TENTH EDITION

POPULATION GEOGRAPHY

PROBLEMS, CONCEPTS, AND PROSPECTS

Gary L. Peters
California State University, Chico

Robert P. Larkin
University of Colorado, Colorado Springs

Karen D. Johnson-Webb
Bowling Green State University

Kefa M. Otiso
Bowling Green State University

Kendall Hunt
publishing company

Book Team

Chairman and Chief Executive Officer Mark C. Falb
President and Chief Operating Officer Chad M. Chandlee
Vice President, Higher Education David L. Tart
Senior Managing Editor Ray Wood
Assistant Editor Bob Largent
Senior Developmental Editor Lynnette M. Rogers
Vice President, Operations Timothy J. Beitzel
Assistant Vice President, Production Services Christine E. O'Brien
Senior Production Editor Mary Melloy
Permissions Editor Carla Kipper
Cover Designer Jenifer Fensterman

Kendall Hunt
publishing company

www.kendallhunt.com
Send all inquiries to:
4050 Westmark Drive
Dubuque, IA 52004-1840

ISBN 978-1-4652-1985-5

Printed in the United States of America
10 9 8 7 6 5 4 3 2 1

To Carol, Jason, Erika, Joyce, Michael, Christine, Tanya W. Webb, Sr., Jackie, Joy, Isaac, and our students

Brief Contents

Contents

Preface

The scientific study of the human population is interdisciplinary, involving demographers, geographers, sociologists, anthropologists, psychologists, political scientists, economists, biologists, physicians, and even philosophers. The dramatic growth of the human population in recent decades compels us to pay attention to its causes and consequences. On the other hand, the increase in the number of countries experiencing slow growth and zero population growth is worthy of study as well.

Our purpose is to provide students with an introduction to population geography, a task that requires drawing upon materials from many disciplines and integrating them into a readable text. We begin with population growth in an effort to generate an interest in the study of population. Following that, we look at demographic data, which is so essential to helping us understand population processes. Population distribution and composition are considered, followed by discussions of theories of population growth and change. After that, the focus turns to the basic demographic processes—mortality, fertility, and migration. The final two chapters examine relationships among population, environment, and food supply.

Acknowledgments

It would be impossible to acknowledge individually each of the people from whom we have learned about population, though citations give credit to many of them. Without the continued research and writing efforts of many people, this book could not have been written. We would also like to thank all of those instructors who continue to use this text; we encourage you to send us your comments.

Though this book has evolved considerably since the first edition appeared in 1979, it continues to reflect our own graduate learning experiences. We would still like to thank Professor Paul Simkins for stimulating our initial interest in population geography during our student years at Penn State. Many of the ideas in this book can probably still be traced back to Paul, and we hope that he approves of the current edition. Of course, we remain responsible for whatever errors and shortcomings this book may have.

Gary L. Peters
Robert P. Larkin

We would like to thank Gary L. Peters and Robert P. Larkin for a job well-done and for bequeathing to us this important work. We recognize that the task of stewarding it is easier because we stand on their giant shoulders. We also acknowledge the input of other "giants" including our own myriad mentors, teachers, and colleagues. We also thank the many students of our Population Geography courses who have also taught us much as we have taught them the material in this book. Like our co-authors, we are responsible for any errors and shortcomings in this book.

Karen D. Johnson-Webb
Kefa M. Otiso

About the Authors

Gary L. Peters (1941–2013) was born in Marysville, California on March 20, 1941. After serving as a radio operator in the U.S. Navy he attended Yuba Junior College and then transferred to Chico State University, where he majored in Geography. He then obtained his Master's Degree and Ph.D. in Geography from Pennsylvania State University. Following the completion of his studies, Gary taught in the geography department at California State University Long Beach before finishing his career at Chico State University. Throughout his career Gary published ten books—including *Population Geography: Problems, Concepts,* and *Prospects and American Winescapes: The Cultural Landscapes of America's Wine Country*—and numerous academic articles. (This necrology is used with permission of the Association of American Geographers.)

Robert P. Larkin is a professor emeritus of geography and environmental studies at the University of Colorado, Colorado Springs. He received his B.A. in Social Studies Education from the State University of New York, Cortland College, his M.A. in geography from the University of Colorado at Boulder, and his Ph.D. in geography from the Pennsylvania State University.

He has been teaching at the college and university level for more than thirty years. His interests are in population geography, gerontology, and the teaching of geography. He has published eight books and a variety of articles on geographic topics.

Karen D. Johnson-Webb is associate professor of geography at Bowling Green State University. She received her B.A and M.A. in Geography at Michigan State University and her Ph.D. in Geography at The University of North Carolina at Chapel Hill. Her research interests include migration and occupational niches. She has published one book, *Recruiting Hispanic Labor: Immigrants in Non-Traditional* Areas as well as other articles on a variety of geographical topics.

Kefa M. Otiso is associate professor of geography and Director of the Global Village at Bowling Green State University. He received his B.Ed. in Geography and Kiswahili from Kenyatta University, Nairobi, Kenya, his M.A. in geography from Ohio University, Athens, and his Ph.D. in geography from the University of Minnesota, Twin Cities.

He specializes in urban, economic and migration processes at various scales. He has published two books and a variety of articles on urban, economic and migration topics.

Introduction

Key Terms

Population Geography | Demography

When the first edition of this book appeared in 1979, the world's population was close to 4.4 billion. In 1999 the earth's population reached 6.0 billion; by the start of 2013 there were more than 7 billion. That number has tremendous implications for people around the world and the natural systems that they depend on.

Geographers are at the forefront of efforts to understand and communicate contemporary human-environment interactions. A geographic perspective is especially valuable in clarifying and interpreting the considerable variations that exist from place to place with respect to birth and death rates, as well as major patterns of population movements. Most of the world's growth is occurring in the less developed countries, whereas most of the more developed countries are growing slowly or not at all; many are even declining. Such demographic disparities are likely to set in motion other processes, especially international migration, which could easily become this century's most important demographic question.

Furthermore, as demographer Geoffrey McNicoll (1999, p. 435) noted, ". . . the demographic relativities of the future will differ from today's in the certain diminution of the West—and, more generally, the North—giving still greater cause for a new international order."

It is hardly surprising to find that more and more attention is being focused on population issues. From scholarly journals to the popular press, a burgeoning literature on population dynamics and population problems challenges us. Business has discovered the importance of demographics as well, as is apparent in publications such as *American Demographics* (now published as *Advertising Age*). Our view of population, the geographic perspective, requires a brief look at the intellectual development of population studies as a geographic concern.

Population Geography

Because of the interdisciplinary nature of population studies, it is not always easy to distinguish **population geography** from other disciplinary contributions. However, it seems that we should briefly consider ways in which some important population geographers have defined their realm of study.

Because of the interdisciplinary nature of population studies, it is not always easy to distinguish **population geography** from other disciplinary contributions.

Population geography is a relatively young subdiscipline, though geographers have long been concerned with population characteristics as a part of broader regional studies. Though the roots of the subdiscipline developed somewhat earlier, most would agree that Glenn Trewartha (1953), a noted climatologist as well as population geographer, provided the first definitive statement on population geography. Therein he noted that population geography had been, and continued to be, a neglected aspect of the discipline. Furthermore, he argued that population was a pivotal element in geography, and that its continued neglect would seriously affect the development of geography.

" . . . the geographer's goal in any or all analyses of population is an understanding of the regional differences in the earth's covering of people. "

Some geographers would still agree with Trewartha (1953, 87) that ". . . the geographer's goal in any or all analyses of population is an understanding of the regional differences in the earth's covering of people. Just as area differentiation is the theme of geography in general, so it is of population geography in particular." Other geographers, as we shall see, would find different approaches to the study of population geography more appealing. Trewartha's ideas were reassessed by Graham (2004) and Pandit (2004).

The response to Trewartha's plea for increased attention to population geography was neither rapid nor overwhelming, at least not at first. In the year following the publication of Trewartha's presidential address, geographer Preston James (1954, 107) wrote that ". . . the fundamental problem of population geography is the search for a systematic method of outlining enumeration areas that are meaningful in terms of the question being asked." On the same page he went on to note that ". . . in the field of population geography the need is to find a way to generalize the distribution of people without obscuring the relationships of man to the other phenomena with which he is already associated." Additionally, James argued that demographic studies could benefit from the geographic perspective, especially from the regional concept and regional methodology. He also provided a useful review of population studies by American geographers, suggested numerous new research directions, and even offered a discussion of training requirements for population geographers.

After an embryonic decade or so, two books on population geography appeared in 1966, one by American geographer Wilbur Zelinsky (1966) and the other by French geographer Jacqueline Beaujeu-Garnier (1966). Zelinsky (1966, 2) commented that a ". . . rich harvest of facts and ideas has not yet been reaped by the population geographer because this is still the period of germination in the history of his discipline." Furthermore, he offered the following definition (Zelinsky, 1966, 5):

> Population geography can be defined accurately as the science that deals with the ways in which the geographic character of places is formed by, and in turn reacts upon, a set of population phenomena that vary within it through both space and time as they follow their own behavioral laws, interacting one with another and with numerous nondemographic phenomena.

Perceptively noting the burden that this definition placed on the reader, he offered a shorter version (Zelinsky, 1966, 5) stating that ". . . the population geographer studies the spatial aspects of population in the context of the aggregate nature of places." He contended that population geography was concerned with three "distinct and ascending levels of discourse": (1) simple description of the location of population numbers and characteristics, (2) explanation of the spatial configurations of these numbers and characteristics, and (3) the geographic analysis of population phenomena. As did Trewartha and James, Zelinsky emphasized areal differences in populations.

Beaujeu-Garnier's concept of population geography was best expressed in the following statement (1966, 3):

> If the demographer measures and analyzes the demographic facts, if the historian traces their evolution, if the sociologist seeks their causes and repercussions by the observation of human society, it is the business of the geographer to describe the facts in their present environmental context, studying also their causes, their original characteristics, and possible consequences.

In her view the geographical study of population had three foci:

1. the distribution of people over the globe,
2. the evolution of human societies, and
3. the degrees of success that those societies have achieved.

In his major work on the topic Trewartha (1969, 1-2) stated that "... the geography of population (or population geography) is concerned chiefly with one aspect of population study—its spatial distribution and arrangements." As he had in 1953, Trewartha again emphasized the importance of population geography within the broader discipline of geography, noting that "... population serves as the point of reference from which all other geographic elements are observed, and from which they all, singly and collectively, derive significance and meaning."

In the following year Demko, Rose, and Schnell (1970, 4) offered yet another definition, conditioned by changes in methodology that were affecting most fields of geography by that time (spatial analysis, logical positivism, quantitative methods):

> Population geography is, therefore, that branch of the discipline which treats the spatial variations in demographic and nondemographic qualities of human populations, and the economic and social consequences stemming from the interaction associated with a particular set of conditions existing in a given areal unit.

Their emphasis was somewhat broader than those of previous definitions, leaning more toward process than simple areal differentiation, more toward hypothesis-testing than simple description.

Soon afterward, geographer John Clarke (1972, 2) wrote that population geography was "... concerned with demonstrating how spatial variations in the distribution, composition, migrations, and growth of populations are related to spatial variations in the nature of places." A few years later Courgeau (1976, 261) added that:

> Demography and population geography are concerned with the same topic: the study of human populations. Basically, they are quantitative disciplines that mainly use statistical data, but they also employ qualitative approaches. The main difference between the two sciences is the fact that the demographer places his emphasis on time whereas the geographer places his emphasis on space.

At least to some degree this distinction remains acceptable.

In the following decade Clarke (1984) noted the success that population geography had experienced as a subdiscipline, especially since the 1970s, and pointed out that since Trewartha's seminal paper in 1953 nearly ten percent of all published geographical papers had dealt with some aspect of population. He also noted the diversity of studies under the population geography umbrella, along with a look at population and political changes, as well as data and advances in quantitative methods. However, he ended up noting that, despite many of the methodological advances of the 1970s and early 1980s, "The search for general methods and laws has often obscured the complexities of reality." (Clarke, 1984, 9)

Demography

Definitions of **demography** also vary, both over time and among practitioners. According to demographer Donald Bogue (1969, 1), "... demography is the empirical, statistical, and mathematical study of populations." In his view "formal" demography focused on: (1) changes in population size, (2) the composition of the population, and (3) the distribution of population in space.

> Definitions of **demography** vary, both over time and among practitioners.

Demographer William Peterson (1975) defined demography as "... the systematic analysis of population phenomena. . . ." However, he differentiated between formal demography and population studies, arguing that formal demography was concerned with the gathering, collating, statistical analysis, and technical presentation of data, whereas population studies were concerned with population characteristics and trends in their social setting. Along this latter line, Thomlinson (1976, 4) stated that "... population study involves the number and variety of people in an area and the changes in this number and variety." More recently, McFalls (2007, 3) wrote that "Demography is defined as the

Describing housing demography, Dowell Myers (1992, 6) commented that, "This theory stresses the detailed interconnections between populations and their housing stocks."

study of human populations: their size, composition, and distribution, as well as the causes and consequences of changes in these characteristics. Populations are never static."

Along with formal demography and population studies, a third dimension deserves mention, namely "housing" demography. Describing housing demography, Dowell Myers (1992, 6) commented that, "This theory stresses the detailed interconnections between populations and their housing stocks." Both urban and population geographers have found this concept useful and have incorporated it into their work (Gober, 1992). Somewhat later Buzar, Ogden, and Hall (2005, 428) noted that "... household-level demographic change lies at the nexus of the cultural, the economic and the urban." They also pointed out that geographers had been working at the household scale now for three decades, even though much of that work has not yet entered mainstream geography journals. Fortunately, that is changing as household demography becomes one more relevant way to examine and interpret urban problems.

Trends in Population Geography

Geographers and demographers share many common and overlapping interests in their respective studies of population. However, population geographers tend to place more emphasis on the spatial patterns of population characteristics and processes, whereas demographers, though often interested in population distributions, seldom make spatial patterns their central interest.

Focus on the spatial dimensions of demography and demographic techniques have been central in much of the literature of population geography. Findlay (1991, 64) noted that, "The 1980s may well be recorded by historians of population geography as the decade in which the subdiscipline became strongly demographic and moved in the direction of being redefined as spatial demography." This trend, exemplified by such works as Woods and Rees (1986) and Congdon and Batey (1989), was especially pronounced in the United Kingdom. Findlay and Graham (1991) also looked at the nature of recent studies in population geography, first noting the emphasis on the spatial perspective, then criticizing population geographers for their neglect of postmodernist debates that took place in geography in the 1980s. "That these issues were not taken up and discussed in population geography in the 1980s," Findlay and Graham (1991, 156) argued, "meant that population geographers were failing to participate in the mainstream methodological debates within geography, but clung instead to the surer, but dated terrain of the 1960s and 1970s." Perhaps even more clearly, Findlay and Graham (1991, 158) commented that "A narrow definition of population geography as spatial demography is quite simply inadequate to face the challenge of what others expect geographers to contribute to the understanding of population." Similarly, Ogden (2000, 628) noted that ". . . population geographers need to forge stronger links with other branches of the discipline, and thereby to renew awareness of the extent to which demographic phenomena underpin social structure and change and how interesting are demographic changes at all scales from the local to the global."

In population geography we need to be more involved in solving key problems in society.

Clearly, in population geography we need to be more involved in solving key problems in society. With respect to geography and the work being done by geographers, Brunn (1992, 2), from his perspective as editor of the *Annals of the Association of American Geographers*, lamented:

> I am sometimes concerned that, unlike other disciplines, we are apparently unwilling to tackle global problems on a global scale . . . we are not prepared to look at global and macroregional problems, processes, and models. . . . The time is propitious for us to get some idea of what we are doing and how we might proceed as a sound and worthy discipline into the next millennium.

Given their roots in geography, population geographers should not lose track of how the subdiscipline fits into, and interacts with, other branches of geography. Whether looking at international flows of refugees or crosstown movements within metropolitan areas, whether considering the relationships between people and global warming or the supply and demand for food in the Sahel, we should be able to find some solid geographic foundations upon which to build. In that process, people, and their relationships to places and environments at all scales, must remain central. As Stoddart (1987, 331) noted, for geographers "The task is to identify geographical problems, issues of man and environment within regions—problems not of geomorphology or history or economics or sociology, but geographical problems: and to use our skills to work to alleviate them, perhaps to solve them." Perhaps the summary by White, et al (1989, 282) provides as clear a picture as any of where we should be going:

> Geographers might disagree about the most significant topics within population geography, the most appropriate research questions, and certainly the most appropriate research methodologies. However, it is clear that despite these honest differences, population-geography research has grown rapidly since 1953, and shows no signs of slowing. Undoubtedly, population geography is pivotal to an understanding of the cultural landscape.

"The argument advanced here is that we must concentrate our efforts on particular issues or areas in which we have some expertise—and then we must augment that work, deepen our knowledge and hone our skills."

Findlay (1993) urged population geographers to pursue more specific tasks. For example, he suggested the need for studies of the demographic disorder that has resulted from the end of the Cold War and the search for a "new world order." Rapid changes in birth rates, for example, have occurred in Russia and other former East Bloc countries. New refugee flows have resulted as well, especially in the region that once was Yugoslavia. By 2000 Ogden (2000) showed that geographers were adding substantially to population studies, especially to migration studies.

In addition, geographer Alan Nash (1994, 386) beseeched population geographers to improve their scholarship (a message often given to other geographers as well), noting that "The argument advanced here is that we must concentrate our efforts on particular issues or areas in which we have some expertise—and then we must augment that work, deepen our knowledge and hone our skills." He recommended, especially, the pursuit of more focused studies of migration, fertility, and gender issues. In a more recent survey of population geography Hugo (2006, 520) concluded that "With almost two-thirds of the world's population, Asia must loom large in any assessment of the dominant global issues—poverty, inequality, sustainable development, climate change, conflict, spread of disease, famine and environmental degradation." In another study Hugo (2007, 85) noted that "The linkage between permanent and temporary migration promises to be a rich area for contemporary population geography research which has yet been little investigated." Ongoing surveys of the work of population geographers and the need for research in different areas are fundamental to the directions that our sub-discipline takes in the years ahead. A more detailed and thorough account of the work of population geographers and the evolution of this sub-discipline can be found in Bailey (2005).

To the extent that geographers continue to be concerned with the earth as a home for humans, we also need to focus more attention on the global and national interactions between growing populations and earth's capacity to support them. The world's population has more than doubled since the end of World War II and it could possibly double again before it stabilizes. As then-Senator Albert Gore (1992, 295) pointed out:

> Human civilization is now so complex and diverse, so sprawling and massive, that it is difficult to see how we can respond in a coordinated, collective way to the global environmental crisis. But circumstances are forcing just such a response; if we cannot embrace the preservation of the earth as our new organizing principle, the very survival of our civilization will be in doubt.

More recently Gore (2006) returned to his writing on global environmental problems, this time with a powerful insistence on the impact that global warming was having, and would have in the future, on planet Earth. Though this book, and the Academy Award-winning film, "An Inconvenient Truth," based on it and directed by Davis Guggenheim, say little about population growth per se, they certainly have awakened a new and broader interest in environmental problems. As we shall see later, it is impossible to remove population growth from the equation because it is so closely related to our consumption of energy of every sort, including fossil fuels.

Whether you agree with Gore's pronouncements or not, the reality is that there are few problems facing humanity today that cannot be more easily solved with a slower population growth. Though the rate of human population growth has been slowing for more than three decades now, its continued slowing is hardly inevitable. It clearly depends, more than ever, on the choices that millions of individuals make about how many children they want to have.

In the 1990s geographers and other social scientists, with an interest in population growth, began to focus more than ever on the importance of women and their roles, both in life and in fertility decisions. We know, for example, that complications from pregnancies and various reproductive disorders rob women of many days of healthy life in less developed countries. We also know that throughout the world millions of women still do not have access to family planning and safe methods of controlling fertility. Violence against women (including rape and sexual coercion) is a worldwide problem and remains a pervasive, if not always widely recognized, abuse of human rights around the world. The practice of female genital mutilation continues to scar millions, both physically and emotionally; and HIV/AIDS has affected men and women almost equally in many parts of the world. Women are also disproportionately represented among the world's illiterate and poor.

One path to slowing world population growth would be to improve the lot of females everywhere.

It is apparent, then, that one path to slowing world population growth would be to improve the lot of females everywhere, so that they could marry and reproduce if and when they want. Gender equality may not guarantee the "survival of our civilization" (which probably means very different things to different people around the world anyway) but it would go a long way toward improving the lives of individuals, families, and communities around the world. Adding a geographic perspective to fertility research would be valuable (Boyle, 2003).

Even as geographers remain concerned about the implications of population growth for the planet's long-term ecological health, a growing number of researchers are looking at the number of places in which populations have ceased to grow, and in some cases are even declining. The cause of such declines is easy enough to articulate: more people die each year than are being replaced by births. More complex, however, are the implications of demographic declines, both in the short and in the longer run. As journalist Phillip Longman observed (2004, 7-8), "In both hemispheres, in nations rich and poor, in Christian, Taoist, Confucian, Hindu, and especially Islamic countries, one broad social trend holds constant at the beginning of the twenty-first century: As more and more of the world's population moves to crowded urban areas, and as women gain in education and economic opportunity, people are producing fewer and fewer children." Geographers are only beginning to focus on population decline on a national or international scale, though they have considered it before in local and regional examples.

For more than half a century now ordinary people on every continent have embraced the notion that every individual has a claim to basic rights. Such rights were enumerated in 1948 when the General Assembly of the United Nations adopted and proclaimed resolution 217 A (III)—The Universal Declaration of Human Rights. The purpose of the declaration was to set a common standard of achievement for all peoples and all nations with respect to recognition of the rights and freedoms of each and every citizen of the planet. The declaration came out of a period following two world wars, a period that included such horrors as the Nazi atrocities. Article 1 (the first of 30) states that "All human beings are born free and equal in dignity and rights. They are endowed with reason and conscience and should act towards one another in a spirit of brotherhood." When all is said and done with respect to trends in population geography, we find it easy to agree with Gober and Tyner (2003, 196), who wrote that "We see the new millennium as a time of great potential and promise for a vigorous expansion of the field of population geography while, concurrently, not losing sight of its strong foundation." More generally we agree with Heyman (2006, 105) when he suggested to geographers that "The work that needs to be done in the discipline is not figuring out how we can bend our research interests to fit agendas set elsewhere; instead, it entails figuring out how we can contribute to shaping social agendas and priorities along more humane and socially just lines." Geographers differ from other social scientists, and we need to heed those differences, especially our use of field experience, to call more attention than we do to the manifold problems that are created at every scale by population growth—its impact on cultures, communities, and environments.

We are members of a species that has expanded from about a billion to over 7 billion in less than 300 years, an unprecedented biological feat. In the process we have found ever more clever ways to utilize Earth's resources, but not without consequences that are now starting to appear at the global scale. For example, most of the hottest years on record have occurred since 1995, glaciers have been retreating almost everywhere, and few reasonable people any longer doubt that Earth is heating up or that humans are at least partly responsible (Barrett, 2012). Population geographers should focus more attention on how communities, regions, and nations might be better served by altering settlement and commuting patterns in order to curb our appetite for burning fossil fuels. We should also look more closely at how the benefits of economic growth might be more equally distributed among regions, at more creative ways to encourage rich countries to help poor ones through the demographic transition and into an era of slower growth, and at whether migration patterns might be better managed by changing investment policies that encourage large companies to circle the globe in search of cheap labor and lax environmental laws.

The Growing Population Literature

Before ending this introduction, it is useful to recognize some of the periodicals in which population articles appear, so that both instructors and students can seek out more detailed and current studies with which to enrich this basic introduction to population geography.

A number of journals are dedicated exclusively to population articles. Among the major ones in English are *Demography, Population Studies, International Migration Review, American Demographics, Population and Development Review, Population Bulletin, Population Research and Policy Review, Family Planning Perspectives, International Journal of Population Geography, Population and Environment,* and *Population, Space, and Place.*

The Population Reference Bureau, located in Washington, D.C., publishes not only *Population Bulletin*, but also a number of other studies of population, as well as numerous materials of interest especially to those who teach population. Additional articles on population topics appear frequently in the following geographic periodicals: *The Geographical Review, Annals of the Association of American Geographers, The Professional Geographer, The Canadian Geographer, Transactions of the Institute of British Geographers, Area, Social and Cultural Geography, Political Geography, Geojournal, African Geographical Review* and *Gender, Place, and Culture.*

Aside from the above demographic and geographic publications, population articles of interest also appear occasionally in *Science, Scientific American, Nature, The Sciences, Urban Affairs Quarterly, International Journal of Health Science, American Economic Review, International Journal of Comparative Sociology, Research on Aging, Journal of the American Medical Association, American Behavioral Scientist, The Gerontologist, Social Forces, Foreign*

Affairs, American Sociological Review, Sociology and Social Research, Annals of the American Academy of Political and Social Science, Economic Development and Cultural Change, Rural Sociology, and Social Science Quarterly, the New England Journal of Medicine, and the *Journal of the American Medical Association.*

Though the periodicals mentioned above are primarily scholarly, numerous population articles appear in more popular publications as well, including *National Geographic, Time, U. S. News and World Report, The Economist, The Atlantic Monthly, Business Week,* and *Forbes,* along with newspapers such as *The Los Angeles Times, The New York Times, The Washington Post, The Christian Science Monitor*, and *The Wall Street Journal.* Almost anywhere that you turn today, you are likely to come across something concerning population, so keep your eyes open and become an informed and critical consumer of demographic information. Ultimately, the study of the human population includes you, so learn all that you can and understand that demography will do much to affect your own life.

CHAPTER 1

Population Growth and Change

Key Terms

population growth
basic demographic equation
rate of natural increase
rate of population growth
doubling time
carrying capacity
population projections
mathematical methods
component methods
genocide
baby boom

Our aim in Chapter 1 is to place the present world population into a broader historical perspective and then to consider current and future population trends. Only after we have seen where we have been, and how we got to where we are, can we begin to ponder our demographic future.

Measuring Population Growth and Change

Any understanding of the ways in which population processes operate to shape or alter the size or composition of a region's population requires a knowledge of various measures of **population growth** and change.

Population growth is the change in population over time, and can be quantified as the change in the number of individuals in a population per unit time.

The Basic Demographic Equation

The most fundamental characteristic of any population is its size. An area's population may be increased either by a birth within the area or by the migration into the area of a person from another area. Similarly, the population may be decreased either by the death of someone within the area or by the migration of someone from the area out to another area. Thus, the primary population processes are births, deaths, and migration. These basic demographic processes may be combined to produce the following equation:

$$FP = SP + B - D + I - O,$$

where FP = final population, some time interval beyond SP,
SP = starting population,

> The most fundamental characteristic of any population is its size.

B = births during the interval,
D = deaths during the interval,
I = in-migration during the interval, and
O = out-migration during the interval.

Primary population processes are births, deaths, and migration, which combine to produce the **basic demographic equation.**

It is easy to see why demographers have sometimes referred to it as the "**basic demographic equation.**"

The Rate of Natural Increase

For any given population the rate of natural increase (RNI) equals the crude birth rate (CBR) minus the crude death rate (CDR). Thus

$$\text{RNI} = \text{CBR} - \text{CDR}.$$

The crude birth rate is the number of births per 1,000 population in a one year period, or

$$\text{CBR} = (\text{B}/\text{P}) \times 1{,}000,$$

where B = number of births in one year and
P = mid-year population.

In 2011 the crude birth rate for the world was about 20 per thousand and for the United States it was approximately 13 per thousand. The crude birth rate is influenced to some degree by the age and sex structure of a population.

The crude death rate is the number of deaths per 1,000 population in a one year period, or

$$\text{CDR} = (\text{D}/\text{P}) \times 1{,}000,$$

where D = number of deaths in one year and
P = mid-year population.

In 2011 the crude death rate for the world was about 8 and for the United States it was around 8. The crude death rate is considerably more affected by the age structure of the population than is the crude birth rate, so comparisons among countries need to be done with caution. This should be immediately apparent when you consider that the world rate and the rate in the United States are the same.

Rate of natural increase is the crude birth rate minus the crude death rate of a population.

Because the crude birth and death rates are both expressed per 1,000 population, it is obvious that the **rate of natural increase** will also be expressed in units per 1,000 population—births minus deaths. Since the crude birth rate for the world was 20 and the crude death rate was 8, the world's rate of natural increase for 2011 was equal to 20 minus 8, or 12 per thousand. Similarly, for the United States the rates were 13 and 8, respectively, and the rate of natural increase in the United States in 2011 was 13 minus 8, or 5 per thousand. Note that the rate of natural increase is not necessarily the same as the rate of population growth because the effect of migration is not included in the former. For the world, the rate of natural increase equals the rate of population growth because migration to and from the earth is currently nonexistent. For the United States, however, the rate of natural increase is well below the actual rate of population growth because of a sizable annual net immigration (which is the subject of considerable debate today).

The **rate of population growth** is a measure of the average annual rate of increase for a population.

The Rate of Population Growth

The **rate of population growth** is a measure of the average annual rate of increase for a population. Barring migration, it is possible to convert the rate of natural increase to the natural rate of population growth by simply converting the rate per 1,000 to an annual percentage rate. For the world the rate of natural increase was 12 per thousand, which is equivalent to an annual rate of population growth of 1.2 percent. Keep in mind, however, that most of the time migration must be considered, so the rate of population growth will usually differ from the rate of natural increase. To some extent the relative effects of natural increase and migration are inversely related to the size of the area under consideration. Whereas at the world scale migration plays no part at all in population growth, at the local scale migration may even be more important than natural increase in determining the overall rate of population growth.

In the United States example the rate of natural increase amounts to an annual growth rate of 0.5 percent. Actually, the 2011 rate of population growth for the United States was closer to 1.0 percent. The difference between these two rates results from net immigration. In other words, 50 percent of the growth in the United States, population in 2011 was due to natural increase and 50 percent was due to net immigration.

Doubling Time

The **doubling time** of a population is the number of years that would be required for a population to double in size, assuming that the population continues to grow at a given annual rate. This growth is analogous to the growth of money in a bank savings account. In both cases the "interest" is compounded. Without going into detail, it is possible to closely approximate the doubling time for a population by dividing the annual rate of population growth into the number 70. Thus, for the world, growing at 1.2 percent annually, the time required to double the present population would be 58 years. This assumes, of course, that the 1.2 percent growth rate continues over the entire period. For the United States, growing at 1.0 percent, the doubling time would be around 70 years.

The **doubling time** of a population is the number of years that would be required for a population to double in size, assuming that the population continues to grow at a given annual rate.

World Population Growth

This section is concerned with the growth of the human population from prehistoric times to the present. Once we get into the twentieth century the focus shifts from the total world population to regional patterns of growth—mainly to the current division of the world into less developed (mainly poor) and more developed (mainly rich) regions—and to the differences between these regions with respect to population growth. We generally use the terms *less developed* and *more developed* regions with reference to levels of economic development—occasionally less developed countries may also be referred to as Third World countries.

Brief Overview of World Population Growth

In order to understand the current world population situation, as well as future prospects for the world's ever-increasing numbers, it seems worthwhile first to discern how it is that the population reached its current level—a world of 7 billion people growing at an average annual rate of about 1.2 percent. Those figures suggest that each year around 84 million people are added to what many already perceive to be an overcrowded planet. Every four years more people are added to the world's population than currently live in the entire United States. Most geographers and demographers believe that this rate of growth cannot continue indefinitely. As Berelson and Freedman (1974, 3) noted more than three decades ago, "The rate of growth that currently characterizes the human population as a whole is a temporary deviation from the annual growth rates that prevailed during most of man's history and must prevail again in the future." This recent period of rapid population growth is unique in demographic history, both in terms of the rate of population growth and in terms of the absolute size of the world's population. The world's population nearly quadrupled during the twentieth century, from 1.6 billion in

1900 to 6.1 billion in 2000. Population geographers and demographers do not expect that to happen again.

For most of human demographic history, population growth was exceedingly slow; the annual rate of increase probably did not reach 0.1 percent (a doubling time of about 700 years) until sometime in the seventeenth century, after which it began to accelerate. This acceleration was gradual at first, but it became more noticeable after 1750.

Our current knowledge of historical populations remains conjectural. Clever demographic detectives, using whatever clues they can uncover (from archaeological excavations to early church baptismal, marriage, and death records), have pieced together the story of the human population's slow but inexorable expansion in both numbers and occupied territories. Our knowledge of historical population sizes and growth rates remains speculative because censuses and other organized and systematic collections of population data were nearly nonexistent before the middle of the eighteenth century. Earlier censuses had been taken in a few places, but their data were controversial at best. Even today reliable statistics don't exist for perhaps half of the world's population.

Even today reliable statistics don't exist for perhaps half of the world's population.

Estimates of population numbers in prehistoric times vary considerably and are generally made on the basis of assumptions about the **carrying capacity** of the land—its capacity to sustain a given human population at a given level of technology—and the distribution of the human population. Cohen (1995a) and others have tried to model human carrying capacity of the earth, but there has generally been little agreement among different approaches, partly because of the complexity of human societies and cultures. Hopfenberg (2003, 109) used food supply data in a logistic model and concluded: "That food supply data adequately fits the logistic model of human population dynamics provides evidence that, consistent with ecological notions typically applied only to nonhuman species, human population increases are a function of increased food availability." This fits well with Cohen (1995b, 35), who earlier observed that "The ability to produce food allowed human numbers to increase greatly and made it possible, eventually, for civilizations to arise."

The **carrying capacity** of land is its capacity to sustain a given human population at a given level of technology.

"The ability to produce food allowed human numbers to increase greatly and made it possible, eventually, for civilizations to arise."

Speaking of numbers, this and other books on population are full of them, so let us digress for a minute and talk about them. On May 30, 2013, according to the Census Bureau, the population of the United States was 315,949,324 and the population of the world was 7,088,560,200. What precision! But don't take these numbers, or others like them, too seriously. Most of the above digits mean nothing, though they demonstrate how nice it is to have computers around to generate them. Round them off to 316 million and 7.1 billion and you have lost no accuracy–the accuracy is only apparent to start with. Furthermore, such numbers can be manipulated easily while forgetting that they represent real lives. We encourage you to remember what Cohen (1995b, 20) pointed out, that "Uncertainty does not render statistical numbers worthless; even with uncertainty, statistical numbers are indispensable. They are often far more informative than verbal descriptions or intuitive hunches. But every statistical number should enter your consciousness with a penumbra of doubt." In other words, don't take numbers for granted–look at them carefully, think about how precise they might be, and keep in mind that there may be considerable uncertainty surrounding them. Numbers are necessary for our discussions, but it is also necessary to view them with a touch of suspicion.

Deevey (1960) estimated that the world's population around one million years ago was about 125,000. According to his estimates, this population grew very slowly to approximately 3.34 million 25,000 years ago and to 5.32 million 10,000 years ago. By *A.D.* 1, Deevey and others estimate, the world's population was in the neighborhood of 250–300 million. At that time the average annual rate of increase was probably on the order of 0.05 percent. At that growth rate it would take about 1,400 years for a population to double, compared to a doubling time of about 58 years for today's population.

The world's population did not reach its first billion until sometime around 1820. By then the annual rate of growth had increased tenfold to roughly 0.5 percent. Though all of human history had been required to reach this first billion, only 110 years were required to add the next billion; by 1930 there were 2 billion residents on our planet. In only 45 years this 2 billion doubled to the 1975 population of 4 billion, and by 1987 the world's population had grown to 5.0 billion. Only twelve years later the next billion had been added. As a species, we have certainly demonstrated our capacity for successful reproduction, but it may well be time for us to restrain ourselves before we find a way to destroy our own ecological niche (though not the earth—it could get along quite well without us, as it did for most of its history). As Ornstein and Ehrlich (1989, 45) noted:

> Increasing numbers is a "goal" of all organisms. But never before has there been an "outbreak" of a single species on such a global scale. Unfortunately it is not yet clear how enduring our unprecedented triumph will be, because it has created an unprecedented paradox: our triumphs can destroy us. As people strive to increase their dominance even further, they are now changing the earth into a planet that is inhospitable to civilization.

Geographer Crispin Tickell (1993, 220) noted that "All previous civilizations have collapsed." Though he recognized variations on the general theme, he suggested that each early civilization suffered from a fatal combination of population, resource, and environmental variables that turned unfavorable at some point. Furthermore, Tickell (1993, 220) added that:

> The prime engine of the recent dizzymaking rise in the human population and change generally is the industrial revolution. We have the misfortune to be perhaps the first generation in which the magnitude of the global price to be paid is becoming manifest.

Cohen (1995b, 367) put it this way: "The human population of the Earth now travels in the zone where a substantial fraction of scholars have estimated upper limits on human population size." A look at how the world population has grown may help us better understand where we are today and how many more people are likely to be added to Earth's population in the decades ahead.

Warning!

The study of population is based largely on the collection and analysis of demographic data. These data vary considerably in reliability, so it is necessary to proceed somewhat cautiously, to develop a healthy skepticism about population information and its interpretation. In an informative article, Bouvier (1976, 8–9) suggested the following warnings that you should certainly heed:

- **Warning 1:** Do not use growth rates to indicate changes in birth rates.
- **Warning 2:** Do not use natural increase to indicate population growth, except in those areas where migration is nonexistent.
- **Warning 3:** Do not confuse numerical growth or decline with rates of population growth or decline.
- **Warning 4:** Do not take population figures as gospel truth, especially if they come from areas with less than adequate data-gathering facilities.

Each of these warnings should be considered carefully; errors in demographic thinking often result from a failure to consider one or more of them.

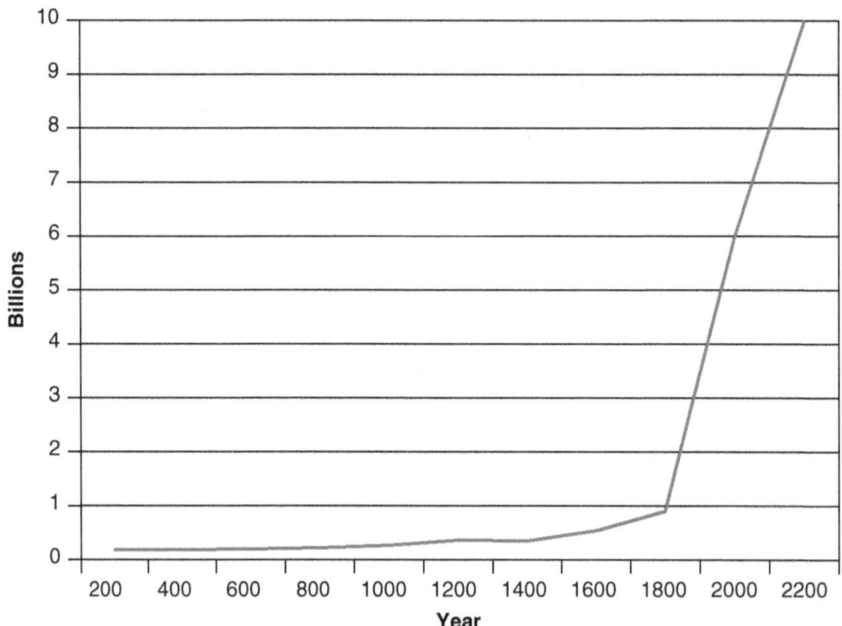

Figure 1–1 Arithmetic Growth of World Population
Source: McEvedy and Jones, 1978; United Nations 2011.

Figure 1–1 provides a graphic illustration of human population history. The slow growth that characterized so much of the early history of humankind gives way, first gradually, then much more rapidly, to increased rates of population growth. These changes in growth rates involved alterations of both birth rates and death rates, with an emphasis on the latter; alterations that were in turn linked to sweeping changes in the socioeconomic fabric of societies.

Figure 1–2 shows past population growth on a logarithmic rather than an arithmetic graph, allowing us to focus more on changes in the rates of increase. The contrast with Figure 1–1 is both striking and suggestive. Rather than a single period of population growth, three periods of relatively rapid demographic increase become apparent, each of them followed by a slowing of growth rates. Deevey (1960) argued that each of these periods of accelerated population growth was a response to a revolution in which the

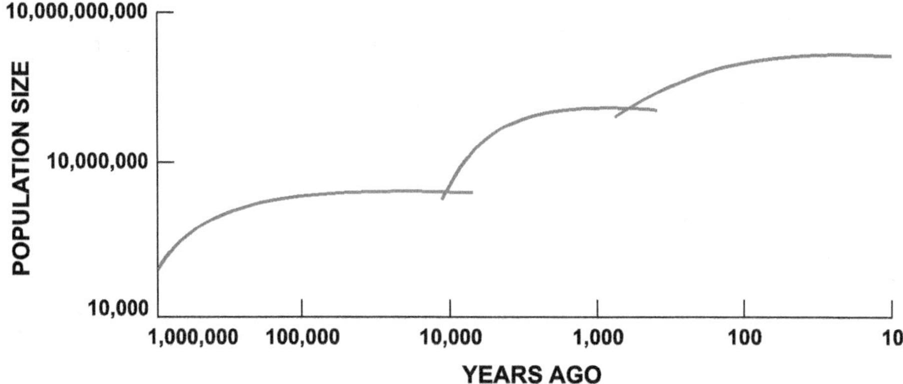

Figure 1–2 Logarithmic Growth Curve for World Population from 1,000,000 Years Ago
Source: Adapted from "The Human Population" by Edward S. Deevey, Jr., *Scientific American*. September 1960.

earth's carrying capacity was dramatically increased. Each of these revolutions in carrying capacity can be viewed as a diffusion process, radiating outward from one or more origins to gradually encompass the inhabited world. The earliest of the three revolutions was the *toolmaking or cultural revolution*; the second was the *agricultural revolution*; and the third was the *scientific-industrial revolution*, which continues today.

The implications of Figures 1–1 and 1–2 for future population growth are dramatically different. Figure 1–1 implies a continuing rapid increase in the size of the world's population (perhaps followed by a catastrophic crash?), whereas Figure 1–2 suggests that the world should experience a tapering off of growth rates as the world population adjusts to the current technological levels and their concomitant limitations on population expansion. Barring another major revolution that would again expand the earth's carrying capacity, Deevey's interpretation of population history implies a slowing of world population growth, one that we are already beginning to see. The world's population growth rate reached a peak sometime in the late 1960s at around 2.1 percent, from which it has dropped gradually to a level of about 1.2 today. Though the direction of change is encouraging, we need to keep in mind that it only represents a change in doubling time from about 33 years to 58 years. Furthermore, given the size and youthfulness of the planet's population, absolute annual population increases (as opposed to the rate of increase) will remain large for many years.

You might wonder, as others have done, how many people have ever lived on our planet. Demographer Carl Haub (2002) estimated an answer. By making a number of "guestimates" about populations, births, and deaths in the past, he concluded that 106,456,367,669 people had been born by 2002. Of those births, about 5.8 percent, 6.125 billion, were still living at the time. By 2008 perhaps another 800 million births had occurred and the total population had passed 6.6 billion. In 2011, it is estimated that 108 billion people have ever lived on earth

The Three Major Periods of Population Growth

As we have already noted, Deevey (1960) proposed that population growth occurred unevenly over time, mainly in conjunction with three major revolutions in human history—the cultural, the agricultural, and the scientific-industrial revolutions. These three major revolutions serve as the basis for subdividing our discussion of population growth into three discrete periods. This is appropriate because each of the revolutions must have opened up new possibilities for population growth, mainly because each one extended the earth's carrying capacity. However, the availability of population *facts* decreases rapidly as we move back in time, hence the following discussion must be approached with a degree of caution.

Population growth occurred unevenly over time, mainly in conjunction with three major revolutions in human history—the cultural, the agricultural, and the scientific-industrial revolutions.

The Cultural Revolution and Population Growth

It was the emergence of primates, perhaps as early as 85 million years ago (Gugliotta, 2002), that set the stage for the gradual origin and spread of our own human population. The earliest primates, which probably overlapped the age of dinosaurs, were small (around two pounds) and ate mainly insects and fruit. At some point in time, perhaps eight million years ago, humans diverged from chimpanzees, our closest living relatives. It was a long road from there to where we are today, however, one still full of mysteries, potholes, ruts, and even evolutionary deadends.

The cultural, or tool-making, revolution occurred in prehistoric time. What knowledge exists of events in that era must be drawn primarily from the archaeological record. Seeking into the origin and evolution of the human population continues to occupy the time and energy of many researchers, mainly anthropologists. Periodically, new finds of

The cultural, or tool-making, revolution occurred in prehistoric time. What knowledge exists of events in that era must be drawn primarily from the archaeological record.

the skeletal remains of early hominids push back the frontiers of our knowledge to yet earlier times. The geographical search for early people has established an African origin.

Several pieces of archaeological evidence have pushed the frontier of hominid history farther back into the dawn of our evolutionary chronology. The earliest hominid on record so far, one similar in many ways to modern chimpanzees, is nearly seven million years old and goes by the name *Sahelanthropus tchadensis*, a name that recognizes its origin in Chad in the Sahel region of Africa. This region is well west of the Great Rift Valley in which most subsequent evidence of hominids have been discovered and suggests that human origins may have been geographically more widespread than has traditionally been appreciated. Around 5.5–5.8 million years ago *Ardipithicus ramidus kadabba* (a new genus) roamed parts of East Africa, as did *Ardipithecus ramidus ramidus* at least 4.4 million years ago. Meave Leakey's (1995) Kanapoi fossils, found not far from Lake Turkana, extended *Australopithicus afarensis* back to about 4.1 million years ago.

Between *Ardipithecus ramidus* and *Australopithecus afarensis*, at an age of 4.2 to 3.9 million years ago, Leakey and Walker (1997) have added a new species, *Australopithe-cus anamensis*, based on further fossil evidence from sites at Kanapoi and Allia Bay.

Previously known skeletal remains uncovered in Africa suggested an age of 2.8 to 3.8 million years for *Australopithecus afarensis*, as exemplified by "Lucy" (Johanson and Edey, 1981) and the "Dikika Baby" (Sloan, 2006), 1.6 to 2.2 million years for *Homo habilis*, and 1.4 to 1.7 million years ago for *Homo erectus*. Bipedalism was probably the earliest characteristic that separated hominids from their nearest relatives, the gorillas and chimpanzees. After that, increasing brain size is one of the primary features that distinguishes each one of the hominid groups from the next. However, as Ornstein and Ehrlich (1989, 37–38) have suggested, ". . . although the human brain appears to have enlarged rapidly (in geological time) in response to the pressures of culture, it does not seem likely that the brain crossed any real physical threshold that suddenly permitted new kinds of cultural activities."

Perhaps our arrogant view of ourselves long colored the way in which we perceived our evolution from ape to hominid to modern people. Like other successful animal species, the line of humans, once separated from the apes five or six million years ago, apparently went through considerable trial and error, made many false starts, and reached a few dead ends. Ian Tattersall (2000) argued convincingly that at least four kinds of hominids lived together within a single landscape in part of what is now northern Kenya about 1.8 million years ago. Though we may never know the degree to which they interacted with each other, Tattersall argues that *Paranthropus boisei*, *Homo rudolfensis*, *Homo habilis*, and *Homo ergaster* (also known as African *Homo erectus*) occupied the same geographical area at the same time. As Tattersall (2000, 61) phrased it, human evolution has not been a simple linear pattern:

> Instead it has been the story of nature's tinkering: of repeated evolutionary experiments. Our biological history has been one of sporadic events rather than gradual accretions. Over the past five million years, new hominid species have regularly emerged, competed, coexisted, colonized new environments and succeeded—or failed. We have only the dimmest of perceptions of how this dramatic history of innovation and interaction unfolded, but it is already evident that our species, far from being the pinnacle of the hominid evolutionary tree, is simply one more of its many terminal twigs.

According to Wenke (1990), two scenarios currently set the stage for debate about the continued search for human origins. One of these scenarios suggests the migration of *Homo* ancestors, most likely *Homo erectus*, out of Africa some 1.5 million years ago, followed by gradual diversification among scattered groups but at the same time a gradual evolution toward *Homo sapiens* among all the groups because of both common genetic inheritances and similar adaptive pressures. The result of this first scenario, as Wenke (1990, 137) described it, was "that they all converged at about 30,000 years ago as one species, *Homo sapiens sapiens*." The other scenario begins in the same way, with the migration of *Homo ancestors* out of Africa some 1.5 million years ago. This scenario differs after the initial migration by suggesting considerably more divergence among groups as they spread out and adapted to different environments. Following the second scenario, according to Wenke (1990, 137), the final result is that "perhaps 140,000 years ago, *Homo sapiens* evolved in one place (probably Africa), and spread across the world, displacing most groups, driving some into extinction, and absorbing a small fraction of the others through intermarriage."

In recent years the latter scenario has received considerable attention, not from archaeological but from genetic evidence. Cann, Stoneking, and Wilson (1987) argued that the maternal lineage of all humans could be traced backed to a single African woman who was alive perhaps 200,000 years ago, a woman now referred to by many as "Mitochondrial Eve." Their data were based on studies of mitochondrial DNA (deoxyribo-nucleic acid), which are only maternally inherited. These DNA then form the basis for a "molecular clock," which in turn can be used to develop a branching tree. As Stringer (1990, 99) has noted, "One attempts to make a molecular clock by comparing genetic differences among various species or varieties within species and expressing their relatedness in a tree . . . then calibrates (or 'dates') the tree by comparing it with another group that diverged from the tree at a known date." Studies since 1987 have tended to confirm the African origin of "Mitochondrial Eve" and have given rise to considerable speculation about what that means for the Neanderthals (Stringer, 1990). Barinaga (1992, 686), citing increasing questions that have been raised about the DNA methodology, noted that ". . . the root of the human tree has been thrown open to question once again." Current criticism is focused primarily on how the mitochondrial DNA have been analyzed and whether or not these data can identify a geographic origin for our species. Excellent summaries of the two opposing views of evolution over the past 200,000 years are Wilson and Cann (1992) and Thorne and Wolpoff (1992). The earliest migration date must be pushed back because of the discovery of fossil remains found in Dmanisi, Republic of Georgia, that date back to 1.75 million years ago and provide clear evidence that *Homo erectus* or something similar had left Africa earlier than 1.5 million years ago.

In a review of these two competing hypotheses about our origin, Tattersall (1997, p. 67) wrote that ". . . my strong preference is for a single and comparatively recent origin for *H. sapiens*, very likely in Africa—the continent that, from the very beginning, has been the engine of mainstream innovation in human evolution. Additionally, Olson (2002, 28) noted that ". . . the genetic evidence available today points to a straightforward conclusion. According to our DNA, every person now alive is descended from a relatively small group of Africans who lived between 100,000 and 200,000 years ago." The oldest fossil of a modern human found so far was discovered in Orno Kifish in Ethiopia, where it lived about 195,000 years ago. Put a different way, Olson (2002, 38) stated that "Everyone alive today is either an African or a descendant of Africans." More recently, Shreeve (2006, 69) wrote that "DNA studies have confirmed this opening chapter of our story over and over: All the variously shaped and shaded people of Earth trace their ancestry to African hunter-gatherers, some 150,000 years ago." He goes on to suggest (2006, 69) that "Perhaps the most wonderful of the stories hidden in our genes is that, when unraveled, the tangled knot of our global genetic diversity today leads us all back to a recent yesterday, together in Africa."

The emergence of anatomically modern humans, *H. Sapiens*, as the *only* surviving members of a long chain of trials and errors occurred recently—probably no more than 25,000 to 28,000 years ago. Before that, we know, for example, that 90,000 to 100,000 years ago in the region that now includes Israel, modern humans lived side-by-side with Neanderthals (whose reputations have been much improved as a result of recent research), though that group seems to have died out without spreading. In 2003 researchers reported some of the oldest known finds of skulls of modern humans. They were approximately 160,000 years old and were found in the proximity of a now-vanished lake in Ethiopia. This evidence puts the fossil evidence in line with the arguments of geneticists—an African origin for *Homo sapiens* between 150,000 and 200,000 years ago.

Despite the find of early modern human fossils in Qafzeh, Israel, that were dated to nearly 100,000 years ago, most evidence today suggests that the successful migration of modern humans out of Africa and around the world began between about 70,000 to 50,000 years ago. As Shreeve (2006, 63) noted, "All non-Africans share markers carried by those first emigrants, who may have numbered just a thousand people." Likely paths were around the north of the Red Sea or across its southern opening (Shreeve, 2006). From there modern humans spread northwestward into Europe, north into what is now Russia, and east through Asia, reaching Australia by around 50,000 years ago. The New World was not populated by modern humans until about 20,000 years ago, or even less, when sea levels were low enough to allow people to cross between Siberia and Alaska.

Between 30,000 and 40,000 years ago modern humans appeared in Europe, where Neanderthals lived already. However, at least according to Tattersall (2000, 61), "Certainly the repeated pattern at archaeological sites is one of short-term replacement, and there is no convincing biological evidence of any intermixing in Europe." In a period of perhaps 10,000 years the Neanderthals disappeared.

Wong (2000), however, identifies some differences of opinion among anthropologists' interpretation of the European confrontation between Neanderthals and modern humans. For example, anthropologist Fred Smith (cited in Wong, 2000, 107) tells us that "The likelihood of gene flow between the groups is also supported by evidence that Neanderthals left their mark on early modern Europeans."

Near the end of 2004 a new skeletal discovery created a stir among anthropologists (Wade, 2004). The discovery was made on the Indonesian island of Flores, about 370 miles east of Bali. The bones were from an adult that would have been around 3.5 feet in height, lived on the island until about 12,000 years ago, and were not pygmy forms of modern people. Rather, they appeared to be downsized versions of *Homo erectus*. This discovery suggests that some of the earliest people to leave Africa may have lived for much longer than has been previously thought. These dwarf humans, assigned the name *Homo floresiensis*, but dubbed Floresians, were still around 20,000 years after Neanderthals disappeared, made stone tools, and lived among giant rats, pygmy elephants, and Komodo dragons. Nicknamed "Hobbit," this small hominid created considerable controversy, with some arguing that it was some kind of dwarf or perhaps a case of microcephaly, but recent testing has ruled out the latter and found in favor of the separate species school.

We are still left with more questions than answers. It seems clearer all the time that the evolution of modern humans was a complex drama that unfolded over a period of some five million years. Trial and error, numerous wrong turns, and dead ends led to the final emergence and dominance of modern humans between 25,000 and 30,000 years ago. Tattersall argues that the final defining characteristic that gave modern humans their edge over the Neanderthals may have been the development of language and an ability to form mental symbols. As he notes (Tattersall, 2000, 62), "We do not know exactly how language might have emerged in one local population of *H. sapiens* . . . . But we do know that a creature armed with symbolic skills is a formidable competitor—and not necessarily an entirely rational one, as the rest of the living world, including *H. neanderthalensis*, has discovered to its cost." Whatever the particular advantages of modern humans were, there is little doubt that McNeill and McNeill (2003, 4) were right when they commented that "What drives history is the human ambition to alter one's condition to match one's hopes."

"What drives history is the human ambition to alter one's condition to match one's hopes."

Though these debates will undoubtedly continue, perhaps it is more important at this point for us to consider Wenke's (1990, 186) summary comment that ". . . we should also take note of the fact that even though we are physically very much like our ancestors of 12,000 years ago, we are enormously different culturally . . . in a cultural sense we are not at all the same species as the human hunter-gatherers of the late Pleistocene . . . to live a life for which evolution has shaped us—we should probably eat a varied diet, live closely in a small group, and walk a lot."

As fascinating as such studies of early *Homo* and earlier ancestors may be, little is known of the numbers involved, though they were undoubtedly small. As Deevey (1960, 5–6) commented, "For most of the million-year period the number of hominids, including man, was about what would be expected of any large Pleistocene mammal—scarcer than horses, say, but commoner than elephants." Not only were the numbers small, but the rate of growth in these numbers must also have been exceedingly small. The growth of *Homo sapiens sapiens* prior to the agricultural revolution remained extremely slow. Hunting, fishing, and foraging provided an existence, though undoubtedly a precarious one. Life was most likely "nasty, brutish, and short."

Population densities on the eve of the agricultural revolution were low, and populations were vulnerable to environmental changes such as climatic fluctuations. Estimates of population size are subject to wide margins of error and must be accepted with reservation; however, the order of magnitude seems reasonable. According to Deevey (1960), the world's population 10,000 years ago was 5.32 million, though others have suggested that it might have been twice that number. Even if we accept 10 million, we're talking about a small base from which we have grown in a relatively short time.

The Agricultural Revolution

An exact date for the beginning of the agricultural revolution is impossible to set, but it is likely that incipient cultivation and domestication developed sometime around 10,000 B.C. in the Near East. A curious point about the beginning of agriculture revolves around the length of time that modern humans had been around (at least 140,000 years) before agriculture began to take root, so to speak. Why did it take so long, we might wonder? One answer, not completely agreed upon, is the better climate that developed at the end of the last Ice Age. Weatherford (1994, 47) noted, for example, "Around the world, humans seem to have switched from foraging to farming because of the whole set of changes produced by global warming." In writing about agriculture and the warming climate at the end of the last Ice Age, Flannery (2005, 61) argued that "It's hard to avoid the feeling that the hostile ice age climate and its savage transition to the interglacial had, until then stymied this great flowering of creativity and complexity." Furthermore, he noted (2005, 63) that "The long summer that has been the last 8,000 years is without doubt the crucial event in human history."

> An exact date for the beginning of the agricultural revolution is impossible to set, but it is likely that incipient cultivation and domestication developed sometime around 10,000 B.C. in the Near East.

A reasonable scenario is that around 12,000 years ago hunter-gatherers began to settle along the shore of the eastern Mediterranean, exploiting local plants and animals, making seasonal hunting trips, and gradually becoming more sedentary. Archaeologists have confirmed that by around 8,000 B.C. Jericho (sometimes described as the "world's oldest town") housed several hundred people, and they were definitely agriculturalists. The rimland around the Fertile Crescent was one of the first areas to experience the agricultural changes that would slowly burgeon into a major revolution. As Weatherford (1994, 111) pointed out, "The transition from rural to urban life first occurred in Mesopotamia with the rise of Uruk, Sumer, and other cities in the area between the Tigris and Euphrates rivers, in modern Iraq." As agricultural practices evolved and diffused, people experienced tremendous changes; they were able to turn from the wandering and tenuous life of hunter-gatherers to the more sedentary and secure life of agriculturalists. However, these changes were gradual. As Cipolla (1974, 22) noted:

> It can be stated however, with a fair degree of certainty that the foundations of settled life in the Old World were first laid in South-West Asia between the ninth and the seventh millennium B.C. This seemingly took place where prototypes of the earliest domesticated animals and plants existed in a wild state and where the concentration on particular species as sources of food was stimulated by the ecological changes that marked the transition to Neothermal climate.

Between perhaps 7,000 B.C. and 5,000 B.C. there was some possible domestication of plants in Mesoamerica. Following 5,000 B.C., a slow but consistent domestication continued there (Diamond, 1997).

The introduction of farming allowed greater population densities to exist and probably produced the first food surpluses people had ever known. In turn, a few people were then freed from the fundamental task of providing food. A multitude of inventions and innovations followed, including the development of village settlements, irrigation, metallurgy, and long-distance trade. These inventions and innovations in turn increased people's capacity to satisfy their needs from the environment and further increased the carrying capacity of the land. Trade, and the convergence of trade routes, certainly affected population distribution and the rise of early cities as well. From today's perspective it seems clear that agriculture brought with it both risks and rewards. Hunters and gatherers lived in small groups, worked together, and did not have a well-defined hierarchical structure. As Weatherford (1994, 50) noted, "The division of the world between farmers and foragers created a permanent tension between two types of subsistence with very different needs." Clearly, with the shift to agriculture new skills and ways of thinking appeared, and, as McNeill and McNeill (2003, 6) pointed out, "Economic specialization and exchange created poverty as well as wealth."

> The introduction of farming allowed greater population densities to exist and probably produced the first food surpluses people had ever known.

> *"Economic specialization and exchange created poverty as well as wealth."*

The demographic response to agricultural and related changes was a gradual acceleration in the rate of population growth. As the agricultural revolution diffused to various parts of the earth's inhabited surface, its impact on population growth became increasingly significant. However, as Deevey's interpretation suggested, once the impact of this revolutionary increase in the earth's carrying capacity had completed its diffusion, the rate of population growth slowed again, and the population became stabilized at a new and higher plateau. Geographic variations in population growth rates existed, however. For example, McNeill and McNeill (2003, 37) noted that "In temperate climates, where diseases were less burdensome than in tropical lands, farming village populations clearly grew much faster than hunting bands had previously done." At the same time, settled people were more vulnerable to certain risks, including infectious diseases.

By the beginning of the Christian Era the earth's population was about 250–300 million. Though this population continued to grow, its numerical progress was slow, and the regional distribution of growth was varied. Within the overall pattern of growth, cyclical changes were typical. The rate of population growth was kept in check by food supplies (often interrupted by famines), by wars, and by epidemics of various diseases.

For the most part famines have been localized, and their impact on the death rate depends on both the severity of the famine and the links between the famine-stricken area and other locations. However, famines have occasionally been devastating. Fourteenth century China may have experienced the planet's first great famine (with deaths thought to be in excess of 4 million); they occurred in many regions over the next few centuries, including the great potato famines in Ireland (1845–1849). Since 1900 famines have been worst in China (1921–1923 and 1928–1929), India (1943–1944 and 1965), and Russia (1932–1934, 1941–1944, and 1947), though they have occurred also in Poland, Greece, Africa's Sahel region, Ethiopia, Bangladesh, Somalia, Nigeria, and Kampuchea. With improved transportation linkages, the effects of local crop failures have gradually diminished.

Wars have directly affected population growth rates at various times, but their impact is not always easy to assess. The simple counting of battlefield deaths alone would underestimate the demographic impact of most wars, because wars also disrupt food supplies and act as diffusion agents for numerous diseases. Of war and the latter, Zinsser (1967, 113) commented, "And typhus, with its brothers and sisters—plague, cholera, typhoid, dysentery—has decided more campaigns than Caesar, Hannibal, Napoleon, and all the inspector generals of history." The greatest losses occurred in World War I and World War II.

Epidemics and pandemics have often been devastating in their impact on regional populations. Extreme examples include the Justinian Plague of A.D. 541–544 and the Black Death of A.D. 1346–1348. The latter may have reduced the European population by 25 percent, and local death tolls reached as high as 50 percent. Recovery of the European population after this decimation was slow, and it was further hampered by the One Hundred Years War. By the sixteenth century, however, Europe had regained her lost population and was beginning a gradual acceleration in the rate of population growth, though there were still localized periods of famine, war, and disease. The rapid growth and distribution of AIDS since the 1980s (discussed in Chapter 4) convinced everyone who might have thought otherwise that epidemics and pandemics are still with us.

After 1492 population declined precipitously in the New World as well. Though we will never know for sure, reasonable estimates suggest that more than 50 million people lived in the Americas when Columbus first sailed westward. As journalist Lewis Lord (1997, 70) noted, "The 150 years after Columbus's arrival brought a toll on human life in this hemisphere comparable to all of the world's losses in World War II." Geographer William Denevan (1996) explored estimates of the Native American population in detail. Diamond (1997, 2005) suggested that 95 percent of the Native American population died as a result of diseases introduced by Europeans into the New World. A more recent study by Livi-Bacci (2006) further supports the idea that new world populations were decimated by contact with early European explorers.

The Industrial Revolution

The Industrial Revolution originated in England in the latter half of the eighteenth century, though its roots may be found in earlier times. At its heart, the Industrial Revolution was a shift from animate to inanimate energy sources, from humans and domesticated animals to steam power generated by carbon fuels–charcoal, coal, then later oil and natural gas. Its impact on humans was vast, fairly rapid, and underwent a geographic dispersion that continues to this very day. As McNeill and McNeill (2003, 248) commented, "Industrialization forever altered the nature of work. From the natural rhythm of the days and seasons that governed farm work, people shifted to schedules controlled by the clock." The shift of populations from rural to urban areas began, and the world became noisier, often dirtier, and undoubtedly warmer as well. From England it diffused rapidly into the countries of Western Europe and to the United States. By the beginning of the twentieth century it had reached Russia and Northern Italy. Japan was the first Asian country to

The Industrial Revolution originated in England in the latter half of the eighteenth century, though its roots may be found in earlier times.

experience this revolution. As countries industrialized, industry replaced agriculture as the major sector of the economy. The Industrial Revolution continues today, and it has so far only partially diffused to the less developed countries. New inventions and innovations continue to pour forth almost daily.

By about 1750 in England and Wales, and soon thereafter in other countries as they began to industrialize, population growth accelerated. Prior to this time, crude birth and death rates had both tended to be high. In average years there may have been more births than deaths, while in bad years the reverse was likely. Death rates undoubtedly fluctuated more widely than did birth rates. High birth rates were deemed necessary in order to overcome the prevailing high death rates, though birth rates were generally not as high as they might have been because of a variety of social constraints.

During both the cultural and agricultural revolutions people increased their capacity to wrest a living from the earth, but it was not until the scientific-industrial revolution that, for the first time, they began to gain control over death. This control over death rates was a result of many changes, mainly changes that probably at first cut off the high peaks in the cyclical fluctuations of death rates. Better agricultural practices and improved distribution systems cut down on the localized effects of famines. Improved sanitary practices and facilities decreased the deaths from some diseases quite early. Then during the nineteenth century major medical advances accelerated the downward trend in death rates.

The innovations associated with each revolution were not contained in their area of origin but diffused outward, as is well illustrated by the spread of the Neolithic farming cultures of Europe. In 6,000 B.C. farming in Europe was primarily limited to a few sites near the Aegean Sea. During the next 1,000 years it spread northward into the Danubian Basin and by 4,000 B.C. to the North European Plain. An even more rapid diffusion of the industrial-scientific revolution has occurred. These innovations were carried by Europeans as they colonized new areas, and in the current century there are few places that have not been touched by industrialization to some degree.

The speed and direction of diffusion was governed by such things as distance, obstacles, nature of the environmental base, and receptivity of various social structures. One critical point is that the spread of new innovations was uneven. Whenever such innovations were introduced into a society there was a traumatic effect that necessitated new forms of organization, new patterns of leadership, and the acquisition of new skills. Rapid population growth often accompanied the changing socioeconomic conditions.

One further point to be emphasized about Deevey's (1960) interpretation of world population growth is the nature of the population growth curve for each revolution. After each rapid spurt in population growth, the growth rates slackened off—the numbers reached a plateau, and then further additions were slow to be achieved. Each revolution therefore removed, partially at least, some pre-existing constraint upon population growth, but it must also have set into motion forces that eventually brought growth under control. Obviously, these forces are of urgent concern in our present circumstance.

In response to the Industrial Revolution the world's population entered a period of rapid and sustained population growth. During the nineteenth century this growth was concentrated in the more developed countries. By the middle of the twentieth century, however, population growth had subsided in the more developed countries and was accelerating in the less developed countries, setting the demographic stage for the new millennium.

Many would have predicted that the rapid population growth of the twentieth century (from about 1.6 billion in 1900 to 6.1 billion in 2000) would have resulted in a world of extreme poverty and economic deprivation as resource scarcities led to higher prices for basic commodities. Such was not the case, however, and we need to keep this in mind as we look ahead to discussions of population growth and economic well-being, food supplies, and environmental concerns. We also need to keep in mind that the near-quadrupling of the world's population during the twentieth century was made possible in large part by the widespread development and use of fossil fuels.

Though no definitive yardstick is available for measuring such things, economists have suggested that the world's material standard of living increased perhaps nine-fold during the twentieth century—a considerable achievement. On the average people live longer, healthier lives now than they did 100 years ago. Geographically, however, the vast improvements in wealth during the twentieth century accrued mainly to the nations of Europe, the United States, and Japan. We enter the new century with vast differences in wealth among the world's nations—a person's place of birth largely determines his or her economic and demographic destiny.

A person's place of birth largely determines his or her economic and demographic destiny.

The Human Population Today

Jane Jacobs (2004, 168) noted, perceptively, that "The world today is a bewildering mosaic of cultural winners, groups of people sunk into old or recent Dark Ages and downward spirals, groups in the process of climbing out, and remnants of preagrarian cultures, as well as remnants of declined empires."

Today's population situation is unique in the world's history; not only is the current rate of increase still fairly high, but the base population (7 billion) is also the largest ever. The historical record shows that the acceleration of world population growth started with the European countries and the lands that Europeans settled overseas, especially the United States, Canada, and Australia. However, the areas that are growing the fastest today are Africa, Asia, and Latin America. These so-called less developed regions have more than three-fourths of humankind and are responsible for more than 90 percent of the world's population growth. Death rates in these regions' countries have been falling during the last twenty-five years, whereas birth rates have remained twice as high as they are in the more developed nations (Goldstein and Schlag, 1999).

Between 1950 and 1987 the world's population doubled from around 2.5 billion to over 5 billion, an increase of over 2.5 billion people in less than forty years. The Cold War ignored growing populations and changing geographic concentrations of people. During those years population growth had been unequally distributed geographically; more than 85 percent of that growth occurred in the less developed countries. In most of the more developed countries today, fertility hovers near or below replacement level (the United States is the major exception), so that an even higher percentage of population growth in coming decades will occur in the less developed countries, those least able to absorb additional people. By the year 2008, 85.4 percent of the world's population resided in the less developed countries; nearly half of them were residents of either China or India (the world's second "demographic billionaire"). It is easy to see why, then, we can expect international migration to flourish in the decades ahead as globalization of the economy brings together capital, which is heavily concentrated in the rich countries, and young people, who are heavily concentrated in the poor ones.

Still another comparison between the more developed and less developed countries can be made by considering the differing age structures of their populations. A country that has a rapidly growing population has a large proportion of its residents in the younger age groups. In the rapidly growing areas of the world—Africa, Asia, and Latin America—high proportions (typically 30–45 percent) of the population are under 15 years of age, whereas in North America, Europe, and Australia and New Zealand there are significantly lower proportions of young people (20 percent or less). A population with a large proportion of old people will have different needs and requirements than a population with a great many young people. Though we don't know exactly what the earth's carrying capacity for humans actually is today, many would argue that we may be approaching it soon. Some would even suggest that we've already passed it.

However, overpopulation is an elusive concept. Though it may seem to you that parts of the world are indeed overpopulated, that doesn't mean either that everyone would agree with you or that, by virtue of some regions being overpopulated, that the world is also overpopulated. If we could define an optimum population, then it would be easier to define overpopulation. It would be any population greater than the optimum. But there is no agreed upon definition of optimum population. Optimum for whom, we might ask, and for what? We might try to relate the optimum population in turn to carrying capacity, but as we've seen already, the carrying capacity of the world has been regularly altered by changes in technology that have allowed us to produce more food, the essential need that we humans have. We will revisit these ideas in more detail in Chapter 9, though we will never resolve the issue to everyone's satisfaction.

Population Projections

So far we have viewed the present population situation mainly in the perspective of the past, but what does the future hold in store for the world's population? One answer to that question, based on a set of assumptions about the dynamics of a population over some time period, is the **population projection**. Young (1968, ix) cautioned us long ago, however, to remember that:

A **population projection** is made on the basis of the population at some date and assumptions about births, deaths, and migration between that date and some future date.

> The projection of a future population from a present growth rate is a hazardous undertaking at best. These rates contain many variables and are sensitive to small changes in these variables. Furthermore, since population growth is cumulative, very slight changes in present rates can make enormous differences when projected 100, 200, or more years into the future.

Always remember that we cannot predict the future, and those who try are doomed to failure. As only one example, in 1949 *Popular Mechanics* predicted that "Computers in the future may weigh no more than 1.5 tons."

Demographers are careful to differentiate between projections and predictions. The *projection* for the size of a population at some future date is based on a set of *assumptions* about the demographic processes that will affect population growth over the time period. The simplest assumption is that the future rate of population growth will be the same as that of today. However, for most situations this is unrealistic. Typically, the projection is broken down into separate projections for the birth rate, death rate, and migration. These may then be combined into a single projection for the future population. What the projection shows is that, *if* the assumptions hold true for births, deaths, and migration, *then* the projection will be accurate. Demographers should not be held responsible if the assumptions are not fulfilled, you see. Often a projection may alter people's reproductive behavior and set into motion events that will assure that the projection will be off the mark. Usually more than one projection is made and quite often a series of projections is made, using different assumptions about future birth, death, and migration rates.

Population Projections: A Brief Overview

First, we should distinguish among the following three commonly encountered terms: projection, forecast, and prediction. A population projection is made on the basis of the population at some date and assumptions about births, deaths, and migration between that date and some future date. As Gibson (1977, 7) noted, "Population projections are 'correct' by definition (except for computational errors) because they indicate the population that would result if the base data (starting) population is correct and if the underlying assumptions about future change should turn out to be correct." The usefulness of projections, then, depends on what assumptions have been made and how well they accord with the actual events.

Demographers prefer to avoid the term *prediction* altogether because it suggests that only one projection has been made and it is considered as an ultimate truth, as occurring with a high degree of certainty. Past projections have often fallen so wide from their marks that the most demographers will venture today is to choose one of a series of projections as a forecast, and that usually only for short distances into the future, and then only reluctantly.

Two broad classes of population projections exist: mathematical and component. **Mathematical methods** are easier to understand and to apply, but **component methods** are preferable for most projections, especially for those beyond the short term, which is usually taken to be five years or less. Whereas mathematical methods employ some mathematical formula to a base population using an assumed rate of growth over the projection interval, component models separately project births, deaths, and migration, then combine the "components" into an overall population projection. These latter projections, of course, are also done mathematically, and the terminology sometimes confuses people. For demographers and population geographers there is no escape from mathematics.

Mathematical methods employ some mathematical formula to a base population using an assumed rate of growth over the projection interval.

Component models separately project births, deaths, and migration, then combine the "components" into an overall population projection.

Within each class of projections different models exist, so a wide range of projection models can be called upon. The model that should be used, of course, depends upon several factors, such as the size of the area for which projections are being made, the assumptions that can or should be made, the types of data that are available, and the length of the projection interval. With respect to scale, for example, national projections require different considerations than do projections for local areas. Understanding and projecting migration is, perhaps, more critical for local area projections than is the projecting of births and deaths. Conversely, projecting immigration at the national level may be much easier than projecting births mainly because immigration is controlled, at least to some extent, by the national government. Also, data for local areas are not always available in sufficient detail for employing component models, thus making mathematical models more attractive, especially for short-term projections.

Population Projections: World and Major Regions

Despite the difficulties, population projections are deemed essential and useful. They can stimulate our thinking about the consequences of population trends, for example. We need to keep in mind, however, what a United Nations study stated:

> What will population trends be like beyond 2050? No one really knows. Any demographic projections, if they go 100, 200, or 300 years into the future, are little more than guesses. Societies change considerably over hundreds of years—as one can readily see if one looks back at where the world was in 1900, or 1800, or 1700. Demographic behavior over such long time spans, like behavior in many spheres of life, is largely unpredictable (United Nations, 2004, 3).

The United Nations' world and regional population projections for the 1950–2050 period are shown in Table 1–1. Among the basic assumptions that the United Nations used to make these projections are the following:

1. At least a minimal degree of social order and control will be maintained.
2. Efforts at maintaining or improving the quality of life will continue and will not be totally frustrated.
3. Regional vital rates will move differently in terms of time, but eventually everywhere mortality and fertility will fall slightly below the lowest levels now observed.

As Table 1–1 shows, the potential for population growth is considerably higher in the less developed areas of the world than in the more developed ones.

Table 1–1 Population of the World, Major Development Groups and Major Areas, 1950, 1975, 2010 and 2050 According to Different Variants

	Population (millions)			Population in 2050 (millions)			
Major area	**1950**	**1975**	**2010**	**Low**	**Medium**	**High**	**Constant**
World	2 535	4 076	6 895	8 112	9 306	10 614	10 943
More developed regions	814	1 048	1 235	1 158	1 311	1 478	1 252
Less developed regions	1 722	3 028	5 659	6 955	7 994	9 136	9 691
Least developed countries	200	358	832	1 517	1 726	1 952	2 434
Other less developed countries	1 521	2 670	4 827	5 437	6 267	7 184	7 257
Africa	224	416	1 022	1 932	2 191	2 470	2 997
Asia	1 411	2 394	4 164	4 458	5 142	5 898	5 908
Europe	548	676	738	632	719	814	672
Latin America and the Caribbean	168	325	590	646	751	869	863
Northern America	172	243	344	396	447	501	444
Oceania	13	21	36	49	55	62	60

Source: Population Division of the Department of Economic and Social Affairs of the United Nations Secretariat (2011). World Population Prospects: The 2010 Revision. Highlights. New York: United Nations.

After getting close to 7 billion in 2010, the world's population is projected to increase by nearly 30 percent over the next 50 years, to a population of slightly over 9 billion according to the medium variant of the United Nation's updated projections. Generally, the future course of fertility is more difficult to project than that for mortality.

Population Projections: The United States

For comparative purposes it is useful to look at what has happened to official population projections for the United States. These projections are extremely important because they serve as the basis for many other projections, including the country's projections of the demand for housing, educational facilities, hospitals, and a myriad of other needs.

Population projections for the United States in the 2000–2100 period are shown in Table 1–2. Though these projections were published in early 2000, they do not include the 2000 census figures, which showed a population about 6 million larger than the 2000 figure used for projections. As a result, these projections are assuredly on the low side. The assumptions underlying the different variants in Table 1–2 are discussed in Hollmann, Mulder, and Kallan (2000). The middle series is considered the "most likely" variant, though with the usual caveat. In a commentary in *Time* magazine, Stengel (2006) wrote about the growing population of the United States as it passed the 300 million mark. He went on to celebrate the nation's growth and pointed out that it would only take about 40 more years to reach 400 million. He wrote that (2006, 8) "In America, we have always done Big well–big cars, big screens, Big Macs; we're the supersize nation. But now we are being challenged to trade Big for Smart." He didn't even mention the possibility of trading Big for Smaller. Even if we were to build "greener" buildings, squeeze more miles per gallon out of cars, and put solar panels on Wal-Marts, another 100 million Americans will further burden Earth's environment.

Given discussions in the United States about immigration rates, it is of particular interest to note the final column of the projections in Table 1–2 because they assume no

Table 1–2 Population Projections for the United States

	Middle Series	Lowest Series	Highest Series	Zero International Migration Series
2000	275,306	274,853	275,816	273,818
2005	287,716	284,000	292,339	280,859
2010	299,862	291,413	310,910	287,710
2015	312,268	297,977	331,636	294,741
2020	324,927	303,664	354,642	301,636
2025	337,815	308,229	380,397	307,923
2030	351,070	311,656	409,604	313,219
2035	364,319	313,819	441,618	317,534
2040	377,350	314,673	475,949	321,167
2045	390,398	314,484	512,904	324,449
2050	403,687	313,546	552,757	327,641
2055	417,478	312,160	595,885	330,991
2060	432,011	310,533	642,752	334,724
2065	447,416	308,716	693,790	338,999
2070	463,639	306,589	749,257	343,815
2075	480,504	303,970	809,243	349,032
2080	497,830	300,747	873,794	354,471
2085	515,529	296,923	943,062	360,026
2090	533,605	292,584	1,017,344	365,689
2095	552,086	287,826	1,097,007	371,492
2100	570,954	282,706	1,182,390	377,444

Source: U.S. Census Bureau. www.census.gov/population/projections/nation/summary/np-t1.pdf. (Feb. 14, 2000)

immigration. As a result, we see that the difference between the middle series variant and the zero immigration variant is about 76 million people by 2050 and more than 193 million by 2100. No matter how we view it, the United States will have a much larger population in the future if current high rates of immigration are sustained, which seems likely. Tables 1–3 and 1–4 show how different age and race groups will be affected by population change.

In summary, selecting an appropriate growth rate for projecting a population requires numerous considerations. We must look not only at past trends and current patterns, but also try to look at ways in which these trends may be altered in the future. Our projections can be no better than the assumptions upon which they are based. In addition, studying projections forces us to confront different scenarios about the consequences of different growth patterns and the possibility of designing policies to affect those patterns (Lee, 2000).

Small Area Population Projections

Though national population projections are of considerable importance, they are not the only ones that are of interest. Political units, from states to counties and cities, need to know something about their demographic futures, as do local school districts, highway

Table 1–3 Projected Population of the United States, by Age and Sex: 2000 to 2050 (In thousands except as indicated. As of July 1. Resident population.)

Population or percent, sex, and age	2000	2010	2020	2030	2040	2050
Population Total						
Total	282,125	308,936	335,805	363,584	391,946	419,854
0–4	19,218	21,426	22,932	24,272	26,299	28,080
5–19	61,331	61,810	65,955	70,832	75,326	81,067
20–44	104,075	104,444	108,632	114,747	121,659	130,897
45–64	62,440	81,012	83,653	82,280	88,611	93,104
65–84	30,794	34,120	47,363	61,850	64,640	65,844
85+	4,267	6,123	7,269	9,603	15,409	20,861
Male						
Total	138,411	151,815	165,093	178,563	192,405	206,477
0–4	9,831	10,947	11,716	12,399	13,437	14,348
5–19	31,454	31,622	33,704	36,199	38,496	41,435
20–44	52,294	52,732	54,966	58,000	61,450	66,152
45–64	30,381	39,502	40,966	40,622	43,961	46,214
65–84	13,212	15,069	21,337	28,003	29,488	30,579
85+	1,240	1,942	2,403	3,340	5,573	7,749
Female						
Total	143,713	157,121	170,711	185,022	199,540	213,377
0–4	9,387	10,479	11,216	11,873	12,863	13,732
5–19	29,877	30,187	32,251	34,633	36,831	39,632
20–44	51,781	51,711	53,666	56,747	60,209	64,745
45–64	32,059	41,510	42,687	41,658	44,650	46,891
65–84	17,582	19,051	26,026	33,848	35,152	35,265
85+	3,028	4,182	4,866	6,263	9,836	13,112
Percent of Total						
Total						
Total	100.0	100.0	100.0	100.0	100.0	100.0
0–4	6.8	6.9	6.8	6.7	6.7	6.7
5–19	21.7	20.0	19.6	19.5	19.2	19.3
20–44	36.9	33.8	32.3	31.6	31.0	31.2
45–64	22.1	26.2	24.9	22.6	22.6	22.2
65–84	10.9	11.0	14.1	17.0	16.5	15.7
85+	1.5	2.0	2.2	2.6	3.9	5.0

Male						
Total	100.0	100.0	100.0	100.0	100.0	100.0
0–4	7.1	7.2	7.1	6.9	7.0	6.9
5–19	22.7	20.8	20.4	20.3	20.0	20.1
20–44	37.8	34.7	33.3	32.5	31.9	32.0
45–64	21.9	26.0	24.8	22.7	22.8	22.4
65–84	9.5	9.9	12.9	15.7	15.3	14.8
85+	0.9	1.3	1.5	1.9	2.9	3.8
Female						
Total	100.0	100.0	100.0	100.0	100.0	100.0
0–4	6.5	6.7	6.6	6.4	6.4	6.4
5–19	20.8	19.2	18.9	18.7	18.5	18.6
20–44	36.0	32.9	31.4	30.7	30.2	30.3
45–64	22.3	26.4	25.0	22.5	22.4	22.0
65–84	12.2	12.1	15.2	18.3	17.6	16.5
85+	2.1	2.7	2.9	3.4	4.9	6.1

Source: U.S. Census Bureau, 2004, "U.S. Interim Projections by Age, Sex, Race, and Hispanic Origin," <http://www.census.gov/ipc/www/usinterimproj/>Internet Release Date: March 18, 2004.

Table 1–4 Projected Population of the United States, by Race and Hispanic Origin: 2000 to 2050 (In thousands except as indicated. As of July 1. Resident population.)

Population or percent and race or Hispanic origin	2000	2010	2020	2030	2040	2050
Population Total	**282,125**	**308,936**	**335,805**	**363,584**	**391,946**	**419,854**
White alone	228,548	244,995	260,629	275,731	289,690	302,626
Black alone	35,818	40,454	45,365	50,442	55,876	61,361
Asian alone	10,684	14,241	17,988	22,580	27,992	33,430
All other races 1/	7,075	9,246	11,822	14,831	18,388	22,437
Hispanic (of any race)	35,622	47,756	59,756	73,055	87,585	102,560
White alone, not Hispanic	195,729	201,112	205,936	209,176	210,331	210,283
Percent of Total						
Population Total	**100.0**	**100.0**	**100.0**	**100.0**	**100.0**	**100.0**
White alone	81.0	79.3	77.6	75.8	73.9	72.1
Black alone	12.7	13.1	13.5	13.9	14.3	14.6
Asian Alone	3.8	4.6	5.4	6.2	7.1	8.0
All other races 1/	2.5	3.0	3.5	4.1	4.7	5.3
Hispanic (of any race)	12.6	15.5	17.8	20.1	22.3	24.4
White alone, not Hispanic	69.4	65.1	61.3	57.5	53.7	50.1

1/ Includes American Indian and Alaska native alone, native Hawaiian and other Pacific islander alone, and two or more races.

Source: U.S. Census Bureau, 2004, "U.S. Interim Projections by Age, Sex, Race, and Hispanic Origin," <http://www.census.gov/ipc/www/usinterimproj/> Internet Release Date: March 18, 2004.

Small area population projections are more difficult to make because local variations in fertility, mortality, and migration may be much wider than those at the national scale.

planners, and urban and regional planning departments. In addition, numerous private corporations are interested in the changing demographics of local areas, so that they can better gauge local and regional changes in demand, marketing strategies, and even changing tastes and preferences. Thus the need for small area, that is, subnational, population projections exists and is growing.

At the same time, as you might expect, small area population projections are more difficult to make because local variations in fertility, mortality, and migration may be much wider than those at the national scale. Especially difficult to project are migration rates for local areas, because changes in an area's socioeconomic characteristics may quickly alter current migration patterns. For example, no one looking at the changes that had occurred in the Asian population in Long Beach, California, during the 1970s would have projected that that city and neighboring Lakewood would have a population of more than 20,000 Cambodians by 2000. At best, small area projections are a tricky business, and changing migration patterns are the major culprit.

The United States Bureau of the Census often publishes state population projections (Table 1–5 is one example). But even there the assumptions that must be made are difficult to choose. Consequently, different projections result primarily from different assumptions about migration patterns. However, the Bureau wisely chooses not to do population projections for areas smaller than states, leaving that task to state and local agencies and to private firms such as Donnelley Marketing Information Systems.

Culture, Population Growth, and Planning

The United Nations designated 1974 as World Population Year and in August of that year convened the World Population Conference in Bucharest, Romania. The purpose of that conference was to focus world attention on problems associated with population growth. At the time it was the largest international population meeting ever held and had representatives from 136 governments around the world.

Though most governments recognized the existence of population problems in their own countries, as well as throughout the world, there was much disagreement and debate about the reasons for the problems and the types of solutions that should be implemented. A number of countries participating in the conference felt that the reason for high birth rates was the lack of social and economic development, so that the emphasis should not be put on population and family planning programs but rather on development. One of the frequently heard slogans at Bucharest was "Take care of the people, and the population will take care of itself." A different position, however, was taken by a number of other countries, including the Western European nations, the United States, and Canada, which felt that reductions in population growth rates would make a substantial contribution to the process of economic development and that what was needed first was a decrease in population growth to induce development.

A more lucid explanation of this relationship between population-family planning and development was expressed by Nortman and Hoffstater (1975, 3):

> Whatever the stance on the political stage, the most ardent family planning advocates recognize that contraception "alone" will not produce housing, schools, or steel mills; and among the staunchest supporters of the "new economic order," many appreciate the demographic value of legitimated and government-subsidized family planning services.

At least three different positions on the population problem can be identified. One position is that population growth is a crisis issue and the problem is so grave that catastrophe is near unless dramatic actions are followed in order to reduce the growth. A second position is held by those who feel that population growth will intensify and multiply other

Table 1–5 Population Projections for Colorado, 2000–2020

	Census 2000				Projection 2020				2000–2020 Change	
	Number			Percent	Number			Percent	Total	
Age Group	Total	Male	Female	Total	Total	Male	Female	Total	Number	Percent
Total	4,301,261	2,165,983	2,135,278	100.0	5,278,867	2,673,752	2,605,115	100.0	977,606	22.7
0–4	297,505	152,353	145,152	6.9	387,617	198,926	188,691	7.3	90,112	30.3
5–9	308,428	158,119	150,309	7.2	374,214	192,310	181,904	7.1	65,786	21.3
10–14	311,497	160,118	151,379	7.2	354,829	182,246	172,583	6.7	43,332	13.9
15–19	307,238	159,971	147,267	7.1	356,206	185,673	170,533	6.7	48,968	15.9
20–24	306,238	162,619	143,619	7.1	348,671	183,260	165,411	6.6	42,433	13.9
25–29	331,795	175,593	156,202	7.7	366,054	191,208	174,846	6.9	34,259	10.3
30–34	332,232	172,898	159,334	7.7	375,798	195,906	179,892	7.1	43,566	13.1
35–39	366,092	185,712	180,380	8.5	362,533	187,561	174,972	6.9	-3,559	-1.0
40–44	370,731	186,634	184,097	8.6	337,217	175,414	161,803	6.4	-33,514	-9.0
45–49	334,855	167,899	166,956	7.8	332,145	172,597	159,548	6.3	-2,710	-0.8
50–54	279,270	140,452	138,818	6.5	314,505	161,697	152,808	6.0	35,235	12.6
55–59	194,722	96,345	98,377	4.5	317,178	160,418	156,760	6.0	122,456	62.9
60–64	144,585	70,739	73,846	3.4	300,997	149,732	151,265	5.7	156,412	108.2
65–69	121,222	57,663	63,559	2.8	253,427	123,143	130,284	4.8	132,205	109.1
70–74	105,088	47,250	57,838	2.4	197,011	93,439	103,572	3.7	91,923	87.5
75–79	85,922	36,043	49,879	2.0	124,550	55,128	69,422	2.4	38,628	45.0
80–84	55,625	21,422	34,203	1.3	81,141	33,357	47,784	1.5	25,516	45.9
85+	48,216	14,153	34,063	1.1	94,774	31,737	63,037	1.8	46,558	96.6
Under 18	1,100,795	565,710	535,085	25.6	1,327,467	682,912	644,555	25.1	226,672	20.6
5–17	803,290	413,357	389,933	18.7	939,850	483,986	455,864	17.8	136,560	17.0
18–24	430,111	227,470	202,641	10.0	494,070	259,503	234,567	9.4	63,959	14.9
25–44	1,400,850	720,837	680,013	32.6	1,441,602	750,089	691,513	27.3	40,752	2.9
45–64	953,432	475,435	477,997	22.2	1,264,825	644,444	620,381	24.0	311,393	32.7
65+	416,073	176,531	239,542	9.7	750,903	336,804	414,099	14.2	334,830	80.5

(*Continued*)

Demographic Indicator	2000	2020	Change
Median Age	34.3	36.0	1.7
Male	33.2	35.2	2.0
Female	35.4	36.9	1.5
Dependency Ratio (1)	61.7	72.8	11.1
Youth (2)	46.0	48.2	2.2
Old Age (3)	15.6	24.6	8.9

Demographic Indicator	2000	2020	Change
Child-Woman Ratio (4)	30.6	37.7	7.1
Sex Ratio (5)	101.4	102.6	1.2
Under 18	105.7	106.0	0.2
18–64	104.6	107.0	2.3
65–84	79.0	86.9	7.9
85+	41.5	50.3	8.8

(1) Dependency Ratio = (Age under 20 + Age 65 and over) / (Age 20–64) × 100

(2) Youth dependency ratio = Age under 20 / Age 20–64 × 100

(3) Old age dependency ratio = Age 65 and over / Age 20–64 × 100

(4) Child-Women ratio = Age under 5 / Female 15–44 × 100

(5) Sex Ratio = Male / Female × 100

Source: U.S. Census Bureau, Population Division, Interim State Population Projections, 2005 Internet Release Date: April 21, 2005

social problems, but that although population is important, it is not everything. A third position is held by those who feel that population is a nonproblem, or even a false problem, with the real problem being development or redistribution of income and power. A small but growing concern among demographers and others is with population decline, which has already begun in many European nations and a few others as well. An unexpected consequence of modernization has been the decline of birth rates to below replacement level. In summary, the very nature of the population problem and the consequences of population growth are under closer scrutiny and examination today than at any time in the past (Hardin, 1999).

The Laissez-Faire Point of View

The general argument stated by those in favor of some form of population control is that individual fertility decisions do not add up to what is socially optimal, or even desirable, hence such decisions cannot be left to individual families. Thus, parents intending to have children may impose a significant part of the cost and responsibility for those children on people other than themselves. These parents are therefore likely to have "too many" children.

On the other hand, however, there are those in favor of a laissez-faire solution. They feel that it is a question of individual choice, because it is the individual who bears the cost and receives the benefits of his own action. In general, the laissez-faire argument regarding population control is essentially the same as the laissez-faire argument in economics. Under proper functioning of the free market, without controls, the prices, both monetary and nonmonetary, that people pay for things reflect the real cost of production; and the prices that they receive reflect the real value of what they produce. Thus, when an individual makes an economic decision he or she bears all the costs and receives all the benefits. If the costs are less than the benefits, a positive decision is made; if the costs are greater than the benefits, a negative decision is made.

Laissez-faire population exponents feel that those best able to determine the costs and benefits of children are those who are contemplating having them. They are the ones who must assume the financial and social responsibility for that child and they are the ones who will benefit from that child. There are some, however, who believe that the costs and benefits for the family are not the same as the costs and benefits derived by the society for each additional child. For the individual family the ideal number of children may be five or six, whereas the ideal family size for the society may be only two.

Laissez-faire population exponents feel that those best able to determine the costs and benefits of children are those who are contemplating having them.

The Question of Cultural Genocide

Somewhat akin to the laissez-faire population exponents are those who feel that population control is a device proposed by more economically developed countries to control the less economically developed countries. Since the former group of countries is primarily the white non-poor nations and the latter group is the non-white poor nations, many believe that population control is a form of "genocide."

"**Genocide**" is a controversial concept with manifold emotional overtones. The United Nations Genocide Convention defined genocide as any of the following acts committed with intent to destroy, in whole, or in part, a national, ethnic, racial, or religious group:

Genocide is the deliberate and systematic destruction of an ethnic, racial, religious, or national group.

- killing members of the group;
- causing serious bodily or mental harm to members of the group;
- deliberately inflicting on the group conditions to bring about its physical destruction in whole or in part;
- imposing measures intended to *prevent births* within the group;
- forcibly transferring children of the group to another group.

The above definition was adopted unanimously by the General Assembly of the United Nations. According to the definition, mass sterilization of a compulsory nature would be considered genocide. Item (D) of the definition, "imposing measures intended to prevent births within the group," is either directly or indirectly related to the family planning programs espoused by the United States and other more developed countries. The important question thus becomes: Do family planning programs, as espoused by predominantly white, wealthy nations represent conscious, deliberate efforts to curtail nonwhite fertility, or do they reflect a genuine concern for the well-being and health of the rest of the world?

In an analysis of population control Darden concluded that, in his opinion:

> The poor and nonwhite should oppose any program which involves institutional limitation of population growth. Why? Because there is no guarantee that relative poverty would decline if the poor and nonwhites accepted such a fertility program. There might be fewer poor people in absolute numbers, but the gap between rich and poor would either get wider or remain constant. . . . In brief, institutionalized, coercive limitation of population growth is a policy aimed directly or indirectly at the poor and nonwhite and is therefore unacceptable as a solution to the problems of hunger, and other social ills in the United States and the world (Darden, 1975, 51).

Genocide is not just a thing of the past, no matter how much human rights are trampled in its path. In 2004 the United States officially recognized genocide in the Darfur region of the Sudan, even as it did little to stop it. Though humanitarian aid was sent to the region, the killing continues. We are constantly reminded of "man's inhumanity to man."

The Ethics of Population Control

As previously mentioned, there are those who believe that any form of coercive population control is unethical. However, a significant number of people feel that in order to solve the population problem and limit population growth it will be necessary to induce people to limit the size of their families. They feel that the hazards of excessive population growth pose such critical dangers to the future of the species, the ecosystem, individual liberty and welfare, and the structure of social life, that there must be a reexamination and ultimately a revision of the traditional value assigned to unlimited procreation and to the increase in population size.

Callahan (1971, 2) outlined some general ethical guidelines for governmental action and presented them in a "rank order of preferences" from the most preferable to the least preferable. They are listed here and are still very much worth thinking about. The government has an obligation to do everything in its power to protect, enhance, and implement freedom of choice in family planning. This means the first requirement is to establish effective voluntary family planning programs.

If it turns out that voluntary family planning programs do not curb excessive population growth, then the government has the right to go "beyond family planning." Callahan felt, however, that before governments take this second step they must justify the introduction of these new programs by showing that voluntary methods have been adequately and fairly tried. Callahan believed that the voluntary programs had not yet failed because they had not been tried in any massive and systematic way.

When the government has to choose among possible programs which go "beyond family planning," it has an obligation to first try those programs which, comparatively, are the least coercive. In other words, positive incentive programs and manipulation of social structures should be resorted to before "negative" incentive programs and involuntary fertility controls are applied. According to Callahan, if it appears that some degree of coercion is required, that policy or program should be chosen which:

- entails the least amount of coercion;
- limits the coercion to the fewest possible cases;
- is not problem-specific;
- allows the most room for dissent of conscience;
- limits the coercion to the narrowest possible range of human rights;
- least threatens human dignity;
- least establishes precedents for other forms of coercion;
- is most quickly reversible if conditions change.

In summary, the ethical considerations associated with population control are complex. Population policies, though they must take into account the interests and needs of particular regions and population groups, should have as their ultimate aim the best interests of the entire human species. Any plan to reduce world population growth to (some would argue even below) zero will have to carefully consider at least the following: economic development (including variations on the "Western model") and its role in reducing family size, the empowerment of women and gender equity, and the extension of family planning services to provide safe and efficient means of preventing unwanted births.

Population Dynamics and the World of Business

Demographic considerations and their spatial or geographic components play an important role in today's business world. Understanding these issues is of vital importance to business executives, who are increasingly turning to demographic experts for answers to a variety of problems. As one writer noted, business executives need to ". . . understand population changes and their impact on basic corporate decisions such as labor supply, location of facilities, the changing nature of markets, and the age makeup of consumer groups" (Hyatt, 1979, 1). *American Demographics* a now discontinued magazine presented succinct looks at the relationship between demographic trends and everything from where to market products to where to retire. Some of the material from *American Demographics* has since been incorporated into another publication called *Advertising Age.*

Marketing

Marketing relies heavily on demographic statistics and their spatial or geographic aspects. Market segmentation and differentiation now play a key role in marketing strategies. According to Francese and Renaghan (1991, 50), ". . . many markets have become too complicated and too unforgiving to rely on just one or two demographic variables." A variety of variables such as race, ethnicity, income, education, and age, have a symbiotic relationship and together form the basis of "database marketing." In order for marketers to succeed in the future they will have to understand the multi-dimensional demographic profiles of their market segment; in the United States multiculturalism itself is becoming a concern for marketing specialists, with growing Latino and Asian populations appealing to many product manufacturers and distributors. Additionally, the aging baby boom is a market segment that will be important in marketing.

The formation of households is an important variable to consider by planners for such utilities as gas, electric, and telephone; population gains and losses certainly shape the demand for power. Changing fertility rates and divorce rates, as well as the increase in the number of late marriages, have a tremendous impact on household formation and thus are important variables when predicting consumer demand for utilities.

African American buying power is expected to reach 1.1 trillion dollars by 2015. Latino buying power was 1 trillion dollars and Asian buying power was 540 billion dollars in 2010 (Humphreys, 2010). The gay market is by and large an affluent one. Average annual income for a gay household was $61,000 compared to $51,914 for the general US population. Gay buying power in the US was projected in 2010 to be 743 billion dollars (Witeck-Combs et al., 2010). Women influence 85% of all consumer purchases and account for 7 trillion dollars in spending. African immigrants are estimated to have 50 billion dollars in buying power (New American Dimensions, 2009). Race and ethnic groups as well as many gay communities reside in geographically distinct regions and locations of the US. This fact has obvious implications for the importance of the geographical analysis of markets.

Business Forecasting

Business people are always interested in the future. They are concerned about next week's sales, next year's profits, future changes in interest rates, and five-year capital investment schemes. Their natural tendency, however, has been to focus on the next fiscal quarter, instead of the next quarter-century. Short-term symptoms always seem to overshadow long-run causes. Worries about short-term problems tend to obscure the longer run, but, equally important, they obscure changes in population variables. For example, a home builder is usually concerned with changes in interest rates, but a shift in the divorce rate or a shift in migration patterns could be equally as important to that industry. Builders have only recently begun to appreciate the impact of the "**Baby Boom**," and the boomers have yet to have their impact on Medicare and Social Security, though it is now beginning.

Baby boom is the period following World War II from 1946–1964 characterized by a rapid increase in fertility rates and in the absolute number of births in the U.S., Canada, Australia, and New Zealand.

During the baby boom period, between 1946 and 1964, there were nearly 80 million births in the United States, 50 percent more than during the preceding fifteen years.

The impact of population change on business forecasting has recently taken on added importance. In a study on changing demographics and the future of business, James Hyatt observed that ". . . while only a few years ago the attention was on worldwide population growth rates, analysts are beginning to understand that for the United States a much more complex set of population shifts are at work. Fast, slow, and no growth are all occurring at the same time in different parts of the country" (Hyatt, 1979, 5). This important geographic dimension of population change is now receiving more attention from business forecasters.

Many of the present and future problems and opportunities to be faced by business have their roots in the most striking demographic phenomena of twentieth-century America: the unique high fertility period that followed the Second World War, the so-called "baby boom." During the baby boom period, between 1946 and 1964, there were nearly 80 million births in the United States, 50 percent more than during the preceding fifteen years. The baby boom or "bulge" has far-reaching effects in housing, employment, retail sales, education, and many other areas of concern to the business community. According to Hyatt (1979, 5), "Business managers would be well advised to keep in mind the location of that bulge from year to year, just as they pay attention to other economic indicators."

The first baby boomers began to turn 60 in 2006, and as Waldrop (1991, 24) suggested earlier, "It will begin a population explosion among affluent, maturing householders. And it will turn the 1990s into peak years for consumer spending." These baby boomers entering midlife brought about changes in a variety of business services. For example, the 1990s saw a rise in consumer demand for bifocal eyeglasses as those midlife boomers underwent inevitable changes due to the aging process. Visual and hearing impairments increase significantly after age forty-five. Increasing attention was paid to health concerns and boomers, in order to stay fit, had to be more selective about food choices. Talk of ice cream and Twinkies gradually gave way to conversations about cholesterol, monounsaturated fats, and triglycerides. This brought about a rise in consumption of food items like fish, poultry, low-fat milk, whole grain cereals and fresh fruits and vegetables. The boomers are also relatively affluent, and in recent years they have been pouring money into mutual funds and other investments, a major factor in the strong stock market performance during the 1990s, despite an economy that was struggling for the most part.

Although much of the attention of the business community has been focused on the young, there has been an increasing awareness of the aged and the aging of America (Soldo and Agree, 1988). Indeed, adults over age 65 now outnumber teenagers and have become one of the fastest-growing population groups in the country. In 1900 only 3.1 million people in the United States were 65 years of age and older, but by 1985 that figure reached 28.5 million and projections for the year 2030 are that there will be 64.6 million. In 1985, one in nine Americans or 11.7 percent of the population was at least 65 years old, but by 2010, because of the maturation of the baby boomers, one in seven Americans will be at least 65 years old (Hooyman and Kiyak, 1988, 22).

As baby boomers become "senior" boomers, ". . . public policy questions how we will support the needs of a growing older population and how we will structure policies to assure a fair and equitable distribution of resources for all age groups" will have to be assured (Bouvier and DeVita, 1991, 27). Government bureaucracies, charged with planning the provision of services for the elderly, will be faced with a variety of decisions with both political and social consequences (Laws, 1991, 32). Some of these problems will be further complicated by the changing ethnic and racial composition of the United States.

Not only will there be more older Americans, but because of their relatively high disposable incomes, they will present a growing market for the business community. England (1987, 8) points out that "Americans over age 65 are the second-richest age group in U.S. society. Only those Americans in the next-oldest age bracket, from 55 to 64, are better off." The aged have assets nearly twice that of the median for the nation. "The spending power of the mature market may be one of the best-kept secrets left in the age of demographic scrutiny," suggests Lazer (1985, 23).

Zero Population Growth and Business

To most entrepreneurs "growth" is a magic elixir; they are naturally attracted to growth, they love growth—be it in profits, sales, or incomes. Investors usually look for firms with solid growth records. Also, to most business managers a growing gross national product (GNP) means a stronger economy with more consumer spending, more employment, and more sales and services. There is, therefore, understandable trepidation among those with businesses when they contemplate the prospects of a slowing, or even cessation, of population growth. They fear reduced demand, which in turn means less profit, smaller dividends, more unemployment, and general economic uncertainty (even though effective demand depends not just on numbers but also on affluence, the ability to pay).

The impact of population growth on the economy of the United States is assumed to be positive because it has continued for so long. Economic theorists have, however, found it difficult to find a direct correlation between population growth and economic well-being in highly industrialized nations. During parts of the nineteenth century population growth was high, yet per capita income growth was modest. On the other hand, in the twentieth century population growth slowed down while the GNP remained high. Among the three economic superpowers—Japan, the United States, and the European Union—only the United States has a significant rate of population growth, and a lack of growth does not seem to have diminished the affluence of its competitors. Nonetheless, concern remains about the connection between economic growth and prosperity, especially as fertility continues to decline and populations grow older. In support of this concern Longman (2004, 41) noted, ". . . for better or for worse, population growth is still the prime driver of economic growth. Increase in population causes new houses to be built, new cars to be manufactured, and new law offices to be built."

After following similar demographic paths for many decades, trends in Western Europe and the United States began to diverge around 1980. Up until then fertility had been getting lower in each region, but after 1980 fertility declines continued in Western Europe whereas fertility began a gradual upward trend in the United States. The increase in American fertility was a result of both more births among native-born citizens and more immigration. Many immigrants have come from higher fertility societies, and their fertility in the United States has continued at a higher level than for the American-born population. This has been especially noticeable for the Latino population and for some Southeastern Asian groups as well. One result of the fertility divergence between Western Europe and the United States is its effect on the age distributions of the two populations; by 2050 the average age in the former is predicted to be around 53 and for the latter around 36. In turn, then, Western Europe's aging population will be more of a burden, whereas America's younger population is likely to be more innovative.

Labor Force

The composition and quality of the labor force are important variables for personnel managers to consider. During the next few decades the nature of the labor force will be considerably different from what it has been in the past. There will be more workers and they will need to be better educated. There will be more minorities in the labor force as well as more females, mothers of young children, and older workers over age 55. Many of these changes in the United States will be direct results of the baby boom.

Among the most important demographic variables in determining the size of the work force is the age structure of the population.

The supply of workers available for businesses will be the result of two separate trends in the labor force. First is the size of the working age population, and second is the labor force participation rate. Among the most important demographic variables in determining the size of the work force is the age structure of the population. The bulge in the young working force cohort is now progressing through the population, so the number of young people entering the work force will be declining, though earlier projections are going to be off because of rising immigration and higher fertility in places such as California.

At the other end of the work force age spectrum, among those over 55 years of age, important changes will take place. Changing attitudes about retirement could have far-reaching impacts on the corporate world. Can business easily absorb those older workers who decide to continue their careers? Will companies find it harder to promote young workers if retirement ages advance? Will older workers continue to be productive? These are all important questions to personnel managers and are underlain by demographic changes in society. During the 1990s corporate downsizing resulted in the elimination of many jobs held by these workers, leaving them floundering to compete for lower-paying jobs with few or no benefits.

Illegal immigration alone has become a critical issue among average Americans, yet politicians and corporate executives do their best to ignore it.

Labor migration in the future promises to be at least as important as population growth, and it will affect the populations of numerous sending and receiving nations. As Wallerstein (1999, 17) observed, "We shall nonetheless see a rise in the real rate of migration, legal and illegal—in part because the cost of real barriers is too high, in part because of the extensive collusion of employers who wish to utilize such migrant labor." Illegal immigration alone has become a critical issue among average Americans, yet politicians and corporate executives do their best to ignore it. As Bartlett and Steele (2004, 58) discovered, "For corporate America, employing illegal aliens at wages so low few citizens could afford to take the jobs is great for profits and stockholders . . . companies are rarely, if ever, punished for it." This issue can only attract more attention in the years to come.

In 2010 there were an estimated 10.8 million undocumented immigrants living in the United States (Hoefer, Rytina and Baker, 2011). Even though it is illegal to hire an undocumented immigrant, most of these millions are working in the US. It is jobs that attract them. Congress, never quick to solve real problems, is struggling to create new legislation that would deal with the problems of undocumented immigration, but at the time we are writing this it doesn't look promising. From building walls to seal off our southern border to giving amnesty to all who have entered undocumented, opinions and emotions run the gamut. There is much sound and fury, but so far little light.

The Education Roller Coaster

The nation's education system, perhaps more than any other social institution, has been greatly affected by fertility variations. The impact of the postwar baby boom children on the schools was dramatic, costly, and painful. In the late 1950s and 1960s school enrollments soared. The elementary-school-age population (5–13 years) grew from 23 million in 1950 to 37 million in 1970, for example. Secondary schools faced similar problems by the early 1960s, and the high school population doubled in size between 1950 and 1975 (Bouvier, 1980, 21). High school enrollments rose 14 percent in 1957 alone, and a critical shortage in classrooms was evident. Administrators projected a need for an additional 750,000 teachers in three years. This demand caused a rapid increase in school budgets.

Changing Enrollment Trends

School planners were totally unprepared for this rapid increase in enrollments. By the time colleges had geared up to graduate enough teachers and massive building programs had produced enough elementary and secondary classrooms, the crest of the baby boom wave was about to leave the K-12 school ages.

The mid-1960s saw enrollments in colleges and universities skyrocket. This increase was a result of both the sheer numbers of baby boomers and the increased proportion of young people going to college. In the 1950s only about 10 percent of those between ages 25–29 had completed at least four years of college. That proportion increased to 16 percent by 1970 and 23 percent by 1980. Between 1957 and 1975 college enrollments increased from 3 million to 11 million, creating a growth industry of its own. The baby boom generation thus became the most highly educated generation in American history (Bouvier and DeVita, 1991, 14).

The major change in school enrollment in the 1970s and 1980s has been a significant increase in nursery and preschool attendance. The percentage enrolled in nursery schools, pre-kindergarten or kindergarten programs, or child-care centers with an "educational" curriculum, increased from 11 percent in 1965 to 38 percent in 1988 (Bianchi, 1990, 30). In 2005 that percent increased to 74.

Another school turnaround occurred. After years of stable or declining enrollments, the school-age population was projected to increase by about 8 percent during the decade of the 1990s. This increase was the result of the baby boom's children filling the elementary and secondary schools across the country (Bouvier and De Vita, 1991, 15); it was also very different from one region to another.

Geography Variations

In order to understand these educational trends and their demographic causes, it is necessary to look beyond national statistics and to understand regional or more local circumstances. The geographic distribution of the population, as well as spatial differences in vital rates, presents a complicated picture. For example, the decline in births was not uniform across the United States in the 1960s and early 1970s. In some areas the peak in births was attained in 1957, whereas in other areas it was not reached until 1960 (Reinhardt, 1979, 10). Also, the decline in births within states was not uniform, and there were significant differences between metropolitan and nonmetropolitan areas.

These geographic differences, as well as other changes in population distribution, mean that some areas have had dramatic declines in the school-age population, whereas others have still been growing. Many rural areas and urban regions losing population have experienced dramatic declines, but many newly developing suburbs have enrollment increases, as do some inner-city areas. Nursery school attendance is more common in the Northeast than other U.S. regions, and among children who live outside the central cities of metropolitan areas (Bianchi, 1990, 30).

Migration has also had a significant impact on school-age populations. Increases can be seen throughout much of the Sunbelt and the West, with declines in the Midwest and the Northeast. An analysis of college-age students also points out significant spatial differences, with many Sunbelt and Western states being net importers of college students. Changes in demographic factors can have a significant impact on school planning policy decisions and should be incorporated into the decision-making process at all levels.

CHAPTER 2

Population Data

Key Terms

de jure	census
de facto	vital registrations
urban	sample surveys

Population studies require an abundance and variety of data—we need numbers and rates in order to better understand both population changes and the ways in which those changes are related to socioeconomic and environmental variables. Demographic changes do not take place in a vacuum.

Populations and their associated characteristics are constantly in flux, so it is necessary to maintain current statistics to the extent that we can, though the costs of doing so are often high. Because of this dynamic nature of populations, however, it is also desirable to look at a full historic range of data, when possible, and not just at information for a short period of time. In turn, then, because of the large masses of statistical information required, coupled with the need for historical data, the population geographer usually obtains information from reports by government agencies or other large organizations. In most cases the individual scholar has neither the money nor the time to gather the required information.

Three broad uses for demographic data can be distinguished (Seltzer, 1973, 5):

1. Planning, policy guidelines, and projections ranging from five years to the long term
2. Monitoring current demographic trends and applied programs
3. Scientific study, at either the micro or macro level, of interrelationships between demographic phenomena and socioeconomic developments

Each of these uses has specific requirements in terms of accuracy, timeliness, topical and geographical detail, and user confidence. The interrelationships of these variables for each data use are outlined in Table 2–1. The information in the table suggests that the greatest difficulties are encountered in monitoring current trends and programs; that the data used for longer-range planning purposes is the easiest to acquire and has the least stringent specifications; and that the information needed for scientific inquiry is immediate.

Table 2-1 Relevance of Specified Criteria for Assessing the Adequacy of Demographic Estimates, by Type of Date Use

Type of Use[a]	Accuracy	Timeliness	Detail: Topical	Detail: Geographical	User Confidence	Index of Total Burden[b]
Planning and projections	Medium	Medium	Low	Medium	Varies	8–10
Monitoring current trends	High	High	Medium	High	Varies	12–14
Scientific study	High	Low	High	Low	High	11

[a] The assessments and the use types are extensions of distinctions developed by Tukey (1960). In terms of his dichotomy, "planning and projections" and "monitoring current trends" fall largely under the domain of decision theory, while "scientific study" would call for conclusion theory.

[b] Sum of scores where high = 3, medium = 2, low = 1 and varies = 1–3.

Source: Reprinted with the permission of the Population Council from Demographic Data Collection: A Summary of Experience, by William Seltzer (New York: The Population Council. 1973), p. 6.

Conceptual Difficulties

At first glance many of the basic demographic measurement concepts seem to be concrete and unambiguous. The concepts of "age," "death," and "live birth" would seem to be straightforward and easy to define. The collection of accurate and reliable demographic data also involves some other concepts such as "city," "place of residence," and "household," which might appear to be relatively ambiguous. However, under close scrutiny, even the seemingly unambiguous terms (age, death, live birth) are difficult to conceptualize.

Age Data

Information on the age structure of a population is essential to many areas of population analysis.

Information on the age structure of a population is essential to many areas of population analysis. Ryder (1964, 449) noted that age is the central variable in the demographic model. It identifies birth cohort membership. It is a measure of the interval of time spent within the population, and thus of exposure to the risk of occurrence of the event of leaving the population, and more generally is a surrogate for the experience which causes changing probabilities of behavior of various kinds. Age as the passage of personal time is, in short, the link between the history of the individual and the history of the population.

Shryock and Siegel (1971, 201) also commented on the value of age data:

> Age is the most important variable in the study of mortality, fertility, nuptiality, and certain other areas of demographic analysis. Tabulations on age are essential in the computation of basic measures relating to the factors of population change, in the analysis of the factors of labor supply, and in the study of the problem of economic dependency. The importance of census data on age in studies of population growth is even greater when adequate vital statistics from a registration system are not available.

Gathering data on the age structure of a population would appear to be a simple process. The United Nations (2008, 135) defined age as "the interval of time between the date of birth and the date of the census, expressed in completed solar years." This type of information can be obtained by asking respondents either their ages as of their last birthdays or their dates of birth. Unfortunately, according to Seltzer (1973, 9), "these questions cannot be answered by populations that take no note of birthdays, and they will produce biased results when asked of those using a calendar other than the 'Western' solar calendar."

Even though the "Western" method of determining age seems logical and straightforward to those who live in countries where it is used, other traditions for determining age prevail in other areas of the world, and if these traditions are not accounted for in the data-gathering process then error and confusion can occur. For example, one of the most common non-Western methods of determining age is that used by the Chinese, who consider a person to be one year old at birth. Thereafter, a person ages by one year each New Year's Day. However, as noted by Hock (1967, 861):

> . . . some of the Chinese respondents . . . reckon their ages according to the Chinese system, some according to the Western system, and . . . others might compute their ages according to Japanese reckoning, by adding one year to their age on the occasion of the Western New Year instead of the Chinese New Year.

Similar results were found in a study of South Korea. Thus, the answer to a seemingly simple question, "How old are you?", can have more than one interpretation.

Total Population

Another seemingly simple question deals with the total population of a place. However, even this concept has its problems. For example, do you assign people to the place that they customarily inhabit regardless of where they happen to be residing on census day (**de jure**) or do you count them wherever they are physically present at census time regardless of their home place (**de facto**)? Different countries make different choices.

In Great Britain the *de facto* census is favored. This method has the advantage of showing exactly where everyone was at a given moment (census day). Its obvious disadvantage is that certain population figures may be increased or decreased because of tourists, traveling salespersons, or other transients. The United States, on the other hand, has traditionally conducted a *de jure* census: people are tabulated according to their permanent places of residence. The *de jure* method provides information that is not affected by temporary or seasonal movements of people; however, it has two principal shortcomings: (1) it involves costly and time-consuming work to transfer people from their places of interview to their places of usual residence, and (2) it is sometimes difficult to be certain where a person's "usual" or "legal residence" is located. Without further complication, the Census Bureau notes that "people should be counted where they live and sleep most of the year."

De jure is an expression that means "based on law."

De facto is an expression that means "in fact" or "in practice" but not spelled out by law.

Definition of Urban

Even more troublesome than defining the total population is defining the term "**urban.**" Almost every country has its own definition, and comparative studies of urbanization may be extremely difficult because of the variety of definitions. Also, within some countries the definition has changed over time. Most urban definitions fall into one of the following four categories: size, administrative function, legal identity, and site characteristics.

In 2010, the US Census Bureau defined an urban area as comprised by a densely settled core of census tracts and/or census blocks that meet minimum population density requirements, along with adjacent territory containing non-residential urban land uses as well as territory with low population density included to link outlying densely settled territory with the densely settled core. An urban area, must encompass at least 2,500 people, at least 1,500 of which reside outside institutional group quarters. Additionally, the Census Bureau identifies two types of urban areas:

1. Urbanized Areas (UAs) of 50,000 or more people;
2. Urban Clusters (UCs) of at least 2,500 and less than 50,000 people.

Urban is described as a geographical area (including cities and towns) distinct from rural areas.

Most urban definitions fall into one of the following four categories: size, administrative function, legal identity, and site characteristics.

Other countries also use the size criterion, but with different sizes. In Denmark, for example, urban places are those with populations of at least 250, whereas in India urban places are those with populations of at least 5,000. Size definitions of urban places are often inadequate because the size of a minor civil division, such as a county, may be used in place of the size of an urban place. Exact definitions are necessary before comparative studies can be made.

Some countries prefer to use administrative units as designations of urban places, as is common in parts of Latin America. In Peru, for example, urban places are defined as ". . . capitals of all departmentos and distritos plus all towns larger than the average size of these administrative centers not possessing rural characteristics." The major problem with this kind of definition is that places with no administration functions will be considered rural, regardless of their population. Another possible criterion for defining urban places is legal identity—towns with legal identity are classified as urban and all others, regardless of size or function, are classified as rural. This definition traces its origin back to the chartered cities of the Middle Ages.

Finally, site characteristics may be used to define an urban place. A place may be considered to be urban if it has certain urban characteristics, such as piped-in water supplies, numbered streets, and lighting. Occasionally both site characteristics and the previously mentioned criteria are applied, and the resulting definition of an urban place can be quite complex.

The United States' definition of urban has changed over time. Early American censuses created no distinction between rural and urban. A distinction was first made in the *Statistical Atlas of the United States* (1874), where urban was defined as towns greater than 8,000; subsequently, the Census of 1900 lowered the size limit to 2,500. Massive suburbanization, resulting in an increased number of people living in urban conditions, but not within incorporated city limits, rendered older definitions inadequate. In 1950 a density criterion for fringe areas was adopted, and an 11.5 percent increase in urban population was registered.

Several types of urban statistical divisions have been used by the Census Bureau. The most recent were agreed to in 2000, as authorized by the Office of Management and Budget. They are as follows: Metropolitan Statistical Areas, Metropolitan Divisions, Micropolitan Statistical Areas, Combined Statistical Areas, and New England City and Town Areas. These designations were updated in 2010 (OMB, 2010).

Migration and Areal Subdivisions

Conceptual problems concerning migration have always been problematic. As Petersen (1975, 41) noted, "We know whether someone has been born or has died, but who shall say whether a person has migrated?" Everyone agrees that when we go to the neighborhood grocery store we have not migrated. Likewise, there is agreement that when we leave our home in one country and establish residence in another we have migrated. However, between these two extremes problems arise. How far does one have to move in order to be considered a migrant? How long must one stay at the new location? The questions of how far and how long, as well as other circumstances, play an important role in the definition of migration (discussed further in Chapter 8).

"We know whether someone has been born or has died, but who shall say whether a person has migrated?"

The geographic subdivisions that are used for collecting and mapping population data present further problems. As Zelinsky (1966, 22) pointed out, "Urban residence and function have generally spilled far beyond the political bounds of the metropolis; and if urban growth is rapid, no amount of technical alacrity can keep census divisions abreast of the actual urban-rural frontier."

In summary, the collection and analysis of demographic data is wrought with conceptual problems, primarily problems of definition, and, unless these discrepancies are dealt with, gross inaccuracies can occur.

Types of Records

Information for demographic analysis generally comes from three sources:

1. census enumerations
2. vital registrations
3. sample surveys

These sources are the most comprehensive records available and are generally provided by national governments. A variety of fragmentary demographic data exists and may occasionally serve useful purposes, but such data are usually collected for selected small populations. Examples include the following: school censuses, church records, records of births and deaths recorded by hospitals, city directories, auto title transfers, and marriage licenses. They are often readily available, and their use should not be over-looked; however, they generally are inadequate substitutes for the data available from the major sources of demographic information.

A **census** is a total count of the population of a specified area, generally a nation, and ". . . a sort of social photograph of certain conditions of a population at a given moment which are expressible in numbers" (Willcox, 1940, 195). In addition to enumerating the population, most censuses collect other data, such as age, sex, place of residence, place of birth, income, occupation, education, and religion. The quantity of information collected is mainly a function of the amount of money available for financing the census. A census is a major and expensive undertaking requiring considerable expertise and planning for it to be successful. Thus, censuses are usually conducted by national governments. The 1980 Census of the United States cost over a billion dollars, nearly five dollars per person enumerated. The 2010 Census cost 13 billion dollars or about 42 dollars per person.

A **census** is a total count of the population of a specified area, generally a nation, and ". . . a sort of social photograph of certain conditions of a population at a given moment which are expressible in numbers."

Vital registrations are the recording and compilation of vital statistics, at or near their time of occurrence. They usually include such events as deaths, fetal deaths, births, marriages, divorces, and at times disease and illness.

Vital registrations are also major sources of demographic data. Vital statistics are recorded and compiled at or near their time of occurrence and usually include such events as deaths, fetal deaths, births, marriages, divorces, and at times disease and illness. Unlike the census, which is a static, cross-sectional view of a population at a specific moment of time, a registration system is a dynamic recording of events that can change rapidly.

Unlike the census, which is a static, cross-sectional view of a population at a specific moment of time, a registration system is a dynamic recording of events that can change rapidly.

The major responsibility for reporting vital events to civil registration authorities, depending upon the country, is given to a local registrar, parents or relatives, or to physicians, midwives, undertakers, or religious officials; persons with special duties relative to births and deaths. Seltzer (1973, 35) suggested that "ideally, a national birth and death registration system obtains a report of each event shortly after it occurs, and statistical summaries of these reports, in conjunction with externally derived population estimates, can be used to compute vital rates." If the vital registration system is functioning effectively, it is capable of producing precise mortality and fertility estimates.

In some countries, such as the Netherlands, Finland, Belgium, and the Scandinavian countries, the vital registration system is so complete that a local registration bureau keeps a card for each individual. Major demographic events such as marriages, divorces, and changes in residence are noted on the card and this information can be available quite readily.

Because of the time and expense involved in a census, sample surveys are often used. Their advantages in terms of quality and cost are now well recognized. A major concern is assurance of a truly representative sample of the population. Many sample surveys dealing with demographic issues are conducted by census bureaus and other governmental and private agencies; sampling is also used within vital registration systems.

Census Enumerations

By authority of the Constitution of the United States, the population of the United States must be enumerated every ten years. In other countries the time span between censuses may be as short as five years, or censuses may occur at infrequent intervals depending upon economic and political circumstances.

The United Nations (2008, 7) regards a census as ". . . the total process of collecting, compiling, evaluating, analysing, and publishing or otherwise disseminating demographic, economic and social data pertaining, at a specified time, to all persons in a country or in a well-delimited part of a country." A census should also have the following characteristics:

- It should be of the population of a strictly defined territory.
- It should include everyone.
- It should be conducted at one time, preferably on one day.
- It should be conducted at regular intervals.

A Brief History of Census Enumerations

The thought of counting an area's population occurred to people in ancient times. Perhaps the oldest account of census taking is that of the Incas, who organized their society according to a decimal system. Though this was not a census in any modern sense, it is at least a variation on the general theme.

Enumerations of various kinds were made in early times in Sumeria, Babylonia, and Egypt. A rather large amount of data, though quite fragmented, were recorded in ancient China. A census of sorts was taken in Rome during the sixth century, B.C.

Examples of early census counts appear in the Bible in the Old Testament. Some passages illustrate both the purpose of the counts and the reasons for resistance to them. Many early population counts were for such purposes as conscription into the military service and the labor force. Other counts were taken for tax purposes. The word census comes from the Latin word, *censere*, which means "to value" or "to tax." Thus, it comes as no surprise to find that early attempts at enumeration were looked upon with considerable suspicion and resentment. For example, two Biblical accounts are given of a census of Israel that was taken by David for military conscription purposes. However, because God found the census to be prideful, he punished David for undertaking it.

According to Carr-Saunders (1936, 14), when the notion of taking a census was first proposed in England in the 1750s, ". . . the dire results of the census which David forced Joab to make were quoted by those who opposed the measure, and it was prophesized that some 'public misfortune' or an 'epidemical distemper' would follow if an enumeration were attempted." The result was a long delay in the first census of England, as well as continued anguish about censuses and census takers.

Modern Censuses

Modern census taking commenced in Scandinavia. Sweden began taking regular population counts in 1749; Norway and Denmark began in 1769. The first United States census was taken in 1790, and the United States was the first country to legislate a continuing, time-specific census. Thus, the practice of regular census taking originated in Western Europe around the middle of the eighteenth century and became widespread in Western

Europe before the middle of the nineteenth century; during that interval enumeration efforts became more sophisticated.

It is not sufficient to know that a country has taken a census; we also need to know whether or not censuses are taken at regular intervals—many are not. For example, China took a census in 1953 and then not again until 1982.

Many improvements are still needed in census taking and demographic knowledge, in spite of the fact that only about two percent of the world's population lives in countries that have never taken censuses. The greatest recent gains in census taking have been made in sub-Saharan Africa.

The Census of the United States

The United States Bureau of the Census is the record keeper for the nation. The major output of the Bureau's data-collection activities are its statistical reports, which not only provide a record of the number of people in the United States but also document changes in characteristics of the people, manufacturing establishments, farms, retail stores, and a host of other business enterprises from which Americans derive their living. The Census Bureau publishes more statistics, encompasses a larger range of topics, and serves a greater variety of statistical needs than any other federal agency. According to U.S. law, the Census Bureau is responsible for taking a variety of censuses; they are outlined in Table 2–2.

From the beginning, however, the primary task of the Census Bureau has been to provide the federal government with an accurate head count of the U.S. population. The count is to be conducted on the first day of April in every year that ends with a zero. The Bureau of the Census is required to notify the President by December of that year as to how many people he presided over eight months earlier.

The first census of the United States was strictly a head count and included only five questions, one of which was dropped in 1870—the one that asked the head of each household how many slaves he owned. It was conducted by 650 U.S. marshals and their deputies, and the results were issued in a single volume, fifty-six pages long. The entire operation cost $44,000, or approximately $.011 or 1.1 cents per person. However, the censuses have changed considerably since 1790. The volume of output has increased from a single short volume to some 15,000 separate publications.

Table 2–2 Types of Censuses in the United States

Census	Periodicity	Recent Censuses
Population	10 years	1960, 1970, 1980, 1990, 2000, 2010
Housing	10 years	1960, 1970, 1980, 1990, 2000, 2010
Agriculture	5 years	1984, 1989, 1994, 1999, 2004, 2009
Business	5 years	1982, 1987, 1992, 1997, 2002, 2007
Construction industries	5 years	1982, 1987, 1992, 1997, 2002, 2007
Governments	5 years	1982, 1987, 1992, 1997, 2002, 2007
Manufactures	5 years	1982, 1987, 1992, 1997, 2002, 2007
Mineral industries	5 years	1982, 1987, 1992, 1997, 2002, 2007
Transportation	5 years	1982, 1987, 1992, 1997, 2002, 2007

Source: United States Bureau of the Census.

Though the United States was one of the first nations to establish a modern census, it lagged far behind in establishing a permanent bureaucracy to continue the task. Originally the State Department had the chore, but in 1830 it was transferred to the Department of the Interior.

A full-time bureau was not set up until 1902, when Congress decided to put the Census Bureau in the Department of Commerce and Labor. Later, when this department was split into two separate departments in 1913, the Census Bureau went into the Commerce Department, where it remains today. However, Congressional proposals to eliminate the Department of Commerce were made in late 1995. If Commerce is eliminated, the Census Bureau would remain, though it was not clear at the time where it would be housed. Appendix A shows the Census Bureau's current regions and divisions.

Many changes in the type and number of census questions have been made since 1790. The largest number of questions ever asked was seventy in the 1970 census, although four-fifths of the people enumerated didn't have to respond to more than twenty. In 1980 and 1990 the number of questions declined slightly. The number of questions in the 2000 census (Appendix B) was also higher than in the 2010 one (Appendix C). Questions have been eliminated in the past because the information was no longer valid, for example, the number of slaves, or because the question was confusing. For instance, in 1960 enumerators were asked to rate the condition of the homes that they visited as either good, dilapidated, or deteriorating. This question was subsequently dropped because of the lack of agreement on the meaning of terms like dilapidated. Whatever problems and shortcomings the Census Bureau might have, however, we will continue to take a decennial census. As Anderson (1988, 240) concluded in her excellent history of census-taking in the United States:

> The census remains both an apportionment tool and a baseline measure for American social science. These dual functions have provided the impetus for technical innovation in the past and will continue to do so in the future. . . Americans will undoubtedly continue to conceive of social issues in terms of census categories and to redistribute the benefits and burdens among the people—"according to their respective numbers."

Vital Registration

Historically the collection of vital registration information has been the responsibility of ecclesiastical authorities. Most cultures have some form of religious ceremony attached to each of the major vital events: baptism and birth, wedding and marriage, funeral and death. The systematic recording of vital events has been traced back to A.D. 710 in parts of Japan, while in 1497 the Archbishop of Toledo (Spain) instructed parish priests to maintain vital records. In England, the Vicar General under Henry VIII instructed the clergy to record the number of baptisms, burials, and marriages. In 1563 the Council of Trent made the keeping of burial and baptismal records compulsory in the Roman Catholic Church.

Most of the first annals were kept by parish priests, mainly to determine the severity of the plague. Many of those early records were destroyed by age, fire, or pests long ago. Gradually, toward the end of the eighteenth century, government agencies began taking over the vital registration function. For example, national civil registration began in Sweden in 1756, in France in 1792, in England in 1837, and in Ireland in 1864. Thus, vital registration data became available at about the same time that census enumerations started appearing.

Vital Registration in the United States

The United States was one of the first countries to begin taking a regular modern census, but one of the last countries to initiate the collection of vital statistics. Early attempts to collect vital statistics were sporadic. As early as 1639 Massachusetts had begun collecting and maintaining records of births and deaths, but this was done only locally. It was the government, however, and not the clergy, that began keeping the vital records, though the model was based on English precedent. Despite all attempts to enforce compliance with the registration system, early record keeping was not effectively accomplished.

The United States was one of the first countries to begin taking a regular modern census, but one of the last countries to initiate the collection of vital statistics.

During the nineteenth century there was an increased awareness of the importance of vital records. Lemuel Shattuck, the founder of the American Statistical Association in 1839, used that association as a forum for lobbying for vital registration laws, and such laws were successfully established for Massachusetts in 1844. In 1850 the United States Census Office invited Shattuck to help develop plans for obtaining data on births and deaths, a venture that was unsuccessful.

In 1880 the Census Office created a Death Registration Area, which included all cities and states that registered 90 percent or more of their established deaths. At that time only Massachusetts and Washington, D.C., qualified. However, there was a gradual increase in death registrations as other areas improved their record collection techniques. It has been estimated that by 1890, 31 percent of the population in the United States was included in the Death Registration Area. Part of the reason for the slow expansion of the Death Registration Area was the lack of a central office to encourage states to register births and deaths. The Census Office was created each ten years for the purpose of taking the decennial census, but following each census it was disbanded. Thus, there was virtually no promotion of the Death Registration Area.

In 1902 the Permanent Census Act created the Bureau of the Census, which was charged with the responsibility of collecting vital statistics along with its other duties. Subsequently the entire record collection process was accelerated somewhat.

In 1915 the Birth Registration Area was established. At that time it contained approximately 33 percent of the population. Table 2–3 shows the additions to the Birth Registration and Death Registration Areas. By 1930 both Areas were 90 percent complete and, with the addition of Texas in 1933, both Areas were complete. In 1957 a Marriage Registration Area was established, and in 1958 a Divorce Registration Area was added.

It was originally the task of the Bureau of the Census to collect vital registration information; however in 1946 the Public Health Service's National Office of Vital Statistics was authorized to collect the data. In 1960 another reorganization occurred and the National Center for Health Statistics was formed. Its National Vital Statistics Division currently collects, publishes, and distributes information on births, deaths, marriages, and divorces. The data are collected from county courthouses, registrars, and halls of records and tabulated and recorded on monthly and annual bases.

Table 2–3 Percent of Population in Birth and Death Registration Areas in the United States

Year	*Birth Registration Area*	*Death Registration Area*
1880	—	17
1890	—	31
1900	—	26
1910	—	51
1920	60	81
1930	95	95
1933	Both areas complete, with addition of Texas	

Vital Registrations throughout the World

Birth and death registration is now virtually complete in most of North America, Europe, and Oceania, and in certain parts of the Caribbean and Latin America. However, as Seltzer (1973, 36) noted, "In view of the basic motivational and administrative obstacles to the establishment of civil registration on a reliable basis, it is not surprising that official registration systems have been estimated as recording 40 percent or less of the live births occurring in Indonesia, Pakistan, India, the Republic of Korea, and the Philippines." Apparently death registrations are less complete than births in Asia also, and in most cases the coverage of tropical Africa is even more deficient than that of Asia. Considerable uncertainty exists in all the less developed countries with respect to the completeness of their registration systems.

Two major problems in the less developed countries relative to vital registrations are the following:

1. A considerable number of births and deaths are never reported.
2. Whole sections or areas of countries are not part of the data-collecting system.

The implications for demographic research are clear—a great deal of uncertainty surrounds the fertility and mortality data of less developed countries.

Sample Surveys

A **sample survey** is a canvass of selected persons or households in a population usually used to infer demographic characteristics or trends for a larger segment or all of the population.

Because of the expense and time involved in interviewing everyone in a country, demographers frequently use **sample surveys** to collect information.

Because of the expense and time involved in interviewing everyone in a country, demographers frequently use **sample surveys** to collect information. The advantages of using a sample over a complete census are that

1. A sample can be taken much faster than a complete enumeration, and if timeliness is important to a particular study a sample may be the only means to collect the needed data.
2. A sample survey is much cheaper to administer than a complete enumeration.
3. The quality and accuracy of a sample can be greater than with a complete enumeration, particularly where there are a limited number of trained personnel to conduct a complete enumeration.
4. There is less paperwork involved, and data handling and processing can be easier.

In some nations, particularly the less developed ones, a sample survey is the only economical method of collecting reliable demographic data. As Bogue (1969, 106) has commented:

> In some nations sample surveys are so widely used and have become so precise that a national census serves less to provide national totals of population composition than to provide data about smaller areas—states, cities, counties, townships, census tracts, and similar minor civil divisions—and to permit more detailed breakdown into refined categories and more elaborate cross-classifications of characteristics than do the sample surveys.

Requirements for an accurate and reliable sample survey include the following:

1. There must be an effective sampling design that will avoid biases in the selection process.
2. The sample design must lend itself to the estimation of sampling errors. It is difficult to develop and execute a sample design that is both economically feasible and scientifically correct without the help of a sampling expert.

Many variables affect the results of a sample survey, including the following:

1. characteristics of the person supplying the information
2. characteristics of the person collecting the information
3. salaries and benefits paid to those conducting the interviews
4. the amount of supervision
5. the collection procedure used
6. the type of questions used
7. the actual wording of the questions used

Multiple-round surveys are becoming more important in demographic data collection. In the multiple-round survey, or longitudinal survey, information about a specific group or problem is collected on a continuing basis. It is a useful method for collecting information, particularly in areas where there is poor census coverage. According to demographer Ralph Thomlinson (1976, 75), "Multiple-round surveys offer the triple advantages of cost reduction through sampling, keeping up-to-date by re-interviewing annually, or if desired, more frequently, and the opportunity to add and subtract questions as desired. In the United States the Bureau of the Census regularly conducts a variety of sample surveys. Three major categories of reports are published, *Current Population Reports, Current Housing Reports,* and the *American Community Survey.*

Quality and Completeness of Population Data

Though there are exceptions, the quality and quantity of the collected demographic data are generally related to levels of economic development. In the poorest countries demographic data are usually either absent or incomplete and of poor quality, whereas in the wealthy countries demographic data are likely to be abundant and of good quality. Censuses and vital registration systems require a considerable outlay of money, and in many of the poor countries money is simply unavailable for such purposes. For this reason sample surveys, often done with the help of the United Nations, are being conducted in many of the less developed countries. They cut down on the cost and, if carefully designed and executed, can provide useful information. Martin (2006) and others are concerned about whether future censuses will contain as much detail as they have in the past.

Completeness of Coverage

Countries differ considerably in the number and types of questions they include on their census forms. Most censuses include questions on age and sex, and for perhaps half of the world's population censuses include questions on education and occupation. While about one-third of the world's people are asked questions on their language and religion, far fewer are asked about their ethnic identity. Though part of the reason for wide variation in census designs is financial, the variations are also affected by the political sensitivity of asking questions that some people may consider an invasion of privacy. Many people still view censuses with suspicion and resentment.

The list of questions asked in the 1990 United States Census was quite lengthy, and about the only topic not dealt with was religion. On the other hand, Gambia's 1963 census consisted of only five questions: place of enumeration, tribe, age, sex, and marital status. Beyond the problem of insufficient coverage, another problem is the lack of geographical detail. Though in the United States some data are available for geographic units as small as census tracts, block groups, and even blocks, in the less developed countries very little data is published for comparable minor civil divisions. Even when such data are available in less developed countries, their use is often limited by a lack of maps showing the exact boundaries of the divisions used.

Another problem of geographic coverage is the failure of census takers to visit isolated areas, either because of the physical difficulties involved in getting to such places or because of the fear of visiting such areas. The lack of adequate roads and communications systems in less developed countries often means that some geographic areas will be either omitted or very poorly covered.

Even for countries that have detailed modern censuses, problems arise with historical studies. Questions often disappear from the census as we go back in time. A problem that often occurs in rapidly growing urban areas is the split tract problem; this happens when a census tract in a given time period is split into two tracts for the next census. Geographer Erick Howenstine (1993) provides help for people who run into this particular difficulty.

Errors and Omissions

Occasionally, census records are purposely falsified, as in the case of the 1940 Guatemalan Census. According to Whetten (1961, 20):

> The 1940 census might have been fairly adequate if the figures had not been deliberately tampered with. President Ubico, who wanted to show a large population for political purposes, issued orders to local authorities to alter the count after the census was taken. Relevant papers in the central files were carefully destroyed, but enough documentary evidence has been found in the offices of local authorities to indicate that the census results were inflated by at least 900,000 inhabitants.

Massive misreporting occurred in Nigeria in the 1962 census, as political parties in different regions attempted to increase their representation in the national government. The 1962 census was officially disavowed. Fortunately, such examples of deliberate falsification of census records are rare. For the most part people attempt to faithfully and honestly answer census questions, though they often give wrong answers when they misunderstand questions or suffer lapses in memory. Most census errors are a result of omission rather than commission.

Errors in Age and Mortality Reporting

Aside from the conceptual problems related to the collection of age data, misreporting of age is believed to be quite common, especially in the less developed countries. Studies of African and South Asian populations confirm this skepticism about age data. In a study in Africa, Blacker (1969, 277) noted that, ". . . the widespread ignorance of age in the sense of numbers of completed years bedevils the collection of accurate data, and studies which have been made of reported African age distributions have revealed massive and deep-seated errors." The age situation in India has received a similar evaluation, and even in more developed countries some data suggest that age is often misreported.

Problems similar to those of age misreporting can be found when examining mortality records. For example, because all mortality reports involve proxy respondents, a poor accounting of deaths is common for persons living alone or living with non-relatives prior to their death.

Mortality in the more developed regions is concentrated among the elderly, whereas in the less developed areas approximately half of the deaths occur to persons less than 15 years old; and many times more deaths occur in the first year of life than among all of those over 65. In the more developed countries, where a larger proportion of older people are found, mortality surveys that omit the deaths of persons living alone or with non-relatives will have a larger impact. Underreporting of deaths of the very young is common in all areas, but it has a greater impact on the less developed countries, where there is a larger proportion of young people.

Errors in the United States Census

In such a large undertaking as the United States census, with more than 635,000 door-to-door follow-up workers in 2010, it is no wonder that errors have been made. Kahn (1974, 44) noted the following:

> While the general public is sometimes surprised to hear that census figures are not beyond question, this is no news to the nation's enumerators themselves. They are among the few bureaucrats in any federal establishment who have consistently and candidly and voluntarily conceded that they make mistakes.

Among the early critics who reminded the census takers of their fallibility, George Washington remarked that:

> Returns of the Census have already been made from several of the States and a tolerably just estimate has been formed now in others, by which it appears that we shall hardly reach four million; but one thing is certain: our *real* numbers will exceed, greatly, the official returns of them; because the religious scruples of some would not allow them to give in their lists; the fears of others that it was intended as the foundation of a tax induced them to conceal or diminish theirs; and thro' the indolence of the people, and the negligence of many of the Officers, numbers are omitted (Fitzpatrick, 1939, 329).

Fortunately, the Census Bureau realized the fallibility of its information and by 1950 started to conduct regular self-analyses called Post Enumeration Surveys. In 1970 they spent over $3 million for self-evaluation; more than the total amount spent on the census a century earlier. In 1980, 1990, 2000, and 2010 the task was both more complicated and more expensive.

Historical undercounts in the United States census are shown in Table 2–4. The 2000 census was unique; it actually showed a slight overcount, indicating that the number of people overcounted actually exceeded the number of people undercounted. It also erased or greatly reduced traditional undercounts for specific subgroups of the population. The undercount for African Americans in the 2000 census was 1.84 percent, well below the 4.57 percent undercount of that population in the 1990 census. Undercounts of other minority populations were reduced to zero, or at least to figures that were not statistically significant from zero. As Graham and Waterman (2005) noted, undernumeration is common for many different populations.

The 2000 Census in the United States

The 2000 Census was full of surprises. For example, the census counted 281,421,906 residents, nearly 7 million more people than the Census Bureau had estimated it would find on April 1, 2000. It also showed the decade of the 1990s to be the greatest period of

Table 2–4 Historical Net Undercounts in Censuses in the United States

Year	*Percent Net Undercount*
1940	5.4
1950	4.1
1960	3.1
1970	2.7
1980	1.2
1990	1.8
2000	0.1

Source: United States Bureau of the Census.

population growth in the nation's history—population grew by nearly 33 million people in a decade (nearly the equivalent of the 2000 population of California). Furthermore, population growth occurred in all 50 states, though hardly at the same rates. Twelve seats in the House of Representatives changed from one state to another, more than had been expected. The big winners (two new seats each) were Texas, Florida, Georgia, and Arizona—New York and Pennsylvania each lost two seats.

After a three-decade decline, the response rate for the 2000 census pegged upward, with an initial return of 67 percent of mailed-out questionnaires. The count was also thought to be the most accurate one that the Census Bureau had produced for some time, an important consideration considering that nearly $200 billion in federal funds are allocated annually on the basis of census figures (Anderson and Fienberg, 1999).

After considerable argument and deliberation, the Census Bureau declined to statistically adjust the new census data in order to adjust for undercounts, which may have been around 3 million people or just over 1 percent (Prewitt, 2000). The comparability of race and ethnicity data has been confounded by the decision to allow people to choose more than one race for the first time ever. Nearly 7 million people took advantage of the idea and identified with more than one race (Skerry, 2000).

Census 2000 also made clear that rapid changes were occurring in the ethnic/racial composition of the United States. The Hispanic population grew much faster than expected (from 22 to 35 million in a decade) and is now larger nationally than the African American population. The Asian population grew rapidly also. As an example of what is ahead for the nation, the non-Hispanic white population in California slipped below 50 percent, leaving the Union's most populous state with a population that has no racial-ethnic majority.

The 2010 Census in the United States

It takes considerable time, effort, and money to complete a national census in the United States. Preparation started in 2003 with a test of response methods and the race question. This was a survey of 240,000 households and focused on trying to find the best way to increase the mail response to the census, hence decrease the expensive process of following up on non-responsive households.

In 2004 two census tests were conducted to test new procedures and technologies, especially in culturally diverse areas. One test site included seven neighborhoods in Queens County, New York, and the other included three rural counties in southwest Georgia, Colquitt, Thomas, and Tift.

In 2005 a test of the design and wording of the census form was conducted, along with a search for methods to increase responses to the census. A sample of 420,000 households was selected nationwide.

In 2006 census tests were conducted in Travis County, Texas, and on the Cheyenne River Reservation, where census enumerators visited each home rather than mail out questionnaires. Among other things, the 2006 tests were used to determine the usefulness of hand-held computers for providing maps to enumerators and for interviewing respondents.

A dress rehearsal, which was the first real test of what had been learned from the previous tests, was conducted in 2008 in San Joaquin County, California, and a nine-county area surrounding Fayetteville, North Carolina. Census workers used hand-held computers and coordinates from the Global Positioning System to up-date its address lists and maps.

The 2010 census used only the short-form (see Appendix C), rather than the two forms that have been used in previous censuses. The more detailed information that was collected on the long-form is now collected using the American Community Survey. This sample survey is done annually, so it will provide better updating of data between censuses. Another change was the use of about half a million hand-held computers for data collection. These were used by census workers to update mailing lists and to conduct follow-up interviews with those who did not mail back a census questionnaire.

The 2011 Census in Canada

A census in Canada is taken every five years. The 2011 census of Canada counted 33,476,688 people in the nation as of May 10, 2011, a gain of 1,160,333 people since the last census, in 2006, a gain of 5.9 percent. This rate of growth was among the highest of the Group of 8 Countries (G-8). Every province and most of its territories gained population between 2006 and 2011.

In Canada immigration was the primary source (about two thirds) of population growth during the five-year period since the last census, whereas the rate of natural increase declined by more than 30 percent over the 1991–1996 five-year period. The number of births decreased during the period, whereas the number of deaths increased somewhat, a consequence of an aging population. The level of urbanization was up slightly for the census period as well, at 80 percent, with the highest urban growth rates in Calgary, Saskatoon, and Kilowna. Slightly over 70 percent of Canadians lived in one of the nation's 27 census metropolitan areas, up from 68 percent in 2006.

CHAPTER 3

Population Distribution and Composition

Key Terms

ecumene
nonecumene
arithmetic density
physiological density
nutritional density
sex ratio
population pyramid
dependency ratio
elderly

The description and analysis of spatial distributions are recurrent themes in geography; where phenomena are located, and the reasons for their particular locations, have traditionally been a central part of the discipline. In that vein, then, in this chapter we look at the "where" of population and analyze some of the reasons behind the spatial distribution of the world's population, the age and sex structure of populations, and examples of the distributions of both the elderly and major race and ethnic populations in the United States.

It is obvious that there have been (and continue to be) temporal changes in the world's population. Though the overall trend in population growth had been one of gradual increase until approximately 300 years ago, when a more rapid rate of increase gradually became perceptible, large fluctuations of numbers occurred in specific areas (China and Europe, for example) throughout most of human history. Uneven growth over time has been matched by a similarly uneven growth over space; that is, growth is asymmetrical geographically because the populations of some countries and continents have grown more rapidly than others.

The number of inhabitants in a given area—be it a city, county, state, nation, or any other political unit—is usually one of the first population facts called for before any analysis of an area is undertaken. The numbers of people in specific geographical units are common elements that are needed for various administrative and research purposes; in many cases first judgments about a country or region are based upon a combination of the extent of its geographical area and the size of its population. More importantly, this information is not just of geographic importance. According to Smith and Zopf (1976, 39):

> Sociologists must have the facts relative to the numbers of inhabitants in various areas before they can compute indexes of criminality, juvenile delinquency, marriage, and so on; administrators must have them in order to determine how state and federal funds for education, agriculture, road construction, and so on, are to be apportioned among the counties, states, or other political divisions with which they are concerned.

Many other disciplines, including economics and political science, as well as many private and public agencies, including city and regional planners, also need accurate population counts.

Global Distribution Patterns

An analysis of the distribution of the world's population, shown in Figure 3–1, reveals some fascinating patterns:

- First, the great majority of the earth's inhabitants live on a small part of the land area. Approximately 90 percent of the people live on about 10 percent of the land.
- Second, over 90 percent of the world's population live north of the equator and less than 10 percent live to the south; however, it must be remembered that over 80 percent of the total land area is located in the Northern Hemisphere as well.
- Third, the earth's population is concentrated on the margins of the continents. Estimates are that approximately 70 percent of the world's population live within 1000 kilometers (600 miles) of the sea, about 67 percent live within 500 kilometers (300 miles) of it, and 50 percent live within 200 kilometers (120 miles) of it (in the United States an increasing proportion of the population lives in coastal areas as well).
- Fourth, population numbers generally decline with altitude. Approximately 56 percent of the world's population live below 200 meters (656 feet), whereas close to 80 percent live below 500 meters (1640 feet) (Clarke, 1972). Except for the

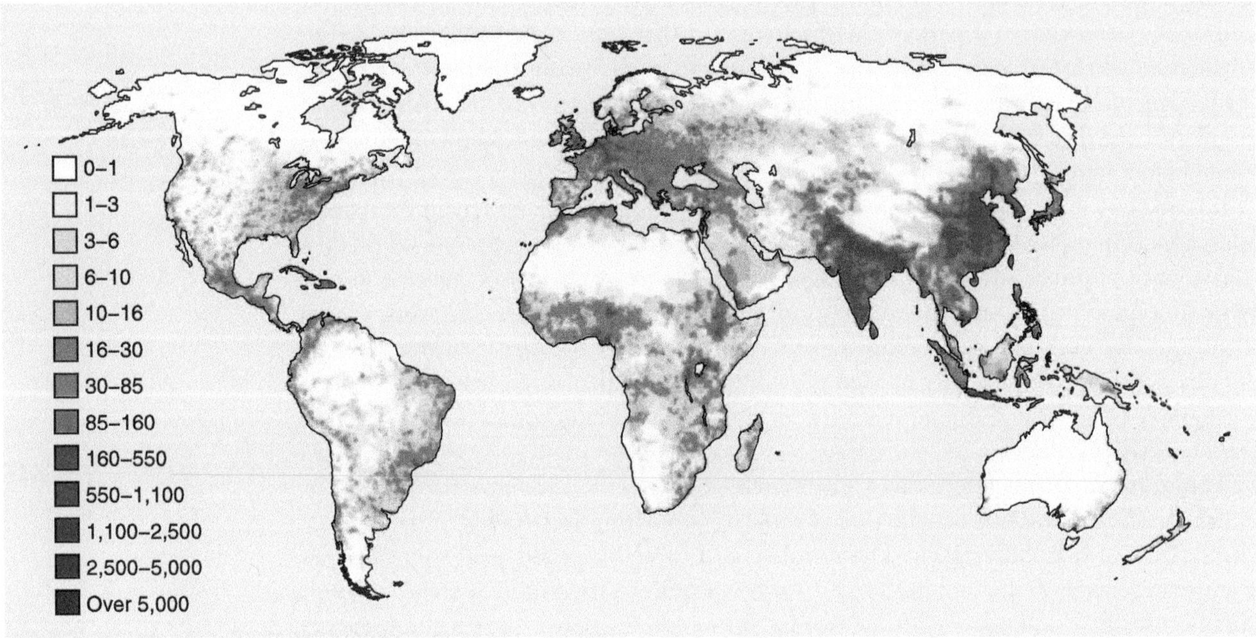

Figure 3–1 World Population Density
Source: Data from the Population Reference Bureau.

attraction of minable natural resources (minerals and fuels, especially) and recreational opportunities (skiing, for example), mountainous areas are limited in their range of economic activities. However, the large number of people in the world who are living in low-elevation coastal zones (< 10 meters; 33 feet) that are susceptible to flooding is of growing concern (Vafeidis et al., 2011; Small and Cohen, 2004).

There are four areas in the world with large population clusters: southern Asia, eastern Asia, western Europe, and east-central North America (the smallest). The southern Asian cluster includes India, Pakistan, Sri Lanka, Burma, Cambodia, and Thailand, along with related areas. The eastern Asian cluster includes China, Korea, Japan, and the Philippine Islands. Together, these two clusters include over one-half of the world's population. The western European cluster includes most of the Commonwealth of Independent States, the United Kingdom, Germany, and the other western European nations. The last and smallest cluster includes urbanized eastern parts of both Canada and the United States.

There are four areas in the world with large population clusters: southern Asia, eastern Asia, western Europe, and east-central North America (the smallest).

Ecumene and Nonecumene

One manner of describing the distribution of the world's population is to divide the world into two groups, the **ecumene** and the **nonecumene**. The ecumene is the permanently inhabited portion of the world, whereas the nonecumene is the uninhabited or virtually uninhabited portion. Though at first glance it may seem relatively simple to divide the world into ecumene and nonecumene, it is actually quite difficult because neither the inhabited nor the uninhabited areas of the world are represented by a continuous distribution.

The **ecumene** is the permanently inhabited portion of the world, whereas the **nonecumene** is the uninhabited or virtually uninhabited portion.

Climate plays an important role in determining these broad spatial patterns. The permanent icecaps are relatively easy to identify as part of the nonecumene, as are the world's most arid realms, but much of the rest of the nonecumene is isolated (some mountain ranges, for example) and either unoccupied or sparsely occupied (Trewartha, 1969). Of the total land area of the earth, approximately 30 percent is barren waste, without any significant human occupance; large areas of forest are also nearly uninhabited by people. The nonecumene, then, consists of perhaps 35–40 percent of the earth's total land area; a substantial portion of the earth's surface is without permanent human occupance. Of course the nonecumene is smaller today than it has been in the past because an established feature of human history has been a progressively expanding ecumene—population growth, technological advances (air conditioning, for example), and the discovery of minable minerals in isolated locations all contribute to that augmentation.

Climate plays an important role in determining these broad spatial patterns.

Further investigation of the ecumene reveals some absorbing regional characteristics. The North-American and western European population concentrations share numerous similar attributes. For example, the North-American cluster was a direct result of European colonization (Native Americans comprise less than one percent of the region's population today) and thus bears the stamp of European culture. Both of these clusters have advanced technology, high per capita incomes, well-developed regional specialization of economic activities, low percentages of their labor forces employed in agriculture, and a large proportion of residents living in cities.

By contrast, the two major Asiatic concentrations, with the exception of Japan, still exist mainly in the less developed world, though rapid industrialization and other signs of economic development are changing the picture. Asian NICs (newly industrialized countries), sometimes dubbed the "Four Tigers," include South Korea, Hong Kong, Taiwan, and Singapore. Since the 1980s the People's Republic of China has been undergoing numerous economic reforms aimed at bettering the Chinese economic system. Soon the Chinese economy will have the planet's largest total GNP (though it will remain much lower in terms of per capita GNP); its growing role in world trade is already apparent. Changing economic circumstances in these countries are resulting in changing demographic pictures as well.

Population Density

There are several ways to measure population density. The most common measure compares the total population of a place to its total area, a measure referred to as arithmetic density.

There are several ways to measure population density. The most common measure compares the total population of a place to its total area, a measure referred to as **arithmetic density**. Table 3–1 depicts the population density for the principal regions of the world for 1960, 1975, 1995, and 2011. Though Europe and Asia are the most densely settled continents, the average density of the less developed countries is more than triple that of the more developed countries (compared to less than double in 1960). During the 1980s the population density in Asia surpassed that in Europe, which previously had been the most densely settled continent.

Although density figures are convenient for analyzing differences in population distribution, they are often misleading. In 2011 there were about 51 persons for every square kilometer on the earth's land surface, but Oceania has only 4, Africa 35, Europe 32 and Northern America 16, whereas the more densely settled continents have considerably higher densities, with 132 for Asia. Densities obviously vary on a continental scale, but they vary to an even greater degree within smaller areas. For example, Australia has a density of only 3 inhabitants per square kilometer, whereas the Netherlands has 402. Other examples are evident in Table 3–2. Smaller nation-states, and particularly island nation-states, often have dense populations, in contrast with larger nation-states such as the United States and Russia, which have densities of 32 and 8.6 respectively. Population density values, however, have the least significance when large areas with marked environmental differences are involved. Though the average density for the former U.S.S.R. was 5 people per square kilometer, it is important to note that the density in the Asiatic portion of that country was only about 10 percent of the density in the European portion. Arithmetic density is widely used because the necessary data are easily obtained; keep in mind, however, that it tells us very little about the relationship between people and resources in a nation. We can find rich nations with high densities (Germany, Japan), rich nations with low densities (U.S., Australia), poor nations with high densities (Pakistan, Bangladesh), and poor nations with low densities (Sudan, Chad).

A somewhat more refined measure of density is the ratio of population to arable land, a measure referred to as physiological or nutritional density.

A somewhat more refined measure of density is the ratio of population to arable land, a measure referred to as **physiological** or **nutritional density**. Arable land is that portion of the earth's land surface that is suitable for tillage, a definition which obviously leaves out some productive areas of nonarable land, such as mining land, natural pasture, forests, and

Table 3–1 Land Area and Population Density in the World and Major Areas, 1960, 1975, 1995, and 2011

Area	Land Area (1,000 km²)	1960	1975	1995	2011
World total	135,779*	22.1	29.4	42.0	51.0
More developed regions	60,907	16.0	18.6	19.2	27.0
Less developed regions	74,872	27.0	39.5	60.5	69.0
Europe	4,936	86.1	96.0	103.1	32.0
U.S. and Canada	21,515	9.2	11.0	13.7	16.0
Oceania	8,509	1.9	2.5	3.3	4.0
South Asia	15,775	54.9	80.4	116.6	331.0
East Asia	11,756	67.0	85.5	122.7	134.0
Africa	30,320	9.0	13.2	23.7	35.0
Latin America	20,568	10.5	15.8	23.4	29.0

*Not including the Antarctic continent.

Source: United Nations and 2011 World Population Data Sheet, Population Reference Bureau.

Table 3–2 Population Density, 2011		
Area	**Per. Sq. Km.**	**Per. Sq. Mile**
Afghanistan	50	118
Egypt	83	212
Chad	9	21.7
Mexico	59	150
Brazil	23	62
Honduras	69	188
U.S.	32	85
Japan	339	874
Australia	3	7.3
Germany	229	591
Netherlands	402	1039

Source: World Population Data Sheet 2011, (Washington D.C.: The Population Reference Bureau); UN Statistical Division, 2011;U.S. Census Bureau International Database, 2011.

scenic regions. Nutritional density cannot account for the great variations in productivity within arable areas that are the result of such environmental characteristics as drainage, soil, and climate. According to Trewartha (1969, 74), nutritional density provides a better indicator than does arithmetic density of the degree of crowding in a region compared with its physical potential for producing food and agricultural raw materials. Japan's arithmetic density in 1960, for example, was 655 per square mile; its nutritional density was an unbelievable 4,680, a fact which suggests the necessity for significant imports of food and vegetable raw material.

Several other density measures have been suggested, including the "standard land unit," which includes varying measures for different modes of land use, and various indices based on the calories produced by a country, the values of national income, and productivity. Some density maps give the ratio of the agricultural population to the amount of arable land, a measure that eliminates nonfarmers. As Beaujeu-Garnier (1966, 32) commented, "All of these methods have their interest in showing up this or that aspect of population density, but none is of undisputed absolute value; and, above all, none escapes from the criticism that, being applied to large and complex areas, it distorts a phenomenon which is essentially localized and variable." Martí-Henneberg (2005) produced an interesting discussion of regional population densities in Europe from 1870 to 2000.

Factors in Population Distribution

An analysis of the distribution of the world's population suggests the importance of certain fundamental factors:

1. the historical evolution and duration of a population in a certain area
2. the natural environment and its associated aspects such as landforms, soils, precipitation, temperature, and natural resources
3. the socioeconomic and technical development of a region

As we shall see, the latter factors have become increasingly important as the Industrial Revolution and modernization in its broadest sense have spread throughout the world. Given an increased technological base, natural environments have less influence on a nation's population distribution. Storm windows and natural gas help mitigate the harshness of cold winters, whereas air-conditioning has turned many desert areas into such delightfully desirable places as Palm Springs, California.

With economic development, employment structures shift from mainly employment in agriculture to increasing proportions of employment in manufacturing and services, both of which tend to concentrate populations in cities. Thus, a major redistribution of population from rural to urban areas usually accompanies modernization. A good historical account of such changes in the United States can be found in Otterstrom (2001).

Two of the basic characteristics of any population are sex and age.

Sex and Age Structure

Two of the basic characteristics of any population are sex and age. Both of these factors are important in relation to other population variables such as mortality, morbidity, marriage, fertility, and migration.

Sex Structure

The sex ratio is defined as the number of males per 100 females.

The sex ratio is one of the simplest measures of population structure, and statistics on sex are usually easy to get and accurate. There is no ambiguity about the meaning of female and male and there is little need to misrepresent such information. The **sex ratio** is defined as the number of males per 100 females. Typically, the ratio lies between 90 and 105. For example, the sex ratio for the population of the United States in 2010 was

$$\text{Sex ratio} = \frac{\text{number of males}}{\text{number of females}} \times 100$$

$$= \frac{138{,}053{,}563}{143{,}368{,}343} \times 100 = 96.29.$$

A sex ratio of 100 indicates that there are equal numbers of males and females in a population; a ratio above 100 indicates that there are more males than females; and a ratio below 100 indicates that there are more females than males. The basic factors that determine the sex ratio are

1. death rate differences between the sexes
2. net migration rate differences between the sexes
3. the sex ratio of newborn infants.

All modern censuses calculate the proportions and numbers of females and males in the population. The first census in the United States determined the sex structure of the white population, and since 1820 the sex structure of all segments of the population has been tabulated. Because of the relative simplicity of the question pertaining to sex, census officials since as early as 1930 have concluded that the data on sex structure is the most reliable of all population data included in their tabulations.

Examining sex ratios and correlating them with other demographic factors such as births, deaths, and migration reveal some basic facts:

1. The sex ratio at birth is relatively high (105) for the United States and for most of the rest of the world.
2. In general, females have lower death rates than males at all ages.
3. Females outnumber males in migration from rural to urban areas and in other short-distance migrations.
4. Males outnumber females in long-distance moves.

The sex ratio of the population of the United States has varied greatly from place to place, time to time, and group to group. Figure 3–2 depicts the changes in the sex ratio of the United States from 1840 to 2010. The sex ratio of 94.5 in 1980 was the lowest in

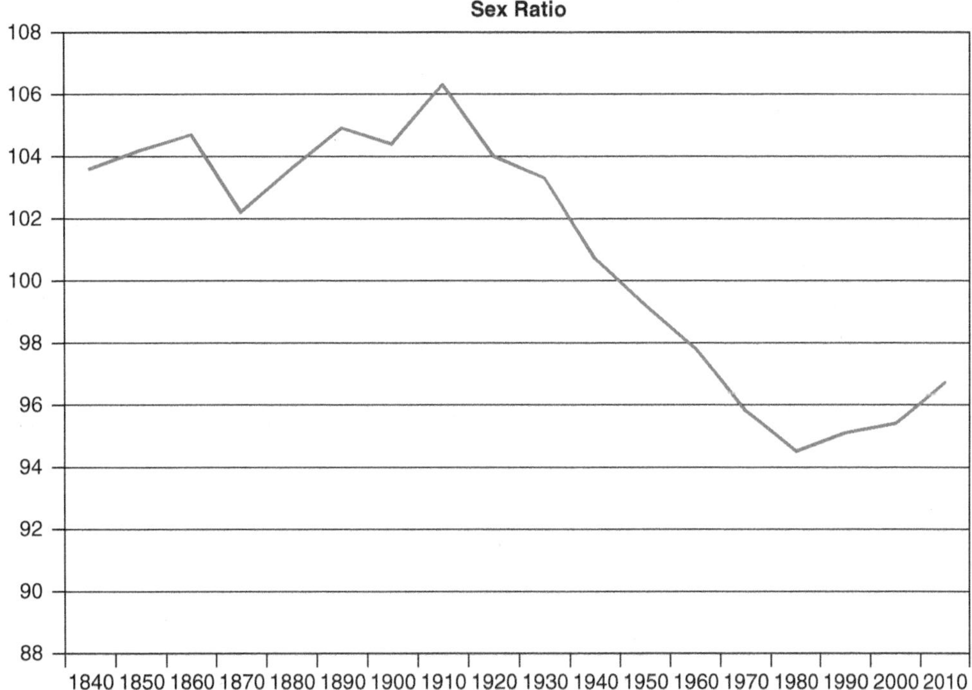

Figure 3–2 Sex Ratios in the United States, 1840–2010
Source: Data from Statistical Abstract of the United States and Census 2012.

American history. The decline in the sex ratio that occurred after 1910 in the U.S. can be attributed to a continuing reduction in female mortality rates (Smith and Spraggins, 2000). Between 1980 and 1990, the upward trend in the sex ratio can be accounted for by the death rates for men which declined faster than those for females and increased immigration to the U.S. which included more men.

Changes in the sex ratio over time in the United States are primarily due to changing immigration patterns. The all-time high of 106.3, which occurred in 1910, was the result of large numbers of immigrants, primarily males. The low sex ratio of 1870 was primarily due to the decline in immigration during the Civil War and the large numbers of males killed in battle.

A state-by-state analysis of the sex ratio of the United States reveals some interesting changes in spatial patterns over time. Figure 3–3 shows sex ratios for each state in 1790. These sex ratios, however, represent only the white population. The sex ratio for all of the states was 103.7. A state-by-state analysis reveals that higher sex ratios were found in what were then considered to be frontier areas, whereas the lower sex ratios were found in the more settled areas of New England. This same pattern also appears in Figure 3–4 for 1850, though these data include both white and nonwhite residents. Once again, the frontier areas are male-dominated, whereas the more densely settled and urbanized areas have a lower sex ratio. The rush of males to the mining areas of California during the Gold Rush of 1849 is apparent in the extremely high sex ratio of that state (1222.9). It is also interesting to note that the lowest sex ratio, 95.0, is found in Washington, D.C., though in general urban areas have lower sex ratios than rural areas.

Figure 3–5 shows sex ratios at the turn of the century. The sex ratio of the United States was still above 100, mainly because of large numbers of immigrants. The opening up of the western frontier brought about large imbalances between the proportions of the sexes. The westward movement was largely dominated by males, making the sex ratios in several western states extremely high, whereas large numbers of females were left

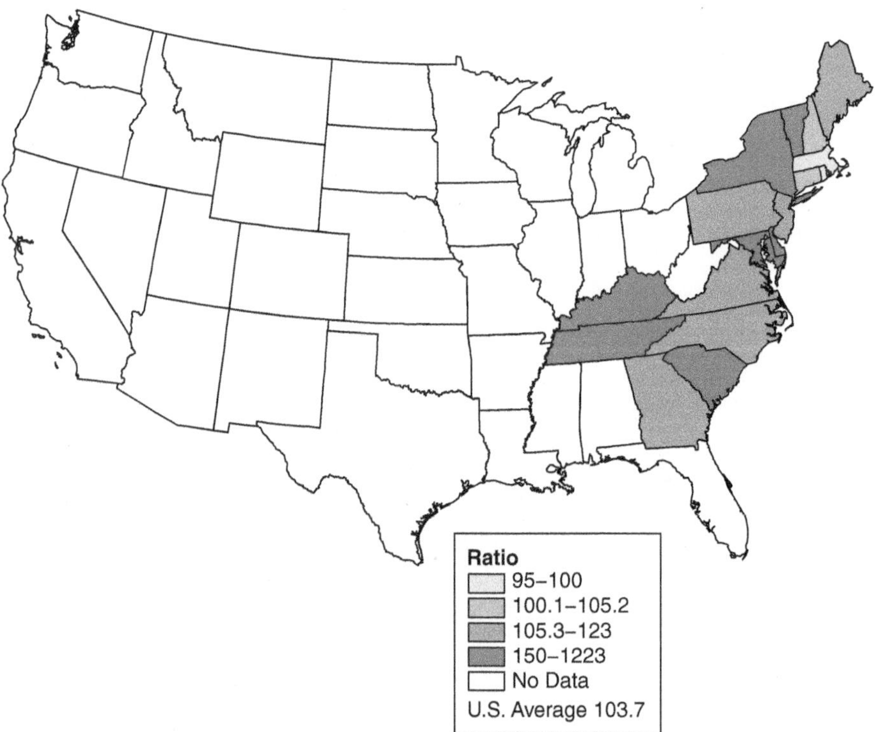

Figure 3–3 Sex Ratios for the White Population, 1790

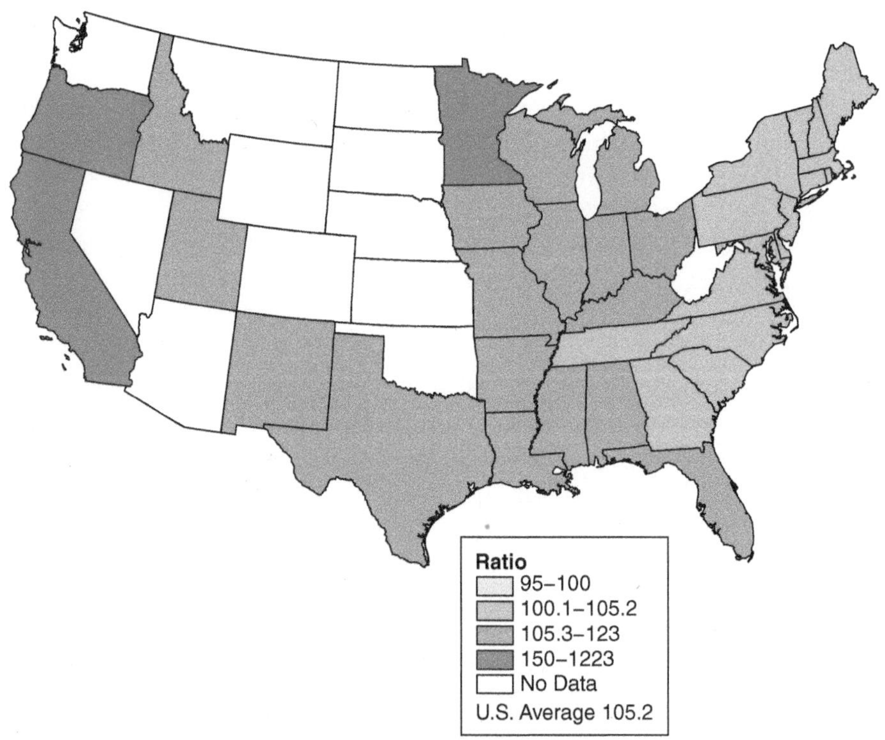

Figure 3–4 Sex Ratios for the Total Population, 1850

behind in the South and East. Washington, D.C., once again had the lowest sex ratio. Utah, although a western state, had a much lower sex ratio than the other western states, undoubtedly because of the unique nature of the settlement of Utah and the role of the Mormon Church. Migration to Utah often involved families, whereas single males were most evident in early migrations to the other western states.

With the passing of the frontier came a reduction in the sex ratio for the United States, as well as a lowering of sex ratios for western states, as is apparent in Figure 3–6 for 1940. Not only is there a lowering of the sex ratios in the West, but also the sex ratios for the various parts of the country tend to cluster close to the national average.

Figure 3–7 illustrates sex ratios in 1970. It is apparent that all sections of the United States are well settled and that only relatively small regional differences exist within the country. The sex ratio for all sections of the country is close to the national average of 94.8.

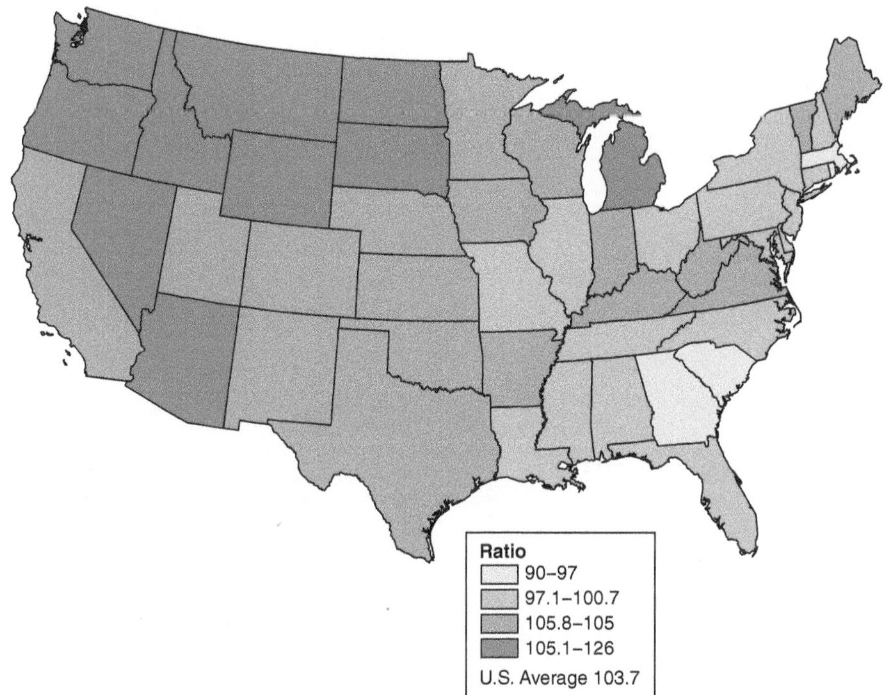

Figure 3–5 Sex Ratios for the Total Population, 1900

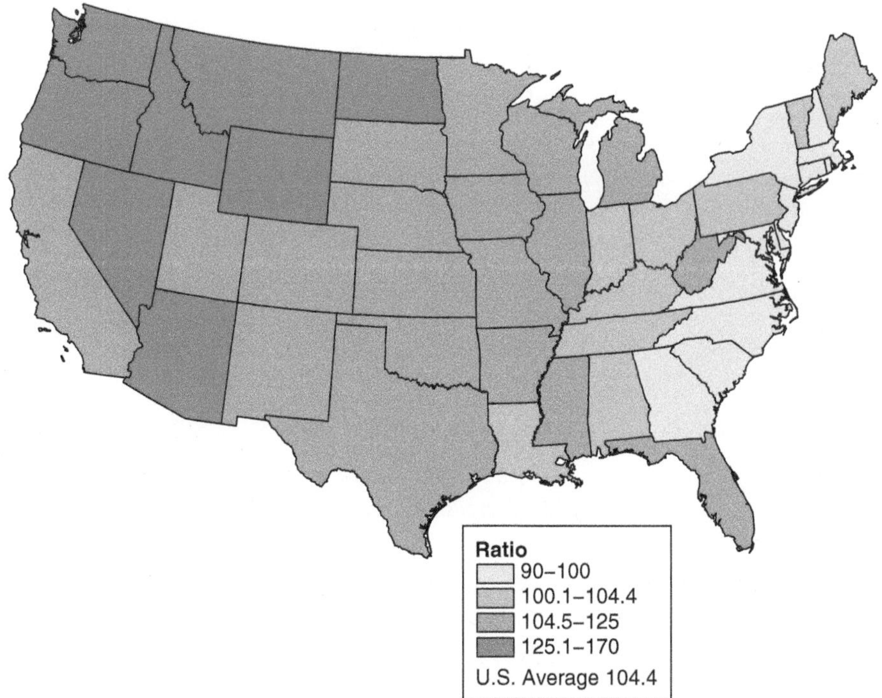

Figure 3–6 Sex Ratios for the Total Population, 1940

The only exceptions are Alaska and Hawaii, with Alaska having the highest sex ratio of all states. Alaska can be considered the last frontier and we would expect it to have a higher proportion of males, again because of the dominance of males in migration to frontier-type areas. Though the sex ratio increased slightly between 1970 and 2000, the spatial pattern of sex ratios in 2000, as shown in Figure 3–8, differed little from the pattern in 1970.

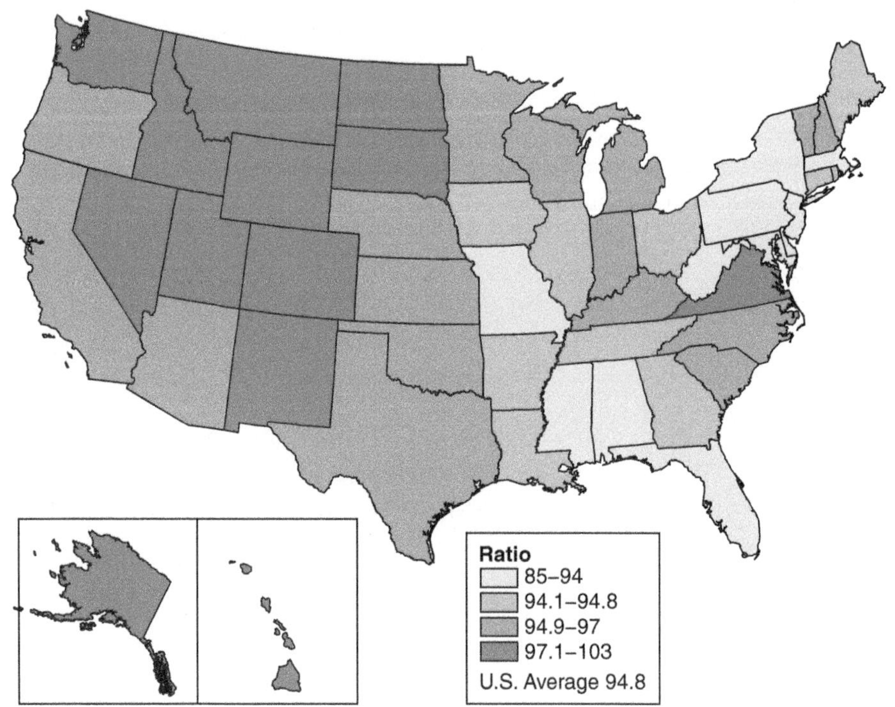

Figure 3–7 Sex Ratios for the Total Population, 1970

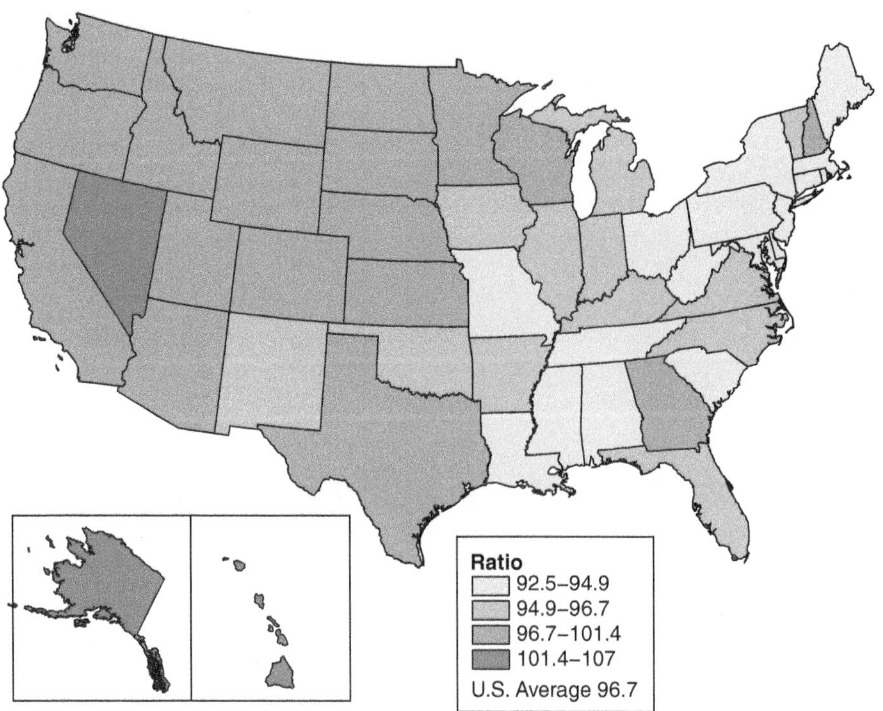

Figure 3–8 Sex Ratios for the Total Population, 2000

International Variations in Sex Structure

Though there are more males at birth, females tend to outlive males, especially in regions where there is good medical care. Thus, we would expect to find lower sex ratios in the more developed countries and higher ones in the less developed areas, where maternal mortality and fertility are still high.

Very few countries in the world have sex ratios higher than 105 or lower than 90, as shown in Table 3–3. The reasons for high sex ratios include such variables as in-migration of males, high female mortality due to poor health facilities, undercounts of females, and high fertility, with its resulting young age composition. In some areas, such as West Malaysia, Hong Kong, and Cuba, migration and the influx of laborers have been important variables determining sex ratios. However, in other areas, such as Libya, Pakistan, India, and Sri Lanka, migration was not important; high sex ratios in these areas are probably due to higher mortality rates for females than for males.

Extremely high sex ratios, such as those found in Qatar, 311, United Arab Emirates, 228, Bahrain, 166, Kuwait, 148, Oman, 142 and Saudi Arabia are due to the massive streams of male immigrant workers to these countries. The countries with low sex ratios are primarily found in Europe and their low sex ratios can partly be traced to large numbers of emigrants, mainly males. Another reason is the loss of males during World War II.

China is unusual because of its high sex ratio at birth, which, according to the 2010 census, was nearly 119, far above the normal range of 103–107. In some provinces, including Hainan and Guangdon, sex ratios at birth exceeded 135. Sex ratios are high in China in part due to the cultural preference for male heirs and to the underreporting of female births. China's One-Child Policy, which is enforced differently in different provinces, has led to an interesting situation in some geographical areas. Because sex ratios have been high for some time now, it has led to two related phenomena, "missing girls" (Riley, 2004) and "bare branches." The "missing girls" refer to those girls who would have been present if sex ratios were at or near the normal range. Major reasons for these missing girls include the underreporting of female births and the use of sex-selective abortions; the latter made possible because of the increasing availability of ultrasound B machines. Though female infanticide is often cited as another reason for high sex ratios, Riley (2004) argued that it is relatively rare and would be difficult to get away with in a society where so many people are aware of each other's daily activities. Attané (2006) has noted that the lack of females will also result in lost childbirths down the road.

Hudson and Den Boer (2004) have studied implications of the "bare branches," young males who are unlikely to find a mate because of the "missing girls." Without mates they will be unable to have children or a stable family life. These researchers suggest that in turn it is at least likely that this excess of men will lead to increases in violent crime. These "bare branches," most numerous in China but found in large numbers also in India, may even represent threats to domestic stability and international security. The world has probably never seen such an excess of males in a society, but the possible implications are dramatic. The Chinese government is aware of the problem and is searching for financial and other incentives that would normalize sex ratios by 2010, though it is not at all clear right now how successful they will be.

Sex and Gender

Though demographers and geographers have long recognized sex ratios as a measurement of the relative distributions of males and females in a population, there has been a growing interest in gender studies as well during the past two decades. Gender studies among geographers and other social scientists are concerned with socially created differences between the sexes (in contrast to biological distinctions). Such social differences between the sexes

Table 3–3 Sex Ratios of 105 or More and 90 or Less

Country	Sex Ratios
Sex Ratios of 105 or More	
Qatar	311
United Arab Emirates	228
Bahrain	166
Kuwait	148
Oman	142
Saudi Arabia	124
Bhutan	113
Western Sahara	112
China	108
Afghanistan	107
Solomon Islands	107
India	107
Samoa	107
Jordan	106
Equatorial Guinea	105
French Polynesia	105
Sex Ratios of 90 or Less	
Latvia	85
Ukraine	85
Estonia	86
Netherlands Antilles	86
Russian Federation	86
Lithuania	87
Belarus	87
Armenia	87
Martinique	88
Georgia	89
Guadeloupe	89
U.S. Virgin Islands	90
Hong Kong	90
Aruba	90
Republic of Moldova	90
Hungary	90

Source: United Nations, World Population Prospects. Table 1, 2010.

(gender roles) typically arise from the differential ways in which roles in a society are assigned to the two sexes; from early educational opportunities in less developed countries to the "glass ceiling" in America, from women as secretaries and nurses to men as construction workers and doctors.

Because such social-typing can vary considerably from place to place, geographers have become interested in spatial manifestations of gender differences. Common grounds for feminist geography (and the geographical study of gender) include at least the following: the portrayal of women's experiences, concern with the quality of life for women, and a political vision of gender equity (Monk, 1994). Additional ideas about geography and gender are included in Monk and Hanson (1982), Bondi (1992), and Bondi (1993); books focusing on various aspects of gender studies include Adepoju and Oppong (1994), Peters and Wolper (1995), and Jones, Nast, and Roberts (1997), Hussain, Khan, and Momsen (2005), Wright (2006), Vernoy (2006), Longhurst (2007) and Samarasinghe (2007). Two brief essays by Rodger Doyle, each accompanied by a world map, have examined female illiteracy (Doyle, 1997) and women in politics (Doyle, 1998). Low rates of female participation in politics and high levels of female illiteracy are characteristic of many nations in Africa, Asia, and Latin America, as is apparent in Chant (2007) and Radcliffe (2006).

Age Structure

Much of a population's economic and social behavior is determined by the proportion of people found in various age groups. Smith and Zopf (1976, 149) suggested the following three reasons why a knowledge of the age distribution of a population is basic in nearly all population analyses:

1. Age is one of the most fundamental of one's own personal characteristics; what one is, thinks, does, and needs is closely related to the number of years since one was born.
2. Groups are determinants of primary social and economic importance in any society.
3. The qualified student of population must possess the technical skills needed to bring out the significant features of the age composition of a population with which he/she is concerned and also those required to make the proper allowances or corrections he/she may attempt.

Because the age structure is so closely related to other demographic characteristics, such as the birth rate, death rate, migration rate, and marriage rate, caution must be used when comparisons and analyses involving these characteristics are undertaken. Andrews and Phillips (2005) is an excellent survey of ageing, age structure, and place.

Factors Affecting Age Structure

Many factors determine the age structure of a national population, but the primary variable is the birth rate. A population with a high birth rate will have a large proportion of young people, whereas a population with a low birth rate will have a smaller proportion of young people. The role mortality plays in determining the age structure of a national population is best described by Coale (1964, 49) as follows:

> Most of us would probably guess that populations have become older because the death rate has been reduced, and hence people live longer on the average. Just what is the role of mortality in determining the age distribution of a population? The answer is surprising—mortality affects the age distribution much less than does fertility, and in the opposite direction from what most of us would think. Prolongation of life by reducing death rates has the perverse effect of making the population somewhat younger.

Migration generally has little or no impact on the sex structure of national populations because most countries have restrictions on international migrations. However, for sub-national populations, for example the populations of states, counties, cities, or even particular districts within cities, migration often has a considerable impact on age structure. Migration is generally age-selective. That is, almost all types of migrations involve a large proportion of young people. An area that has experienced a great deal of out-migration usually has fewer young people in the age group 20–40, whereas a region with a significant amount of in-migration may show an excess of young adults. A somewhat different argument has developed with respect to how the age distribution of places may impact the age composition of out-migrants from an area (Little and Rogers (2007).

Catastrophes such as famine, pestilence, or war may also influence the age structure of a region. For example, war has a twofold effect on the age structure. First, many young adults are combatants in a war and are killed; and second, there is usually a sharp decline in the birth rate during a war.

Among older adults, activity patterns are now being viewed as much more complicated than some earlier studies and stereotypes have suggested. Furthermore, there are considerable geographic variations in those activity patterns, as noted by Fortuijn, et al (2006). Their research is dismissive of the notion that older adults are inactive, immobile and unhealthy, all things that governments worry about when they look at the consequences of aging populations. Danson and Hardill (2006) look at similar themes and also focus on the changing nature of places in Europe as elderly folks retire to sunnier climes while immigrants fill increasing gaps in local labor forces. As they note (2006, 317), "The interaction of such demographic changes with migration flows, traditionally based on selective mobility of skilled and dynamic younger people but now of more diverse groups, is painting complex patterns across the continent." Throughout most of the more developed world life expectancies are rising, birth rates are declining, and migration is increasing.

Population Pyramids

A **population pyramid** is a useful aid in examining the age structure of a population and it also provides information about the social attributes of the population concerned.

A **population pyramid** is a useful aid in examining the age structure of a population and it also provides information about the social attributes of the population concerned. As illustrated in Figure 3–9, a population pyramid is really two bar graphs placed back-to-back, with the vertical center line representing zero. The number or percentage of people in each age group, by sex, is indicated with horizontal bars; and the vertical axis gives the different age groups, usually in five-year intervals. The shapes of population pyramids vary according to the country or community and the time period. A pyramid for a particular country represents the demographic history of that country over the past two or three generations – a generation is roughly 30 years long.

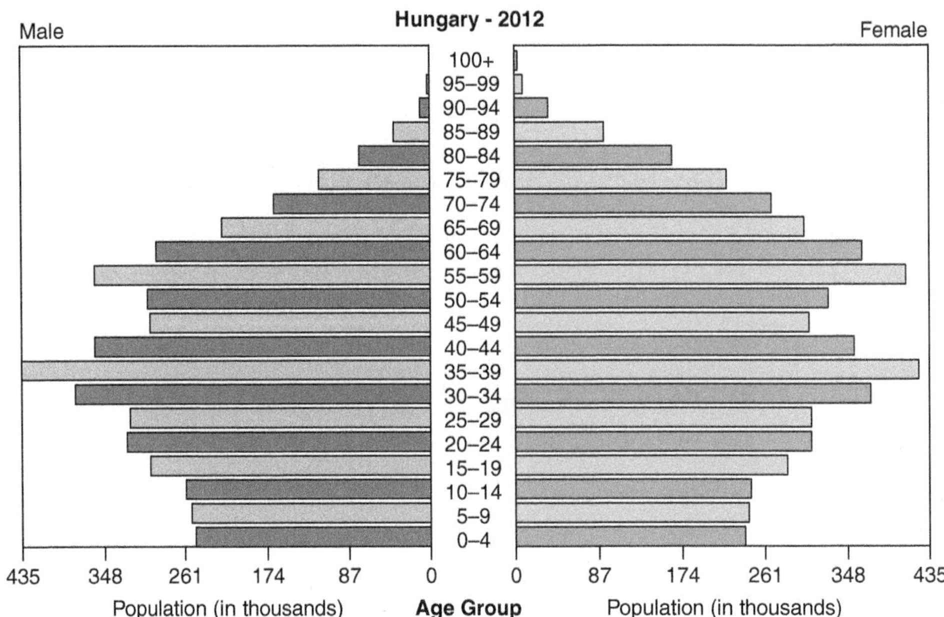

Figure 3–9 Population Pyramid for Hungary, 2012

Source: U.S. Census Bureau, International Database.

A national population with a large number of children, as the result of a high birth rate, would have a pyramid with a wide base, like that shown in Figure 3–10; whereas the pyramid for a population with a low birth rate would have a small base, like the pyramid for the Japan in Figure 3–11. The pyramid is also affected by death rates, though usually to a lesser extent. In wartime the death rate for certain age groups may rise considerably, and the bands representing those age groups will be shorter. Finally, selective migration can also change the shape of the pyramid.

It is possible to recognize certain broad categories of population structure by comparing the shapes of pyramids for different countries. Four types of population structures can be recognized, each coinciding with a particular cultural development. The first

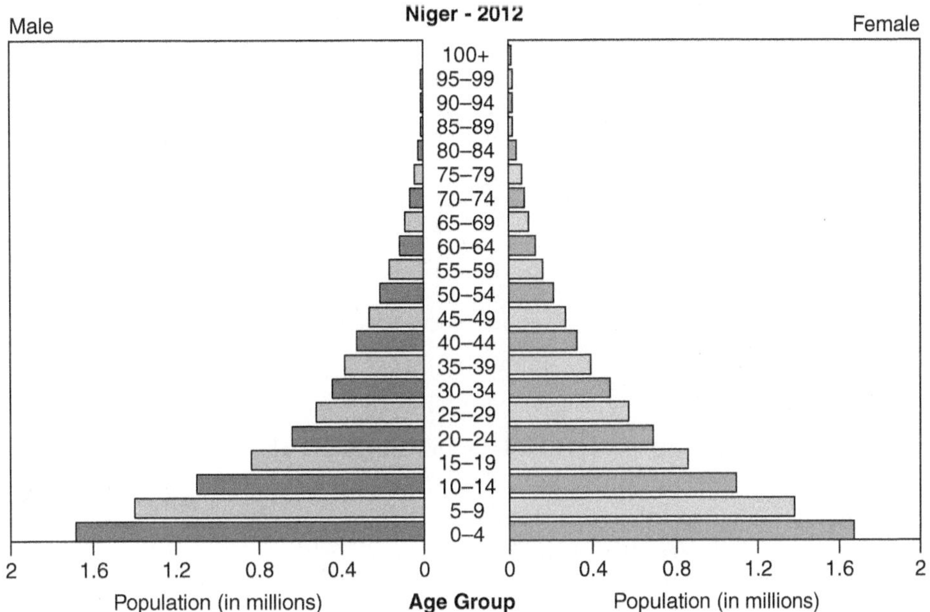

Figure 3–10 Population Pyramid for Niger, 2012
Source: U.S. Census Bureau, International Database.

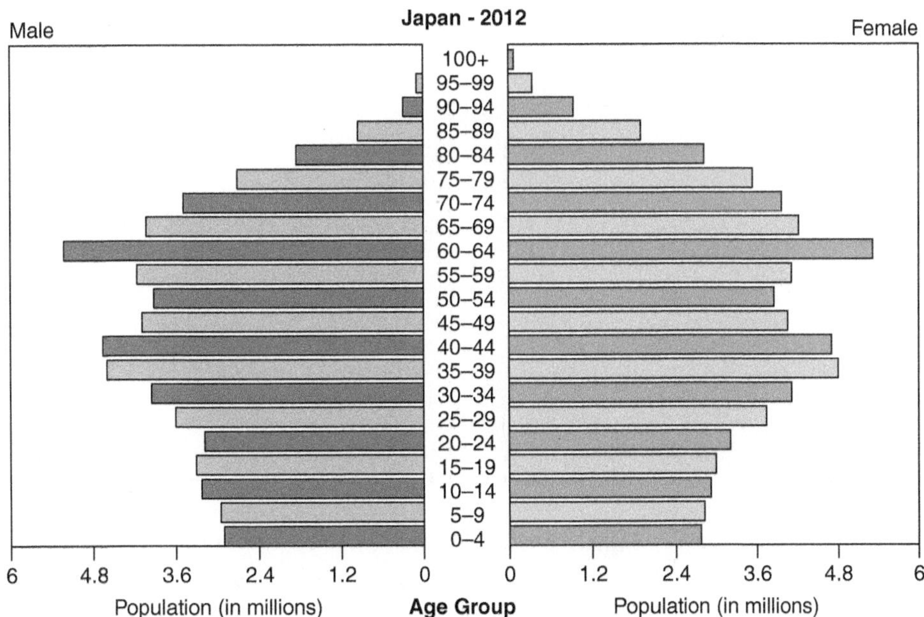

Figure 3–11 Population Pyramid for Japan, 2012
Source: U.S. Census Bureau, International Database.

type (Figure 3–12) is a pyramid with a regular triangular shape, typical of a country with high birth and death rates. Few countries today have this type of pyramid, but it was common for most countries during the seventeenth and eighteenth centuries. The second type of population pyramid has a narrow top, wide base, and concave sides (Figure 3–13); it represents a country with a falling death rate, particularly in the youngest age category, and a high birth rate. This pyramid is typical for most of the less developed countries in Africa, Asia, and Latin America. A beehive shape characterizes a third type of population pyramid (Figure 3–14) and indicates a relatively stable population with low birth rates, low death rates, and a high median age. Most European countries would have pyramids of this type. England, Wales, and Sweden are good examples. The final pyramid type (illustrated in Figure 3–15) represents a country that has had a rapid decrease in fertility, thus giving a tapered shape to the bottom of the pyramid.

Dependency Ratio

The **dependency ratio** is a simple statistic that measures the role of age composition on the productive activity of a population by comparing the proportion of the population in the nonproductive ages with those in the working ages.

The **dependency ratio** is a simple statistic that measures the role of age composition on the productive activity of a population by comparing the proportion of the population in the nonproductive ages with those in the working ages. It assumes that people between ages 20 and 64 are the productive segment of the population, whereas those under 20 and over 64 years are the dependent segment of the population. It may be calculated as follows:

$$\text{dependency ratio} = \frac{\text{population under 20 plus population 64 years and over}}{\text{population 20–64 years}} \times 100$$

The purpose of the dependency ratio is to measure the number of dependents that each 100 people in the productive years must support. Thus, some nations have a high dependency ratio because of a large number of elderly, whereas other nations have high ratios because of a large number of children.

In the absence of comparable international data on economic activity, the dependency ratio is often used to reflect variations in economic dependency, even though some people in the dependent age range are producers, whereas many persons in the productive age range, particularly in Western societies, are economically dependent. For example, in the United States it is currently possible to begin drawing social security

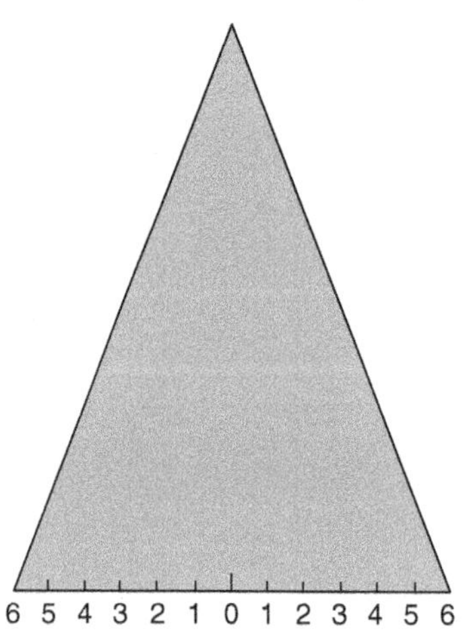

Figure 3–12 Population Pyramid with Regular Triangular Shape

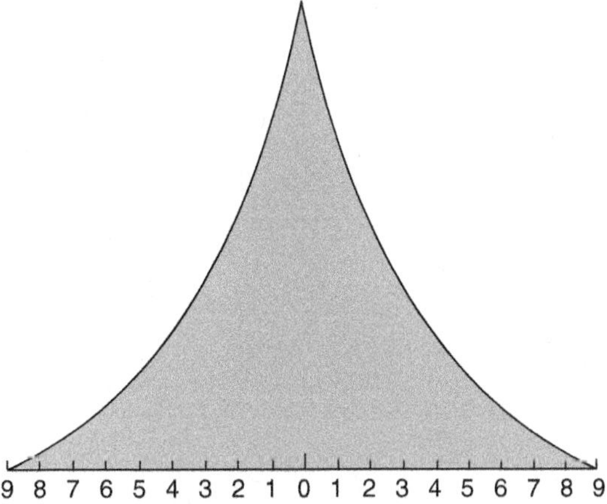

Figure 3–13 Population Pyramid with Concave Sides

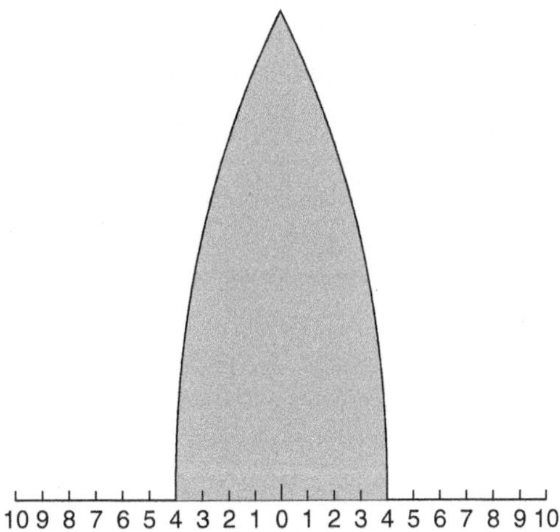

Figure 3–14 Population Pyramid with Beehive Shape

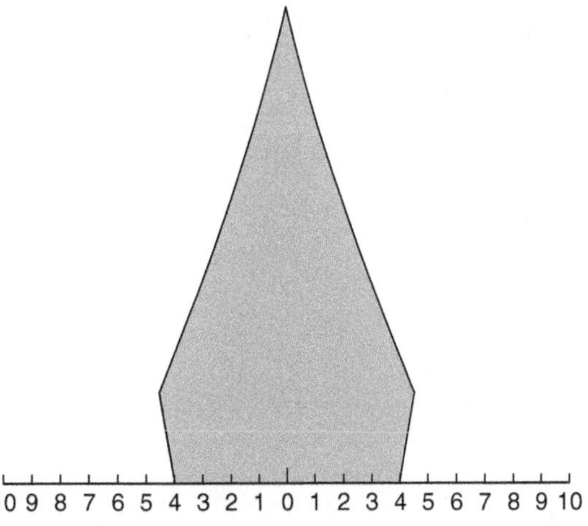

Figure 3–15 Population Pyramid with Tapered Base

benefits at age 62, and many do. In Italy the percent of the population that receives retirement benefits is nearly half again as large as the population 65 and older. At the same time the working age population doesn't consist 100 percent of people who are working. Many in that age group don't work, either because jobs are not available or because they don't have to. In considering the burden of retirees, a better figure is the ratio of pensioners to workers, which is currently around 4.5 in the United States and 3.0 in Germany.

World Patterns of Age Structure

A world view of age structure reveals some interesting patterns, as we see in Table 3-4 and in Figures 3–16 and 3–17. The world's youngest populations are found in the less developed nations of Africa, Asia, and Latin America, where almost half of the population in many countries is under 15 years of age. The elderly comprise only 4–6 percent of the entire populations of Africa, Asia, and Latin America. For example, in Tanzania, only 3 percent of the population is over 64, while 44 percent is under 15. In 2011 the following were among those with the highest proportions of persons under 15: Niger (52 percent), Uganda (48 percent), Angola (48 percent), Mali (47 percent), Afghanistan (46 percent), Democratic Republic of the Congo (46 percent), and Zambia (46 percent). High birth rates in these countries ensure a steady supply of new parents for the next generation, even if those parents in turn decide to have smaller families.

In 2010, the population of the world was 29.2% elderly. Europe ranks as the world's most elderly continent with 16 percent of its population aged 65 and over. The proportion of elderly is even higher for several individual nations of Europe. For example, in 2011 the countries with the highest proportion of persons 65 and over included: Monaco (24 percent), Germany (21 percent), Italy (20 percent), Greece (19 percent), Sweden (18 percent), Austria (18 percent), Finland (18 percent), Bulgaria (18 percent), Portugal (18 percent), Belgium (17 percent), United Kingdom (16 percent), France (16 percent), Switzerland (16 percent), Norway (15 percent), and Denmark (15 percent). As you might imagine, birth rates (and percentages of young people) are quite low in these countries.

Table 3–4 Countries with Oldest and Youngest Populations, 2010

Oldest	Median Age	Youngest	Median Age
Japan	44.7	Niger	15.5
Germany	44.3	Uganda	15.7
Italy	43.2	Mali	16.3
Channel I.	42.6	Angola	16.6
Finland	42.0	Afghanistan	16.6
Hong Kong	41.8	Timor-L'Este	16.6
Austria	41.8	Congo (Dem. Rep. of)	16.7
Slovenia	41.8	Zambia	16.7
Bulgaria	41.6	Malawi	16.9
Croatia	41.5	Chad	17.1

Source: United Nations Population Division, World Population Prospects—The 2010 Revision: Highlights.

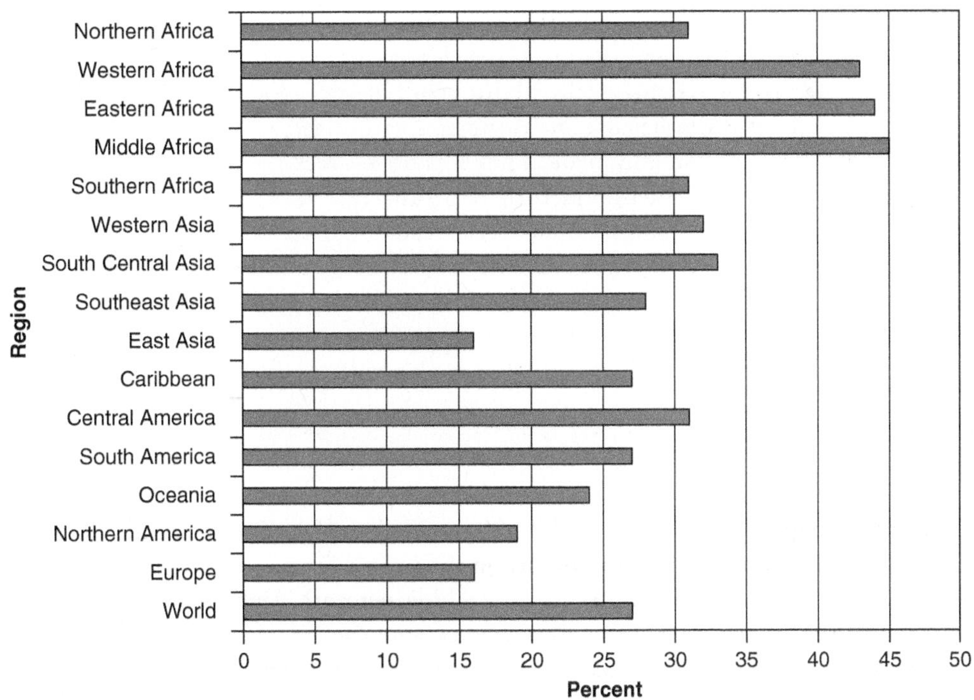

Figure 3–16 Population Under Age 15—World and Major Regions
Source: Data from 2011 World Population Data Sheet. (Washington, D.C.: The Population Reference Bureau)

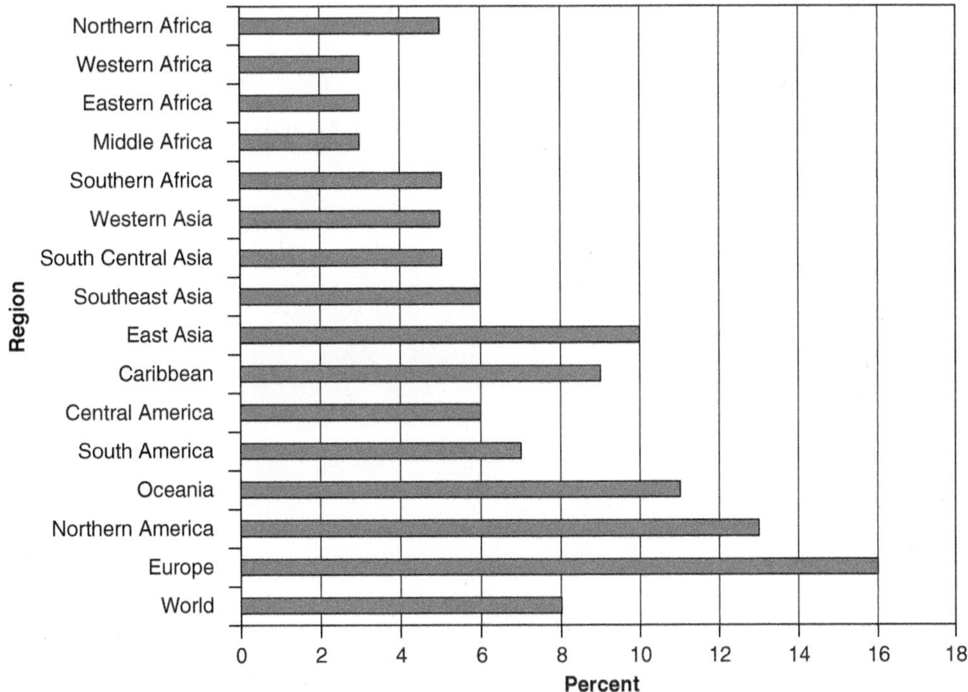

Figure 3–17 Population Over Age 65—World and Major Regions
Source: Data from 2011 World Population Data Sheet. (Washington, D.C.: The Population Reference Bureau)

Aging populations in European countries are forcing many governments to reconsider the generosity of medical and other benefits for the elderly.

Figure 3–18 shows the spatial distribution of elderly populations. The dilemma for modern countries is simple enough: as the population ages, the number of workers available to support pension and medical benefits for retirees declines. The choices are stark: raise the age at retirement, cut benefits to the elderly, or increase the tax burden on workers. The problem, of course, as is readily apparent in the United States, is that no leader wants either to increase taxes or to reduce benefits, but the time of reckoning is upon us. Worse yet, even as people are living longer, more are choosing to retire early. Few even recognize that the very idea of retirement, of life after work, is itself a product of the twentieth century and the abundance created by the Industrial Revolution.

Baby Boomers in the United States

Following World War II, the United States, as well as most European and many other countries, experienced an increase in fertility. In the United States the cohort that is generally recognized as the "Baby Boomers" includes children born between 1946 and 1964. There are about 76 million boomers; those born in 1946 turned 65 in 2011. The US Census projects that by 2022, those 50 years and older will number 122 million (compared to those between the ages of 18 and 49; 143 million). In 1946 there were 20 percent more babies born in the United States than in the previous year, a not unexpected result of the return home of missing husbands and lovers. However, one year was apparently not enough to make up for the war years, and Americans went on something of a fertility binge, totally unexpected by demographers and other social scientists (keep in mind that the best of economic, demographic, and other projections have been known to fall far

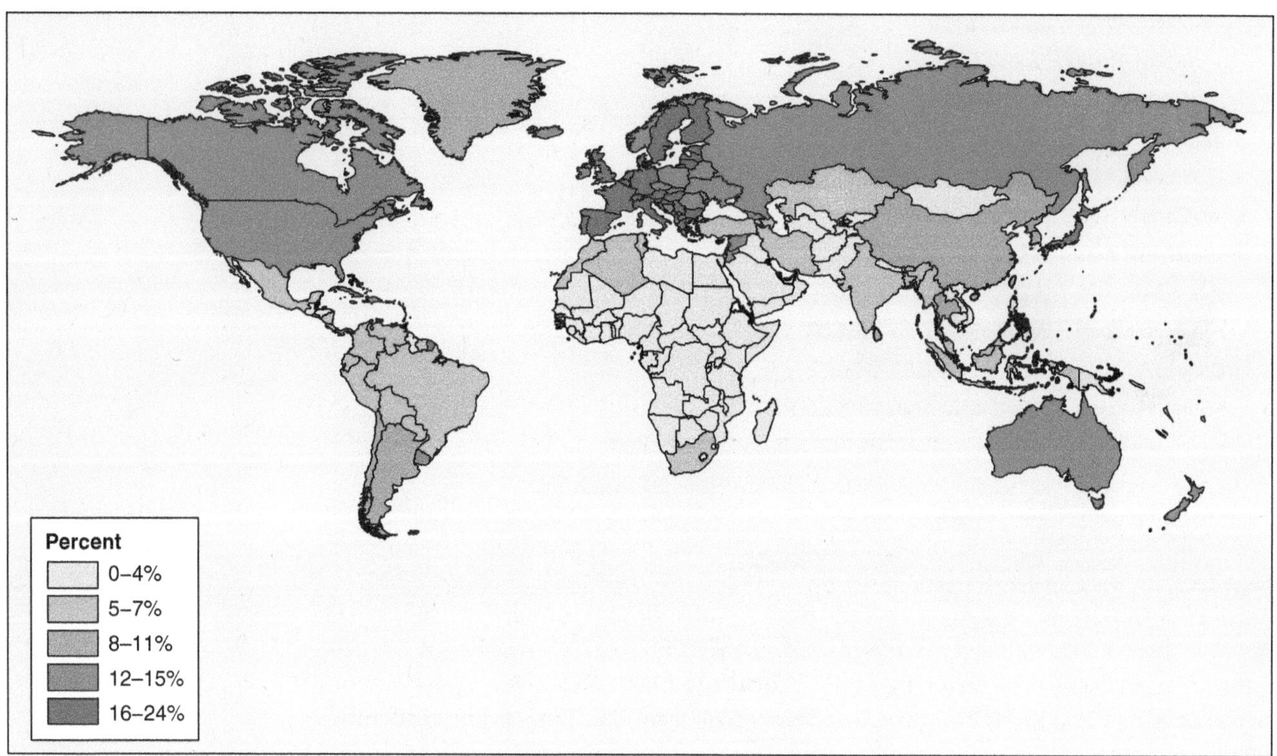

Figure 3–18 Percent of Population that is 65 Years of Age and Over
Source: Data from the Population Reference Bureau, 2011 World Population Data Sheet.

from the mark). At the same time, the baby boom was not a return to the high fertility of earlier periods; rather, it was a result of a significantly larger percentage of women marrying and having two or three children.

The sustained period of earlier marriages and higher fertility, so different from what occurred in the depression years of the 1930s, was at least partly a response to better economic conditions after World War II. Bouvier and De Vita (1991, 7) also noted that "Because men's wages were rising during this period, women felt little economic pressure to enter a male-dominated work world and thus were likely to be ambivalent about the desirability of 'yet another child' and apt to 'relax contraceptive vigilance.'" Interestingly, the end of the baby boom coincided rather well with the introduction of oral contraceptives in the mid-1960s.

By the time the leading edge of the baby boom cohort began to reach public schools in the early 1950s it was apparent that this cohort was going to create unprecedented problems, from the need for more schools as they entered kindergarten to their retirement needs.

Having been supplemented over the years by immigration, as if their numbers were not sufficient already, in 1990 the baby boomers numbered around 80 million, or nearly one-third of the population of the United States. At that time they ranged in age from 26 to 44 and formed a prime component of everything from the workforce to the housing market. As demographer William Frey (2010, 29-30) has noted:

> The aging of the baby boom generation is noteworthy not only because of its size, but also because its members' social and demographic profile contrasts sharply with earlier generations. Baby boomers are more highly educated, have a higher percentage of women in the labor force, are more likely to occupy professional and managerial positions, and are more racially and ethnically diverse than their predecessors.

We don't want to completely endorse the idea that "demography is destiny." Nevertheless, the baby boomers and their impact on American society are clear evidence that demographics cannot be set aside. Changing age, sex, and ethnic compositions can, and will, have impacts on the economic and social systems within which they are found. However, Friedland and Summer (2010, v) have stated,

> "Society's future is not determined solely by demographic changes. Focusing on the anticipated growth in population by age group is just too simplistic an approach. Rather, the future is shaped by the choices made—or not made—individually and collectively, bounded by the limits in resources and, in particular, knowledge. Knowledge is at the heart of gains in productivity, economic growth, and the advances in medical care, agriculture, communication, transportation, and the environment."

Demographers don't think boomers are preparing adequately for retirement (Russell, 1998), for example. In fact, baby boomers will probably have to work five to seven more years than they had originally planned (Hyman, 2012). Hyman (2012, 32) comments, "According to most analyses, about one in three Boomers, or roughly 25 million, are prepared financially to retire, one in three will never be prepared, and one in three are in the middle—and things aren't looking good for them."

In the early 2000s, after air was let out of the stock market bubble, a new bubble in housing began in earnest in the United States, and it didn't start to sag until probably sometime in 2006. The recession of 2008 hit many baby boomers hard. For baby boomers, their retirement money was tied to the equity in their homes. As the housing bubble burst, homeowners have found themselves upside down in their mortgages. If the housing market improves, and housing equity is restored, the plight of the baby boomers in terms of retirement may be ameliorated.

The Elderly in the United States

In the United States, being **elderly**, or old, is usually defined according to the Federal Social Security Program as having reached 65 years of age.

As we approach the twenty-first century most Americans can expect to attain at least the Biblical age of three score years and ten; life expectancy today is over 77 years, with males nearing an average of 75 and females approaching an average of 80. However, an analysis of the elderly in the United States over the past 90 years shows that only recently have most Americans been able to reach such advanced ages, as is illustrated in Figure 3–19.

In the United States, being **elderly**, or old, is usually defined according to the Federal Social Security Program as having reached 65 years of age. The elderly population of the United States has risen consistently since 1900, when slightly more than 3 million Americans were aged 65 and over. The elderly population increased nearly threefold to 9 million by 1940 and then to nearly 40 million in 2010 (Table 3–5). In 90 years there was almost a 1,100 percent increase in the elderly population, compared with just over a 370 percent increase in the total population, from 76 million to 281 million, during the same time period.

The rapid increase in the elderly population in the United States is the result of three major factors.

1. High fertility rates occurred during the late nineteenth and early twentieth centuries, and because of this large birth cohort there has been an increased number of people in the 65- to 75-year age bracket.
2. There has been a marked decline in mortality over the past 75 years, mainly because of advances in sanitation and medicine, which allowed more people to live to age 65. Presently, around 75 percent of newborn children are expected to reach age 65, whereas in 1900 only 39 percent of newborns were expected to reach that milestone.
3. Large increases in the elderly population in recent years were due to the high level of immigration prior to World War I. These migrants were primarily young adults, and thus a large cohort of that age group was added to the population.

Table 3–5 Population 65 and Older in the United States: 1900–2010

Year	Population (thousands)	Population Increase from the Preceding Decade (thousands)	Percent of Population Increase from the Preceding Decade
1900	3,099		
1910	3,986	887	28.6
1920	4,929	943	23.7
1930	6,705	1,776	36.0
1940	9,031	2,326	34.7
1950	12,397	3,366	37.3
1960	16,659	4,262	34.4
1970	20,156	3,497	20.4
1980	24,830	4,674	23.2
1990	29,835	5,005	20.2
2000	34,991	5,156	17.3
2010	40,267	5,276	15.1

Source: Leon Bouvier, Elinore Atlee and Frank McVeigh. (1975) "The Elderly in America," *Population Bulletin,* 30(3):4, courtesy of the Population Reference Bureau, Inc., Washington, D.C. Newest figures from the United States Bureau of the Census.

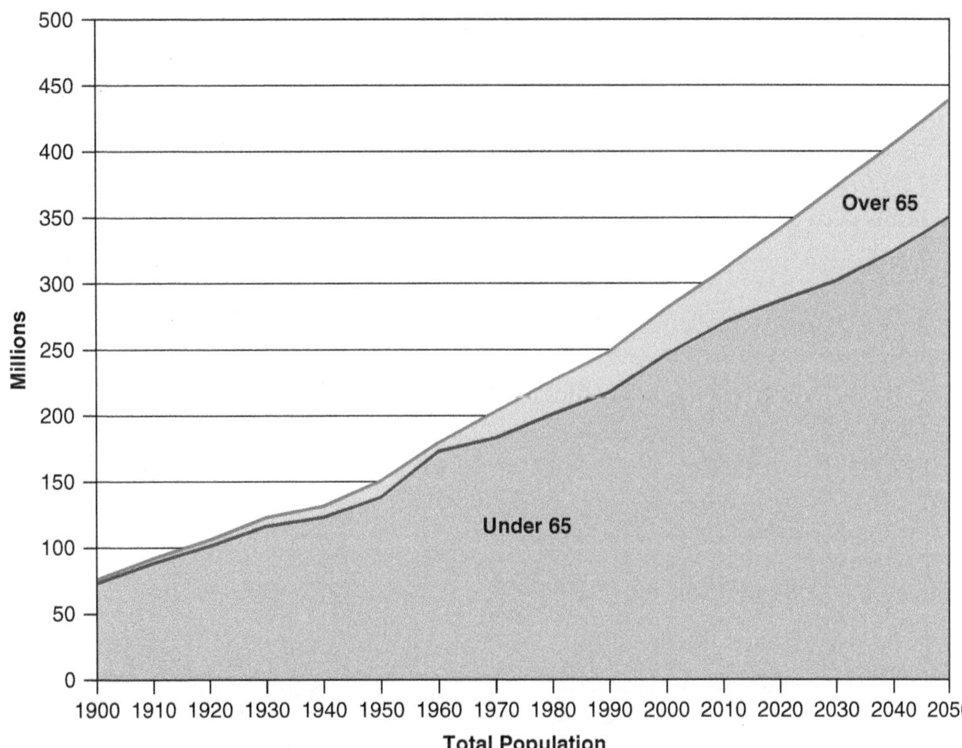

Figure 3–19 Population in the United States Over and Under 65 Years Old, 1900 Through 2010 and Projected to 2050

Source: 1900 to 2000 data: Hobbs, Frank and Nicole Stoops, U.S. Census Bureau, Census 2000 Special Reports, Series CENSR-4,Demographic Trends in the 20th Century, U.S. Government Printing Office, Washington, DC, 2002; 2010 to 2050 projected data:Table 12. Projections of the Population by Age and Sex for the United States: 2010 to 2050 (NP2008-T12), Population Release Date: August 14, 2008, Division, U.S. Census Bureau.

Population projections by the Bureau of the Census forecast a continuing increase in the numbers of elderly in the population until at least 2030. There are expected to be 53.7 million by 2020, and 77.2 million elderly by the year 2040, and 89.8 million by 2060. These figures should be reasonably accurate projections because the future elderly have already been born, though immigration may have a significant impact on the later figures.

Not only has the total number of elderly been increasing rapidly, but the proportion of the elderly in the United States has also increased. In 1900 about one out of every 25 Americans was 65 or older, whereas in 1990 about 1 out of 8 Americans was elderly. The proportion of the elderly in a population is affected by variations in the numbers of people in other age categories. The increase in the proportion of the elderly is primarily attributed to declining fertility rather than to the lowering of the death rate. Changes in the proportion of the elderly over the past 60 years are illustrated in Figure 3–20.

One major reason why Americans are living longer now than ever before has been the substantial decline in heart attack deaths that has occurred since about 1950. However, this greater longevity is to some extent a mixed blessing at best. For example, Crimmins (2001, 5) pointed out that "Improvement in health does not necessarily accompany an increase in life expectancy since only about half of the disability and functioning loss at older ages is caused by lethal diseases. The other half is caused by conditions that are not linked to mortality trends, including arthritis, vision loss, and Alzheimer's disease." More attention needs to be focused on the quality of life for elderly folks, rather than simply the extension of life. There is hope, of course, that stem cell research and other biotechnology innovations may aid in healthier lives for the elderly, but their promise remains in the future.

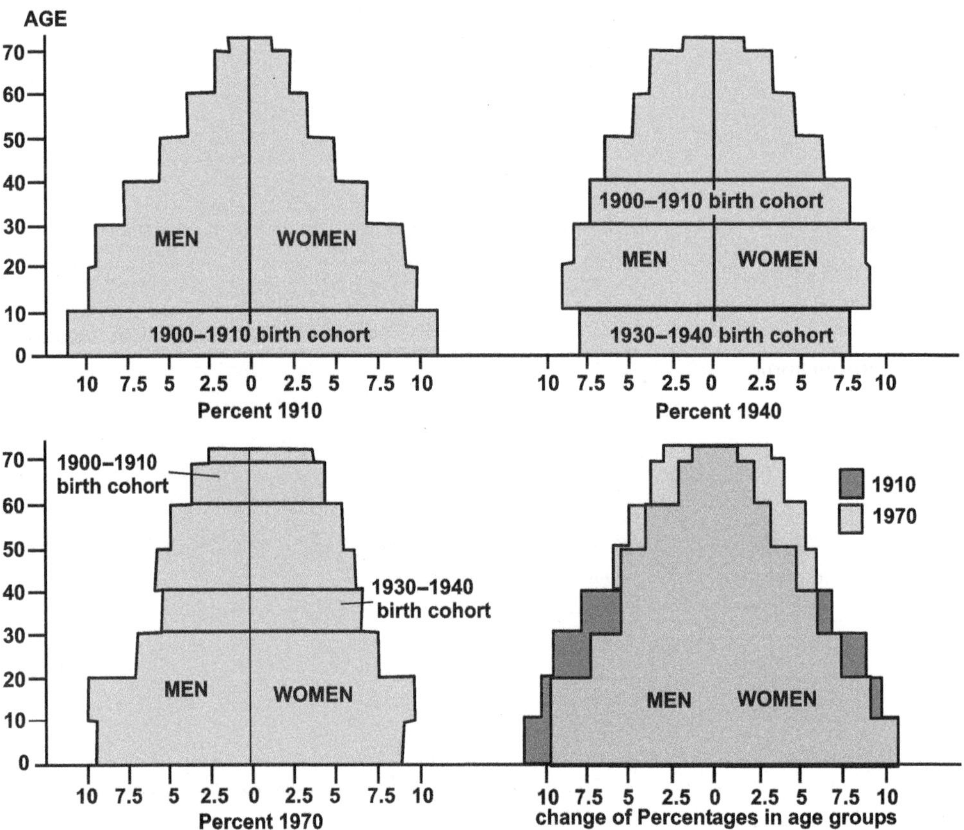

Figure 3–20 Age Pyramids, or Percentages in Age Groups, for the United States in 1910, 1940, and 1970

Source: U.S. Bureau of the Census, General Population Characteristics, Final Report PC(1)-B1 United States Summary (U.S. Government Printing Office, 1972), p. 1–276.

International Migration and Demographic Change

The end of the Cold War, major economic and demographic trends, and global communications systems are converging at the end of the millennium to produce unprecedented human population movements. In turn, these movements are going to reshape demographic trends within both sending and receiving countries; they are already reshaping American demography, for example, though not without considerable debate, much of it emotionally charged (Isbister, 1996; Brimelow, 1995).

As the Cold War reached its sudden and unexpected end, euphoria swept the nations of Europe and elsewhere. On November 9, 1989, celebrants drank Champagne atop the Berlin Wall, hugged and danced in the streets, and cheered as the Wall came tumbling down. Then-President George H.W. Bush proclaimed the coming of a "New World Order." However, fading euphoria and reality checks on several continents suggest that we may be much closer to a new world "disorder."

Regional conflicts, often strongly nationalistic in origin, have flared up from Bosnia and Chechnya to Rwanda and Haiti, generating streams of frightened and battered refugees in their wake. Women, children, and the elderly are suffering disproportionately as uncertainty and instability replace the old order—any new order seemed distant and indefinable, especially when the major powers resisted interfering in places such as Bosnia. From Vietnam, Laos, and Kampuchea in the 1970s, Afghanistan in the 1980s, and Colombia, Rwanda, Burundi, Bosnia, Chechnya in the 1990s, and Somalia and Sudan in the 2000s, streams of refugees have sought safe havens elsewhere, generally in different parts of their own country or in neighboring countries.

At the same time, worldwide demographic and economic changes are generating ever larger streams of international migration, most of which are focused on the more developed countries of North America and the European Union. At least three important trends converged as the millennium ended: (1) demographic—rich countries have slow-growing or even stagnant populations, whereas poor countries have been growing rapidly, (2) economic—the global economy (victory in the Cold War has been accepted by many as "proof" that unfettered capitalism must prevail) is bringing capital and labor together in unprecedented ways and amounts, and (3) communications—global networks (examples include CNN and the Internet) are tying the world together in such a way that nearly everyone recognizes the great disparities that exist between rich and poor nations (the richest 20 percent of the world's population generates about 85 percent of the world's GNP).

As a consequence of these converging trends, both capital and people are going to be on the move as never before. Capital, concentrated in the more developed countries, will go in larger amounts to poor countries, but not fast enough to stem the flow of great numbers of people who will move to rich countries, where the capital is. Both of these trends will accelerate, as we are seeing from California to Belgium and Germany. In turn, these trends are disturbing the status quo, creating social and political backlashes that must be dealt with in rational ways if societies everywhere are to provide better lives for future generations.

International migration and the changing racial and ethnic compositions of individual nations are inextricably linked together. A few examples in the United States serve to focus attention on internal changes that cannot be separated from increasing flows of immigrants.

Since the 1930s immigration to the United States has been increasing steadily (Figure 8–6), and since the 1960s the origin of these immigrants has shifted increasingly from Europe to Asia and Latin America. Today there are vast differences in rates of growth for major racial and ethnic groups within the United States. Rapid growth in the Asian and Hispanic populations results from both increased immigration rates (including many refugees from Southeast Asia) and the increased fertility of many recent arrivals, especially among the Latino population. In California—the new Ellis Island and recipient of more immigrants since 1980 than any other state—for example, the total fertility rate for Latinas is slightly more than twice that for Anglo women, though it is starting to decline.

At every level, from the national to the local, rapid changes in the racial and ethnic compositions of populations are occurring. For example, the composition of California's population has changed considerably in the past twenty-five years. For the first time in recent decades Asians (13.6 percent) outnumber African Americans (6.6 percent), though the two groups together are about 40 percent smaller than the Latino population (38.1 percent). If current trends continue, then Latinos will become the largest minority. This already is the case in Los Angeles where the Latino population comprises 48.5 percent of the total population.

(continued)

International Migration and Demographic Change (*continued*)

Thus, at every geographic scale the impact of increased immigration is becoming apparent—the nature of places is being altered. Without doubt, growing numbers of immigrants are generating concern, both in North America and in the European Union. Less apparent, however, is how political units—from cities to nations—are going to cope with such rapid changes in population composition. Immigrant bashing has become more common, as have attacks on immigrants and even citizens who might look like immigrants—civil society cannot accept such acts.

Yet, here and elsewhere, the volume of both international migration and refugee movements will continue to increase in the decades ahead. Most countries, especially those in the more developed world, are going to experience growing ethnic diversity. With luck, stronger international linkages will develop between immigrant nations and those that are providing large numbers of immigrants. Successful policy changes are going to have to be based on the assumption that more immigrants are going to be seeking permanent residence. Finally, social and political adjustments are likely to lag behind economic and demographic changes well into the next century.

In the meantime care for the elderly in the United States is in critical condition (DeFrancis, 2002). Nursing homes, for example, have experienced an array of problems, including understaffing and bankruptcies. Medicaid, which pays nursing home costs for about two-thirds of nursing home residents, is in deep trouble, and neither Medicare nor private health insurances are likely to pick up the slack. Staffing problems reflect labor shortages for nurse's aides and others; those shortages in turn reflect low wages throughout the nursing home industry. In some states workers in fast food restaurants are paid nearly as much as nurse's aides.

Americans are living longer, and the "oldest old," people 85 and over, is the fastest growing portion of the elderly population. In 2010 there were about 5.7 million people over 85 in the United States, and that number is projected to increase to 19 million by 2050. As the "baby boomers" enter old age, there is going to be significant growth in all categories, but especially in the "oldest old," which will put severe pressure on the nation's medical resources. The leading edge of the baby boom generation began to turn 65 in 2011, and because social security benefits are available starting at age 62, 2008 was the beginning of significant changes not only for America's elderly but also for all Americans. As sociologist Christine Himes (2001, 37) noted, "Our society is just beginning to face the complex issues involved in an aging society. We cannot anticipate the changes needed by looking backward; the population aging we are facing in the next several decades is unprecedented. We must look forward."

Spending on the elderly in the United States has increased considerably since the establishment of Social Security (1940) and Medicare (1965) (Lee and Haaga, 2002). As one analyst observed, "Medicare spending is growing faster than the economy. On top of that, the Baby Boomers have started to retire: in 2031, when the last of the boomers reach age 65, there will be 77 million people on Medicare, compared to 47 million today" (Roy, 2011, 48). The aging of the Baby Boomers will continue to have important implications in the decades to come.

Spatial Distribution of the Elderly

Though the elderly population in the United States has increased dramatically within the past century, these increases have not been evenly distributed within the country. According to the US Census Bureau, the South region had the largest number of elderly people, over 12.4 million and the Northeast had the largest proportion, 13.8% (see Appendix A for a census region map). The West has the fastest growing elderly population. States with smaller proportions of the elderly either have populations with relatively high fertility (Idaho, Alaska, and Utah) or have received large numbers of relatively young in-migrants from other areas of the country (Maryland, Colorado, and Nevada). Immigration plays a part in the age structure of states as well (Texas and California). States with over 12 percent of their populations classified as elderly include Arkansas, Nebraska, Iowa, South Dakota, Missouri, Arizona and Florida. Many states in the Great Plains have lost a considerable number of young people due to out-migration, whereas others have an influx of retirees, which increases the proportion of the elderly. In both cases the age distribution becomes older. South Dakota and Florida are states that have similar proportions of the aged but for very different reasons. Bouvier, Atlee, and McVeigh (1975, 8) noted that in 1970, 14.6 percent of the population of Florida and 12.4 percent of Nebraska were 65 years or over. Between 1960 and 1970, however, the aged population increased by almost 80 percent in Florida but by only 12 percent in Nebraska. The high proportion of the elderly in Florida is explained by in-migration of the old; in Nebraska it is caused by the exodus of the young in addition to the longevity of the state's population (Figure 3–21).

Within America's metropolitan areas the distribution of the elderly is also of interest. In 2010, 78.9 percent of persons 65 years and older lived in metropolitan areas. Cowgill (1978), for example, discussed "gray ghettos" in many of the older core areas of major cities. Goodman (1987) used Lorenz curves and Gini coefficients to study the distribution of the elderly in three American cities: Baltimore, Philadelphia, and Pittsburgh. Though Pittsburgh had the largest elderly population, he found that the elderly were more concentrated in the central cities of Philadelphia and Baltimore than they were in Pittsburgh. They were most concentrated in Baltimore.

However, as Soldo (1980, 13) earlier noted, "This does not reflect intentional discrimination; rather, while their children have moved to the suburbs, the elderly have stayed behind." In small towns and rural areas the elderly are also often present in relatively high proportions. In 2010, 17.2 percent of rural dwellers were elderly. "Contrary

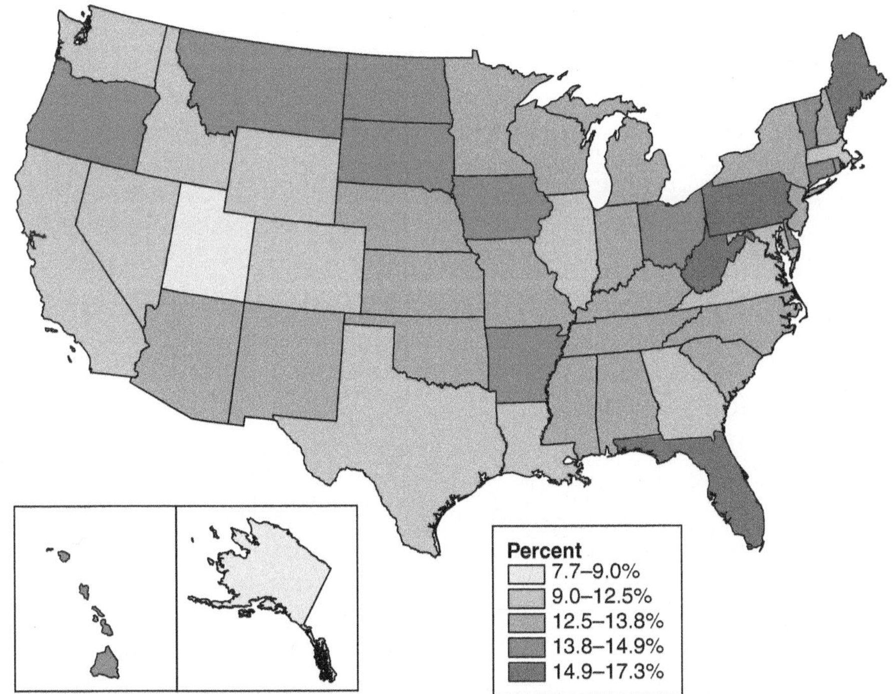

Figure 3–21 Percent Aged 65 or Older, 2010
Source: U.S. Census Bureau, 2010.

to popular opinion," Soldo (1980, 13) went on to say, "relatively few Americans move during their retirement years." Most who do move do not move very far. Among the longer-distance movers the primary destinations are in the Sunbelt, especially in California, Arizona, and Florida. Even these migration streams are selective, however, because most of the elderly people involved are the "young" and more affluent elderly, those most able to make such moves (Soldo, 1980). General problems of an aging population are discussed in Peterson (1999) and Wilmoth (2006).

Race and Ethnicity in the United States

We are well aware that biologists have thoroughly discredited the idea of separate races, and we certainly agree with Olson (2002, 5) that "Human groups are too closely related to differ in any but the most superficial ways . . . the cultural differences between groups could not have biological origins." Frazier et al. state that "[t]he concept of race in the United States has taken on a socially constructed meaning that transcends the biological distinctions of group, with implications for how those groups are perceived and treated by others within the larger society" (2003, 5). However, as Hirschman (2004, 385) noted, "Although modern science has discredited race as a meaningful biological concept, race has remained as an important social category because of historical patterns of interpersonal and institutional discrimination." As Frazier et al. contended, "race is significant in America because Americans have made it significant on all levels" (2011, 5). Ethnicity is another term that deserves examination. Frazier et al. define it as "a group-constructed identity using one or more of its cultural attributes" (2011, 9). These attributes include language, religion, nationality, unique customs and practices and race. Race and ethnicity are often associated with locational and other types of inequities for minority race and ethnic groups, not only in the U.S. but in other regions of the world as well. The legacy of slavery and institutionalized and legal discrimination against minorities is that many members of these groups must grapple with the inequities that are the result of this history. Many of these inequities appear in poor educational and health outcomes and well as reduced overall life chances.

Geographers Allen and Turner (1988) provided the most ambitious look at ethnicity in America that has ever been undertaken. They noted, importantly, that "Differences in values, occupational directions, lifestyles, religious identities, and patterns of socializing based on ethnic background may sometimes be obvious but are more likely to be subtle and hidden from public view . . . ethnicity has played a major role in shaping social structure, patterns of political and economic competition, religion, lifestyles, food preferences, and the processes of cultural change in America" (Allen and Turner, 1988, 205). Furthermore, such changes can be observed at a variety of scales, from the national to the local. Though here we treat examples only at the national level (observing variations among states), geographers and others have been interested in other scales (and other nations) as well; good examples are the study of the social geography of Canadian cities by Bourne and Ley (1993), the study of rapid changes in a New York neighborhood by Alba, et al. (1995), and Logan, et al. (2004).

The United States has always been shaped by changing racial and ethnic patterns, and its future will be shaped by them as well. Though less detailed than the work of Allen and Turner, McKee (1985) also provided some useful geographic perspectives on ethnic patterns in modern America. Aside from these other works on racial and ethnic groups in the United States by geographers, sociologists and others have much to offer as well. Examples include Fuchs (1990), Gonzales (1990), Gonzales (1991), Lamphere (1992), Roberts (1993), Murdock (1995), Ingoldsby and Smith (1995), and Gonzalez, (2000).

The issue of racial identity became even more complex when the 2000 census for the first time allowed respondents to check more than one box when they identified their racial identity. About 7 million people, half of them under age 18, chose more than one racial category, leaving demographers and other social scientists with some new things to think about. As Gregory Rodriguez (2001, M1) noted, "While most Americans still chose only one box last year, the government's official recognition of hybridity has not only muddled the statistical portrait of the nation, it has also undermined the popular belief that race and culture are immutable."

Between 1990 and 2000 high immigration rates and relatively high birth rates resulted in Latinos and Asians being the two fastest-growing racial/ethnic groups in the nation. Both of these groups tend to have high rates of intermarriage with other groups, including non-Hispanic whites, resulting in ever more blending of future generations. In fact intermarriage is on the increase for all groups in the United States. As Rodriguez (2001, M6) also noted, "The newly released census data will not so much resurrect the melting-pot concept as broaden it to include cross-racial and not only white-ethnic mixing."

Though distinctions can be made at many scales, the Census Bureau data and classifications are used for the discussions that follow, and we need to keep in mind that imperfections exist in both the data and the categories. Distributions of racial and ethnic groups are important and need to be studied at various scales, from national to intraurban, but space constraints allow us here to consider only the national distributions of the three largest such groups in the United States: African-Americans, Hispanics, and Asians and Pacific Islanders. The overall racial and ethnic composition of the United States in 2010 is shown in Figure 3–22; Waldman (2009) has written a detailed account of the North American Indian population.

> *"Race and ethnicity are often associated with locational and other types of inequities for minority race and ethnic groups, not only in the U.S. but in other regions of the world. Many of these inequities appear in poor educational and health outcomes as well as overall reduced life chances."*

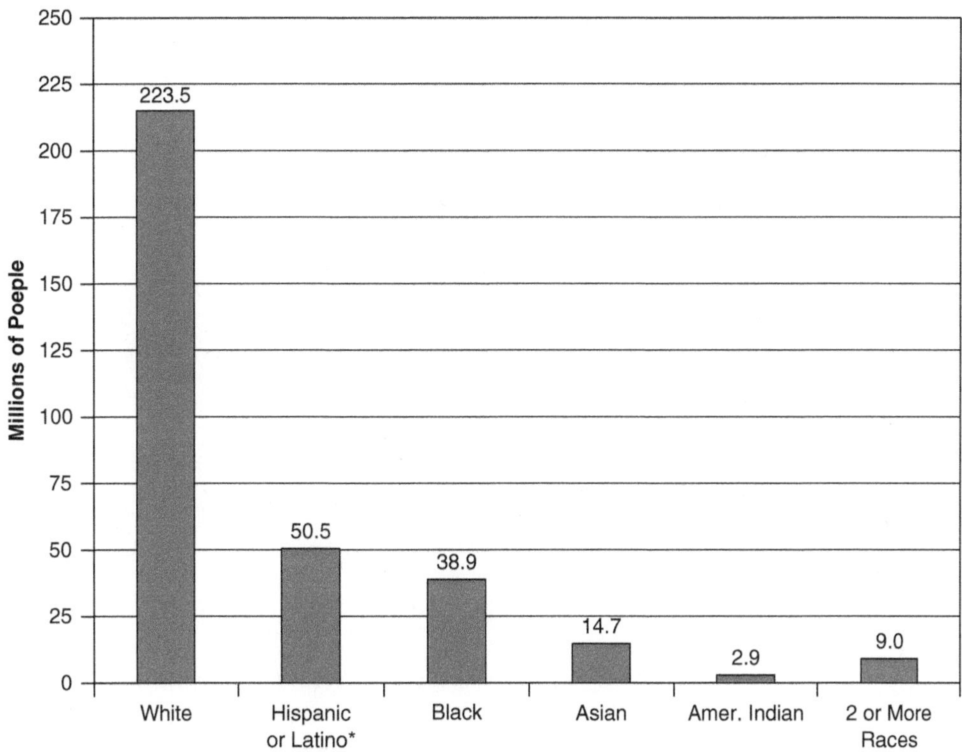

Figure 3–22 Population Composition of the United States, by Racial and Ethnic Groups

Source: Redistricting Data from the U.S. Census, SF P1 and P2, July 1, 2010.
*Hispanics or Latinos come from many racial backgrounds.

African-Americans

According to the Census Bureau there were 38.9 million African-Americans in the United States in 2010. They comprised 13.6 percent of the nation's population. Their share of the United States population has changed over time, from a high of 19.3 percent in 1790 to a low of 9.7 percent in 1930. Hispanics have now surpassed African-Americans as the nation's largest minority. In all likelihood by the middle of this century Anglos will become the largest minority in a nation that will no longer have a majority racial or ethnic group, a situation already apparent in California.

Figure 3–23 shows the distribution of African-Americans in 2010. Slightly over half of all African-Americans, almost 55 percent, still live in the South, around 57 percent live in the central cities of metropolitan areas, and their struggle for equality in America has still not been fulfilled. As Weeks (1992, 280) noted, "Being of African origin in the United States is associated with higher probabilities of death, lower levels of education, lower levels of occupational status, lower incomes, and higher levels of marital disruption than for the white population."

Though Mississippi currently has the highest proportion of African-Americans (36.7 percent), New York, Florida, California, Texas, and Georgia have the largest absolute populations of African-Americans—over 3 million in each. The largest metropolitan concentrations of African-Americans are found in New York, Chicago, Philadelphia, Detroit, and Houston. Washington, D.C. had the largest proportion (52.2 percent) of the African-American population in 2010.

It is interesting to note that the African American population in the United States is becoming more diverse. The share of the black population that is foreign-born has grown from only 1.3 percent in 1970 to 7.8 percent in 2007. Over one-half of black immigrants were born in the Caribbean and over one-third on the continent of Africa. In 2005, one-third of the black foreign born population lived in the New York metropolitan area, 14 percent in the Miami, Florida area, and 6 percent in the Washington, D.C. area (Kent, 2007).

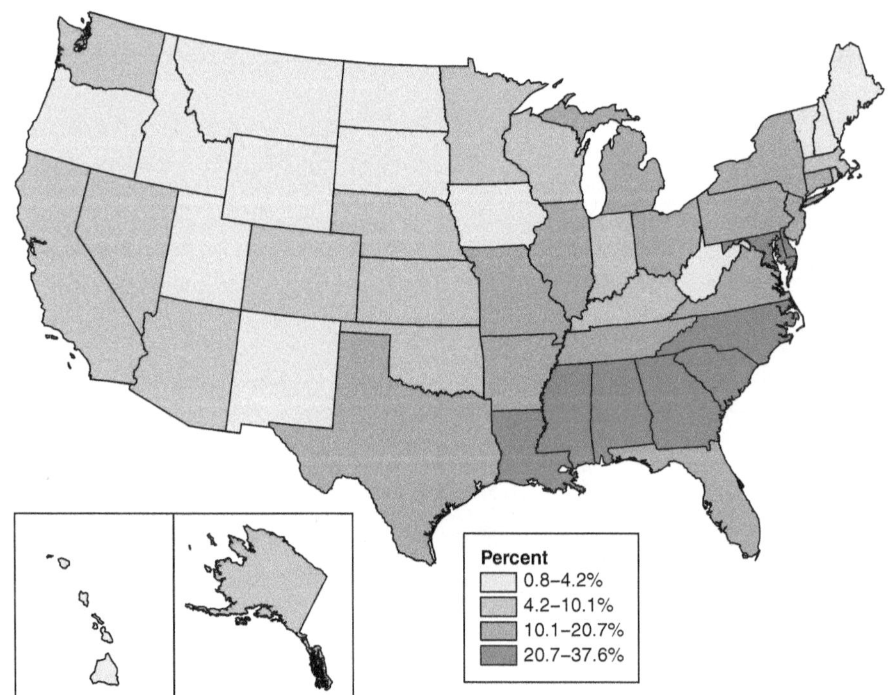

Figure 3–23 Resident Black Population, 2010
Source: U.S. Census Bureau.

Hispanics

In 2010, there were 50.5 million persons of Hispanic origin in the United States. This comprised about 16.3 percent of the nation's population. Though these persons were primarily of Latin American origin, as a group they are far from being homogeneous. More than 32 million are of Mexican ancestry, whereas nearly 4.6 million were Puerto Rican, 1.8 million Cuban, and about 13.4 million from a variety of Central and South American backgrounds. Every Spanish-speaking country in the world is represented in the count, including Spain. Distance from the Mexican border is a major factor in explaining the geographic distribution of Latinos. In addition, metropolitan areas such as Chicago have long presented employment opportunities for Hispanics. Furthermore, long-standing migration streams exist between New York and Puerto Rico and between Miami and Cuba.

A higher fertility than that of non-Hispanic populations and high rates of immigration (both legal and illegal) together raised the number of Hispanics in the United States by close to 100 percent between 1980 and 2000 and by 43 percent between 2000 and 2010. The geographical distribution of Hispanics, shown in Figure 3–24, differs considerably from that of African-Americans. Three-quarters of the U.S. Latino population are in the West and Southwest (primarily of Mexican origin), New York (primarily of Puerto Rican origin), and Florida (primarily of Cuban origin). California and Texas together contain more than half of all Hispanics in the United States (Figure 3–25), with over one in every three in California alone. The major metropolitan concentrations of Hispanics are in Los Angeles, New York, and Houston; the cities with the largest Hispanic populations are New York, Los Angeles, Houston and San Antonio, TX.

As an example of how perspective changes as we change the geographic scale, Figure 3–26 shows the distribution of the Hispanic population within California. The strong concentration in Southern California reflects proximity to the Mexican border and the availability of jobs in urban manufacturing (the garment industry, for example) and

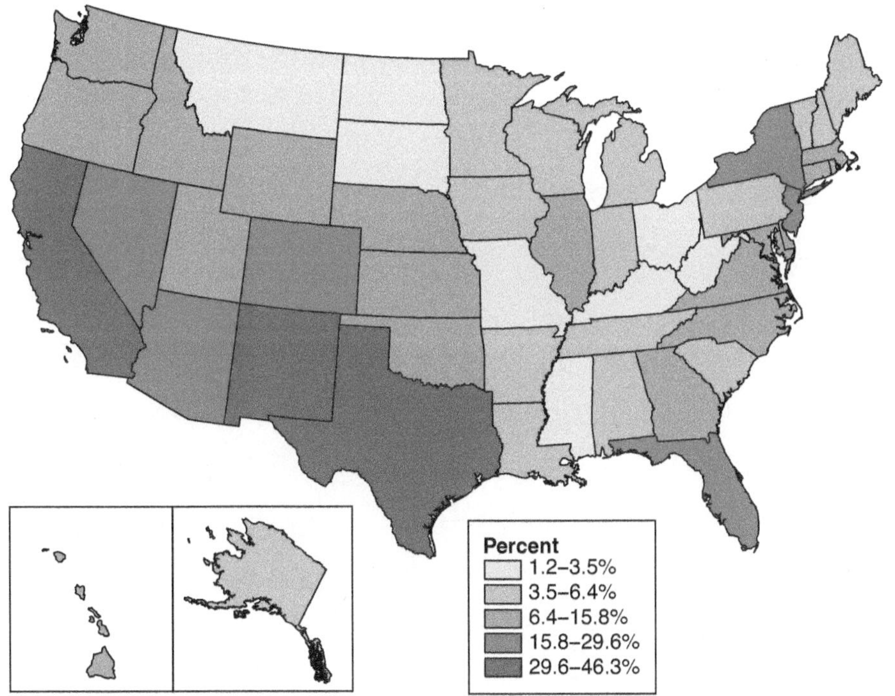

Figure 3–24 Resident Hispanic Population, 2010
Source: U.S. Census Bureau.

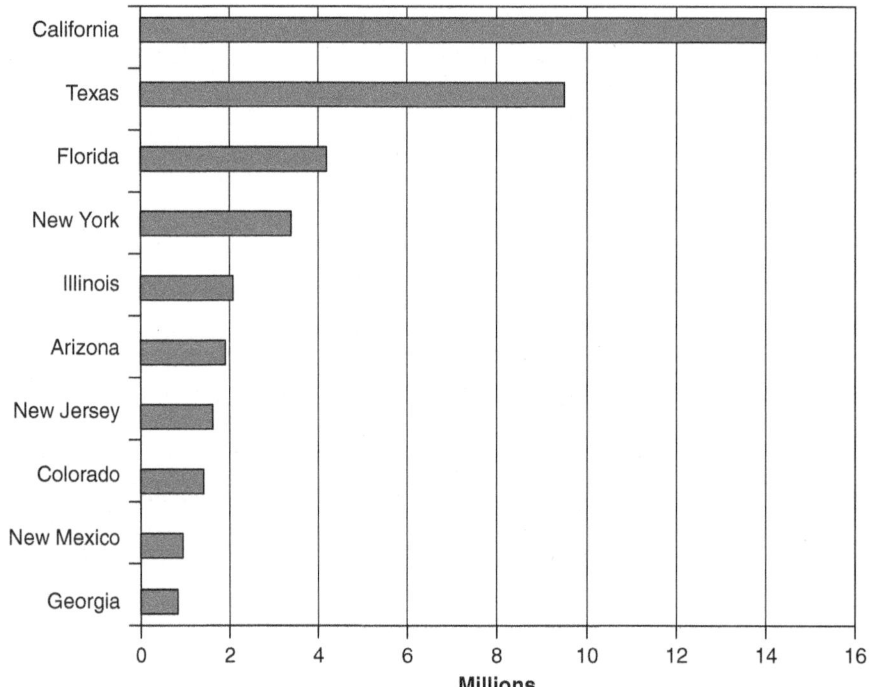

Figure 3–25 States with Largest Hispanic Population, 2010
Source: Data from U.S. Bureau of the Census.

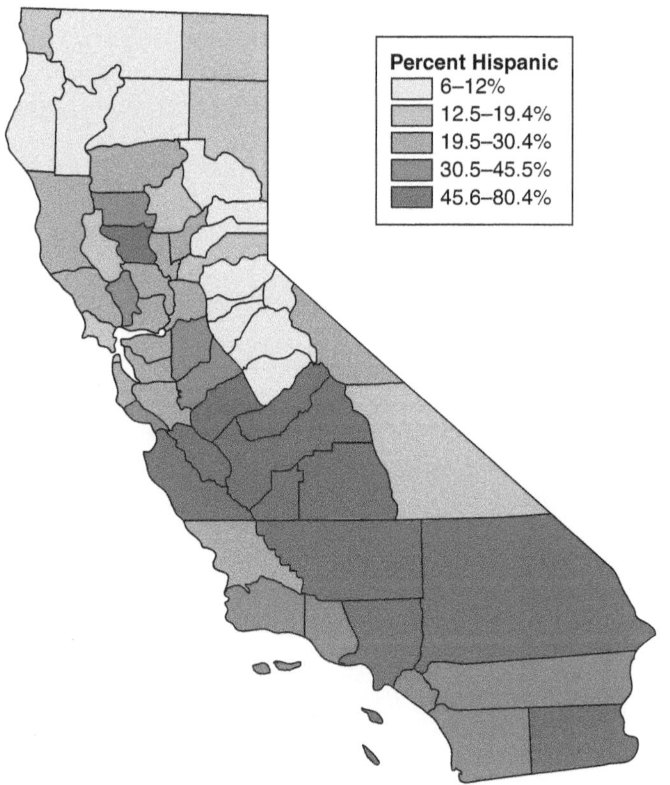

Figure 3–26 Hispanic Population in California, 2010
Source: Data from U.S. Census Bureau.

services (foods and hotels, for example). Latino concentrations in the San Joaquin Valley, on the other hand, reflect their continued employment in California's vast agricultural lands.

Average incomes for Hispanics are higher than those for African-Americans and Latinos are closing the education gap. In 2010, Latinos for the first time outnumbered blacks in college enrollments at both 2-year and 4-year institutions. Latinos still lag behind, however, in terms of completion of bachelor's degrees (Fry, 2011). Hispanic fertility is above that for African-Americans, and continued immigration from Latin America is likely to sustain that differential for the foreseeable future. There are now more Hispanic children than African-American children in the United States and Spanish is the nation's second biggest language.

Asians and Pacific Islanders

Asians are the fastest growing group in the US. According to the Census Bureau, there were 17.3 million Asians and Pacific Islanders in the United States in 2010, and they comprised 5.6 percent of the total population. As with Hispanics, Asians and Pacific Islanders hardly form a homogeneous group. Chinese, Filipino, and Japanese, respectively, are the most numerous Asians in the United States.

Figure 3–27 shows the distribution of Asians and Pacific Islanders in the United States. Geographically, 46 percent of the US Asian population is found in the West, with the largest metropolitan concentrations in Los Angeles, New York, and San Francisco. In terms of metro areas with the largest percentages of Asian population, Honolulu, HI ranks first (43.9%) followed by San Jose, CA (31.1%), San Francisco (23.2%) and Los Angeles (14.7%). Distance again plays a role in explaining this distribution; California is the port-of-entry for most Asian immigrants today. The distant concentration in New York reflects both the cosmopolitan nature of that city and its size, which creates many employment opportunities for them.

As a group, Asians and Pacific Islanders have the highest median family income in the United States, along with high educational attainment and low unemployment. However, within the group there are considerable differences; Japanese, Chinese, and Koreans tend to be at the high end, whereas Indochinese refugees and some Pacific Islander groups tend to be at the low end, often living below the poverty line.

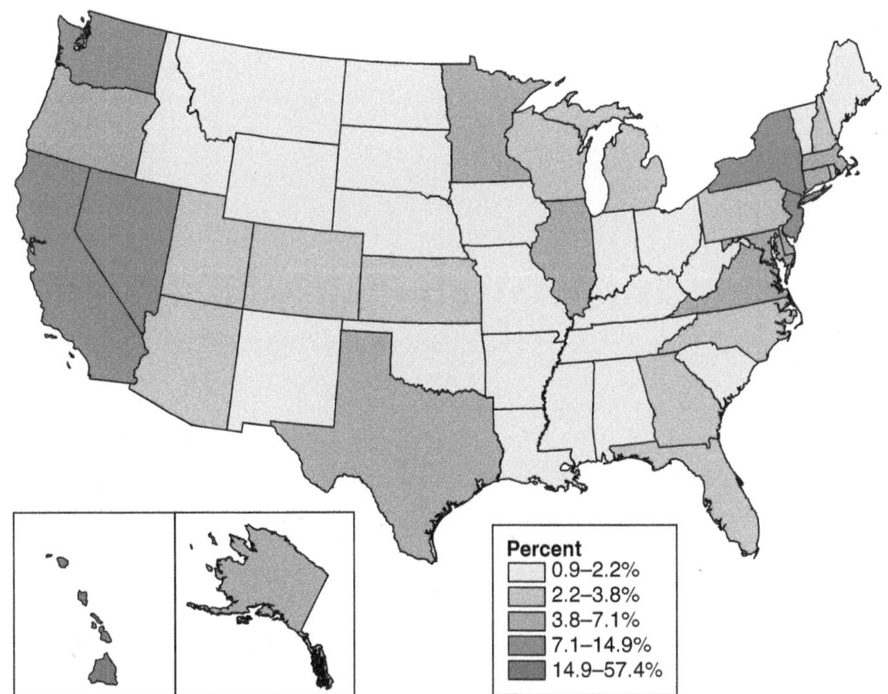

Figure 3–27 Resident Asian Population, 2010
Source: Data from U.S. Census Bureau.

Measuring Diversity and Segregation

Though we have been looking only at the distribution of one group at a time, and only for states, we may also want to know about the relative importance of different groups at different scales. Much is written today about the growing diversity of people in the United States, but less is said about how to actually measure diversity or about measuring a related idea, segregation. Yet, as we look at populations, whether nationwide, statewide, or within our vast metropolitan areas, we are all in agreement that demographic changes are underway, often with considerable rapidity. Furthermore, as Brown and Chung (2006, 125) noted, "Segregation is inherently geographical."

Urban data from the 2010 census showed that the largest 100 cities in the United States have undergone considerable demographic change since 1990. Non-Hispanic whites' share of the population of these cities decreased from 71 percent to 57 percent. Hispanics gained in their share of these metro areas from 11 percent to 20 percent. By 2010 minorities gained majority status in 22 of the largest cities in the US. This is up from 14 cities in 2000 and from 5 in 1990. Among these emerging "minority-majority" cities are New York, Washington DC, San Diego, Las Vegas, and Memphis (Frey, 2011).

"The housing market and discrimination sort people into different neighborhoods, which in turn shape residents' lives—and deaths. Bluntly put, some neighborhoods are likely to kill you."

As demographer Michael White (1986, 198) has pointed out, "Diversity is variety . . . . Segregation in a population is indicated by the unevenness of the distribution of its members across places or categories . . . the most common analysis of segregation is that which measures the racial unevenness across urban neighborhoods or schools."

Diversity can be measured with the entropy index (Shannon index) or the interaction index (Simpson index). Segregation can be measured by the index of dissimilarity, the Gini index, the entropy index, interaction or exposure indices, contingency table measures, and analysis of variance. After surveying a variety of different diversity and segregation indices, White (1986, 216) concluded that "Overall, the entropy or information-based statistic is particularly appealing. It may be the best general measure of unevenness in population distribution." Geographers James Allen and Eugene Turner (1997) found the index of dissimilarity and the entropy index particularly useful in their splendid study of demographic diversity in Southern California. Johnston, Poulsen, and Forrest (2006, 389) found that among major metropolitan areas in the United States, "That the size of a minority group—both absolute and as a percentage of the metropolitan population—is significantly linked to segregation levels and changes in them. There is a clear convergence taking place, reflecting increasing segregation among recently-arrived Hispanics and some desegregation among Blacks." It is important to measure segregation because it is an integral part of US society. Segregation levels impact the geography of opportunity for residents who are relegated to inferior neighborhoods and schools (Squires and Kubrin, 2005). In the US, the racial composition of neighborhoods is connected to the quality of housing, schools, safety, and useful contacts. As John Logan (2003, 33) so astutely put it,

> The housing market and discrimination sort people into different neighborhoods, which in turn shape residents' lives—and deaths. Bluntly put, some neighborhoods are likely to kill you.

CHAPTER 4

Theories of Population Change

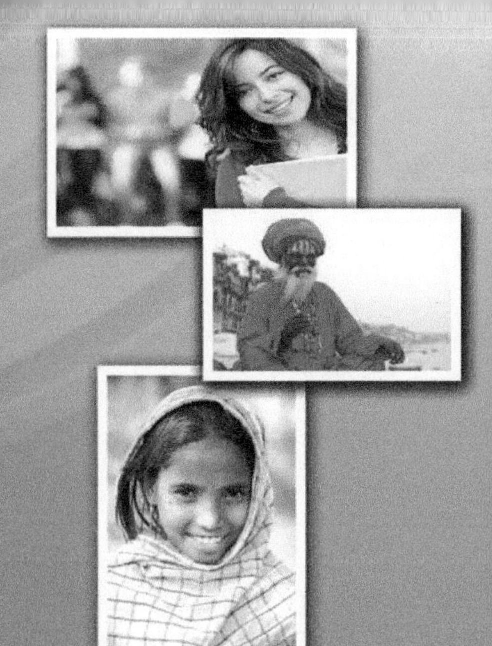

Key Terms

Malthusian theory
Neomalthusians
Optimists Marxism and critical theory
demographic transition theory
demographic inertia
demonstration effects
multiphasic response theory

It has been more than 60 years since Rupert Vance in his presidential address to the Population Association of America asked "Is Theory for Demographers?" (1952). The discipline of demography and indeed the subdiscipline of population geography have long been criticized for their focus on models and techniques, and difference and diversity between people at the expense of theory building (Burch, 2003; White and Jackson, 1995; Findlay and Graham, 1991). It is essential to move beyond descriptions if the causes and consequences of population processes are to be better understood. Additionally, if projections of the future course of demographic events are to be made, then as much insight as possible into the ways that population variables interact with each other and with other nondemographic variables must be gained. Furthermore, more understanding must be gained about how such interactions may differ from time to time, from place to place, and at different scales. Demographic events and systems are situated within broader social, cultural, economic and political contexts. This complicates the search for explanations that might be universal.

Demographic events and systems are situated within broader social, cultural, economic and political contexts. This complicates the search for explanations that might be universal.

In this chapter we consider theories of population change, the general value of these theories, and the value of these theories in helping us to understand future demographic changes, especially in the less developed countries.

Many population geographers and demographers would consider these to be important questions about population:

- What factors, systems, events, etc. cause population growth or decline?
- What will stop or slow population growth?
- When will it stop or slow?

Underlying these questions are fundamental assumptions about populations and their growth. If it is assumed that population growth must stop (an assumption that does not receive universal support!), one of three things must occur. The birth rate must decline, the death rate must increase, or some combination of the two must occur. Trying to predict what will stop rapid population growth, and when it will stop, is fraught with difficulties. Many alternative views of future population growth have been suggested. One alternative suggests a rapid rise in population, followed by a leveling off of population growth before the earth's carrying capacity is reached. Another envisions a rapid rise in population that overshoots the carrying capacity. This overshoot leads to an increase in the death rate because of the lack of food; the increased death rate then leads to a decline in population numbers below the carrying capacity. A similar alternative to the previous one would be an overshoot followed by a long period of continuing fluctuation of population above and below the carrying capacity. Perhaps the most catastrophic consequence of rapid population growth would occur in an alternative in which a rise in population growth leads to serious degradation of the carrying capacity, which in turn brings about a drastic reduction in population size. One final alternative (favored by some economists) suggests a continual increase in the earth's carrying capacity, which keeps ahead of population growth most of the time. In a few instances, for relatively short periods of time, population growth may overtake the carrying capacity, but in a short time new innovations increase the carrying capacity again. Which of these alternatives will be most appropriate for the future is certainly a matter of speculation. However, several theories and models about population growth have been formulated to explain changes in birth and death rates. After reviewing several different demographic theories and models, the chapter ends with detailed examples of demographic change in Cuba and China.

The Role of Theory

A theory is a conceptual framework that explains existing observations and predicts new ones. Theories are well substantiated by evidence, and are developed from extensive observation, experimentation and creative reflection (National Academy of Sciences, 1999). When compared with the physical sciences, population geography, demography, and other social sciences have been relatively unsuccessful in developing theories, though theory building has rapidly accelerated in the past several decades. Geographers, demographers, and others interested in population dynamics share this concern for theory-building because it is valuable as a guide in helping them explain existing phenomena and in predicting as yet unobserved phenomena. Achieving such goals, however, has not been easy. As Teitelbaum and Winter (1989, 3–4) believed:

> Explanation entails causality, which in turn requires comment on a range of variables hard to define or measure. These variables of necessity include the differential resource endowment across nations; and the character of social conflict along gender, ethnic, national, religious, or class lines. When such issues enter the argument, so does political and philosophical language, replete with moral assumptions and a priori beliefs. Since the discussion of population and resources must entail explanation as well as description, the distinction between ideology and science has never been easy or clear. This is inevitable given the nature of the issues involved and the inordinate difficulty in isolating them from . . . questions of all kinds.

A theory is a conceptual framework that explains existing observations and predicts new ones.

Despite the continuing emphasis on models and theories in population geography, geographers in recent years have become more critical of logical positivism and the notion that there are "truths" that can be discovered and enumerated. Social scientists are being confronted with ideas that are forcing fundamental rethinking of the nature of research in the discipline. However, so far such rethinking, aside from feminist theory at least, has had little impact on demography or population geography.

Two thoughtful commentaries on the hazards of predicting future trends were provided by Smil (2005a and 2005b). In the first of these two articles Smil (2005a, 201) made the point that "Modern civilization is subject to gradual environmental, social, economic and political transformations as well as to sudden changes that can fundamentally alter its prospects." Trend forecasts are common, were discussed in the section on population projections in Chapter 2, and are likely to be wrong, partly, according to Smil (2005b, 605), because they may be ". . . punctuated by surprising, sometimes stunning, discontinuities." These discontinuities also bothered Taleb (2007, xx), who wrote that "We produce thirty-year projections of social security deficits and oil prices without realizing that we cannot even predict these for next summer—our cumulative prediction errors for political and economic events are so monstrous that every time I look at the empirical record I have to pinch myself to verify that I am not dreaming." Even though we search for better explanations for demographic processes, and would like to find ways to make better predictions about our demographic future, we are left believing that the wisest comment we have seen on making such predictions was made by Sir Kenneth Clark (1969, 344–5), who wrote that "We have no idea where we are going, and sweeping, confident articles on the future seem to me, intellectually, the most disreputable of all forms of public utterance." Great care must be taken when trying to peer into the future.

Population Theories

It is not surprising that few major philosophers paid much attention to the topic of human population; after all, they lived in a world that was far less crowded than ours. Most early views on population and the factors affecting fertility originated as folklore. People often thought, according to Eversley (1958, 281), that "high living causes sterility . . . intellectual pursuits and gallantry diminish powers of procreation . . . and idiots breed like rabbits." Notions such as these were held by large portions of humankind throughout history, and similar notions are still held today by a substantial number of people.

Perhaps the most common belief about population prior to Malthus' time was that population growth was good and decline was bad. As Tomaselli (1989, 8) commented in writing about population concerns in the eighteenth century, "Playing on the fear of population decline in attacking any given social, legal, or political practice or, conversely, emphasizing the extent to which a reform would bring about an increase in population were such standard rhetorical ploys, that it is often difficult to assess the real importance which was given to the issue of population in and of itself." Wealth and material progress were largely dependent on manual labor, and a growing labor force was viewed as a way to prosperity. With the high mortality rates that prevailed throughout most of human history, it was necessary to have many children in order to overcome the negative effects of the infant and child mortality. Hernandez (1974, 146) pointed out that:

> Another important legacy from premodern times has been the persistence of the utopian dream: an optimum population in an ideal state, according to Plato and Aristotle; medieval notions of a harmonious balance of the material and spiritual; the Renaissance belief in the perfectability of art and life; and the modern scientism that offers unlimited horizons for human betterment.

It was partially in response to such a utopian dream that Malthus wrote his famous essay on population.

Malthus

Perhaps more than any other population theorist, Thomas Robert Malthus (1766–1834), both an economist and a clergyman, has had a profound impact on attitudes and ideas concerning population growth. His name still appears in most discussions of population prospects and seems ineradicably associated with the topic of population. However, demographers do not all agree that Malthus made a significant contribution to the scientific study of population; at the very least he elevated it to a higher level as a topic for serious discussion (Petersen, 1999).

Malthus was born in England in 1766 into a rather well-to-do family and lived during a time of revolution, both political and industrial. For most of their history Europeans had been conditioned to expect, and to accept, a bitter lot and a short life. This fatalism was conditioned by a belief that there were not enough material resources to go around. The major economic thought of Malthus' day was mercantilism, an economic system with strong government policies directed toward the accumulation of wealth. Poverty was associated with the merits of man and the will of God, and the earth was viewed as a place of testing and punishment. Mercantilist population policies were mainly pronatalist because more births meant more workers, hence more aggregate wealth. At the same time, as Wrigley (1989, 31) noted, Malthus and the two other great classical economists, Adam Smith and David Ricardo, were in complete agreement that, "The secular tendency of real wages was likely to be flat, if not tending downward, because any increase in the funds available to pay wages would be matched by a proportional rise in the number of wage earners."

However, during Malthus' time industrialization was accelerating and ideas were rapidly changing; the acceptance of deprivation was increasingly being challenged. Liberty, equality, and fraternity became the new watchwords. Rapidly increasing production and increased trade suggested that resources might not be so limited after all. Rather, abundance might be possible. Out of this background a rash of utopian schemes, wild dreams, and visions of the future were forthcoming. The future was seen as a time when poverty, misery, vice, want, greed, and even death itself might be eliminated. With a little bit of luck, and a few changes, the perfect society could evolve. Malthus' father was caught up in these new ideas concerning the perfectibility of man and society. He invited many utopians to his home for discussions, but within these discussions the young Malthus took a position opposite to that of his father.

The first edition of Malthus' famous population essay, published in 1798, was entitled: *An Essay on the Principle of Population as it affects the future improvement of society; With remarks on the speculations of Mr. Godwin, M. Condorcet, and other writers*. It was written neither as a text in demography nor as an exposition of some new law of population growth. Rather, its intent was to refute some of the utopian ideas that were then gaining currency. His chief targets were the authors mentioned in his title. Commenting upon Condorcet, Malthus (1798, 3) wrote: "I have read some of the speculations on the perfectibility of man and of society with great pleasure. I have been warmed and delighted with the enchanting pictures which they hold forth. I ardently wish for such happy improvements. . . ." Unfortunately, Malthus then argued, the road to utopia would always be blocked because he believed that man would always press up against the limit of subsistence.

Malthus' first edition of the Essay was rather simple. His "principle" of population was the result of two postulates and one assumption. The first postulate was that "food is necessary to the existence of man," and the second was that "the passion between the sexes is necessary and will remain in its present state." Malthus assumed that population would always tend to increase at a geometric rate, whereas food production could only be increased at an arithmetic rate. In his own words:

> Population, when unchecked, increases in geometric ratio. Subsistence increases only in an arithmetical ratio. A slight acquaintance with numbers will show the

> immensity of the first power in comparison of the second . . . the human species would increase as the numbers 1, 2, 4, 8, 16, 32, 64, 128, 256, and subsistence as 1, 2, 3, 4, 5, 6, 7, 8, 9. In two centuries the population would be to the means of subsistence as 256 to 9; in three centuries as 4,096 to 13; and in two thousand years the difference would almost be incalculable.

According to Malthus, population growth would always press against the means of subsistence, unless it was prevented by some very powerful and obvious checks.

According to Malthus, population growth would always press against the means of subsistence, unless it was prevented by some very powerful and obvious checks. These checks, according to Malthus, could be categorized as either *positive* or *preventive.*

Preventive checks were those that affected the birth rate, and in his view moral restraint was the only acceptable one, though he recognized that "vices" such as homosexuality and birth control could also have an impact. The positive checks were those that affected the death rate, including misery, disease, famine, and war. Thus, Malthus felt that there was no way to escape the positive checks and that Godwin's argument that things would become perfect if we could just get rid of government and private property was wrong. The "Principle of Population" was inflexible, inexorable, and inescapable. No exercise of reason could remove its effect, he argued, producing his own dismal view of the future of humankind and helping to saddle economists forever with the description of their discipline as the "dismal science."

From today's perspective a number of shortcomings can be identified in Malthus' thinking. He emphasized land as the major limiting variable on expanding food supplies, yet other inputs such as improved crop rotation patterns, fertilizers, and new hybrid strains of seeds have been, and continue to be, major factors in increasing food supplies. Furthermore, food is not our only necessity. Clothing, shelter, and other items are essential to at least some degree, and industrialization, well underway during his era, certainly increased the availability of many of these items more rapidly than the rate of population growth. He also failed to consider the major changes that transportation systems would undergo, as well as the expansion of trade that would accompany such changes. Because of his clergical leaning, he also failed to foresee the possibility that contraceptives would be widely accepted, or to even imagine that couples would decide to limit the sizes of their families in response to changing socioeconomic conditions.

Though considerable discussion and controversy appeared after the publication of Malthus' book in 1798, his theory dropped from favor by the middle of the nineteenth century, as the Industrial Revolution found ways to reduce the pressure of population, including emigration to the New World, and to raise wages above the subsistence level for more workers. However, the rapid increase in population during the twentieth century, coupled with considerable malnourishment in many areas, revived an interest in Malthus' ideas. Many "Neo-Malthusians" continue to feel that population growth will still outrun the food supply and that the world will not be able to continue supporting a growing population. Strangeland (1904, 356) summarized Malthus' contribution as follows:

> Malthus' work was a great one, written in an opportune time, and though it cannot lay claim to any considerable originality as far as theories presented are concerned, it was successful in that it showed more fully, perhaps more clearly, and certainly more effectively than had any previous attempt, that population depends on subsistence and that its increase is checked by want, vice, and disease as well as by moral restraint or prudence.

The indelible mark that Malthus has left on population studies has been largely negative and pessimistic. Yet, as geographer Vaclav Smil (2000, xxvii) perceptively noted with respect to Malthus, ". . . the final message of that famous English cleric was not one of despair, but one of cautious hope." Smil (2000, xxvii) went on to say that "When a man, shortly before his death, deliberately changes the conclusion and the emphasis of his previously much-publicized work, then it is this final version, rather than the initial product, that should be seen as the intellectual bequest to be associated with his name." At the end of his 1803 book, Malthus wrote:

> On the whole, therefore, though our future prospects respecting the mitigation of the evils arising from the principle of population may not be so bright as we could wish, yet they are far from being entirely disheartening, and by no means preclude that gradual and progressive improvement in human society. . . . And although we cannot expect that the virtue and happiness of mankind will keep pace with the brilliant career of physical discovery; yet, if we are not wanting to ourselves, we may confidently indulge the hope that, to no unimportant extent, they will be influenced by its progress and will partake in its success.

Of course it was not this cautiously hopeful, and often ignored, conclusion that led to the rise of neomalthusianism, but rather the more despairing earlier work of Malthus discussed above.

Neomalthusians

Neomalthusians see the consequences of rapid population growth not just in mass poverty but also in the deteriorating quality of the earth itself as a home for humans. At the same time they go far beyond Malthus' "moral restraint" as a means of controlling population growth, supporting family planning, contraceptives of every sort, and even abortion.

Neomalthusians continue to argue that population growth is problematic. Not only do they argue, as Malthus did, that population growth creates problems relative to food supplies, but they go beyond that to relate population growth to environmental problems as well. They see the consequences of rapid population growth not just in mass poverty but also in the deteriorating quality of the earth itself as a home for humans. At the same time they go far beyond Malthus' "moral restraint" as a means of controlling population growth, supporting family planning, contraceptives of every sort, and even abortion.

Two major voices for neomalthusianism are biologists Paul Ehrlich (1932–) and Garrett Hardin (1915–2003). Both generated considerable public interest in the problems of population growth in the late 1960s, and both have pursued those themes since then. Ehrlich's 1968 book, *The Population Bomb*, set forth the central themes of rapid population growth, uncertain and inadequate food supplies, and environmental degradation. Ehrlich and Ehrlich (1990, 225) revisited the population issue and its relationship to environmental and other problems, and concluded that "A central problem facing us now is finding ways to convince national and international leaders and the world's people that opportunities for action to assure global environmental security are fast slipping away . . . it's the top of the ninth and humanity has been hitting nature hard . . . we must always remember that nature bats last!" The writings of Ehrlich, Hardin, and other neomalthusians are discussed in more detail in Chapter 9.

For now we would note that in the last two hundred years the world's population has grown from slightly less than one billion to more than 7 billion. This growth has been so rapid that it is not at all clear whether it has been made possible somehow, thanks to fossil fuels and technology, to be able to permanently increase the planet's carrying capacity, or whether there has been what biologists would label a vast "overshoot" of that carrying capacity. Even as economists happily see population growth as adding both more labor and more consumers to the planet, ecologists worry that we have created a population that is in the long run unsustainable. Few economists want to concede that the economic system that engulfs humanity ultimately must function within Earth's ecosystem. At the other end of the spectrum renowned astrophysicist Stephen Hawking suggested in 2006 that humans should spread out into space in order to assure survival of the species because life here on earth will be increasingly threatened by disasters of one sort or another from global warming to diseases.

Optimists

In 1965 Danish economist Ester Boserup proposed an argument about the relationship between population growth and food supply that was in many ways directly the opposite from Malthus'. Rather than arguing that population growth "depends" on agriculture, treating it as a dependent variable as Malthus did, she began her investigation by stating that, "Population growth is here regarded as the independent variable which in its turn is a major factor determining agricultural developments." (Boserup, 1965, 11) In essence, she argued that population growth and critical population densities could stimulate agricultural innovation and change. As a starting point, she suggested replacing the distinction between cultivated and uncultivated land with the concept of frequency of cropping. Given definitions of land use and patterns of tool adoptions in agrarian societies that she outlined, she was able ". . . to define the concept of intensification in agriculture in a new way, namely as the gradual change towards patterns of land use which make it possible to crop a given area of land more frequently than before" (Boserup, 1965, 43).

Whereas Malthus would argue that death rates would rise to curtail the population growth and restore equilibrium, Boserup would see another possibility, namely an intensification of the agricultural system.

Faced with population growth in an agrarian society, people would be confronted with a threat to their standard of living. Whereas Malthus would argue that death rates would rise to curtail the population growth and restore equilibrium, Boserup would see another possibility, namely an intensification of the agricultural system. She argued that by working more hours and adopting more intensive ways of growing crops, people could cope with a growing population. The primary trade-off that they would be forced to make is one of leisure for work. She envisioned a succession of increasingly more intensive agricultural practices as population growth continued, ranging from forest-fallow to bush-fallow, through short-fallow and annual cropping, to multi-cropping, the most intensive of the five systems, and noted that "Under the pressure of increasing population there has been a shift in recent decades from more extensive to more intensive systems of land use in virtually every part of the underdeveloped regions" (Boserup, 1965, 16).

In a subsequent work Boserup (1981) broadened her arguments somewhat but continued to argue that population growth acted as a stimulant to technological change, at least under some conditions. However, she recognized also that high rates of population growth could overload systems. Agricultural intensification has continued to occur throughout the less developed world, but in some cases, despite total growth in agricultural production, per capita growth has remained constant or even declined as populations grew at unprecedented rates. For example, such conditions have occurred in various parts of Africa in recent decades.

Though Boserup provided some compelling arguments for treating population growth as an independent variable that in turn drives agricultural intensity and technological innovation, she is not without her critics as well. One argument against her work focuses on population growth itself and asks why, if food was scarce to begin with, a population would begin to grow, especially since slow or no growth has been typical for most of human history. Another criticism is that she treats agrarian societies as closed systems, whereas migration may act as a safety valve for population growth in many cases, though that role has certainly become more limited with the rapid expansion of population in the twentieth century.

Marxism, Critical Theory, Feminist Theory

Down through the years several theorists have disagreed with the basic Malthusian ideas. Foremost among Malthus' critics was Karl Marx, who argued that no such thing as overpopulation existed. Marx believed that Malthus was a bourgeois, chauvinist clergyman who upheld the established order of social inequality. Furthermore, Malthus was accused of copying the writings and ideas of the classical school of economic theorists, who proposed unrestricted competition, self-interest, private property, and individualism as solutions to Europe's population explosion. Malthus was thus branded as an easy target for revolutionary criticism and an archenemy of the worldwide socialist movement, since he was considered a literary oppressor of the underprivileged and an apologist for the exploiting class.

Marx believed that Malthus was a bourgeois, chauvinist clergyman who upheld the established order of social inequality.

In the *Communist Manifesto* and in other works, Marx expounded in class struggle theory that an uprising by the proletariat would eventually lead to a classless society without private property, and that an egalitarian ethic would prevail. He believed that the output of food and resources was outstripped by population growth not because there was a natural law that subsistence could not keep pace with population growth, but because poor people, under the capitalist system, were denied access to the control of food and other resources. Furthermore, he argued, if poor people were given control over the means of subsistence—equipment, knowledge, land, and an adequate share of the wealth—their production of goods and services would far surpass the growth of population. Exploitation by the rich and private ownership of the means of production blocked this solution. In essence, Marx felt that there was no population problem; the problem was a maldistribution of resources.

Marxism is a strand of critical theory. Critical theorists believe that it is necessary to understand the lived experience of real people in context. Research that is grounded in critical theory interprets the acts and the symbols of society in order to understand the ways in which various social groups are oppressed. Critical approaches examine social conditions in order to uncover hidden societal structures. Critical theory teaches that knowledge is power. This means that understanding the ways one is oppressed enables one to take action to change oppressive forces. In this way critical theorists work to bring about change in the conditions that affect people's lives.

Feminist theory, in the most simplistic terms can be understood as a response to the proposition so aptly stated by Simone de Beauvoir (in English), "Representation of the world, like the world itself, is the work of men; they describe it from their own point of view which they confuse with absolute truth" (1972, 161). Feminist theory is the extension of feminism into theoretical or philosophical discourse. Feminist theorists view gender as a social construct, and attached to this construct are a variety of lived experiences that may not be captured in research not grounded in feminist theory. In most societies gendered oppression is structural and the effects of this must be considered in any inquiry of men's and women's lives. One of the aims of feminist theory is to understand the nature of gender inequality and the subordination of women. Researchers in a variety of disciplines and fields examine women's social roles, experiences, interests, etc. More generally, it provides a critique of social relations. Therefore the notion of a value-free orientation in research is rejected by feminist theorists.

During the development of research based in feminist theory, several major research paradigms have emerged to challenge the monopoly of men in knowledge production, research methods, and methodologies, and thus, made way for production of feminist knowledge.

Demographic Transition Theory

Malthus' theory and demographic transition theory have been the two most influential theories in population geography. The original demographic transition theory, or model, was based on historical observations of demographic changes in western European countries. The gap between the birth and death rates in Europe was closed by a declining birth rate, not by a rising death rate as Malthus had predicted. A new kind of demographic stability was reached, first in northwestern Europe and then elsewhere in the more developed world. This transition from a relatively stable demographic system with high birth rates and death rates to a similarly stable system with low birth rates and death rates is known as the **demographic transition**.

This transition from a relatively stable demographic system with high birth rates and death rates to a similarly stable system with low birth rates and death rates is known as the **demographic transition**.

The basic demographic transition model postulates a deterministic causal link between modernization on the one hand and both mortality and fertility reduction on the other. Conceptually it can be viewed as an idealized sequence of stages through which a given population passes, and the end result is a stable demographic system with low birth and death rates. The data upon which the transition model is based came primarily from the demographic experience of northwestern Europe, principally England and Wales. Figure 4–1 illustrates the changes in vital rates that have taken place in England since 1700, compared with those in the United States, and provides an idealized view of this descriptive model.

Prior to its industrialization in the eighteenth century, England generally experienced high birth and death rates, though not as high as those currently being experienced in many less developed countries. The result was a relatively stable population in which births were approximately balanced by deaths. A decisive break occurred around the middle of the eighteenth century, when the death rate began a long-term decline. The reasons for the initial decline are still not completely understood, though most experts agree that it was probably due to unusually good harvests at the time. The good harvests were a result of such agricultural improvements as crop rotation and advances in animal husbandry. There was also a slight rise in the death rate in the early nineteenth century, probably associated with the growth of factory towns and their squalid, unhealthy living conditions. However, after a few decades the death rate resumed its decline, which has continued until today.

In England after 1700 the birth rate initially increased slightly as a result of the Industrial Revolution. With industrialization and the disappearance of the apprentice system, unskilled young people could get factory jobs and did not have to endure long apprenticeships before they received compensation. Thus, young people could marry earlier and start families at younger ages. Improved living conditions may also have added slightly to the birth rate—healthier mothers would be more likely to experience full-term pregnancies that resulted in proportionately more live births and produced healthier babies.

Fertility in England stayed relatively constant for the next one hundred years and then started its major decline around 1870. The primary reason for the decline was not a change in marriage patterns, but a redefinition of the ideal family size within marriage. Several reasons for this decline in ideal family size have been suggested. First, the contribution of children to family welfare was declining. In the preindustrial period children contributed a great deal to family welfare. Because the overwhelming majority of people were involved

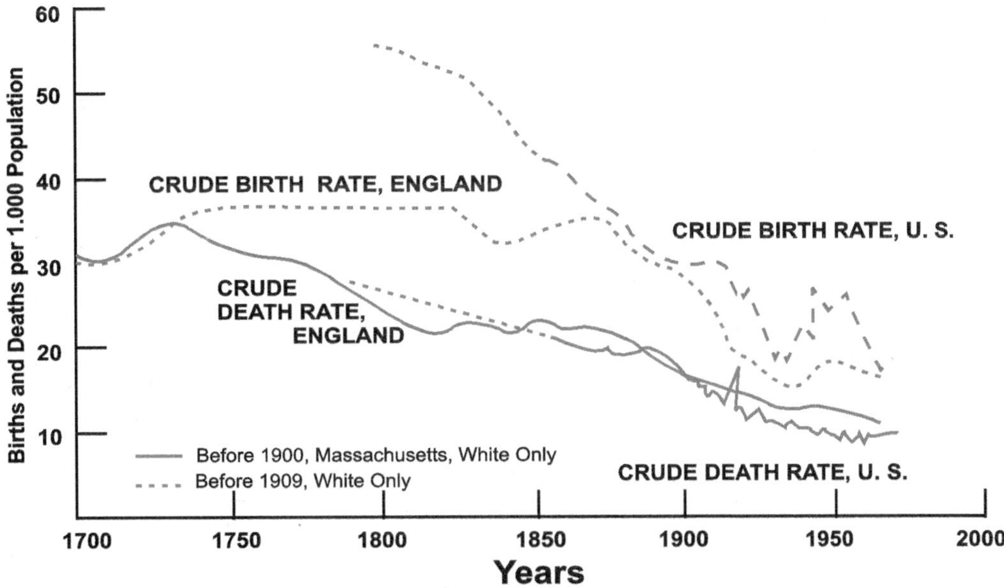

Figure 4–1 Mortality and Fertility Transition in the United States and England

Source: Abdel R. Omran, "Epidemiologic Transition in the United States," *Population Bulletin*, Vol. 32, No. 2 (May, 1977, 15). Courtesy of the Population Reference Bureau, Inc., Washington, D.C.

in agricultural pursuits, children were very useful in weeding crops, fetching water, and gathering wood. In the early textile factories and mines, children served an important economic function because they worked at a much cheaper rate than adults. Children were also a form of social security; they were often the only means of support for their parents when the parents became unable to work. However, as more restrictions were put on child labor practices in the middle and late nineteenth century, and as populations generally became more urbanized, children became more of a burden to parents. The state began to take on a welfare function and to provide for older citizens. With increasing industrialization the number of people engaged in farming dropped significantly and the role of children as producers was further diminished.

A second reason for the redefinition of ideal family size had to do with rising expectations. The Industrial Revolution produced a tremendous volume of goods and services, so people could increasingly shift from satisfying needs to satisfying wants. Luxuries sometimes became viewed as necessities; social mobility and the attainment of riches were easier if one had fewer children. With the growth of knowledge came more favorable attitudes toward family planning, resulting in a significant and sustained decline in the birth rate.

Though Figure 4–1 illustrates the demographic experience of England, general trends appeared to be similar for most of the Western world, with the exception of France, where fertility declines began much earlier, especially within some social classes (van de Walle and Muhsam, 1995). The timing of the transition varied from place to place, but the results were essentially the same, at least when viewed at the national scale. The new balance of birth and death rates represented an improved condition of human efficiency and health, as well as a level of material well-being that had never before been achieved.

The general pattern of the demographic transition can be associated with different areas of the world today. Figure 4–2 represents an idealized demographic transition model. Stage A indicates an area with a high birth rate and a high death rate, equivalent to the preindustrial era in Europe. Few major world regions remain in this stage, though a few African countries, such as those of the Sahel, probably still do. Stage B represents a population with a high birth rate and a declining death rate. Many of the African nations are now in this stage. Parts of Asia, excluding Japan, China, Hong Kong, South Korea and Taiwan, also fit into this category. Stage C includes nations with relatively high birth rates and low death rates, such as many of those in tropical Latin America and some African nations, where populations are growing at some of the fastest rates in the world. Kenya, for example, had a crude birth rate of 35 and a crude death rate of 8 in 2012, yielding an annual population growth rate of 2.7 percent. Countries in stage D have declining birth rates and low death rates. Examples include countries in temperate Latin America, such as Argentina and Uruguay, as well as China, Taiwan, and South Korea. The final category, Stage E, depicts a population with low birth and death rates, and includes the countries of the more developed world.

If there is acceptance that the demographic transition model and therefore agreement in a causal link between modernization and a decline in birth and death rates, then the obvious solution to rapid population growth is to modernize the world as rapidly as possible. However, if there is progressive step-wise movement from Stage A to Stage E, then the speed at which nations move through these stages is critical. The nations in Stage D are heading in the "right" direction and in a short time should reach equilibrium between births and deaths. Though the nations in Stage C are experiencing a rapid increase in population, they are going in the "right" direction and should soon experience a decline in birth rates and a concomitant decrease in population growth. The critical nations are those found in Stages A and B. They have yet to reach their greatest rates of demographic increase and still must pass through Stage C, the period of most rapid population growth.

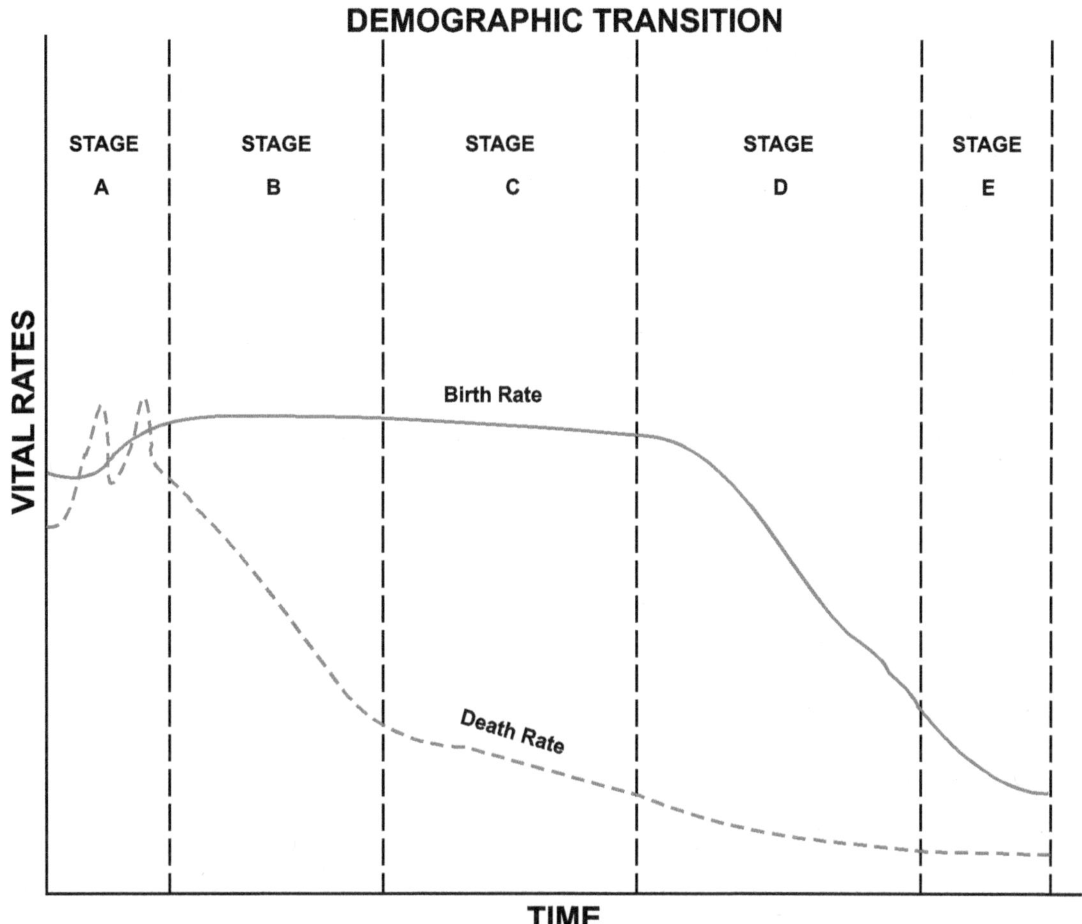

Figure 4–2 Stages of the Demographic Transition

Africa and Asia contain the largest populations, and if they follow the European experience then they would experience a tremendous increase in numbers.

The Relevance of the Transition Model

Though the demographic transition model has received considerable attention, its relevance seems increasingly in doubt. Whether or not the transition model is useful in explaining the population changes that are occurring in the less developed countries is an important issue. Conditions in today's less developed countries differ considerably from those that prevailed in the more developed nations as they moved through the demographic transition. Policy formulation in less developed countries often depends on the concepts of the transition model, and to the degree that these concepts are validated they will be useful in guiding policy development. However, questions have been raised about the degree to which the less developed countries will duplicate the experience of the more developed countries upon which the demographic transition model is based.

Critics of demographic transition theory assert that it is based entirely in modernization theory. Graham notes (2000, 262)

> Like both modernisation theory and dependency theory, [demographic transition theory] provides a general picture of change over time which is predicated on a linear view of history in which the West sets the standard and is further along a given path of progress than the Third World. Like modernisation theory, demographic transition theory . . . denies the Third World a history and assumes that progress consists of achieving the characteristic conditions of the West.

Moreover, demographic transition theory is criticized for its philosophical underpinnings. Graham further states (2000, 265)

> Three points are worth noting about the claims of the transition theorists: firstly, the centrality given to the notion of progress and its implication of improvement, i.e. change for the good; secondly, the measurement of progress against a European yardstick (i.e. for non-European countries, change for the better is change as Europe experienced it); thirdly, the causes of change in non-European populations are thought to emanate from Europe, with non-European populations progressing by adopting the characteristics of European populations. The last two of these points seem closely associated with claims of cultural superiority and the theory has, quite rightly, been condemned for its Eurocentrism.

Population Growth

As we have already seen, population growth in the less developed countries is occurring at unprecedented rates. Africa is the world's fastest growing major region with a 2.5 percent rate of natural increase in 2012. By comparison, when most European countries were undergoing their demographic transitions very few countries experienced annual growth rates of more than 1.5 percent.

One of the obvious problems associated with rapid population growth is the population-related growth in demand for goods and services and the growth in demand for social investment, especially in education. Whereas the European countries were not too much affected by these difficulties, most of the less developed countries today face severe problems and find that population growth increases the difficulty of raising the standards of living (Kindleberger and Herrick, 1977).

The tendency for current population dynamics, such as growth rate, to continue for a period of time, thereby creating a delayed population response to gradual changes in birth and death rates is **demographic inertia.**

To make matters even worse, it probably will be much more difficult to halt rapid growth in the less developed countries than it was to slow down the growth in European countries during their demographic transitions. The younger age structures in the less developed countries, the result of much higher fertility, add a great momentum for further growth. This produces a **demographic inertia**, the tendency for current population dynamics, such as growth rate, to continue for a period of time, thereby creating a delayed population response to gradual changes in birth and death rates. Even if the fertility rate in the less developed countries declined to replacement level within the next decade (a highly unlikely event) they would still continue to grow for another 60 years or more and would nearly double in size. For a country such as India, even a doubling of its current population would present considerable challenges.

Mortality Declines

One of the major reasons for rapid population growth in the less developed countries is the rapid decline in mortality coupled with sustained high rates of fertility. These mortality declines have occurred in response to the geographical diffusion of modern medicines and public health programs. The European countries experienced a more gradual decline in mortality that was closely related to the economic and social forces of industrialization and economic development. In many less developed countries today mortality declines have been much more rapid. As a consequence, the less developed countries have mortality levels that are considerably below those that prevailed in Europe during the early stages of industrialization. Further mortality declines seem likely, especially in Sub-Saharan Africa; however, such declines may be slowed as a result of the spread of the HIV/AIDS epidemic on that continent and the increasing social and political instability of some African countries.

Fertility Levels

The fertility levels of most of the less developed countries today are higher than those that prevailed in Europe before the European countries began their demographic transitions. Many African countries had crude birth rates of 40 or more in 2000, for example. By comparison, the crude birth rate in early nineteenth century Britain was estimated to be around 35. The major reason for the difference is marriage patterns. Extensive nonmarriage and late marriage were typical in nineteenth century Europe, whereas the practice of early and nearly universal marriage in less developed countries has been more widespread.

Although high birth rates prevail in many less developed countries as a result of the prevailing marriage patterns, coupled with relatively low rates of contraceptive usage, signs of declining fertility can been seen. It might be possible to reduce fertility in the less developed countries today by altering the marriage patterns as well as by introducing more effective contraceptive practices. Raising the legal marriage age would decrease the number of years females are exposed to the possibility of conception, for example. This, however, would be a tremendously complex and difficult solution to put into practice in many less developed countries. However, part of the recent fertility declines in the People's Republic of China have resulted from such efforts, mostly due to the influence the government can exert over its population.

Of course, motivation toward smaller families is going to be necessary before significant declines in fertility are recorded in most of the less developed countries. The "demonstration effect" (effects on the behavior of individuals caused by observation of the actions of others and their consequences) of the European transition and the predominance of small families in Europe and the United States have recently diffused to most of the less developed world as a result of expanded trade and better communication systems.

The role of governments and various international agencies may also aid in bringing about rapid fertility declines. Family planning programs are most likely to be effective if they are supported by the government. Governments today are more likely to be involved in many forms of planning than were their nineteenth-century counterparts. Furthermore, many less developed countries today have social scientists and planners who actively participate in the formulation and execution of national policies for economic development and social change. Such planning was almost nonexistent in the European countries during their periods of demographic transition and industrialization. Also, financial and technical assistance for both development and family planning programs is available from such organizations as the United Nations and the United States Agency for International Development.

Encouraging signs of fertility declines in many less developed countries occurred in the 1990s, and these declines continued throughout the 2000s. Family planning programs, communications about family planning, and the widespread availability of modern, effective contraceptives continue to have significant effects throughout the Third World. As Phillip Longman (2004, 36) commented, "The global fall in fertility, even if it does not continue to deepen and spread, is creating a world for which few individuals, and no nations, are prepared." Vallin (2002) suggested that the end of the demographic transition was also the end of the paradigm upon which it had been based, leaving many more questions than answers about future demographic changes in post-transition countries.

Migration

Migration undoubtedly played a significant role in the stabilization of population in parts of nineteenth-century Europe. International migration operated as a "safety valve" and softened the effect of rapid population growth. Millions of European citizens moved to

Oceania and the Americas during the nineteenth century. However, this potential "safety valve" for releasing population pressure is no longer available for large numbers of people because of current economic and political realities. During the European demographic transition the growing rural population was able to find opportunities for gainful employment in urban areas as industry rapidly expanded. New skills and occupations were acquired as people moved from rural to urban areas. In the less developed countries today, however, industrialization is unable to create jobs in the cities fast enough to absorb the rapidly growing labor force. In some countries attempts are being made to slow down the rate of rural-urban migration.

Education and Economic Development

Economic development and modernization in many less developed countries has proceeded more rapidly than in nineteenth-century Europe. This suggests the possibility of a more rapid fertility decline in the less developed countries and a quicker completion of the demographic transition.

However, education also has an impact on fertility; in general, lower fertility is associated with higher educational levels (this is especially true for females, who are at a considerable disadvantage for educational opportunities throughout much of the Third World). Rapid population growth in the less developed countries, together with the large proportions of people in the younger age groups, means a rapidly increasing demand for educational facilities. The less developed countries usually cannot keep up with this demand, and the result is most likely to be deferment of the goal of universal education along with its potential effect of lowering fertility.

Of course, care must be taken in trying to forecast the demographic consequences of modernization. Caldwell (1976, 358) commented that:

> The major implication of this analysis is that fertility decline in the Third World is not dependent on the spread of industrialization or even on the rate of economic development. It will of course be affected by such development in that modernization produces more money for schools, newspapers, and so on . . . But fertility decline is more likely to precede industrialization and to help bring it about than to follow it.

The specific interactions between socioeconomic and demographic changes are far from being completely understood. Undoubtedly other differences between the transition experience of the European countries and the likely course of demographic events in the less developed countries could also be identified. Changes in the composition of the labor force in the less developed countries would have a downward effect on fertility if more women found occupations outside the home, for example. Labor force participation among females tends to encourage them to have smaller families. Once females have outside incomes, a part of the cost of having children will be the income foregone (opportunity cost) during and following pregnancy. This opportunity cost may be enough to discourage having as many children.

Additional Considerations

The demographic transition model has been in a respectable position for many years. As outlined above, its validity has been increasingly questioned, especially with respect to its applicability to less developed countries. From around the middle of the eighteenth century onward in Europe, marital fertility underwent a sustained decline; it was mainly associated with the modernization that was sweeping that continent. At a general level the transition model appears compatible with the events that occurred there. However, when causal links are established in a model and that model is then used to predict the course of fertility decline in today's countries, its success may be quite limited. Most likely the course of events will differ significantly from the earlier patterns.

Furthermore, the validity of the demographic transition model for explaining the demographic changes that occurred in Europe during the transition has been questioned as well. Freedman (1979, 64) wrote that, "detailed empirical work has been unable to establish combinations of development variables at specific levels which were systematically related to the European fertility declines." In some countries fertility declines began in less advanced regions, rather than in the more advanced ones. Also, Freedman (1979, 64) noted that "detailed study of the European fertility transition has shown that many areas that were culturally similar, for example, in language or ethnicity, also demonstrated similar fertility patterns, without prime reference to socioeconomic developmental indices critical to transition theory." Though it appears that spatial and temporal variations in fertility decline were probably more complicated than we have believed, the new evidence does not necessarily invalidate the demographic transition model. It does suggest, however, that a closer look at the model is warranted especially with respect to its predictive ability. Neat causal relationships cannot be specified. For example, it is not possible to show that a 10 percent rise in a nation's level of urbanization will result in a corresponding specified decrease in fertility.

Studies of the demographic transition in Europe have highlighted another problem, one associated with the way that fertility is usually measured. The typical way to chart the transition is to look at crude birth and death rates. However, as van de Walle and Knodel (1980, 15) pointed out, "France and Ireland around 1900 had almost the same crude birth rate." But, they went on to say, "this result was obtained in France by family limitation (within marriage) and in Ireland by high proportions of single persons in the reproductive ages." Thus, historical demographers are now breaking overall fertility down into the various components that comprise it. The major component, of course, is marital fertility. Related data for selected countries are shown in Table 4–1.

Table 4–1 Index Numbers of Marital Fertility[1]: Selected European Countries, 1850–1910

	Date						
Country	**1850**	**1860**	**1870**	**1880**	**1890**	**1900**	**1910**
Belgium	109	109		100	89	71	59
Denmark	99	94	96	100	97	89	77
England-Wales	100	99	102	100	92	82	69
Germany			103	100	96	90	73
Ireland			99	100	101	100	101
Italy		105	100	100	99	98	95
Netherlands	100	98	102	100	97	90	78
Norway			93	100	98	93	90
Scotland		101	103	100	95	86	—
Sweden	96	101	101	100	96	91	80
Switzerland		107	102	100	96	91	76

Source: Etienne van de Walle and John Knodel. "Europe's Fertility Transition. New Evidence and Lessons for Today's Developing World." *Population Bulletin* 34, No. 6 (February, 1980, 18).

[1] Index of marital fertility is the ratio of the actual number of births to the expected number of births to married women during a specific time period. For this table, 1880 is set to 100, so values above 100 indicate that fertility was above the 1880 level and values below 100 indicate it was below that level.

After reviewing several studies of European fertility decline, van de Walle and Knodel (1980, 20–21) provided the following useful summary:

1. The past was largely characterized by natural fertility, that is, the deliberate practice of family limitation was mostly absent (and probably unknown) among most of the population prior to a fairly recent time. This was true, even though a substantial proportion of births may have been unwanted.
2. The transition from high to low fertility represented a shift from natural fertility to family limitation, occurred rapidly, and was an irreversible process once underway.
3. The onset of long-term fertility decline was remarkably concentrated in time and took place under a wide variety of socioeconomic and demographic conditions.
4. Differences in the start and in the speed of the fertility decline seem to have been determined more by the cultural setting than by socioeconomic conditions.

During the 1980s considerable attention was given to refining and reinterpreting the European experience of the demographic transition. One shift in focus has been from the role of individual choice in fertility decline to that played by relatives, friends, and neighbors in influencing local fertility behavior (Coale and Watkins, 1986). Watkins (1990, 242) argued the need for paying more attention to the role played by ". . . members of the community with whom individuals interact on a day-to-day basis, as well as the members of . . . 'imagined communities' . . . I assume that in the end it is individuals who act in the privacy of their bedroom; I propose, however, that even when the couple is literally alone in the bedroom, the echoes of conversations with kin and neighbors influence their actions."

Watkins (1990) focused on the diversity of marriage and marital fertility both within and among the countries of Western Europe during two different time periods, 1870 and 1960, and found that between the two time periods diversity increased to a maximum and then decreased to below what it had been in 1870. She also discovered that variations in fertility during the former period were closely related to linguistic diversity. She found that national social integration ultimately diminished demographic diversity within nations, though at the same time demographic diversity among nations became more pronounced. She identified the following variables as most important in reducing within-nation demographic diversity:

1. the integration of national markets
2. the expansion of state functions
3. nation-building

Within Western Europe today fertility rates are the lowest in the world; most European nations are below replacement level. Interestingly, in nations such as Italy and Spain fertility levels have dropped so low that a few demographers are starting to talk about a "second" demographic transition, one that leads to sustained negative population growth rates. Whether such trends materialize or not, time will tell. It is clear, however, that young European women (with a few exceptions, such as the Irish) are not currently inclined toward having more than one or two children. Young Japanese women are also opting for very low reproductive rates.

The so-called second demographic transition has gained considerable recognition in recent years. It raises questions about the original demographic transition that are worth thinking about. First, why would any nation necessarily reach a balance between crude birth and death rates and then stay at that point? It further illustrates the argument that knowing the future is impossible, so models that predict a certain set of outcomes should be suspect in the social sciences. Second, why would women, given more social and economic freedom than ever before, seek to have two children, especially when the very gains that they have made have resulted in increasing the opportunity costs of having children at all? Third, why do most economists argue that population growth is a good thing and that demographic decline will lead to economic decline as well? This suggests that demographic behavior is on the one hand conditioned by the cultural milieu within which people live but on the other hand it is a result of very private and individual decisions. Even as increasing concerns are raised about the demographic decline of Europe, some European women, including those in France, seem to be having slightly more children than they were a few years ago.

Before leaving the subject of demographic change, one more theory needs to be discussed. This theory enhances our understanding of past demographic changes and suggests some additional factors to consider when projecting future demographic change.

Multiphasic Response Theory

As we have seen, most countries that have achieved high levels of modernization have also experienced a decline in birth rates and death rates. The decline in birth rates, with the possible exception of France, has lagged behind the decrease in death rates. What is not made explicit is the causal mechanism, the direct linkage between modernization and the declines in birth and death rates.

Modernization is a complex process involving sweeping changes in the socioeconomic, cultural and political fabric of societies as they are transformed from primarily rural-agricultural societies to primarily urban-industrial societies. Somewhere in this complex of changes there were reasons for the observed declines in vital rates, but their specific causes are not clear. The search for better explanations continues, even as new patterns may be emerging in the less developed countries.

Demographer Kingsley Davis argued that the demographic transition model was perhaps too simplistic, mainly because it considered only changes in birth and death rates. In so doing, the transition model ignores other possible demographic responses that a society could make when confronted with population growth. Davis (1963, 345) stated that "The process of demographic change and response is not only continuous but also reflexive and behavioral—reflexive in the sense that a change in one component is eventually altered by the change it has induced in other components; behavioral in the sense that the process involves human decisions in the pursuit of goals with varying means and conditions." Thus, the subject of demographic change is complex. Davis attempted to incorporate these complexities in an analysis of demographic change in the industrialized, or more developed countries which he called **multiphasic response theory**.

Multiphasic response theory is the incorporation of reflexive and behavioral responses (rather than solely the continuous responses and changes) to demographic change as well as the changes these induce in turn.

Kingsley looked at Japan in the 1950s as a case study of multiphasic response. He argued that one of the major demographic responses to modernization in Japan was abortion, which caused a rapid drop in the birth rate. He noted that abortions have also been common in many other areas of the world, but their numbers have not been so well documented. Other demographic responses in Japan included the increased use of contraceptives, sterilization, emigration, and postponement of marriage. For example, in 1920, 17.7 percent of females aged 15–19 were married. By 1955, however, this figure had dropped to 1.8 percent. Of the Japanese situation Davis (1963, 349) stated:

> It is the picture of a people responding in almost every demographic manner then known to some powerful stimulus. Within a brief period they quickly postponed marriage, embraced contraception, began sterilization, utilized abortions, and migrated outward. It was a determined multiphasic response, and it was extremely effective with respect to fertility.

The stimulus to this multiphasic response was not population pressure and the prevalence or threat of dire poverty. Japanese industrial output grew more rapidly between 1913 and 1958 than did that of Germany, Italy, and the United States. Rather, according to Davis (1963, 352), the stimulus was "in a sense the rising prosperity itself, viewed from the standpoint of the individual's desire to get ahead and appear respectable, that forced a modification of his reproductive behavior." Carlson and Omori (1999) also found dramatic increases in uptake of contraceptives and use of abortion among Bulgarian women between 1976 and 1995. The stimulus, in that case, a declining economy, pushed Bulgarian women to limit their fertility.

Others have studied agrarian communities and their multiphasic responses to demographic change. Dahal (1983) studied a Nepalese hill community and found a multiphasic response to population pressure that did not trigger lower fertility. He found that farmers intensified their food production using technological means that did not damage the environment and that produced labor shortages. These shortages, combined with traditional cultural values, increased resistance to family planning. With government help they were also able to increase arable land for farming.

A Longer View of Demographic Change

Too often demographic theories and models have been applied to short-term situations, attempting either to understand a particular set of demographic changes or to predict future population trends based on one or a few variables. In any case John Caldwell (2004) has made a strong case for the need to understand demographic change within a much broader context. He accepts the value of demographic transition theory as a general explanation for historical, and possibly future, changes in birth and death rates, but only providing that two conditions are met: (1) people's perceptions of their condition and ability to influence the direction of that change must be considered part of the process and (2) greater attention is paid to the mode of production in a society and to continuing economic growth and its implications for how individuals respond demographically.

According to Caldwell (2004), the world is still in the early stages of a transition from an agricultural mode of production to an industrial one, and, in his view, the demographic transition is a response to that change in the mode of production. The "traditional" family is essential for agricultural societies for production, he argues, whereas industrial societies have no particular need for such families. In fact, Caldwell goes so far as to suggest that industrial societies place no particular demands on individuals or couples to reproduce at all. If they do not, however, and extremely low fertility in many industrial societies today suggest that many are not interested in reproduction, then such societies may have an interest in stimulating fertility. As recently as the 1950s no one foresaw the sweeping changes in fertility and family patterns that have subsequently occurred.

A major element in the shift from agricultural to industrial production has been the movement of women to join men in the workforce. This in turn gives women more equality with men and discourages them from reproducing, especially because the burden of housework and the added work of raising children fall mainly on them and not on their male partners. The shift from agricultural to industrial modes of production, at least in the West, has also brought with it a new egalitarian ethic. In turn, this has engendered the need for more skills in the labor market, the women's movement, the struggle for gay and lesbian rights, and substantial increases in the educational levels of females.

Even as individuals in industrialized countries find that the costs of reproduction outweigh the benefits, hence opting for small families or even no children at all, down the road societies may have different problems. Below-replacement fertility levels cannot long prevail before populations begin to decline, so it is at least fair to think that countries may want to stimulate at least enough births to balance deaths and maintain a semblance of demographic stability. Caldwell (2004) suggests a number of ways that this could be accomplished, including the following: subsidized child care, more flexible working hours for mothers, after school programs that take care of children, paid maternity leave, and access to housing for young married couples.

All this change may not lend itself ultimately to a stable demographic future for the world. Caldwell (2004) suggests, for example, the possibility of an oscillation of world population if the demographic transition is completed worldwide, perhaps in a range of eight to ten billion. What this would mean for individuals, for couples, and for society remains to be seen, but the longer view suggested by Caldwell is a major addition to how demographers and population geographers need to think about demographic trends, both past and future. More will be said about this in Chapters 9 and 10, where we confront relationships between population, food supply, and the environment.

Revolutionary Change: The Case of Cuba

Cuba has received more attention from policymakers, the mass media, and scholars than most developing countries. Although it only ranks eighth in size among Latin American nations, it is the most populous nation in the Caribbean. In area, about 46,000 square miles (119 square km) it is about the size of Pennsylvania, and its 2012 population of 11.9 million is nearly as large as that of Pennsylvania.

The importance of Cuba, and the reason for its receiving so much attention, is clearly not because of its size; it is due to its unique history of social and economic change. Cuba was one of the last American nations to sever ties with Spain, and it had a long and close relationship with the United States. When the revolutionary movement led by Fidel Castro overthrew the government of Fulgencio Batista in 1959, it became the first socialist nation in the Western Hemisphere, only 83 miles from Florida.

The social and economic revolution has had a profound impact on the socioeconomic fabric of the nation. Along with, and partly because of, this change in the social and economic order, the past few decades have brought about significant demographic changes. Among today's less developed countries, Cuba probably has one of the lowest fertility rates (1.7) and one of the highest life expectancies (78 years).

Demographic changes in Cuba have been significantly different from those in most less developed countries. It underwent its demographic transition before World War II, a time when most other less developed countries had not started their transitions (Collver, 1965). The death rate started to decline in the early part of the twentieth century and birth rates started to fall in the 1920s. Migration also played an important role in Cuba's demographic development, with a large influx of African slaves and Chinese indentured servants in the nineteenth century, along with hundreds of thousands of Spaniards and other Europeans. As illustrated in Table 4–2, and quite unlike other developing countries, Cuba experienced a decline in its population growth rate from the beginning of the century to the immediate post-World War II period, and then, as with other developing countries, growth accelerated because of a notable decline in mortality.

Mortality Changes

Demographic trends in mortality prior to the revolution were primarily shaped by economic circumstances. The introduction of sanitary reforms during the United States' occupation of 1899–1902 and a vigorous economy fueled by large-scale foreign investments in the sugar industry during the first quarter of the twentieth century brought about a significant decline in mortality (Diaz-Briquets, 1977). With the depression of the 1920s and 1930s, and its disastrous consequences for the Cuban economy, mortality rates stabilized. With the advent of modern drugs and insecticides following World War II, Cuba's mortality rates, like those of most other less developed countries, experienced an accelerated decline.

Demographic trends in mortality prior to the revolution were primarily shaped by economic circumstances.

On the eve of the communist revolution in 1958, Cuba was already one of the most demographically advanced developing nations, with a life expectancy of over 60 years.

Table 4–2 Population Growth in Cuba: 1899–2010

Year	Population	Growth (Percent)	Net International Migration
1899	1,572,797[a]	33.9	127,257 (1900–1909)
1907	2,048,980	29.1	233,535 (1910–1919)
1919	2,889,004	26.6	268,062 (1920–1929)
1931	3,962,344	15.9	−147,963 (1930–1944)
1943	4,778,583	21.1	−21,920 (1945/49–1953)
1953	5,829,029	22.1	
1970	8,569,121	47.0	−250,000
1975	9,299,000	8.5	−190,000
1980	9,724,000	4.6	−160,000
1985	10,098,000	3.8	−256,502
1990	10,694,000	8.7	−66,501
1995	10,998,500	2.9	−114,004
2000	11,100,000	0.9	−155,801
2005	11,611,000	4.6	−142,000
2010	11,925,000	2.7	−190,123

Source: Sergio Diaz-Briquets and Lisandro Perez. (1981) "Cuba: The Demography of Revolution," *Population Bulletin,* 36(1):5; Main source: census data and figures from 1990–1998 according to "Anuario estadístico de Cuba", ed. 1998. File created on 1997–07–20; revised on 2000–04–03; and last modified 2003–11–11 by Jan Lahmeyer. World Bank (citing: United Nations Population Division, World Population Prospects.)

Among less developed countries of the Western Hemisphere, this life expectancy was surpassed only by those of Jamaica, Puerto Rico, Argentina, and Uruguay. Since the revolution, an enormous effort has been put into the development of medical facilities, and currently, virtually free or very low-cost medical coverage is available across Cuba through a system of polyclinics and rural, regional, provincial, and national hospitals (Danielson, 1979).

As the result of the development of medical programs, Cuba has now achieved excellent health standards with life expectancy at 78 years (2012), nearly equal to that of the United States. Infant mortality was down to 4.5 per thousand live births by 2012, compared to 6 per thousand in the United States. Cuba's crude death rate in 2012 was 8, the same as that of the United States. This figure for Cuba reflects both the young age of the population and significant health improvements. As in more developed countries, the leading causes of death have shifted primarily from infectious to degenerative diseases.

Fertility Changes

Information about pre-revolution fertility is scarce, but the limited data suggests a decline in births during the depression years of the 1920s and the 1930s. This was the result of delayed marriages and more widespread adoption of abortion as a fertility limitation method (Gonzalez et al., 1978). Following World War II the crude birth rate stabilized in the low 30s and by the eve of the revolution it had dropped to the mid- to upper-20s.

Though fertility has fluctuated since the revolution, the general trend has been significantly downward (Figure 4–3). A few years before the revolution the birth rate was around 26 per thousand. With the advent of the revolution in 1959 the birth rate rose from

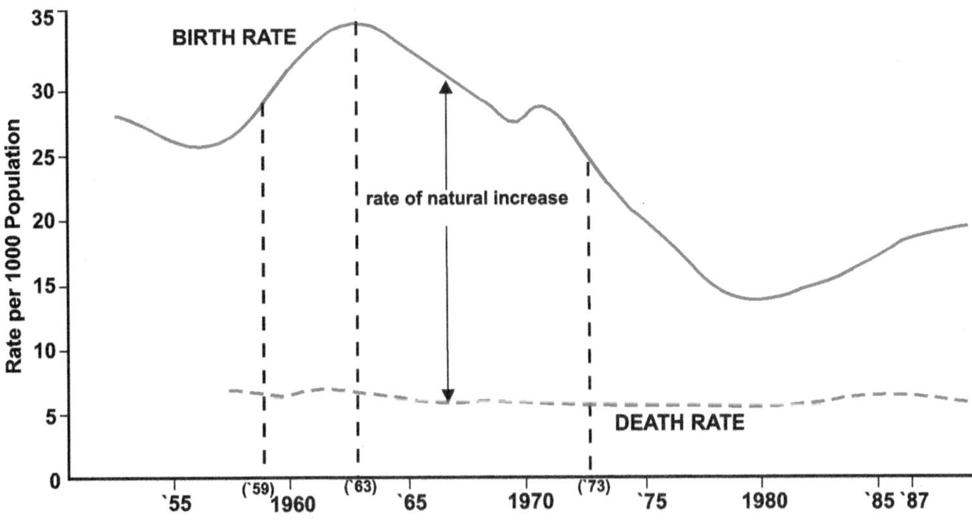

Figure 4–3 Cuba—Birth and Death Rates, 1953–1992

Source: Sergio Diaz-Briquets and Lisandro Perez. (1981) "Cuba:The Demography of Revolution," *Population Bulletin* 36(1):13 and 1992 World Population Data Sheet.

26.1 in 1958 to 35.1 in 1963. This was the highest birth rate in the post-revolutionary period and similar to the level of the 1920s. After 1963 the birth rate fell, with only a brief rise in 1971, a rise attributed to the disruption of normal activities in 1970 in an all-out effort to produce 10 million tons of sugar cane. Thousands of workers were moved from the city to rural areas to help with the harvesting, and the rise in 1971 might have been a reflection of the births averted in 1970 during the harvest. Also, the late 1960s and early 1970s saw a marked increase in marriages.

A dramatic decline occurred in the birth rate from 1973 to 1979, a decline of 40 percent. Overall, there was a 58 percent decline, 20.3 points in the birth rate, between 1963 and 1979; in 2007 the crude birth rate was 11. Cuba's 2007 rate of natural increase, 0.3 percent, less than that for the United States, makes Cuba's rate among the lowest in the less developed world.

Though the age structure of the Cuban population tends to exaggerate this fertility decline, as depicted in the crude birth rate, other fertility measures also point toward a significant decline. Between 1970 and 1978 the total fertility rate dropped by nearly half, from 3.7 to an estimated 1.9 births per woman; by 2007 it was down to 1.5, well below replacement level.

The increase in births in the early 1960s can be directly attributed to the social, political, and economic changes brought about by the revolution. The primary factor behind the rise was probably the growth in real income among the poorer groups, who then felt that the future was more promising and that they could afford more children. At the same time, marriage rates went up and the average age at marriage declined. There was also a shortage of fertility limitation options—partly the result of the economic blockade of Cuba by the United States—and the disruption of contraceptive supplies as well as governmental restrictions on abortion.

An analysis of this birth rate surge shows distinct geographical differences (Figure 4–4) within Cuba. Birth rates increased most in urbanized provinces such as Havana, where the rate went up almost 60 percent between 1958 and 1963. In the least urbanized or modernized provinces, like Oriente, the rate rose only 17 percent.

The causes of fertility decline since 1963 are many and varied. The proximate determinants of the decline—changes in trends of abortion, marriage, divorce, and contraception use—are relatively easy to trace. Restrictions on abortion were eased in 1964 as part of changes in health laws. Since that time free hospital abortions have become available on request to women aged 18 and over during the first ten weeks of pregnancy.

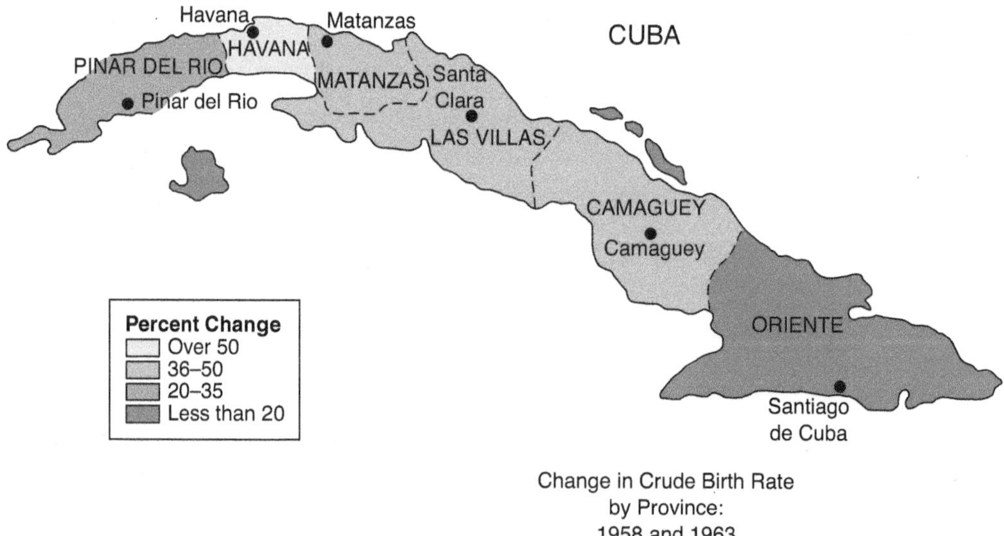

Figure 4–4 Change in Crude Birth Rate by Province: 1958 and 1963
Source: Compiled by authors from statistical data published by the Republic of Cuba.

Later abortions are also available with a physician's permission. There has been a drop in abortion rates since Cuba's high in 1974 of 69.5 abortions per 1000 women aged 15–44. The drop suggests an increase in the use of contraceptives.

Though national statistics on contraceptive use are not available, evidence reviewed by Hollerbach (1980) suggested that contraceptive use has been widespread. The IUD is the most prevalent form of contraception, with condoms second in usage.

The marriage rate has fluctuated since the revolution; the rate of 6.2 per thousand population in 1978 was considerably less than the 10.8 per thousand marriage rate in the United States. This decline could have contributed to the fertility decline, as could the fivefold increase in divorce rates since the revolution. Many socioeconomic changes since the revolution have also played a key role in fertility decline. One thing is certain: the decline was not the result of official antinatalist policies. The government's ideological position is that overpopulation is not one of the causes of poverty. The prevailing view is that changes brought about by the revolution have eroded societal norms favoring childbearing. A good summary of this viewpoint is presented in Hollerbach (1980, 100):

> The decline in fertility, especially rapid since 1973, has not been achieved through antinatalist policies (such as those of China), nor through the creation of demographic targets, which are characteristics of policies in some developing nations. Rather a variety of economic and political factors are responsible, the most significant of which have been increased educational levels . . . adult education programs and expanded enrollment in higher education; the urbanization of rural areas through the concentration of social services and development projects there; construction of small urban communities, and reduction of disparities between urban and rural income levels; and more recently, the incorporation of women into the labor force.

Hollerbach's main emphasis, as with those of other demographic analyses of Cuba, is that modernization has been primarily responsible for these demographic changes. Cuba's experience, therefore, has valuable lessons which might be exported to other less developed nations throughout the world (Harrison, 1980). Other demographic analysts, however, believe that this view—that Cuba's population has been affected uniformly and almost exclusively by modernization sparked by the revolution—needs to be complemented by looking at other factors. They believe that modernization explains the demographic changes for the most disadvantaged groups in Cuba, but that economic setbacks, particularly in recent years, have also had an impact. Their central argument is that different sectors of Cuban society have limited their fertility for different, though overlapping, reasons. According to a study by Diaz-Briquets and Perez (1981, 22):

> Cuba's fertility decline since the mid-1960s has been a response to difficult economic conditions as well as to the undoubted progress made in many social areas. This more comprehensive explanation makes it questionable that poor, high-fertility countries around the world might draw a lesson from Cuba's experience, as has often been asserted. The political, historical, and social context that produced the Cuban revolution is unique in many ways, and the fertility response appears to be just as unique.

Consequences of Fertility Changes

Unlike other less developed countries, Cuba experienced a baby boom and bust similar to those in many more developed nations such as Canada and the United States. Cuba's population pyramid (Figure 4–5) illustrates the large number of people in the 35–44 year age group, a result of the baby boom in the 1960s and the marked decline in infant mortality rates at that time. The narrow base of the pyramid reflects the more recent fertility decline.

Like other countries that have experienced large changes in fertility rates, Cuba has had, and will continue to have, problems trying to adjust to the varying sizes in age cohorts. Cuba has had problems similar to those in the United States. Perez (1977) has pointed out the problems of suddenly needing to accommodate large numbers of children in the school system and of having large numbers of people attempting to enter the labor force at a time when job prospects are dim.

Migration

> Migration has always played an important role in Cuba's population dynamics.

Migration has always played an important role in Cuba's population dynamics. During the nineteenth century large influxes of African slaves, Chinese indentured servants, Spaniards, and other Europeans came to Cuba under various conditions. During the first

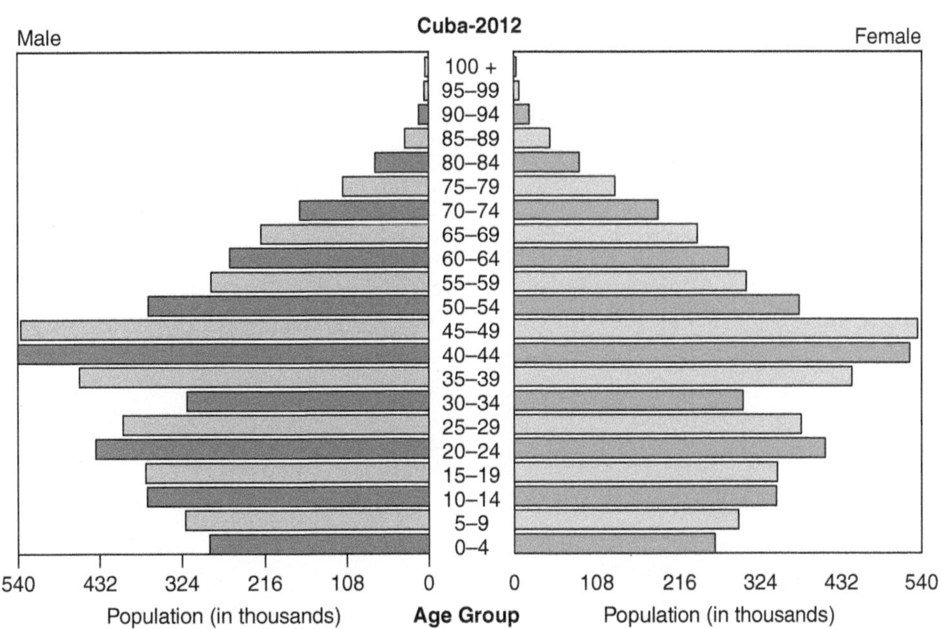

Figure 4–5 Population Pyramid for Cuba: 2012
Source: U.S. Census Bureau, International Data Base.

three decades of the twentieth century, migrants from Spain predominated, though there were also significant numbers of migrants from Haiti, Jamaica, and other Latin American countries. Migration dwindled as an important factor by the early 1930s, then took on added importance again in the early 1960s. Between 1959 and 1980 approximately 800,000 people left Cuba for the United States (Table 4–3). Many of these emigrations

Table 4–3 Cuban Migration to the United States, 1959–2011

Year	Number
1959 (Jan. 1–June 30)	26,527
Year Ending June 30	
1960	60,224
1961	49,961
1962	78,611
1963	42,929
1964	15,616
1965	16,447
1966	46,688
1967	51,147
1968	55,945
1969	52,625
1970	49,545
1971	50,001
1972	23,977
1973	12,579
1974	13,670
1975	8,488
1976	4,515
Year Ending September 30	
1977	4,548
1978	4,108
1979	2,644
1980	122,061
Total, January 1, 1959–September 30, 1980	793,856
Total April 1–December 31, 1980	125,118
1981–1984	38,600
1985	20,300
1989–2003	210,344
2004	20,488
2005–2011	289,946

Source: United States Department of Justice, Immigration and Naturalization Service (INS). (1980) "Cubans Arrived in the United States, by Class of Admission: January 1, 1959–September 30, 1980," October 1980, mimeo; and INS, Statistics Branch and Statistical Abstract of the U.S., 2007; Diaz-Briquets, 2006. Data from the U.S. Department of Homeland Security (2005–2011).

were involved in the Freedom Flights carrying about 344,000 Cubans to the U.S. between 1966 and 1973 (Diaz-Briquets, 2006).

Since then, a very large wave of emigrants came to the United States in the so-called Mariel sealift of April–September 1980, which brought about 125,000 Cubans to the U.S. This post-1960 emigration from socialist Cuba is a reversal of the immigration pattern established in the nineteenth and early twentieth centuries. The Mariel migration to the United States had a considerable impact on both Cuba and specific areas of the United States (Bach, 1980, 40), especially Florida.

Between 1959 and 1980 approximately 800,000 people left Cuba for the United States.

Since the change in U.S. migration policy called the Cuban Migration Agreements of 1994 and 1995, the U.S. has committed to allowing at least 20,000 Cuban arrivals per year. In fiscal year 2008, 49,500 Cubans became legal permanent residents and Cuba was ranked fifth in countries sending migrants to the U.S. (U.S. Congressional Research Service, 2009). Diaz-Briquets (2006) estimates that the U.S. admitted 210,344 Cubans immigrants between 1989 and 2003. According to the U.S. Department of Homeland Security between 2005 and 2011, 289,946 Cubans were granted legal permanent resident status (U.S. Department of Homeland Security, 2012).

China: The World's First Demographic Billionaire

At the end of the thirteenth century, before the exploration of the New World had begun, the population of all of Europe was estimated at 75 million people. At the same time, the population of the Sung Dynasty numbered 100 million. In the next six centuries the population increased, reaching about 540 million in 1949. However, following the 1953 census, which recorded a population of 583 million, there was an attempt to limit population growth. This early campaign was slow to get underway and was disrupted in 1958 by the Great Leap Forward. The Great Leap Forward aimed to rapidly change China from an agrarian economy to an industrialized economy. This program produced massive grain shortages and resulted in many tens of thousands of deaths. Following the Great Leap Forward, in 1962 the Chinese Government again began to promote family planning, but it lapsed into unpopularity again during the Cultural Revolution (1966 to 1969), a movement designed to enforce communism in China and root out capitalism. Since 1971 China has become what the demographer Tien (1983, 3) calls "a born-again advocate of population control."

In the early 1970s, the Chinese government recognized the implications of population growth and its effects on economic development goals. The state family planning agency established a policy known as *wan-xishao*, which meant "later-longer-fewer." This policy encouraged Chinese couples to marry later, preferably later than the average age of 20, to increase the time between births, and to have fewer children overall. Another slogan associated with this family planning program was, "One is not too few, two is good, and three is too many." Targets were set by the government for both family size and age at marriage. Recognizing the economic and cultural differences between the rural and urban areas, rural residents were to have no more than three children after delaying marriage until age 25 for men and 23 for women. For urban dwellers only two children were encouraged after delaying marriage until age 28 for men and 25 for women. In 1979 a new slogan was adopted. It said, "One is best, at most two, never a third."

The One-Child Policy

In 1979 China launched the One-Child Policy to stabilize population growth by limiting families to only one child. Then-Communist Party Chairman Hua Guofeng emphasized that coercion was not permissible, but that the country would rely on publicity and persuasion. Hua spoke out at the same time against overpopulation and bureaucratic obstacles

In 1980 China launched the One-Child Policy to stabilize population growth by limiting families to only one child.

to modernization. The One-Child Policy had a stated goal of keeping the total population "within 1.2 billion" when China had a total fertility rate of around 2.3. In order to meet this objective the one-child policy would require full implementation.

Though some variation in the targets of the policy existed from place to place, a few specific family-planning regulations were well established early in the 1980s. Students and apprentices were not allowed to marry. Incentives were offered to couples that had only one child, whereas disincentives were used to discourage larger families. Incentives were also provided for sterilization after the birth of the first child. A monthly allowance was paid to couples that had only one child, and that child's medical and educational fees were waived. However, if a second child was born, a couple had to relinquish all privileges that had been accorded the one-child family and were required to repay all cash awards that had been received. The third child was denied free education, subsidized food, and housing privileges. That child's parents were penalized by a 10 percent reduction in wages.

In 1981 China's Family Planning Commission was given cabinet status and a variety of scholars, including statisticians, system analysts, economists, and others, were called on to develop and set population policies (Song, et al., 1985). The impact of China's policy has been substantial. According to Tien (1988, 7), "At the time of the program's start China's birth rate was nearly 34 and its total fertility rate exceeded 6 (the 2012 total fertility rate was 1.5) . . . the successes of China's initiative have been, and continue to be, impressive."

This one-child family initiative was an unprecedented attempt to change the reproductive behavior of an entire nation. The success of this program in lowering the rate, according to Jacobson (1991, 282), can be attributed to a variety of aspects of Chinese society:

> The population is nearly homogeneous with a 93 percent Han majority. The closed nature of the political system, the Confucian tradition and the strong sense of family it fostered, and a 3,000 year history of allegiance to authority made the one-child family program succeed where it might otherwise have failed.

Another important variable, according to Greenhalgh (1990, 80), was that the transition to socialism in the 1950s made a fundamental change in the relations between the state and society, thus state-dominated institutions could influence even the most private of decisions. Nonetheless, by 2006 the Chinese population had exceeded 1.3 billion.

The one-child family policy has been criticized in a variety of ways. First, the compulsory nature of the program, at least in some areas, has been criticized and may, as some believe, produce a backlash like that which took place in India. Second, with fewer children, the next generation of elderly will have fewer laborers to support it, and third, most of the immediate economic benefits will not be realized in rural areas.

Rural-Urban Differences

China's family planning policies were most successful in urban areas. While the total fertility rate (Table 4–4) declined precipitously in both urban and rural areas between 1970 and 1988, urban fertility fell a decade earlier and reached a lower level measuring 1.1–1.4 in the early 1980s, well below the level of 2.5–3.0 in the rural areas at the same time. In 1985 the "In-Depth Fertility Survey" showed that "Whereas the vast majority of both urban and rural couples wanted two children, 80–90 percent of urban couples pledged not to have a second child, while fewer than 20 percent of rural couples signed the one-child pledge (Feng, 1989)."

There are several reasons for this disparity. First, large proportions of the country's ethnic minority groups live in rural areas and have been very resistant to the one-child policy. Furthermore, rural reforms in the early 1980s, which dismantled the collective farms, also reestablished the family as the principal unit of rural life and freed many farmers

Table 4–4 China Population Size: Birth, Death, Natural Increase Rates and Total Fertility Rates, 1970–1990, 2000 and 2010

Year	Year-End Population (10,000)	Crude Birth Rate	Crude Death Rate	Natural-Increase Rate	Total Fertility Rate		
					Country	Urban	Rural
1970	82,992	33.43	7.60	25.83	5.81	3.27	6.38
1971	85,299	30.65	7.32	23.33	5.44	2.88	6.01
1972	87,177	29.77	7.61	22.16	4.98	2.64	5.50
1973	89,211	27.93	7.04	20.89	4.54	2.39	5.01
1974	90,859	24.82	7.34	17.48	4.17	1.98	4.64
1975	92,420	23.01	7.32	15.69	3.57	1.78	3.95
1976	93,717	19.91	7.25	12.66	3.24	1.61	3.58
1977	94,974	18.93	6.87	12.06	2.90	1.57	3.12
1978	96,259	18.25	6.25	12.00	2.70	1.55	2.97
1979	97,542	17.82	6.21	11.61	2.80	1.40	3.20
1980	98,705	18.21	6.34	11.87	2.70	1.80	3.00
1981	100,072	20.91	6.36	14.55	2.50	1.50	2.80
1982	101,590	21.09	6.60	14.49	2.90	2.00	3.20
1983	102,764	18.62	7.08	11.54	2.60	1.80	2.80
1984	103,876	17.50	6.69	10.81	2.30	1.60	2.50
1985	105,044	17.80	6.57	11.23	2.30	1.50	2.66
1986	106,529	20.77	6.69	14.08	2.30	1.60	2.60
1987	108,073	21.04	6.65	14.39	2.60	1.80	2.80
1988	109,660	20.78	6.58	14.20	2.40	1.70	2.70
1989	112,704	21.58	6.54	15.04	2.30	1.60	2.50
1990	114,333	23.9	7.60	14.39	2.00	1.20	2.30
2000	126,743	17.7	8.14	7.58	1.82	n.a	n.a
2010	134,091	11.9	7.11	4.79	1.54	n.a.	n.a.

Source: From "Socialism and Fertility in China" by Susan Greenhalgh, in *Annals of the American Academy of Political and Social Science*, 510(1), pg. 73–86. Copyright © 1990 by American Academy of Political and Social Science. Reproduced with permission of Sage Publications, Inc. in the format republish in a book/journal via Copyright Clearance Center; Tu Ping, 2000; National Bureau of Statistics China.

from the enforcement of the one-child policy by rural political cadres. In 1984 physical coercion was prohibited and could no longer be used by rural birth-planning cadres. With decollectivization, labor benefits of children also increased. Most rural dwellers are not included in a state-funded pension system and therefore have to rely on their children, especially their sons, to take care of them in old age.

This preference for sons among a large sector of the Chinese population is a major problem in the acceptance of one-child families in rural areas. Sons are perceived as more important because they can do heavy farm work and are providers of social security. Boys get preferential treatment in health care, food rations, and schooling. Recent research suggests that higher than normal sex ratios among reported live births in China are due mainly to female child underreporting and sex selective abortions (Goodkind, 2011). Johansson and Nygren (1991) also concluded that large percentages of adopted children are females.

Birth Rate Changes in the 1980s

After declining in the early 1980s and reaching a low in 1984, according to the State Statistical Bureau (1988), China's crude birth rate increased by 20 percent between 1984 and 1987. Sixteen of the 30 provinces, municipalities, and autonomous regions experienced an increase exceeding 10 percent between 1981 and 1987, while 26 experienced an increase of over 10 percent between 1984 and 1987. A variety of explanations have been offered for this increase including relaxation of the one-child policy in rural areas, the weakening of local administrator's authority relative to family planning concerns and economic incentives to have more children (Feeney, et al., 1989 and Zeng, 1989).

The crude birth rate is a joint function of three variables—the population age distribution, marriage patterns, and age-specific marital fertility. Zeng, et al. (1991) decomposed the recent changes in China's crude birth rate into these three components in order to assess and compare their contributions to China's increase in its crude birth rate in the 1980s. This decomposition of the crude birth rate into its three components was done for the country as a whole and for each province. Zeng, et al. (1991, 441) concluded that:

> It is thus clear that for the country as a whole the decrease in age at marriage and the changing age structure are the first and second most important factors accounting for the increase in the crude birth rate between 1981 and 1987. The decrease in marital fertility between 1981 and 1987 largely offset the upward pressure of the changing age structure and age at marriage on the crude birth rate.

China's population growth is not "out of control" as some suggested and the increase then experienced in the crude birth rate did not imply a significant change in marital fertility.

The 1990 Census

On July 1, 1990 the Chinese government conducted a nationwide census. According to the National Bureau of Statistics China (1990), the total population for the People's Republic of China was 1,143,330,000 people; this figure did not include Hong Kong, Macao, or Taiwan. This was the fourth census taken since the coming to power in 1949 of the communist government and most experts believe it was their most accurate census. Future plans are to conduct a census every ten years. This total population figure was larger than had been estimated, probably because of previous underreporting, particularly in rural areas where increasing numbers of couples with one female child were allowed to have a second child or more (Tien, 1990).

A number of interesting spatial patterns and variations were revealed by the census. Most of the Chinese people lived in the eastern half of the country, where more intensive agriculture is practiced due to more favorable climatic conditions and better land resources. Although there had been migration to the interior and far west regions, they remain relatively sparsely populated. A doubling of the population since the 1950s significantly increased the density in the eastern part of the country in the 1980s.

As mentioned before, ethnic diversity is also found in China, although the vast majority of people (93 percent) are of the Han ethnicity. There are fifty-five recognized ethnic minority groups that accounted for approximately 8 percent of the population in 1990. This figure was up from 6.7 percent in 1982. A more liberal government policy on family size for these minority groups was probably responsible for this increase. Most of these minority peoples are found in the sparsely populated interior regions that border foreign countries. Conflicts with the Han majority have taken place, with tensions the highest in Tibet due to the ruthless repression and occupation of that country by the Chinese in the 1950s.

As part of the 1990 census, Chinese officials developed a realistic definition and index of the urban population. The urban population in 1990 was 26.2 percent of the total population, an increase of almost 6 percent from the 20.6 percent level in 1982.

Dynamic changes occurred in Chinese cities in the 1990s. China had three of the world's largest cities in 1990; Shanghai, with a population of over thirteen million, Beijing, with more than ten million, and Tianjin, with approximately nine million residents. Internal migration of redundant rural workers to neighboring urban centers dramatically increased in the 1990s. This migrant population increased the total urban population in several of China's largest cities by more than 10 percent (Goldstein and Goldstein, 1991). As a result of this migration, the cities of China began to have urban problems similar to those of other less developed countries.

The regional pattern of urbanization indicated that the highest levels of urbanization were concentrated in the coastal and northeastern regions. According to Pannell and Torguson (1991, 315), "This pattern reflects the high degree of industrial activity and the comparatively well-established transportation networks in the regions in the aftermath of twentieth-century development."

The 2000 Census

According to the 2000 census, China's population for that year was 1.266 billion—an increase of about 124 million over the 1990 census. The 2000 census also showed an urban population of 446 million (about 36 percent of the population); an average family size of 3.44 (down from 3.96 in 1990); and a population of ethnic minorities that comprised about 8.4 percent of the population, up slightly from 1990.

Though Chinese censuses in the past have been considered quite accurate, demographers are concerned that the 2000 census may have been less accurate than previous ones, including 1990. Possible errors are thought to have occurred because of an under-count of children (especially girls) and because of the increasing mobility of the Chinese population. More details about these errors and other problems with the census can be found in Kennedy (2001). As Riley (2004, 4–5) stated, "While some of the reasons for the deteriorating quality of Chinese population data are likely tied to changes in survey design and to what some say is insufficient financial support, other reasons are tied to changes in Chinese society." That society in 2000 was freer, more mobile, and perhaps even less cooperative than it was in 1990.

The 2010 Census

The Sixth National Census of China (2010) enumerated 1,370,536,875 persons however, that includes 7,097,600 in the Hong Kong Special Administrative Region (SAR), 552,300 in the Macao SAR, and 23,162,123 in Taiwan, (the Republic of China claims Taiwan as one of its provinces). This is an absolute change of 73.9 million people (the size of the populations of Turkey or Iran) and a 5.8 percent change rate between 2000 and 2010. The annual growth rate was 0.57 percent down from 1.7 percent in the 1990s.

The sex ratio of the overall population declined from 106.7 in 2000 to 105.2 in 2010, however, the sex ratio at birth remained quite high at 119 male to 100 female births. The size of the population aged 0–14 years (16.6 percent of the total) declined by 6.3 percent from the 2000 figure. The population aged 60 and over in 2000 grew by 2.9 percent in the 2000s and comprised 13.3 percent of the population. In the 2000s, the Han ethnic group decreased slightly to 91.6 percent of the total population. Ethnic minority groups grew by 0.11 percent since the 2000 census.

In the 2010 census, the urban proportion of China's population was measured at 49.7 percent of the population, the highest proportion ever. The rural population comprised 50.3 percent of the population. The Eastern region of the country was the largest

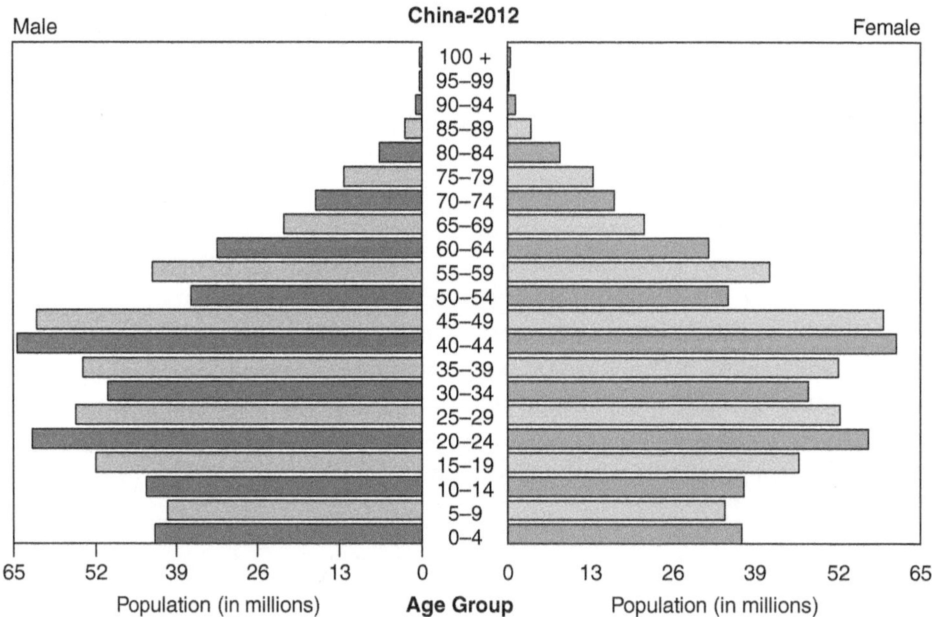

Figure 4–6 Population Pyramid for China, 2012
Source: U.S. Census Bureau, International Data Base.

region in terms of population size (38.0 percent). Additionally, it was the only region that grew (2.4 percent); the remaining regions experienced population decrease in the 2000s mostly due to out-migration to the economically vibrant Eastern region. The Chinese population also became more mobile during the 2000s. The number of those changing cities or towns increased by 81 percent (117 million persons) compared to the 2000 census (National Bureau of Statistics China, 2010).

China remains the most populous nation on the planet. Put somewhat differently, one out of every five persons on earth lives in that nation. During the last five decades or more China has undergone many demographic changes, to say nothing of its many social and economic upheavals. Fifty years ago the crude death rate was 14 per thousand; in 2012 it was around 7 per thousand. Influenced both by government family planning programs and economic changes, fertility declined substantially as well; the 2012 crude birth rate was 12 per thousand, from more than three times that figure fifty years ago. The 2012 total fertility rate of 1.5 was below replacement level, though China's relatively young age structure may still result in a substantial growth of its population. Perhaps another 100 million people will be added before the population stabilizes.

Changes in birth rates and death rates have altered the shape of China's population pyramid as well. Fifty years ago the pyramid had a broad base and narrow top, typical of a less developed country. Today, that pyramid (Figure 4–6) reflects the decline in fertility that has been occurring for two decades or more, and is especially pronounced during the last decade. As a result, China's population is aging quite rapidly. As Riley (2004, 21) noted, "One effect of such rapid aging is that dependency ratios will change; there will be fewer young people to support the country's growing elderly population." This, of course, is the same dilemma that is already being faced by most of the world's wealthy nations. No matter what happens, the future of Chinese demographic change will be felt not just internally but internationally.

Changes in birth rates and death rates have altered the shape of China's population pyramid.

CHAPTER 5

Mortality Patterns and Trends

Key Terms

mortality	life tables
age-specific death rate	morbidity
J-shaped mortality curve	case fatality rate
infant mortality rate	epidemiologic transition
neonatal mortality rate	medical geography
postneonatal mortality rate	HIV/AIDS

Of the three basic population processes, **mortality** is in many ways the easiest to begin with. Death still occurs only once, though modern medical science has been able to retrieve some from the precipice, and it is usually clearly defined, though recent medical technologies have complicated the picture in many ways. For example, physicians use terms such as "brain dead" to describe people who are still able to breath with the help of a respirator. Remember the case of Terri Schiavo, the Florida woman who suffered brain damage, depended on a feeding tube for sustenance, and lived in a vegetative state while her parents, her husband, the courts, and finally Congress and President George W. Bush got involved in her case. She was severely injured in 1990 but kept alive until 2005. Moral and legal issues surrounding the definition of death have drawn more attention in the last decade or so, as organ transplants have become commonplace. A central issue is whether or not to artificially keep someone alive in order to preserve a vital organ until it can be transplanted.

Mortality Deaths as a component of population change. The frequency with which deaths occur.

Measures of Mortality

The crude death rate was introduced earlier and was defined as the annual number of deaths per 1,000 people in the midyear population. The rate is considered to be "crude" because the total population is included in the denominator, whereas the probability of dying in a particular time period is not equal for everyone. Thus, more refined mortality measures attempt to correct for the age structure of the population.

Age-Specific Death Rate

Age-specific death rate Rate of mortality obtained for a specific age group

The probability of dying within a given time interval is closely related to age. The **age-specific death** rate, then, takes age into consideration and may be defined as

$$ASDR = \frac{Da}{Pa} \times 1000,$$

where D_a = number of deaths of people in the age group a, usually either a one-or five-year age group, and
P_a = midyear population in age group a.

J-shaped Mortality Curve A curve depicting the rate of death and how these increase with age.

Generally, death rates are lowest for adolescents and young adults. The typical **J-shaped mortality curve** is shown, separately for males and females, in Figure 5–1.

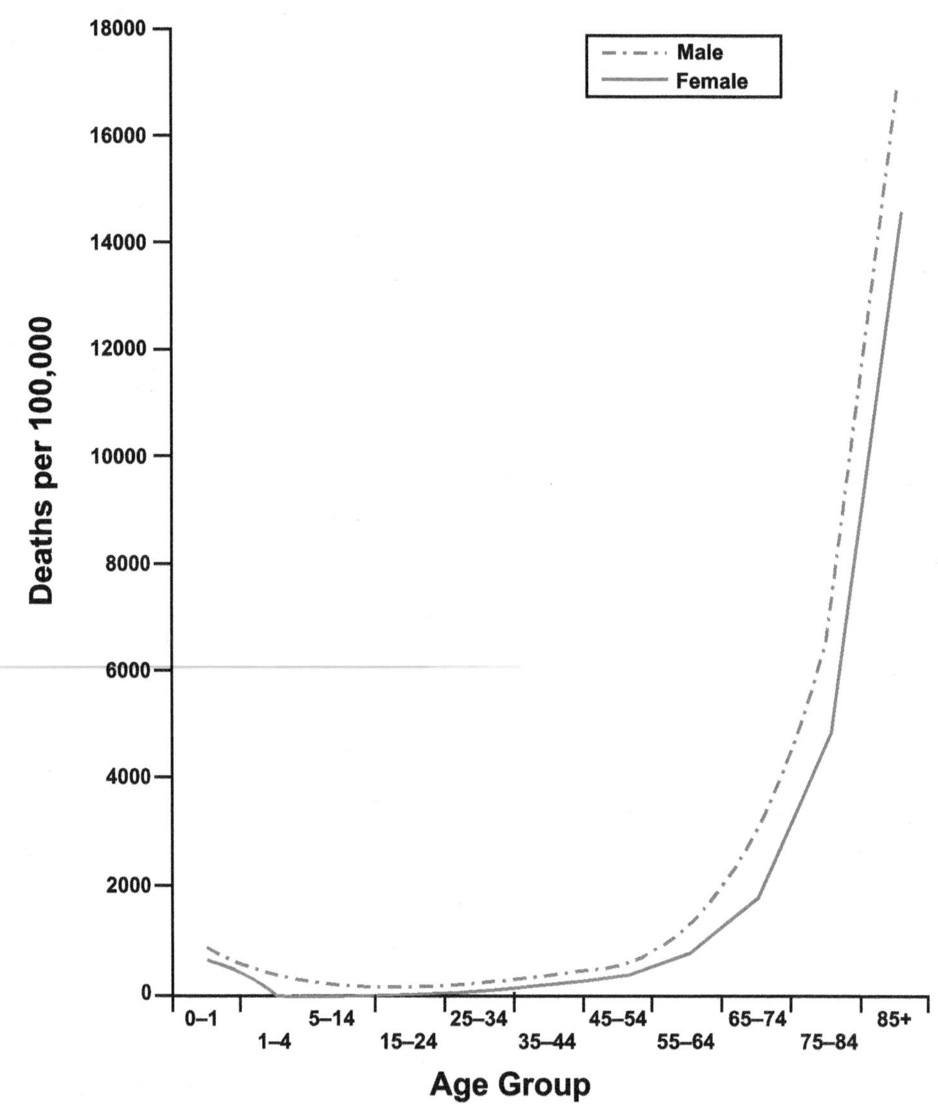

Figure 5–1 Death Rates by Age and Sex: United States, 2010
Source: Data from U.S. Department of Health and Human Services, National Center for Health Statistics, 2012. National Vital Statistics Report..

Infant Mortality Rate

The infant mortality rate is defined as

$$IMR = \frac{D}{B} \times 1000,$$

where D = annual number of deaths of infants between birth and age one year, and
B = annual number of births.

Infant mortality is unevenly distributed throughout the first year of life, with most infant deaths occurring in the first six months. In areas where infant mortality rates are low, a high portion of infant deaths occur within the first 28 days of life. These early infant deaths often result from congenital defects or injuries at birth, deaths that modern medicine may be able to do little to stop.

Infant mortality is unevenly distributed throughout the first year of life, with most infant deaths occurring in the first six months.

Neonatal and Post-Neonatal Mortality Rates

The neonatal mortality rate reflects the influence of congenital and birth-related problems on infant mortality and is calculated

$$NMR = \frac{D_j}{B} \times 1000,$$

where D_j = number of deaths to infants between birth and 28 days of age in a given year, and
B = number of births in that year.

The **neonatal mortality** rate reflects the influence of congenital and birth-related problems on infant mortality.

As Thomlinson (1976, 157) noted, "Neonatal mortality . . . varies remarkably little from country to country, and has shown little susceptibility to reduction under pressure from modern medical science." One important contribution to early infant deaths is the proportion of low-birth-weight infants. Though they have various causes, including multiple births, it is clear that low birth weight is risky for survival. In similar fashion the post-neonatal mortality rate relates the number of deaths that occur to infants between ages 28 days and one year divided by the number of births in that year.

The **post-neonatal mortality rate** relates the number of deaths that occur to infants between ages 28 days and one year divided by the number of births in that year.

The Life Table

Though various mortality rates are useful in describing mortality conditions, they tell us nothing directly about life expectancy at different ages nor about the numbers of people who will survive to different ages. Yet such information is extremely valuable. Life tables, however, are based on observed mortality, usually by both age and sex, and provide us with information on life expectancy and survivorship. The following discussion considers only an abridged form of the life table and its associated functions, as shown in Table 5–1.

Life tables are based on observed mortality, usually by both age and sex, and provide us with information on life expectancy and survivorship.

An abridged life table is calculated from a more detailed life table by deriving life table functions for five-year rather than single-year age intervals. The first column contains the critical life table mortality rate, the probability of dying between age x and age x + n, where n = 1 in the first row, then n = 5 for the remaining columns except for the last one. This is a period life table that incorporates a beginning cohort of 100,000 people, so you can quickly see how many of that cohort survive to age x, how many die between age x and age x + n, person-years lived between age x and age x + n, the total number of person-years lived above age x, and expectation of life at age x. This latter figure is often of interest to demographers and population geographers and is commonly called average life expectancy, or the expectation of life after age 0 (birth).

Table 5–1 Abridged Life Table for the Total Population: United States, 2011

Age	Probability of Dying between Ages x to x+1 q_x	Number Surviving to Age x l_x	Number Dying between Ages x to x+1 d_x	Person-Years Lived between Ages x to x+1 L_x	Total Number of Person-Years Lived above Age x T_x	Expectation of Life at Age x e_x
0–1	0.006761	100,000	676	99,406	7,793,398	77.9
1–4	0.001140	99,324	113	397,023	7,693,992	75.5
5–9	0.000683	99,211	68	495,870	7,296,969	73.6
10–14	0.000839	99,143	83	495,562	6,801,099	68.6
15–19	0.003089	99,060	306	494,626	6,305,537	63.7
20–24	0.004906	98,754	485	492,591	5,810,911	58.8
25–29	0.004956	98,269	487	490,128	5,318,320	54.1
30–34	0.005524	97,782	540	487,600	4,828,192	49.4
35–39	0.007251	97,242	705	484,547	4,340,592	44.6
40–44	0.011003	96,537	1,062	480,214	3,856,045	39.9
45–49	0.016870	95,475	1,611	473,601	3,375,831	35.4
50–54	0.025217	93,864	2,367	463,734	2,902,230	30.9
55–59	0.035858	91,497	3,281	449,712	2,438,496	26.7
60–64	0.052469	88,216	4,629	430,149	1,988,784	22.5
65–69	0.077792	83,587	6,502	402,523	1,558,635	18.6
70–74	0.119029	77,085	9,175	363,858	1,158,112	15.0
75–79	0.191303	67,910	12,991	308,631	792,254	11.7
80–84	0.297772	54,918	16,353	234,712	483,622	8.8
85–89	0.441837	38,565	17,040	149,648	248,911	6.5
90–94	0.612543	21,526	13,185	72,247	99,263	4.6
95–99	0.778938	8,340	6,497	22,835	27,016	3.2
100 years +	1.000000	1,844	1,844	4,181	4,181	2.3

Source: *National* Vital Statistics Reports, Vol. 59, No. 9, September 28, 2011, p. 60.

The Major Determinants of Mortality

Mortality varies both from time to time and from place to place. In order to understand these variations it is necessary to consider the determinants of mortality. Ideally, vital registration data on mortality (in which deaths would be recorded by age, sex, and cause) would be most useful. In reality, however, most information from less developed countries is far less complete.

Progress has been made regarding the use of sample surveys to ascertain both birth and death rates. Though there are still gaps and limitations to mortality data, international agencies like the World Bank and the United Nations Population Division have been able to publish estimates of vital rates for most countries of the world.

Lifespan and Life Expectancy

Jeanne Calment died in her native France in 1997 at the age of 122, having lived longer than anyone else whose age had actually been confirmed. How does this relate to our discussion of life span and life expectancy? First, there is clearly a difference between the two terms. Life expectancy has long been compiled by insurance companies, was mentioned in our discussion of life tables, and is basically the average number of years that a person can expect to live. On the other hand, life span "represents maximum longevity—the absolute number of years any human could hope to survive" (Hopkin, 2004, 12). Though not all researchers agree on an exact figure for the human life span, most put it in the neighborhood of 120–125 years. As Hopkin (2004, 12) points out, "Our maximum life span may have become set during evolution, because there is really no need for any creature to live beyond its reproductive years. . . . Humans escape this seemingly cruel contract, generally speaking, because we have no natural predators." Nonetheless, as Bongaarts (2006, 605) noted, "One of the most notable achievements of modern societies is a large rise in human longevity. Since 1800, life expectancy at birth has doubled from about 40 years to nearly 80 years." It is noteworthy also that some societies have reached long life expectancies without much if any modernization. Buettner (2005) focused on the long lives lived by Sardinians, Okinawans, and Adventists. Though these three groups are different in many ways, they do have some commonalities, including the following: little or no smoking, put family first, stay active, keep socially engaged, and eat lots of fresh fruits and vegetables, along with whole grains.

For every living thing aging is a fact of life, but only humans seek to alter the natural process. However, as Olshansky, Hayflick and Carnes (2004, 98) warned us, "no currently marketed intervention—none—has yet been proved to slow, stop or reverse human aging, and some can be downright dangerous." The same authors note (2004, 99) with respect to aging that "Various definitions have been proposed, but we think of aging as the accumulation of random damage to the building blocks of life—especially to DNA, certain proteins, carbohydrates and lipids (fats)—that begins early in life and eventually exceeds the body's self-repair capabilities." This ongoing breakdown at the molecular level assures us that, no matter what we do to control one disease or another, humans will continue to age and ultimately to die.

Mortality, Morbidity, and the Epidemiological Transition

Morbidity, or sickness, may or may not result in death. The **case fatality rate** is a measure of the proportion of sick persons who die. Basically, diseases may be classified as communicable or degenerative. The role of these diseases in overall death rates has changed over time, leading to the development of an "epidemiologic transition" theory, first described by Abdel Omran (1977, 4) as follows:

> This theory focuses on the shifting web of health and disease patterns on population groups and their links with several demographic, social, economic, ecologic, and biological changes. Many countries have experienced a significant change or transition from high to low mortality accompanying either social development (as occurred in the West with the Industrial Revolution) or a combination of medical development and early social change (which was the story in many developing countries when antibiotics, insecticides, sanitation, and other medical technology were introduced after World War II). Common to all these countries is a clear shift in the kinds of diseases and causes of death that are prevalent at a fixed point in time: the Old World epidemics of infection are progressively (but not completely) replaced by degenerative diseases, diseases due to stress, and man-made diseases. Thus typhoid, tuberculosis, cholera, diphtheria, plague, and the like decline as the leading diseases and causes of death, to be replaced by heart diseases, cancer, stroke, diabetes, gastric ulcer, and the like, together with increased mental illness, accidents, disease due to industrial exposure, and, now, diseases which can be traced to a deteriorating environment.

Morbidity is the frequency of disease and illness in a population.

Case fatality rate is the proportion of deaths within a selected population of cases over the course of a disease.

Morbidity, or sickness, may or may not result in death.

Epidemiological transition is the transition that a society goes through in which death rates decline and the major causes of death shift from communicative diseases to degenerative diseases.

Accompanying this **epidemiologic transition** has been a significant rise in the average life expectancy of those involved. For the world, average life expectancy rose from about 30 years in 1900 to around 66 years in 2000. Worldwide, however, there are still great variations from nation to nation, as is apparent in Figure 5–2. Life expectancy at birth today ranges from a low of 47 years in Sierra Leone to a high of 83 in Japan and Hong Kong. More than 300 million people live in nations where the average life expectancy is less than 50 years. In many of these countries one of every ten newborns dies before age one. Life expectancy tends to be related to the level of economic development, measured as per capita GNP. AIDS has now reduced life expectancy in several African nations. Within the United States long-term trends in life expectancy were examined by Manton, Gu, and Lamb (2006) and Smith and Bradshaw (2006).

The gains in life expectancy in the past four decades have not been shared equally among the world's nations. There is still a wide gap between the life expectancies of the less developed nations in Africa, Asia, and Latin America and the more developed nations.

Within the less developed world there have also been divergent trends. Eastern Asia, dominated by China, had a remarkable gain of 26 years in life expectancy from 1950–1990, surpassing both the Commonwealth of Independent States and Latin America (Mosely and Cowley, 1991, 7). With a gain of only 14 years in life expectancy, Africa showed the least improvement over this forty year period; though even here many variations exist by country. There are also great disparities among more developed countries, with Japan making the greatest gain among them.

In order to better understand the importance of the epidemiological transition theory it is useful to return to Omran's study, in which he introduced four concepts. One is especially important at this point and it was stated by Omran (1977, 9) as follows:

> During the transition, a long-term shift occurs in mortality and disease patterns whereby pandemics of infection are gradually displaced by degenerative and

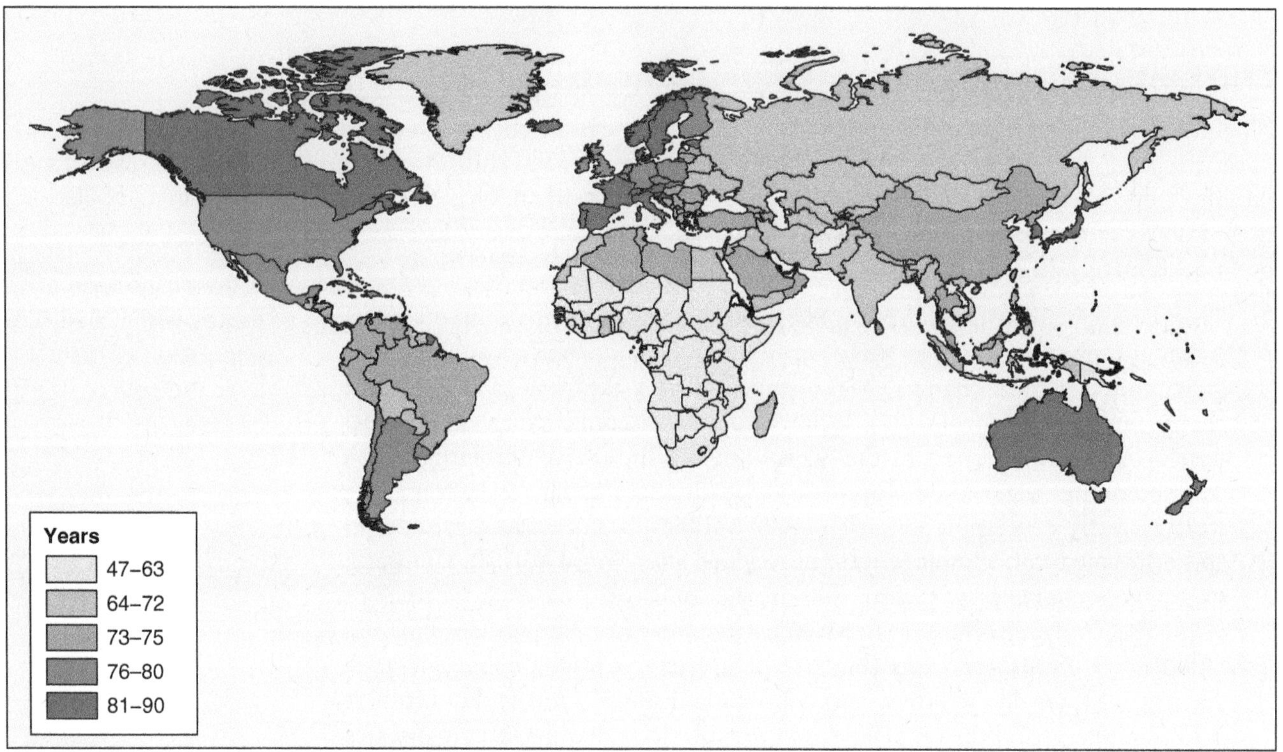

Figure 5–2 Life Expectancy at Birth in 2012
Source: CIA World Fact Book, January 1, 2012.

man-made diseases as the chief forms of morbidity and primary causes of death. Typically, mortality patterns distinguish three major successive stages of the epidemiologic transition: 1. *The Age of Pestilence and Famine*, when mortality is high and fluctuating, thus precluding sustained population growth. In this stage the average life expectancy at birth is low and variable, vacillating between 20 and 40 years. 2. *The Age of Receding Pandemics*, when mortality declines progressively; the rate of decline accelerates as epidemic peaks become less frequent or disappear. The average life expectancy at birth increases steadily from about 30 to about 50 years. Population growth is sustained and begins to take off exponentially. 3. *The Age of Degenerative and Man-Made Diseases*, when mortality continues to decline and eventually approaches stability at a relatively low level. The average life expectancy at birth rises gradually until it exceeds 70 years. It is during this stage that fertility becomes the crucial factor in population growth.

Furthermore, Omran suggested that, because of variations in the timing of the transition, there are three basic models of the epidemiologic transition:

1. the classic or Western model
2. the accelerated model
3. the delayed model

The *classic or Western model* describes the transition as it occurred in Western societies over the past 200 years, a transition in which crude death rates decreased from around 30 to under 10. The mortality decline occurred gradually in response to a mosaic of economic, social, and environmental improvements. In the early stages of this mortality decline medical practices were relatively unimportant, while today downward shifts in mortality are strongly influenced by modern medicine (McKeown and Brown, 1969).

The *accelerated model* describes the mortality transition as it occurred in Japan, Eastern Europe, and the Soviet Union. The transition was socially determined at the outset, but it also benefited from the revolution taking place in medical science.

The *delayed model* may be applied to the observed conditions in most less developed countries today, where rapid declines in mortality have occurred since the end of World War II. Mortality declines in the less developed countries have been the result of modern medicine, the use of insecticides, and organized disease eradication programs; aid from the more developed countries and the United Nations have also had a considerable impact.

Subsequent to Omran's work a fourth stage—the *Hybristic Stage*—has been incorporated into the epidemiologic transition model (Rogers and Hackenberg, 1987, and Olshansky and Alt, 1986). The derivation of the term hybristic is the Greek word *hybris*, which means a feeling of invincibility or extreme self-confidence. This stage is one in which personal behavior and lifestyle strongly impact patterns of disease incidence and injuries. For example, in the United States homicide, suicide, cirrhosis of the liver, and AIDS remain major killers, and tuberculosis and other communicable diseases associated with poverty and poor lifestyle choices are on the rise (Rockett, 1999). In the United States, and a few other rich countries as well, obesity is increasingly threatening societal well-being, and geographers are beginning to grapple with studies of obesity and health in different ways (Evans, 2006). Obesity rates are highest in the American South, but have grown virtually everywhere. Blame for this "epidemic" is aimed at everything from "fast food" and drinks sweetened with "high fructose corn syrup" to video games, television, and the couch-potato syndrome. After a lengthy review of more than a decade's worth of studies on obesity and diet, Ritchie, et al. (2007, 117) noted that the "Dietary factors that protect against excessive weight gain include the consumption of dietary fiber, fruits and vegetables, and calcium and dairy. . . ." Nutritionist Marion Nestle (2006) noted another problem, "The foods that sell best and bring in the most profits are not necessarily the ones

In the United States, and a few other rich countries as well, obesity is increasingly threatening societal well-being.

"The foods that sell best and bring in the most profits are not necessarily the ones that are best for your health."

that are best for your health, and the conflict between health and business goals is at the root of public confusion about food choices." Her advice on nutrition, weight control, and health is simple, easy to understand, but not being followed by a majority of Americans and growing numbers of people elsewhere as well. Eat less, exercise more, choose plenty of fruits and vegetables, and minimize consumption of junk foods. People here and elsewhere spend billions on diet books, but Nestle's advice, and the determination to follow it, are all most people need to live longer, healthier lives. Michael Pollan (2006, 2) made yet another observation: "I wonder if it doesn't make more sense to speak in terms of an American paradox—that is, notably unhealthy people obsessed by the idea of eating healthily."

Some research even suggests that there may be genetic effects arising from exposure to certain chemical compounds in the womb. These compounds, dubbed "obesogens," may interfere with the body's normal hormonal processes. Wherever research leads us, it is not a stretch of the imagination to argue that sitting in front of a television set for several hours a day, munching various salty snacks and washing them down with high-calorie sodas, will lead to weight gains that will be unhealthy, and ultimately costly to both individuals and societies. One of those costs, noted by Patricia Crawford (2007, 98) is that "Our health care system, already stretched, may collapse under the "weight" of this situation [growing obesity]."

Another development that does not fit well with the standard interpretation of the epidemiologic transition is the rise of new viral and bacterial diseases. Over the last three decades at least 30 previously unknown diseases have emerged, some with major consequences. Included among these newly-emergent diseases are rotavirus, parvovirus B19, ebola, hanta virus, human t-lymphotropic virus 1 (HTLV-1), human immunodeficiency virus (HIV), human herpes virus-6, hepatitis C and E, sin nombre virus, sabis virus, HHV-8, and severe acute respiratory syndrome virus (SARS) on the viral side along with such bacterial examples as *legionella pneumophilia, campylobacter jejuni, staphylococcus aureus, helicobacter pylori, and streptococcus pyogenes*. Even in modern countries, then, new threats from communicable diseases cannot be disregarded. HIV/AIDS, for example, has developed into a major pandemic. Kolivras (2006) studied the risk of dengue in Hawaii, using GIS techniques to model that risk. Smil (2005) and others have warned us about the likelihood of future influenza epidemics, a concern that has strengthened with development of bird flu, identified as the H5N1 bird flu strain. Whether this influenza virus or another one ultimately evolves to do its nasty work, most researchers agree that another influenza pandemic is not an unlikely event and is well worth planning for. Osterholm (2007, 55) observed that "The interconnectedness of the global economy today could make the next influenza pandemic more devastating than the ones before it." He goes on to point out that we are so far doing a less-than-adequate job of planning for such a potential disaster. Keil and Ali (2005) suggested that we could learn from the SARS outbreak in Toronto in 2003 and apply that knowledge to preparations for a possible avian flu pandemic. Gibbs and Soares (2005, 54) pointed out that "Never before has the world been able to see a flu pandemic on the horizon or had so many possible tools to minimize its impact once it arrives."

Arguments about the relationship between mortality and economic development have attracted considerable attention. Some argue that advances in mortality conditions in less developed countries today are not as closely related to economic development, as they have been in the past; others argue that economic development is still a major determinant of mortality conditions. In a study of health care in the Third World, Phillips (1990, 51) argued that, "the epidemiological transition is therefore apparently a concept that needs to be applied with caution in many Third World countries. Only . . . where health services are on the whole more uniformly available and data more reliable, can the concept be used with much confidence for present and future health care planning." Salomon and Murray (2002) even questioned the model's usefulness today in the United States.

At first glance it seems that a nation's per capita income is a very important factor in determining mortality levels. The United States and the nations of Northern and Western Europe, with per capita Gross National Income (in Purchasing Power Parity) around $33,000 or more, have a life expectancy of 78 years, whereas many nations in Africa, with per capita incomes around $700 per year, have a life expectancy of 58 years. Although most wealthy countries have high life expectancies, there are a number of low-income countries that have life expectancies approaching that of the wealthier countries. Conversely, there are some countries with relatively high per capita incomes that rate relatively low on measures of health and survival (Mosley and Cowley, 1991, 31). In 2007 Michael Moore's documentary film, *Sicko,* raised numerous questions about the health care system in the United States. However one feels about either Moore or his film, it is true that the United States spends much more per capita on health care than do most European nations, yet life expectancy is higher in most of those countries than in the United States. Insurance companies and pharmaceutical companies hire fleets of lobbyists to discourage any major changes to the health care system in the United States, especially any change that might move the system toward a "single-payer" model that would threaten insurance industry profits.

As Ginter and Simko (2010, 217) note,

> . . . in 2006, the USA was number 1 in terms of health care spending per capita but internationally it ranked 39th for infant mortality, 43rd for adult female mortality, 42nd for adult male mortality, and 36th for life expectancy. A baby born . . . in 2009 will live on average 78 years, ranking America as the 50th worldwide. . . These facts have fueled a question now being discussed in academic circles, as well as by government and the public: Why does the US spend so much to get so little?

However, there still is at least some tendency toward a relationship between income and mortality, though it is apparently weaker now than it was during the epidemiologic transition in the Western nations. According to a study by Franz and FitzRoy (2006, 263), the authors noted that "We confirm the importance of female literacy in explaining both fertility and mortality, and also find a measure of consumption for the poorest share of the population to be significant, while controlling for nutrition, health expenditure, and income distribution."

Causes of Death

The epidemiologic transition helps explain shifts in causes of death from various diseases, though other causes of death must also be considered. Table 5–2 shows leading causes of death in the United States for 2010. Diseases of the heart were the major single cause of death, followed by malignant neoplasms. Geographically, heart disease death rates are highest in the South and lowest across the northern states from Minnesota to Washington. These variations result from a combination of factors including the quality and availability of health care, emergency medical care, diet, and even race and ethnicity. Newer views of the causes of heart disease have shifted from blood vessels to genetic and other risk factors. However, as Kahn (2007, 56) commented, "Until there are tests, genetic or otherwise, that give a clearer measure of risk, everyone would be advised to exercise, watch their diet, and take statins for elevated cholesterol-the same advice doctors gave when the clogged-pipes model of heart disease reigned unchallenged." In 1900 the leading causes of death in the United States were influenza and pneumonia, but today almost 50 percent of all deaths in the United States result from two causes, heart disease and malignant neoplasms.

Table 5–2 Deaths and Death Rates for 2010 and Age-adjusted Death Rates and Percentage Changes in Age-adjusted Rates from 2009 to 2010 for the 15 Leading Causes of Death in 2010: United States, Final 2009 and Preliminary 2010

[Data are based on a continuous file of records received from the states. Rates are per 100,000 population; age-adjusted rates per 100,000 U.S. standard population based on the year 2000 standard; for explanation of asterisks (*) preceding cause-of-death codes, see "Technical Notes." Figures for 2010 are based on weighted data rounded to the nearest individual, so categories may not add to totals]

Rank[1]	Cause of death (based on the international Classification of Diseases, Tenth Revision, 2004)	Number	Death rate	Age-adjusted death rate		
				2010	2009[2]	Percent change
. . .	All causes	2,465,932	798.7	746.2	749.6	–0.5
1	Diseases of heart (100-109,111,113,120-151)	595,444	192.9	178.5	182.8	–2.4
2	Malignant neoplasms (C00-C97)	573,855	185.9	172.5	173.5	–0.6
3	Chronic lower respiratory diseases (J40-J47)	137,789	44.6	42.1	42.7	–1.4
4	Cerebrovascular diseases (160-169)	129,180	41.8	39.0	39.6	–1.5
5	Accidents (unintentional injuries) (V01-X59,Y85-Y86)	118,043	38.2	37.1	37.5	–1.1
6	Alzheimer's disease (G30)	83,308	27.0	25.0	24.2	3.3
7	Diabetes mellitus (E10-E14)	68,905	22.3	20.8	21.0	–1.0
8	Nephritis, nephrotic syndrome and nephrosis (N00-N07-N17,N25-N27)	50,472	16.3	15.3	15.1	1.3
9	Influenza and pneumonia (J10-J18)	50,003	16.2	15.1	16.5	–8.5
10	Septicemia (A40-A41)	37,793	12.2	11.9	11.8	0.8
11	Intentional self-harm (suicide) (U03,X60-X84,Y87.0)	34,843	11.3	10.6	11.0	–3.6
12	Septicemia (A40-A41)	31,802	10.3	9.4	9.1	3.3
13	Chronic liver disease and cirrhosis (K70,K73-K74)	26,577	8.6	7.9	7.8	1.3
14	Essential (primary) hypertension and hypertensive renal disease (110,112)	21,963	7.1	6.8	6.5	4.6
15	Parkinson's disease (G20-G21)	17,001	5.5	5.1	4.9	4.1
. . .	All other causes (residual)	48,954	158.5	. . .	. . .	. . .

. . . Category not applicable.

[1] Rank based on number of deaths; see "Technical Notes."

[2] Rates are revised and may differ from rates previously published; see "Technical Notes."

Source: National Vital Statistic Reports, Vol. 60, No. 4, January 11, 2012

Physicians, however, have long recognized that the causes of death that are listed on death certificates themselves have causes. Mokdad, et al, (2004) identified "actual causes" of death in the United States. According to them, the leading actual cause of death in 2000 was tobacco, responsible for 435,000 deaths, followed closely by poor diet and physical inactivity, responsible for 400,000 deaths. Beyond those two major actual causes of death were the following: alcohol consumption (85,000 deaths), microbial agents (75,000 deaths), toxic agents (55,000 deaths), automobile crashes (43,000 deaths), firearms incidents (29,000 deaths), sexual practices (20,000 deaths), and illicit drug use (17,000 deaths). The authors also note, quite correctly in our view, that when causes are viewed in this way it is clear that the U.S. health care and public health systems need to focus much more than they do on prevention rather than treatment. Less smoking, better diets, and more physical activity would make a considerable dent in the nation's health problems, and may very well help to lower costs in the health care system as well.

British epidemiologist Richard Doll and his colleagues completed a long-term study of the effects of cigarette smoking on life expectancy. This study, conducted on a 35,000 person sample, comprised mainly of doctors, over a period of five decades, concluded that a life of cigarette smoking will be about 10 years shorter on the average than one without cigarette smoking (Kaufman, 2004).

With respect to obesity—which results primarily from a combination of too little physical activity and the consumption of too many calories—and death, there is again little argument. Studies have shown that adult overweight and obesity are linked to significant decreases in life expectancy (Wart, 2004). One study found that people who were overweight at age 40 would have their life expectancies decreased by 6–7 years if they were non-smokers and by 13–14 years if they were smokers. However, according to Wart (2004, 1) "the news might not be completely bleak. Just as you can reduce the long-term effects of smoking by stopping and staying stopped, you can also reduce the risks associated with excess weight." Another study found that in the U.S., obesity has reduced life expectancy at 50 years by up to 1.54 years for females and 1.85 years for males (Preston and Stokes, 2011).

Although the health risks associated with smoking and obesity are well-known, even if they are easily ignored, other influences on our health and general welfare are less obvious. A growing body of literature suggests that the suburban lifestyle aspired to by most Americans may also contribute to higher death rates and lower life expectancies. For example, sprawling suburbs have more fatal automobile accidents than do most central city areas. Walking or riding bicycles in suburban areas may be good for your physical well-being, but it can also lead to unexpected collisions with speeding cars, an all-too-common experience in suburban areas. Suburbs are car-dependent, so most children and adults get less exercise than they would get if they lived in places where kids could easily walk to school and parents could walk at least to local stores and shops, if not to work. Coupled with poor diets, this encourages overweight and obesity. Furthermore, suburbs also lead to more social isolation, especially for the elderly, and social isolation is unhealthy, perhaps carrying risks comparable to cigarette smoking, high blood pressure, and obesity (House, Landis, and Umberson, 1988).

Mortality conditions in the United States prior to 1900 are not well documented, though it is safe to assume that mortality was well above what it has been in most of the twentieth century. Most likely, mortality conditions during the eighteenth and nineteenth centuries in the United States were similar to those experienced in Western Europe during the same period. Wars, famines, and epidemics caused fluctuations in year-to-year death rates. Early data for Massachusetts suggest that some improvement in mortality conditions probably occurred in the nineteenth century, as is apparent in Table 5–3. These data suggest that improvements were affecting mainly infant and childhood mortality, since major gains in life expectancy appear only for life expectancy at birth. Undoubtedly, during

Table 5–3 Life Expectancy at Specified Ages, Massachusetts, 1850 to 1900–1902

Year	At Birth		Age 20		Age 40		Age 46	
	Male	Female	Male	Female	Male	Female	Male	Female
1850	38.3	40.5	40.1	40.2	27.9	29.8	15.6	17.0
1855	38.7	40.9	39.8	39.9	27.0	28.8	14.4	15.6
1878–82	41.7	43.5	42.2	42.8	28.9	30.3	15.6	16.9
1890	42.5	44.5	40.7	42.0	27.4	28.8	14.7	15.7
1900–02	46.1	49.4	41.8	43.7	27.2	28.8	13.9	15.1

Source: United States Bureau of the Census (1975) Historical Statistics of the United States, Colonial Times to 1979. Part I (Washington, D.C.: U.S. Government Printing Office), p. 56.

this period considerable regional variations in death rates could be observed, as well as differences for whites and nonwhites. Life expectancy during the twentieth century also changed dramatically in the United States, as is apparent in Table 5–4.

The major causes of death in Canada, shown in Table 5–5, are similar to those found in the United States or in other industrialized nations. Cancer and heart disease are the major killers, accounting for more than half of all deaths. Half of all cancer deaths in Canada result from tumors in four sites: lung, colorectal, female breast, and prostate. An interesting study by Torrey (2004) compared Canadian and United States mortality. She found that the shorter life expectancies in the United States were mainly a result of higher death rates from circulatory diseases, and suggested that obesity in the United States was a contributing factor.

Table 5–4 Life Expectancy at Birth by Race and Sex, U.S., 1900–2010

Year	White Male	White Female	Nonwhite Male	Nonwhite Female
1900	46.6	48.7	32.5	33.5
1910	48.6	52.0	33.8	37.5
1920	54.4	55.6	45.5	45.2
1930	59.7	63.5	47.3	49.2
1940	62.1	66.6	51.5	54.9
1950	66.5	72.2	59.1	62.9
1960	67.4	74.1	61.1	66.3
1970	68.0	75.6	61.3	69.4
1980	70.7	78.1	65.3	78.6
1985	71.8	78.7	67.2	75.2
1990	72.7	79.6	67.5	75.7
1995	73.4	79.6	67.5	75.8
2000	74.1	79.5		
2005	75.4	80.4		
2010	76.2	81.1		

Source: National Center for Health Statistics. Vital Statistics of the United States, 1970. Vol., 2, Mortality, Part A, Table 5–5 and Statistical Abstract of the United States, 2003. (Washington, D.C.: U.S. Government Printing Office), Table 105. National Vital Statistics Report, Vol. 50, No. 15, September 16, 2002; National Vital Statistics Reports, Vol. 60, No. 4, January 11, 2012.

Table 5–5 Top Ten Leading Causes of Death, Canada 2009			
	2009		
	Number	%	Total Rate[1]
Total, all causes of death...........(A00-Y89)	238,418	100.0	706.8
Malignant neoplasms................(C00-C97)	71,125	29.8	210.9
Diseases of heart..(I00-I09, I11, I13, I20-I51)	49,271	20.7	146.1
Cerebrovascular diseases.............(I60-I69)	14,105	5.9	41.8
Chronic lower respiratory diseases.(J40-J47)	10,859	4.6	32.2
Accidents (unintentional injuries) (V01-X59, Y85-Y86)	10,250	4.3	30.4
Diabetes mellitus.....................(E10-E14)	6,923	2.9	20.5
Alzheimer's disease........................(G30)	6,281	2.6	18.6
Influenza and pneumonia.............(J09-J18)	5,826	2.4	17.3
Intentional self-harm (suicide) (X60-X84, Y87.0)	3,890	1.6	11.5
Nephritis, nephrotic syndrome and nephrosis (N00-N07, N17-N19, N25-N27)	3,609	1.5	10.7

[1]Age-standardized mortality rate per 100,000 population.
Source: Statistics Canada, Catalogue no. 84-215-X.

Diet and Longevity

Observers have long suspected that relationships between diet and longevity were more complicated than they had often been treated in the medical literature. Though it goes without saying that inadequate diets and malnutrition shorten life expectancies, it is less clear what and how much people should eat in order to live longer (assuming that is their goal, of course).

Japan and Hong Kong have the longest life expectancies in the world today (83 years), closely followed by Israel, Singapore, Macao, Iceland, Sweden, France, Switzerland, Italy, Spain and Australia (82 years). Only slightly lower are Canada, Martinique, South Korea, Norway, Austria, the Netherlands, Malta and New Zealand (81 years). Though a variety of factors contributes to longevity in these countries—including relatively homogeneous populations, good access to medical care, and strong social safety nets—it seems likely that diet also makes a contribution. As it has often been said, "You are what you eat."

Considerable attention has been paid in recent years to the longevity of people who live in Mediterranean Europe; countries such as Spain and Greece are among the least affluent in the European Union, yet long lives are still common. Heart attack death rates in the Mediterranean lands, as well as in Japan, are well below what they are in such wealthy nations as the United States and Germany, leading researchers to look at diet as a possible explanation.

Acceptance of the so-called Mediterranean Diet (by, among others, the World Health Organization, Oldways Preservation and Exchange Trust, and the Harvard School of Public Health) has evolved as one explanation (and recommendation) for longevity. The Mediterranean Diet is based on observations of what typical residents of Mediterranean Europe were consuming in the 1960s.

The "pyramid" for this diet provides recommended foods, to be supplemented by both regular exercise and regular, but moderate, wine consumption (popularized by segments on wine and health on "60 Minutes" in 1991 and 1995). At the top of the pyramid, hence the most restricted item in the diet is red meat, to be consumed only a few times a month. Only slightly less restricted are sweets and eggs, followed by poultry and fish (to be eaten only a few times per week). Cheese, yogurt, and olive oil (monounsaturated) are the first items that are recommended for daily consumption, though not in large quantities. The bottom two "layers" of the pyramid really form the bulk of the diet. At the bottom are grains and potatoes, forming the base of the diet; just above them are fruits, vegetables, beans, and nuts.

What we end up with, of course, is a diet that is naturally low in cholesterol; rich in vitamins, minerals, and fiber; and likely to discourage, or at least prolong the onset of, heart disease and some forms of cancer. Interestingly, though Japan is far from the Mediterranean, the Japanese diet is similar in many ways to the above—cereals, fruits, and vegetables comprise the bulk of the diet, supplemented with smaller quantities of fish, and even smaller quantities of red meat.

Is this the perfect diet? Further research is needed before anyone can answer that with certainty, but the Mediterranean Diet or something similar would seem to be a prudent guide for us all. Other dietary items may prove to have particular health benefits; for example, soy is being studied right now—its prevalence in the Japanese diet makes it a potential candidate.

Interestingly enough, despite what you would expect from Omran's transition model, deaths from infectious diseases have been rising in recent decades in the United States. Between 1980 and 1992, for example, deaths from infectious diseases increased by 58 percent. Though much of that rise was accounted for by the growing number of AIDS deaths, deaths from blood and respiratory diseases are also increasing. This raises serious questions about public health policy in the United States, and about the growing numbers of people (nearing 51 million) who have no health insurance. While some of these will soon be able to access affordable health insurance through the Patient Protection and Affordable Care Act of March 23, 2010 (popularly known as Obamacare), it will be a while before all Americans get healthcare insurance.

Though mortality and morbidity have been studied by geographers for quite a long time, most of those studies have been at the national and international scales. However, the study of disease in cities, though not entirely new, seems to be gaining in popularity. In one case detailed maps were used to study an outbreak of cholera that occurred in London during the 1850s (Stamp, 1964). It was found that a close correlation existed between sources of contaminated water and the incidence of cholera. Pyle and Rees (1971) used factor analysis to study health statistics, housing, and social composition in Chicago. The incidence of several diseases was found to be linked to poverty and to population density.

Condran and Cheney (1982) studied the decline in mortality that occurred in Philadelphia between 1870 and 1930. They found that explanations of mortality decline were often dependent on definitions. "Medical treatment," according to Condran and Cheney (1982, 119), "will account for much less of the decline in mortality if it is defined as chemotherapy, specific immunization, or surgery than if the definition is broadened to include the work of public physicians and nurses in isolating patients or in feeding sick infants during the summer months." Such emphases, they argue, have underestimated the impact of medical activities on declines in death rates. Condran and Cheney (1982, 119) noted also that, "the declines of typhoid fever in Philadelphia and elsewhere emphasize the fact that great progress has been made against disease not only by isolating the disease-producing organism, but also by identifying the environmental sources of the disease." Finally, they argued that no single factor alone could explain the dramatic decline that occurred in Philadelphia's mortality. This generalization certainly is also true of other cities, regions, and nations. However, as they also argue (1982, 120), "future research on the origins of the mortality transition will profit from careful attention to individual cities and the changes occurring in small areas within cities."

In another study of urban mortality Cotter and Patrick (1981) focused their attention on the spatial pattern of cholera in 1849 in Buffalo, New York. They found that ethnicity, socioeconomic status, and neighborhood all influenced the direction and timing of the spread of the disease, as did the urban physical environment. They also found that diffusion of the disease was limited by cultural and physical isolation of various segments of the urban population, especially immigrants, who suffered the worst during the epidemic.

"Cholera moved in limited circles," wrote Cotter and Patrick (1981, 49), "because interaction within groups was more frequent and of greater intensity than was interaction between groups." Poor immigrants, mainly Irish, were the first to be affected by the epidemic, because their circumstances favored both entry and spread of the disease. Because many other neighborhoods did not interact with the Irish ones, cholera diffusion was restricted. In the city's Fifth Ward only a few people contracted cholera because conditions were not conducive to the disease's spread. According to Cotter and Patrick (1981, 49), "Not only did topography, low population density, and superior economic status combine to protect this ward, so also did its lack of interaction with the immigrant groups among whom cholera circulated." Thus, the diffusion of cholera in Buffalo was greatly affected by neighborhood locations and their established patterns of interaction.

Crimmons (1981) suggested that American mortality between 1940 and 1977 could be divided into three distinct periods characterized by differences in both age-specific and course-specific mortality rates. From 1940 until the mid-1950s American mortality declined rapidly. From then until 1968 the pace of mortality decline leveled off, leading many to argue that mortality may be about as low as it can get. However, beginning in 1968 mortality began another rapid decline. As Crimmins (1981, 251) noted:

> Much of the recent decline is accounted for by decreases in cardiovascular diseases. This is the first time in history that mortality declines have been dominated by decreases in diseases of old age or degenerative diseases. Because this is so, we may be beginning a new era of mortality decline, which, if it continues, will lead to large increases in life expectancy at older age.

Geographers and others have expanded their research in recent years to other countries and scales. Examples include a study of the relevance of social, cultural and environmental factors in explaining child mortality in North-East India (Ladusingh and Singh, 2006), and an intriguing study of variations in blood lead levels among children in Mumbai (Niranjan and Madhusudana, 2006). Others have begun to look at health and migration, including Antecol and Bedard (2006) and Thomas (2007). Thomas (2007, 40) found that "In the poorest socioeconomic quintiles immigrants had a greater likelihood of child mortality than the native-born while in the wealthiest quintiles child mortality was greater among the native-born."

Finally, Naím (2007) has suggested that there is now a worldwide crime pandemic. According to Naím (2007, 1) "Crime rates are on the rise almost everywhere, and these statistics typically are distinct from the death and mayhem that comes with terrorism, civil war, or major conflict. The data reflect the booming number of civilians assaulted, robbed, or murdered by other civilians who live in the same city, often in the same neighborhood. Frequently, the victims are as poor as the criminals."

> The risk of death differs with the mortality differentials of age, sex, race, income, and rural versus urban residence.

Mortality Differentials

Not everyone is equally likely to die within a given time interval. The risk of death differs with the mortality differentials of age, sex, race, income, and rural versus urban residence.

Age Differentials

As noted previously, the typical J-shaped curve shows that under normal circumstances the probability of dying is relatively high for infants under age one, is very low for children, adolescents, and young adults, and then begins to rise more and more rapidly in the older ages. We've noted as well that improvements in mortality tend to be concentrated heavily in the infant and early childhood age groups, though other age groups also have benefited.

There are also significant differences between the less developed and the more developed countries. In the less developed countries children under age five account for half of all deaths, primarily caused by infectious diseases. Only 5 percent of deaths in these nations are among the elderly older than age 75. Some African villages, for example, have reported deaths among infants and young children at ten times the rate of the aged.

In the more developed nations, on the other hand, only 2 percent of deaths are among children under age 5, whereas almost two-thirds occur to those over 75 years of age.

Infant Mortality in the United States and Canada

Infant mortality in the United States has consistently decreased; however, IMRs have leveled off since the 1990s.

Infant mortality in the United States has consistently decreased, from a level of 100 or more at the turn of the twentieth century (as is apparent in Figure 5–3) to 6.0 in 2012. Figures for infant mortality rates during the eighteenth and nineteenth centuries are unavailable, but most likely they were always in excess of 100 and may at times have been twice that. Some policy makers are concerned that the current infant mortality rate of the United States is higher than the Organization for Economic Cooperation and Development average of 4.6 per 1000. Furthermore, even though the 2012 infant mortality rate for the United States was 6.0, infant mortality rates have leveled off since the 1990s in the US (Heisler, 2012).

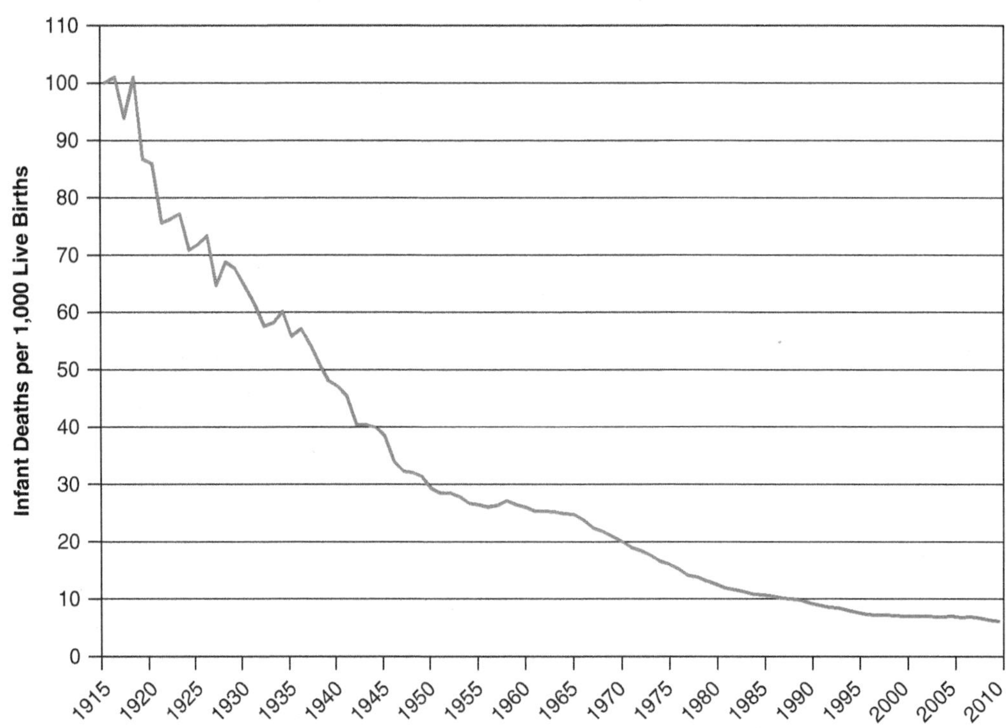

Figure 5–3 Infant Mortality Rate: United States, 1915–2010

Source: Data from Vital Statistics of the United States, Vol. II, Mortality Part A. Section 2, Table 2.1; Health United States 2005; Health United States 2010, National Center for Health Statistics; World Population Data Sheet, Population Reference Bureau.

Significant disparities in infant mortality rates exist within the United States today. These disparities vary geographically as well. A study reported that the infant mortality rate among blacks was 12.7 per 1,000, which was more than double the white rate of 5.5 per 1,000 (Heisler, 2012). However, Hispanic infant mortality rates tend to be lower than both white and black infant mortality rates in the United States. Social disadvantage is often associated with the dismally low infant mortality rates among blacks. However, Hispanics, particularly those of Mexican ancestry, experience many of the same deprivations, yet experience lower infant mortality rates. This has been labeled an "epidemiological paradox."

The major causes of infant mortality, low birth weight and short gestational age, also vary geographically in the United States. Infant mortality rates are highest in the southeastern United States and lowest in New England and the West. Low birth rate and short gestational age have been linked to two broad factors: race, education and economic status of the mother, and the mother's health and her access to health care.

Canada's infant mortality rate declined from 6.3 in 1993 to 5.1 in 2012. Within Canada, however, there are considerable geographic variations in infant mortality. Whereas Ontario, Alberta, Quebec, British Columbia, Nova Scotia, and Prince Edward Island have infant mortality rates at or below 5.0, the Northwest Territories, and Nunavut have rates above 14.0.

Infant Mortality in Other Parts of the World

Despite the affluence and advanced technology in the United States, infant mortality conditions there are not as good as they are in a number of other countries (see Table 5–6), though they are improving. You can't look at this table without shaking your head and asking how the world's wealthiest nation can be satisfied with a health care system that can't do better for all of its citizens.

Worldwide the pattern of infant mortality is as shown in Figure 5–4. Infant mortality rates range from a low of 0.9 in Iceland to a high of 129 in Afghanistan. Among major world regions infant mortality is highest in Africa, with an overall rate of 67. Asia is next highest with an overall rate of 37. Europe is the lowest with a rate of only 5, though Northern America with 6 is close behind.

Infant and child mortality is primarily a problem associated with the less developed countries. In the mid-1990s approximately 98 percent of the deaths among children under five years of age occurred in the less developed countries. The tragedy, according to UNICEF estimates, is that 95 percent of the estimated 14.5 million infant and child deaths in less developed countries in 1990 were preventable (Grant, 1990). Even in the more developed countries, where only 300,000 children under age five died (mostly in Eastern Europe and the former Soviet Union), more than sixty percent of those deaths were potentially preventable as well.

> Infant and child mortality is primarily a problem associated with the less developed countries.

It has been estimated that 24 percent of the world's newborns died before age five in 1950; by 1990, however, the infant and child mortality rate (ICMR) had declined more than 55 percent to 10.5 percent. Perhaps the most impressive decline occurred in China, where the rate went from 26.6 percent to 4.1 percent, an 85 percent reduction. Japan also had an impressive drop from 7.5 percent to 0.8 percent, a drop of 89 percent.

In 2010, 7.6 million children under five years of age died. This represents a 35 percent decrease in the under-five mortality rate from 1990. Less developed regions have reduced their under-five mortality by 50 percent since 1990. However, under-five deaths are increasingly concentrating in Sub-Saharan Africa and Southeast Asia. In these regions, 1 in 8 and 1 in 15 children, respectively, die before the age of five. Globally, close to half of the under-five deaths occur in just five countries: India, Nigeria, Democratic Republic of the Congo, Pakistan and China. Under-five deaths in India and Nigeria constitute 22 percent and 11 percent of total child deaths, respectively (UNICEF, 2011).

Table 5–6 Countries with Lowest Infant Mortality, 2012

Country	Deaths per 1000 Live Births
Iceland	0.9
Hong Kong	1.9
Singapore	2.0
Sweden	2.1
Liechtenstein	2.2
Japan	2.3
Finland	2.4
Norway	2.4
Portugal	2.5
Czech Republic	2.7
Luxembourg	3.0
Slovenia	3.0
Andorra	3.1
San Marino	3.1
Spain	3.2
South Korea	3.2
Estonia	3.3
Israel	3.4
Italy	3.4
Germany	3.4
France	3.5
Denmark	3.5
Ireland	3.5
Belgium	3.5
Channel Islands	3.6
Austria	3.6
Greece	3.8
Netherlands	3.8
Switzerland	3.8
Australia	3.9
Belarus	4.0
Taiwan	4.2
United Kingdom	4.3
Lithuania	4.3
Croatia	4.4
Cuba	4.5
New Zealand	4.7
Poland	4.8
Brunei	5.0

New Caledonia	5.0
Canada	5.1
Malta	5.5
Latvia	5.7
U.S.	6.0

Source: Data from 2012 World Population Data Sheet. (Washington, D.C: The Population Reference Bureau).

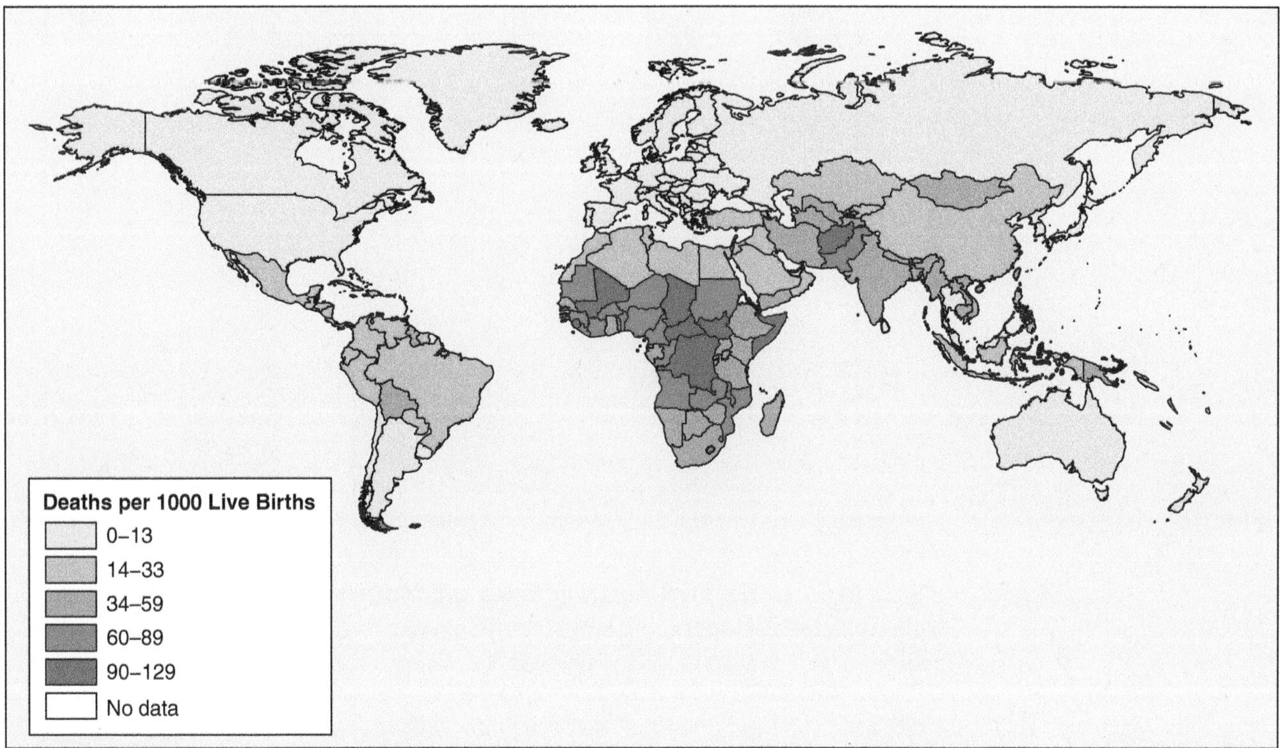

Figure 5–4 Infant Mortality in 2012
Source: Data from 2012 World Population Data Sheet. (Washington, D.C: The Population Reference Bureau).

Researchers have identified several factors that are thought to be related to infant mortality risks. High risk babies include those born to mothers who are adolescents, over age 40, or have had more than seven births, and when the interval between births is less than seven years (Mosley and Cowley, 1991, 10). In the United States the number of babies born to drug-addicted mothers has become increasingly problematic. National policies aimed at discouraging early marriage and childbearing, along with aggressive family planning programs, can prevent births to high-risk mothers and substantially reduce infant and child mortality (Rutstein, 1991).

Figure 5–5 depicts changes in under-five mortality rates as it relates to mother's education for selected countries. According to Mosley and Cowley (1991, 11), "Maternal education appears to affect child health in a number of ways. . . . The more educated a mother is, the more likely she is to use maternal and child health services—specifically prenatal care, delivery care, childhood immunizations, and oral rehydration therapy for diarrhea." A 2010 study found that half of the worldwide reduction in child mortality in 2009 could be attributed to increased educational attainment of young women (Gakidou et al., 2010). This amounts to 4.2 million deaths that were prevented by increased educational attainment of women of reproductive age.

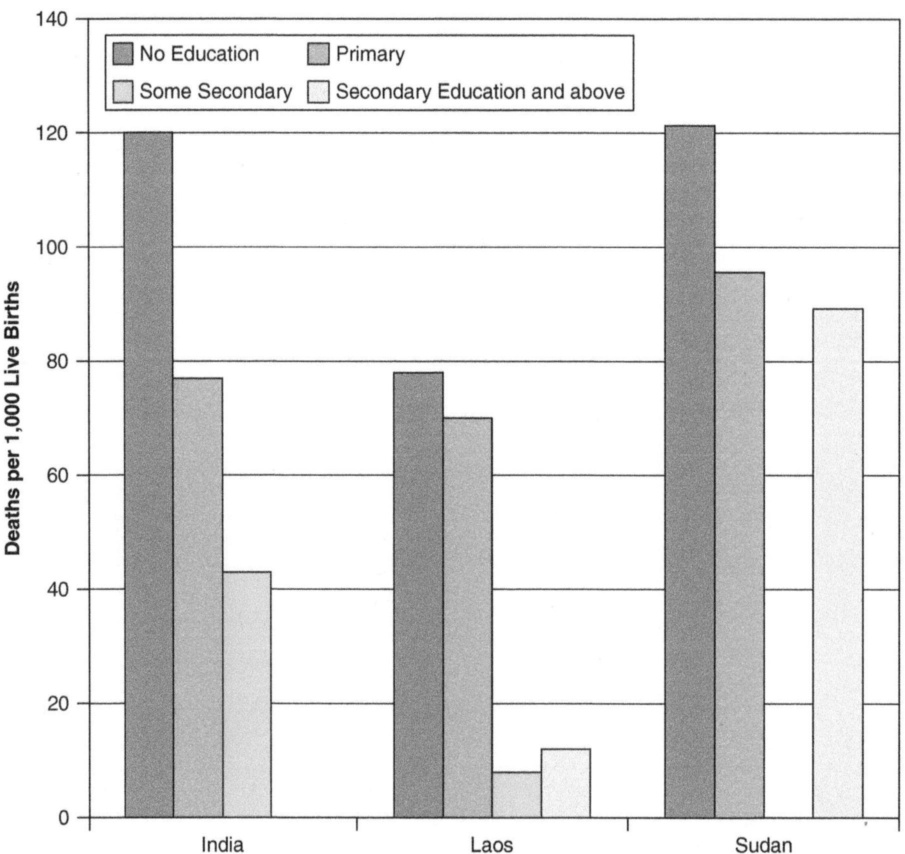

Figure 5–5 Under-Five Mortality Rates and Mother's Education. Results from Selected Demographic and Health Surveys

Source: Lance, P. et al. 2008; Ministry of Health Republic of Indonesia, 2004; National Bureau of Statistics (NBS) 2007; Pandey et al., 2005; Committee for Planning and Investment (CPI), 2007; Ministry of Health, Government of Southern Sudan, 2006.

A few major diseases are responsible for a large proportion of deaths among children under five years of age (Table 5–7). Fifty percent of all under-five deaths are caused by infectious diseases and pneumonia is the number one killer. Diarrheal disease and malaria are also major causes of under-five mortality worldwide. Relatively simple and low cost interventions could sharply reduce these preventable deaths. These interventions include: antenatal care, skilled care during birth and in the weeks after childbirth, early initiation of breastfeeding, exclusive breastfeeding up to 6 months of age, immunization, appropriate use of antibiotics, oral rehydration therapy and zinc, insecticide treated bednets, and anti-malarials, while bolstering nutrition (WHO, 2012).

Malnutrition contributes to more than one third of under-five deaths. There is little doubt that malnourished children are at a disadvantage; their chances of success in life are already diminished. Physically, they are hindered by stunted growth, a reduced resistance to infections, and a variety of other health problems; ultimately, they are more likely to die before reaching adulthood. Mentally, malnourished children suffer both direct consequences—slower brain development, for example—and indirect consequences, which include persistent tiredness and lack of energy—making learning more difficult. Improved nutrition for children needs to be a high priority everywhere, not just in the less developed countries. As Brown and Pollitt (1996, 43) concluded:

> Enriched programs for children in economically impoverished communities can often ameliorate some of the problems associated with previous malnutrition . . .

Table 5–7 Causes of Death of Children Under Age 5, 2010

Causes of Death	Number (thousands)	Percentage of Deaths
Pneumonia	1368	18.0
Diarrheal diseases	836	11.0
Malaria	532	7.0
Injuries	380	5.0
AIDS	152	2.0
Meningitis	152	2.0
Measles	76	1.0
Other neonatal	2660	35.0
Other postneonatal	1368	18.0
Total	7600	100.0

Source: UNICEF, 2012.

> it seems clear that prevention of malnutrition among young children remains the best policy—not just on moral grounds but on economic ones as well . . . . Steps taken today to combat malnutrition and its intellectual effects can go a long way toward improving the quality of life—and productivity—of large segments of a population and thus society as a whole.

Sex Differentials

The risk of death appears to be greater for males at all ages.

Sex differences in mortality have been of considerable interest to demographers. The risk of death appears to be greater for males at all ages. Even among fetal deaths males are more likely to die. No simple explanation of these sex mortality differentials has been accepted, and they may be converging in the United States (Preston and Wang, 2006).

Some scholars have argued that the female is biologically superior to the male. Support for this argument comes from reported sex differentials favoring females among other species, including rats, and from the higher fetal mortality for males in humans. Madigan's study of mortality among religious teaching orders of brothers and sisters also supported the idea of female biological superiority (Madigan, 1957). His sample of residents in religious communities was an attempt to control for sociocultural differences that influence mortality and to view mortality under conditions that were as similar as possible for both sexes. Both sexes were found to have greater life expectancies than the general population (partly as a result of lower stress levels), but the differences in life expectancy between the sexes was as apparent for this sample as it was for the entire population. That demographers are not generally convinced of this argument is apparent in the following comment from demographer Thomlinson (1976, 148):

> Although some persons believe that women are biologically superior to men, this conclusion cannot be properly deduced from the factually supportable premises. One piece of contrary evidence is the higher mortality rates among underdeveloped areas and in some nonliterate societies. Although the evidence is insubstantial, what we can learn from excavations of ancient burial sites implies a higher mortality among women.

However, much of the higher mortality experienced by women in less developed countries stems from high fertility rates; repeated pregnancies and childbirths increase the probability of dying from complications related to them.

Evidence suggests that biological differences between males and females may contribute to the sex differences in mortality. Genetics and hormones may play a part. Estrogen protects the heart and blood vessels while testosterone has been shown to promote hypertension and decrease immunity to infections (Kalben, 2000).

A different explanation of mortality sex differences is to attribute the differences to differentials in occupation, status, and role. Males are employed in more hazardous occupations than are females, for example, and military deaths are primarily male. Males are believed by some to be under greater stress than females, though this is changing in most more developed countries as more females move into higher-stress occupations. Males generally smoke more, drink more, and even drive more than females; they are also more often murdered. Some researchers have even suggested that a certain "macho" factor contributes to higher mortality for males. Men are more likely to ride motorcycles, for example, and more likely to be struck by lightning or die in flash floods. They are also less likely to go to doctors for regular checkups, more likely to medicate themselves, and, generally, more likely to abuse their bodies. Even childhood experiences may have different effects on male mortality (Hayward, et. al., 2004). All of these are likely to raise male mortality rates. At the same time, as societies move through the demographic transition, women may benefit differentially because fewer childbirths undoubtedly increase the probability of living longer. However, both these social and biological differences probably contribute to observed sex differentials.

Indications of mortality sex differentials, as well as changes in these differentials, may be seen in Tables 5–8 and 5–9. Figure 5–6 shows how male and female life expectancies in the United States have changed.

Race and Class Differentials

Many, if not most, racial and ethnic mortality differentials reflect differences in socioeconomic status. No convincing evidence of significant biological differences in the resistance to diseases can be identified among different race and ethnic groups, though a few diseases are largely restricted to certain groups. Probably the best example of such a disease is sickle-cell anemia, which is mainly restricted to blacks.

Crude death rates for the United States in 2009 for whites and blacks appear in Table 5–10. They suggest heavier mortality rates among whites, and especially among white females. However, keep in mind that the crude death rate can be deceptive. Age-specific death rates reveal a different picture entirely, as is apparent in Table 5–11. For most groups black mortality is considerably above white mortality. Table 5–11 also illustrates sex differentials of mortality.

In most countries social Class and mortality are related.

In most countries social class and mortality are related, though the degree of the relationship varies in different types of political and economic systems. Several measures of social class have been used in research studies of mortality differentials including: occupation, educational attainment, and income and wealth (Elo, 2009).

However, such class differentials have probably increased as countries passed through the epidemiological transition. As Pressat (1970, 37) commented:

> It is certainly true that, until recent times, social inequality did not too seriously aggravate the risks of mortality among the lower classes. Because of the ineffectiveness of medical care, the poor lost very little by not being able to take advantage of it. The great epidemics spared no one and the standard of hygiene at the court of Versailles was probably no higher than in country districts. The crisis of subsistence was probably the only plague which did not affect the rich, whose dietary habits could be disastrous anyway.

Table 5–8 Selected Life Table Values; 1979 to 2007

Age and sex	Total							White							Black						
	1979–1981	1985	1990	1995	2000	2005	2007	1979–1981	1985	1990	1995	2000	2005	2007	1979–1981	1985	1990	1995	2000	2005	2007
Average Expectation of Life in Years																					
At birth:																					
Male	70.1	71.1	71.8	72.5	74.3	74.9	75.4	70.8	71.8	72.7	73.4	74.9	75.4	75.9	64.1	65.0	64.5	65.4	68.3	69.3	70.0
Female	77.6	78.2	78.8	78.9	79.7	79.9	80.4	78.2	78.7	79.4	79.6	80.1	80.4	80.8	72.9	73.4	73.6	74.0	75.2	76.1	76.8
Age 20:																					
Male	51.9	52.6	53.3	53.8	55.3	55.9	56.4	52.5	53.2	54.0	54.5	55.8	56.3	56.8	46.4	47.1	46.7	47.3	50.0	51.0	51.7
Female	59. 0	59.3	59.8	59.9	60.5	60.7	61.2	59.4	59.8	60.3	60.3	60.9	60.1	61.5	54.9	55.2	55.3	55.5	56.6	57.4	58.1
Age 40:																					
Male	33.6	34.2	35.1	35.6	36.7	37.3	37.8	34.0	34.7	35.6	36.1	37.1	37.7	38.1	29.5	29.8	30.1	30.6	32.3	33.2	33.8
Female	39.8	40.0	40.6	40.7	41.2	41.4	41.9	40.2	40.4	41.0	41.0	41.5	41.7	42.1	36.3	36.4	36.8	37.0	37.8	38.5	39.1
Age 50:																					
Male	25.0	25.5	26.4	27.0	27.9	28.5	29.0	25.3	25.8	26.7	27.3	28.2	28.8	29.2	22.0	22.1	22.5	23.1	24.2	24.9	25.4
Female	30.7	30.8	31.3	31.4	32.0	32.2	32.7	31.0	31.1	31.6	31.7	32.2	32.4	32.8	27.8	27.8	28.2	28.5	29.1	29.8	30.4
Age 65:																					
Male	14.2	14.5	15.1	15.6	16.2	16.8	17.2	14.3	14.5	15.2	15.7	16.3	16.9	17.3	13.3	13.0	13.2	13.7	14.2	14.9	15.2
Female	18.4	18.5	18.9	18.9	19.3	19.5	19.9	18.6	18.7	19.1	19.0	19.4	19.4	19.9	17.1	16.9	17.2	17.2	17.7	18.2	18.7

Source: United States Life Tables, National Vital Statistics Reports.

Table 5–9 Men per 100 Women by Age and Race, 1900–2010

Age and Race	1900	1930	1960	1970	Projection 1990	2000	2010
All Races:							
Over 65	102.0	100.4	82.6	72.1	67.5	60.2	75.8
Over 75	96.3	91.8	75.1	63.7	57.8	58.2	64.3
White:							
Over 65	101.9	100.1	82.1	71.6	67.7	70.6	64.1
Over 75	97.1	92.0	74.3	63.2	58.4	58.4	65.3
Black:							
Over 65	102.9	105.7	30.1	79.8	68.2	61.4	64.7
Over 75	89.6	89.6	87.6	74.7	62.6	50.8	52.8

Source: Leon Bouvier, Elinore Atlee, and Frank McVeigh. (1975) "The Elderly in America," *Population Bulletin,* 30(3):9. Courtesy of the Population Reference Bureau, Inc., Washington, D.C.; U.S. Census Bureau, Summary File 1.

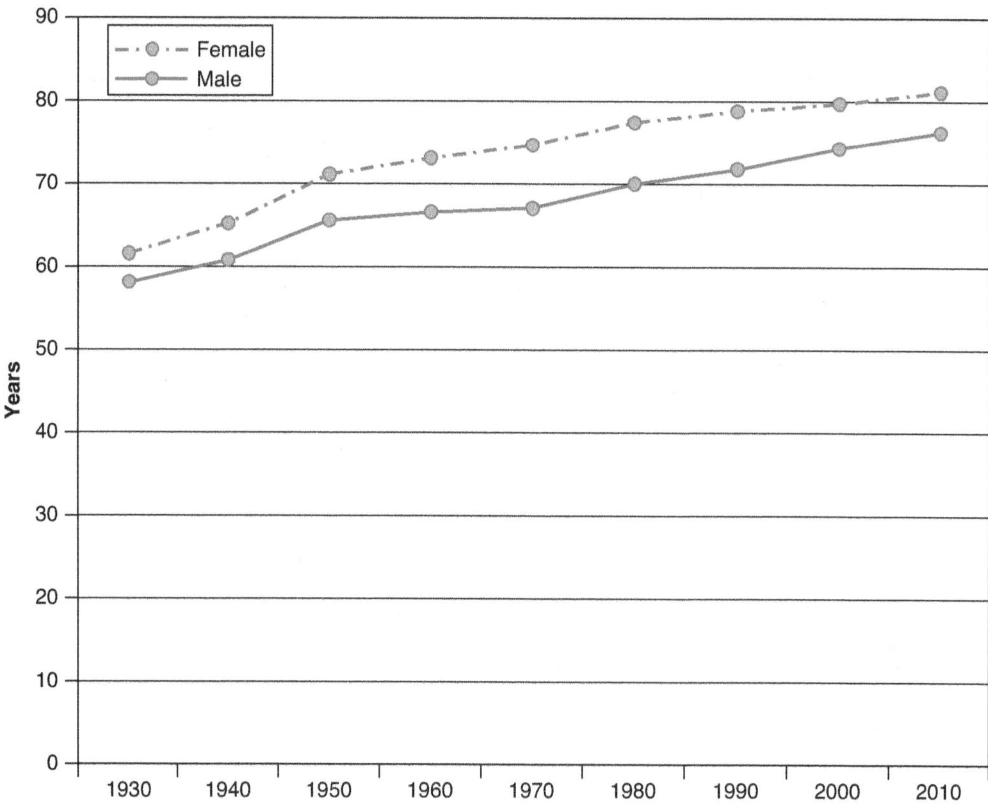

Figure 5–6 Changes in Male and Female Life Expectancy in the United States, 1930–2010

Source: Data from the National Center for Health Statistics.

Table 5–10 Crude Death Rates by Race and Sex, United States, 2009

	Crude Death Rate	
Population	**White**	**Black**
Total	9.6	7.3
Males	9.7	7.8
Females	9.6	6.8

Source: Centers for Disease Control and Prevention, National Center for Health Statistics. Underlying Cause of Death 1999–2009 on CDC WONDER Online Database, released 2012.

Today many differences in mortality may be attributed to differences in socioeconomic factors, but many other differences are also important and they cannot be overlooked. For example, in many places there are differences in death rates between rural and urban areas (Reher, 2001; Gehrmann, 2002) as is shown in Table 5–12 for the United States.

Medical Geography

Medical geography, the study of the spatial aspects of health and illness.

Medical geography, the study of the spatial aspects of health and illness, is an area of geographic inquiry that has seen a considerable amount of recent research activity. Its focus has been twofold:

1. on disease ecology, the "manner and consequences of interaction between the environment and the causes of morbidity and mortality" (Birdsall, 1991, 392) and,
2. on the health care delivery system and its spatial aspects.

More generally, geographer Robin Kearns (1993) has argued the necessity of placing geographic concerns with public health issues more directly within the context of place. Such an orientation toward place in medical geography, he argued, would aid both our understanding of illness and health service provision systems and the collective experience of place in particular communities—health services are clearly related to quality-of-life issues. Kearns emphasizes health, rather than illness, and argues that (1993, 145):

> A central requirement for a geography of health will be models of society as well as of health that recognize the contingent relations that pertain for individuals and groups at particular locations. . . . Only a further integration of issues of place, identity, and health into medical geography will firmly (re)locate the field within social geography.

As Kearns and others have noted, geographers would do well to incorporate more consideration of gender, race, ethnicity, and place into the practice of medical geography. Smyth (2005) reviewed progress that has been made in medical geography, especially as it has separated somewhat from medicine and focused more on geographical factors, especially therapeutic landscapes. As she notes (2005, 488), "The notion of therapeutic landscapes has been embraced by geographers of health and yielded a significant, and growing, body of recent research."

The mapping of diseases has received much attention and studies such as those on the techniques of mapping cancer by Boyle, et al. (1989) and Clayton and Kaldor (1989) have provided useful geographic insights into that disease, as has a detailed historical study of the diffusion of smallpox in Finland by Wilson (1993). A more traditional atlas and compendium of analytical techniques was developed by Cliff and Haggett (1988), and Koch (2005; 2011) added two others.

Table 5–11 Comparison of White and Black Death Rates by Age and Sex, United States, 2009

Race and Sex	Under 1 Year	1–4 Years	5–14 Years	15–24 Years	25–34 Years	35–44 Years	45–54 Years	55–64 Years	65–74 Years	75–84 Years	85 Years and Over
White											
Total	5.1	0.2	0.1	0.7	1.0	1.8	4.1	8.5	19.3	49.0	134.6
Male	5.7	0.3	0.1	0.9	1.4	2.3	5.1	10.6	23.3	58.6	153.6
Female	4.6	0.2	0.1	0.4	0.7	1.4	3.1	6.5	15.7	42.2	126.1
Black											
Total	12.5	0.4	0.2	1.0	1.6	2.8	6.6	14.0	27.0	52.6	137.3
Male	13.9	0.4	0.2	1.5	2.2	3.6	8.2	18.6	35.0	64.9	158.2
Female	11.0	0.4	0.2	0.6	1.0	2.2	5.3	10.4	21.3	45.4	129.5

Source: Centers for Disease Control and Prevention, National Center for Health Statistics. Underlying Cause of Death 1999–2009 on CDC WONDER Online Database, released 2012.

Table 5–12 Urban and Rural Death Rates in the United States, 2009	
Level of Urbanization	**Death Rate**
Large Central Metropolitan Area	6.9
Large Fringe Metropolitan Area	7.2
Medium Metropolitan Area	8.1
Small Metropolitan Area	8.6
Micropolitan Area (Nonmetropolitan)	9.9
Non-Core (Nonmetropolitan	11.0

Source: Centers for Disease Control and Prevention, National Center for Health Statistics, Underlying Cause of Death 1999–2009 on CDC WONDER Online Database, released 2012.

The analysis of geographical variations in health care provision and consumption is another area of concern to geographers. Analytical models in this area have focused on both the individual level (Senior and Williamson, 1990) and the national level (Mohan, 1988). According to Jones and Moon (1991, 443), writing in a review about progress in medical geography, "Sophisticated modeling of this nature, together with informed qualitative work, provides medical geography with a means to complement its traditional regional-scale analyses. In disease mapping it allows the assessment of environmental effects taking into account individual characteristics. In health-care geography it provides a basis for in-depth analyses of the relationship between need and provision/consumption."

Hemorrhagic Fever

Late in 1995 American travelers were warned that dengue fever, a virus transmitted by mosquitoes, was spreading through Mexico, Central America, and the Caribbean. By October of that year more than 35,000 cases of dengue fever had been reported, including some 500 cases of its most severe form, hemorrhagic dengue. This dengue epidemic began in 1994; few doubt that there has been a resurgence, and some argue that the pace of the current epidemic is accelerating. Before 1970, only nine countries had experienced severe dengue epidemics. Dengue is now endemic in more than 100 countries in Africa, the Americas, the Eastern Mediterranean, Southeast Asia and the Western Pacific. Southeast Asia and the Western Pacific regions are the most seriously affected.

Cases across the Americas, Southeast Asia and Western Pacific exceeded 1.2 million cases in 2008 and over 2.2 million in 2010. In 2010, 1.6 million cases of dengue were reported in the Americas and 49,000 cases were severe dengue.

In May, 1993, a young couple in New Mexico died mysteriously just a few days apart—high fevers, headaches, violent coughs, and muscular cramps marked their final days. Similar cases were recorded in Colorado and Nevada. The Centers for Disease Control (CDC) in Atlanta analyzed samples from several comparable cases, finally determining that they had all died of an unknown type of hantavirus.

Other viral outbreaks have been identified as well in various parts of the world: Ebola in Congo (in Yambuku in 1976 and most recently in Kikwit in 1995), Marburg in Germany, Rift Valley fever in Egypt, Lassa fever in Nigeria, Machupo in Bolivia, and Junin in Argentina are examples. What all of these viruses have in common is membership in a broad class of hemorrhagic fever viruses. Most are not new; rather, they are often mutations or recombinations of known viruses termed "emerging pathogens." They can be deadly, and we are likely to hear more about them in the decades ahead.

Hemorrhagic fever includes several "families" of viruses, including the following: arenavirus, filovirus, animal filovirus, flavivirus, hantavirus, and bunyavirus (Rift Valley fever).

According to virologist Bernard Le Guenno (1995, 58):

> The primary cause of most outbreaks of hemorrhagic fever viruses is ecological disruption resulting from human activities. The expansion of the world population perturbs ecosystems that were stable a few decades ago and facilitates contacts with animals carrying viruses pathogenic to humans.

Donovan (2008) wrote about the role of Navajo medicine men at the epicenter of the 1993 hantavirus outbreak. They had heard tales of what happens when the rodent populations are extremely high. Generation after generation had been warned of the devastation and death that would result in these circumstances. The medicine men held a meeting in Window Rock, Arizona to discuss the deaths that had occurred and to issue a warning to the Navajo people to try to avoid the numerous prairie dogs and deer mice that had flourished due to the wet winter and the plentiful food sources available to them.

Hantavirus has recently reappeared in Yosemite National Park, California in 2012. The Centers for Disease Control and Prevention reported 9 cases and 3 deaths among visitors to the park between June and September of that year.

In his best-selling book, *The Hot Zone*, journalist Richard Preston (1994) provided a graphic, compelling description of the Ebola virus at work in the human body:

> The liver bulges up and turns yellow, begins to liquefy, and then it cracks apart . . . . The kidneys become jammed with blood clots and dead cells . . . the blood becomes toxic with urine. The spleen turns into a single huge, hard blood clot. . . . The intestines may fill up completely with blood. The lining of the gut dies and sloughs off into the bowels and is defecated along with large amounts of blood.

Much remains unknown about the various hemorrhagic fever viruses; scattered outbreaks have not led to the identification of hosts in most cases, yet the viruses must be surviving somewhere. As with HIV (which is discussed next), much remains to be learned about how to prevent or minimize such viral outbreaks in a world in which population growth and ecosystem disturbances are certain to continue.

In her book, *The Coming Plague: Newly Emerging Diseases in a World Out of Balance*, journalist Laurie Garrett (1994) warned us that these viruses are our predators and that we had better improve our understanding of where they come from, where they hide, and how they can be dealt with—we have been too comfortable with the notion that "plagues" are a thing of the past. Beyond medical science and forays into microbial ecology, Garrett argues that we must consider changes in social behavior as well.

Geographers could focus more attention on the human-environmental relationships that exist in places where viral outbreaks have occurred. We could also focus more attention on the severe consequences of another doubling of the world's population between now and perhaps the year 2050. Most of that growth will occur in the countries of the Third World, where rapid urbanization is already creating massive centers of urban deprivation, high population densities, and growing anarchy.

Acquired Immune Deficiency Syndrome (AIDS)

AIDS is the defining epidemic of the late Twentieth Century and has become the first modern pandemic. A physically and psychologically debilitating syndrome, it also carries with it devastating cultural and social effects. Caused by the human immunodeficiency virus (HIV), which was first identified in 1981, by the year 2000, it had claimed over

20 million lives globally. HIV/AIDS is now the fourth leading cause of deaths worldwide. More than 30 million people have died of AIDS since the beginning of the epidemic (Pisani, 2011; Vatanoglu and Ataman, 2011).

HIV/AIDS is now the fourth leading cause of deaths worldwide.

Although the exact geographic origin of the AIDS virus has not been determined, most researchers believe that the retrovirus responsible for AIDS originated in Central Africa. Six years before its identification in the United States, the disease had been isolated in central Africa. The earliest specimen of the virus was taken in 1959 from a man in the former Belgian Congo. Since its initial recognition in Africa around 1972, the disease has diffused to most other parts of the world.

AIDS is the defining epidemic of the twentieth century and has become the first modern pandemic.

Currently, researchers have identified two serotypes of HIV; HIV-1, the most common type, which is found worldwide, and HIV-2, found almost exclusively in West Africa, largely within former Portuguese colonies. Though transmission routes and risk factors are similar for both forms of HIV, the latency period for HIV-2 appears to be longer.

The human immunodeficiency virus (HIV) attacks the immune system, complicating the body's ability to combat diseases and infections. It is also the virus that causes acquired immunodeficiency syndrome (AIDS), which is the final and most severe form of HIV. HIV is spread via direct contact with bodily fluids, primarily blood, semen, and vaginal fluids. As a result, sexual contact and shared needles among intravenous drug users are the major agents of transmission.

The AIDS epidemic peaked globally in 1999, and globally, it is estimated that 33.3 million people are living with HIV (Table 5–13). Since 1999, however, the number of new HIV infections has declined by 19 percent globally. This is due to the expanding distribution and use of antiretroviral treatments as well as prevention efforts. However, about 10 million people are still in need of these essential treatments (UNAIDS, 2010). The incidence of HIV has declined between 2001 and 2009 in many countries most affected by the epidemic and 22 of these countries are in Sub-Saharan Africa. However, in several countries in Eastern Europe and Western Asia, HIV infections are increasing.

According to Pisani (2000, 63)

> While the headlines continue to scream of a 'global pandemic', the developments of recent years actually suggest something rather different. Yes, HIV has reached every corner of the globe. Yes, it continues to spread disproportionately fast in marginalised populations in most countries. Yes, all populations should remain vigilant against it. But as the 20th century drew to a close, the 'global pandemic'

Table 5–13 Adults and Children Living with HIV, 2009

Region	Number
Sub-Saharan Africa	22,500,000
South and Southeast Asia	4,100,000
Northern America	1,500,000
Central and South America	1,400,000
Eastern Europe and Central Asia	1,400,000
Western and Central Europe	820,000
East Asia	770,000
Middle East and North Africa	460,000
Caribbean	240,000
Oceania	57,000
Total	**33,300,000**

Source: UNAIDS Report on the Global AIDS Epidemic, 2010.

was—in public health terms at any rate—looking more and more like two distinct epidemics, one global and one regional. Controversial as it may be to say so, the story of AIDS in the 21st century is likely to be dominated by heterosexuals in Africa and injecting drug users around the world.

AIDS in the United States

The first five cases of what is now known as AIDS were reported in the United States in 1981. The majority of early cases were diagnosed in previously healthy homosexual males. Since June 1981, over 600,000 people have died of AIDS in the United States. The initial appearance of AIDS had three prominent geographical foci—the metropolitan areas of San Francisco, Los Angeles, and New York City. According to a study by Dutt, et al, (1987, 457–458) "Prior to 1983, 67 percent of the AIDS victims in the country were confined to these areas." The map of cumulative AIDS cases for 2010 (Figure 5–7) shows that in the intervening period, Florida has also emerged as a focal point of infections. Several northeastern states as well as Illinois, Texas and Georgia also had high numbers of AIDS cases in 2010. These data reflect the strong urban influence of the foci as well as the large populations of these states.

In 1981 only 318 persons were diagnosed as having AIDS. In 2010 that number was approximately 33,000.

The number of people diagnosed with AIDS has risen sharply since its first identification (Figure 5–8). In 1981 only 318 persons were diagnosed as having AIDS. In 2010, newly diagnosed AIDS cases had fallen to about 33,000 from its peak in 1995 at about 71,000 cases.

An analysis of AIDS cases between 1981 and 2001 by age and gender points out some interesting relationships (Table 5–14). Most AIDS cases were between the ages of 20 and 49, but almost half of them were in the 30–39 year age cohort. AIDS cases were still predominantly male, but the disease was affecting an increasing number of females as well (though they were about four times less likely to be infected than males). In the mid-1980s only 7.4 percent of U.S. AIDS cases were female.

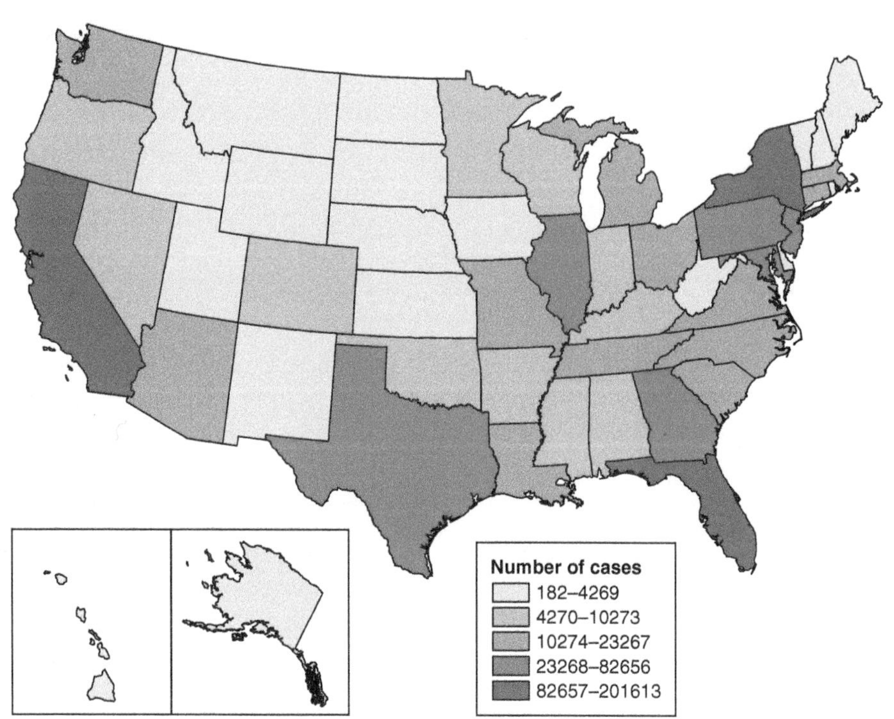

Figure 5–7 Cumulative AIDS Cases, 2010

Source: Centers for Disease Control, Special Data Report, July 2010.

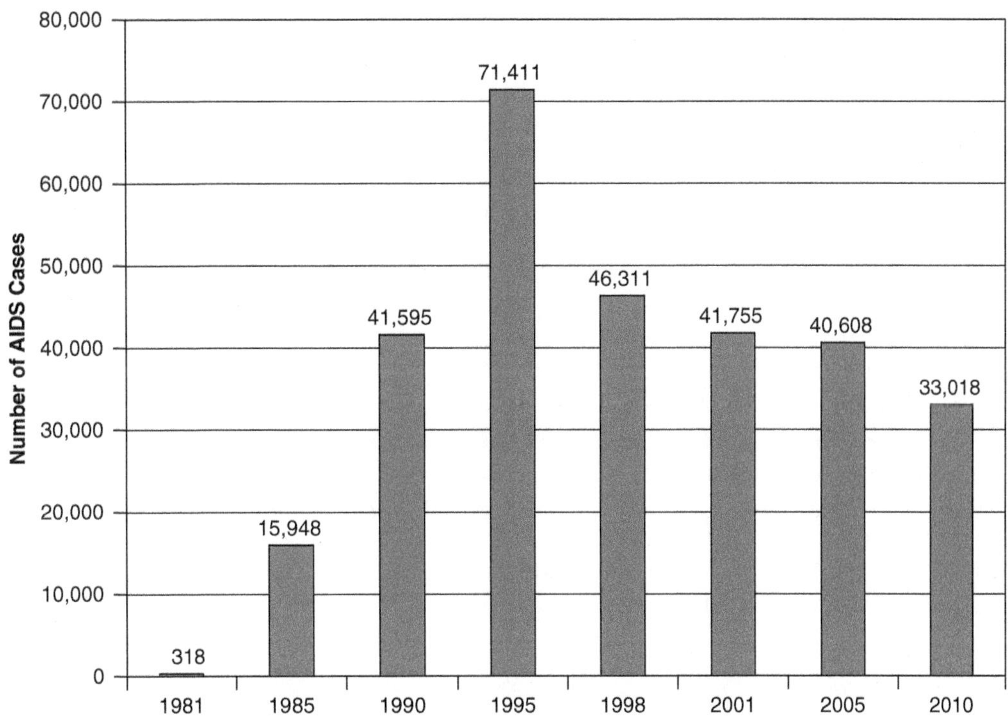

Figure 5–8 AIDS Cases in the United States

Source: HIV Surveillance Reports, Centers for Disease Control and Prevention.

Until the mid-1990s AIDS was a disease that was fatal to almost everyone who contracted it. Between 1981 and 1995, estimated AIDS deaths among those ≥ 13 years increased from 451 to 50,628. AIDs deaths declined by 45% between 1993 and 1998 and by 63% between 1995 and 1998. In 1999, declines in AIDS deaths began to level off and remained stable at an average of 38,279 AIDS diagnoses and 17,489 AIDS deaths per year between 1999 and 2008. From 1998 to 2008, the estimated number of persons living with AIDS more than doubled (CDC, 2011).

By the end of 2008, it is estimated that 1,178,350 persons aged ≥ 13 years were living with HIV in the United States (Table 5–15). Seventy-five percent of those infected were male and 65.7% of those males were men who have sex with men. Blacks had HIV prevalence rates which were eight times the rate among whites. Hispanics had HIV prevalence rates that were 2.5 times the rates of whites (CDC, 2011). The Centers for Disease control estimates that 50,000 new persons are infected every year. Half of these are men who have sex with men and half are black.

An analysis of the racial nature of mortality shows that mortality has been higher among Hispanics and African-Americans than among whites. These differences are probably due to more and better health-care facilities being available to whites than to African-Americans and Hispanics, though the latter group is now contracting HIV in numbers far above their proportionate population size (Brink, 2004).

Although AIDS first appeared in the metropolitan areas of San Francisco, Los Angeles, and New York, it has now spread to other large metropolitan areas as well as to small cities and towns. According to Dutt, et al., (1987, 471) "Once confined to the peripheries of the country, the disease has now spread to its heartland. AIDS is also becoming more prevalent among heterosexuals, who constitute the overwhelming bulk of the sexually active Americans. If this trend continues, the disease will intensify and no part of the country, indeed no group in it, will be AIDS-free in the near future. AIDS has no racial, political, social, or physical barriers. Without the development of a curative drug or preventive vaccine, the potential for

Table 5–14 AIDS Cases Reported by Characteristic: 1981 to 2001

[Provisional. For cases reported in the year shown. Includes Puerto Rico, Virgin Islands, Guam, and U.S. Pacific Islands. Data are subject to retrospective changes and may differ from those data in Table 1190]

Characteristic	1981–2001 total	2001	Characteristic	1981–2001 total	2001
Total[1]	816,149	43,158	Transmission category:		
			Males, 13 years and over	666,026	31,901
Age			Men who have sex with men	368,971	13,265
Under 5 years old	6,975	(NA)	Injecting drug use	145,750	5,261
5 to 12 years old	2,099	(NA)	Men who have sex with men and		
13 to 19 years old	4,428	(NA)	injecting drug use	51,293	1,502
20 to 29 years old	133,725	(NA)	Hemophilia/coagulation disorder	5,000	97
30 to 39 years old	362,021	(NA)	Heterosexual contract[2]	22,914	2,213
40 to 49 years old	216,387	(NA)	Heterosexual contact with		
50 to 59 years old	66,060	(NA)	Injecting drug user	9,821	549
60 years old and over	24,453	(NA)	Transfusion[3]	5,057	105
Sex			Undertermined[4]	57,220	8,909
Male	670,687	31,994	Females, 13 years and over	141,048	11,082
Female	145,461	11,164	Injecting drug use	55,576	2,212
Race/ethnic group:			Hemophilia/coagulation disorder	292	9
Non-Hispanic White	343,889	13,237	Heterosexual contact[2]	35,660	3,205
Non-Hispanic Black	313,180	21,031	Heterosexual contact with injecting drug user	21,736	937
Hispanic	149,752	8,209			
Asian/Pacific Islander	6,157	430	Transfusion[3]	3,914	113
American Indian/ Alaska Native	2,537	194	Undetermined[4]	23,870	4,606

NA Not available. [1]Includes persons with characteristics unknown. [2]Includes persons who have had heterosexual contact with a bisexual male, a person with hemophilia, a transfusion recipient with human immunodeficiency virus (HIV) infection, or an HIV-infected person, risk not specified. [3]Receipt of blood transfusion, blood components, or tissue. [4]Includes persons for whom risk information is incomplete (because of death, refusal to be interviewed, or loss to follow up), persons still under investigation, men reported only to have had heterosexual contact with prostitutes, and interviewed persons for whom no specific risk is identified.

Source: U.S. Centers for Disease Control and Prevention, Atlanta, GA, HIV/AIDS Surveillance Report, Volume 13, No. 2.

destruction from the spread of AIDS cannot be measured or predicted. It is not an overstatement to assert that humanity is in danger." With respect to prospects for a vaccine for HIV/AIDS, physician Richard Horton (2004, 53) noted that, ". . . contrary to the predictions and promises of most AIDS experts, the signs are that a vaccine to prevent HIV infection will not be found for, at the very least, several decades to come—if at all."

Considerable resources have been marshaled in the struggle to understand and treat AIDS, both in the United States and elsewhere. By the mid-1990s one encouraging

Table 5–15 Estimated Number and Rate of Persons Aged ≥ 13 Years Living with HIV Infection by Characteristic, 2008

Characteristic	Number	Rate[1]
Total	1,178,350	469.4
Age Group (years)		
13–24	68,600	134.1
25–34	180,600	440.9
35–44	357,500	846.3
45–54	385,400	871.3
55–64	147,700	439.3
≥65	38,400	99.0
Sex		
Male	883,450	719.5
Female	294,900	230.0
Race/ethnic group:		
Non-Hispanic White	406,000	238.4
Non-Hispanic Black	545,000	1,819.0
Hispanic	205,400	592.9
Asian/Pacific Islander	16,750	147.0
American Indian/Alaska Native	5,000	268.8
Men who have sex with men	580,000	(NA)
Injecting drug use		
Male	131,600	(NA)
Female	73,900	(NA)
Men who have sex with men and injecting drug use	55,200	(NA)
Heterosexual contact[2] (male)	110,900	(NA)
Heterosexual contact (female)	217,400	(NA)
Other[3]	9,350	(NA)

NA Not available. [1]Per 100,000. [2]Heterosexual contact with a person known to have, or to be at high risk for, HIV infection. [3]Includes hemophilia, blood transfusion, perinatal exposure, and risk factors not reported or not identified.

Source: Centers for Disease Control and Prevention, 2011.

development was that of protease inhibitors—powerful antiviral AIDS drugs. These drugs (the first approved in 1995 under the names inverase or saquinavir) are different than earlier drugs such as AZT, DDI, DDC, D4T, and 3TC. Success of the earlier drugs was only temporary; HIV gradually mutated, escaping their effects. Protease inhibitors have been developed to interfere with a different part of the HIV replication cycle; theoretical results suggest that replication attempts will result not in new HIV viruses, but rather in new and harmless viruses instead. Today the benefits of highly active retroviral therapy (HAART) are well documented, primarily in the United States and Europe. These antiviral drugs work by blocking the action of enzymes that aid in the replication and functioning of HIV; they include transcriptase inhibitors and protease inhibitors. As Lamptey, et. al. (2002, 29) noted, "Therapy is more effective when antiretroviral drugs from different classes are used in combination because the virus can develop resistance if only one drug is used." Though these drugs have been successful in managing HIV/AIDS, increasing both the length and quality of life for patients, they are expensive, which limits their availability outside of the wealthiest nations.

In an effort to understand the spatial dimensions of the diffusion of AIDS Gould, et al. (1991, 80) developed modeling techniques to portray the geographic dimensions of AIDS diffusion and attempted to "predict the next maps, not just numbers down the time line." A time-series of maps was produced for the State of Pennsylvania (Figure 5–9) and, according to Gould, et al. (1991, 86), "If you know your human geography of Pennsylvania, you can see the hierarchical diffusion, the jumping from city to city; you can see the spatially contagious diffusion, the wine stain on the tablecloth; and you can

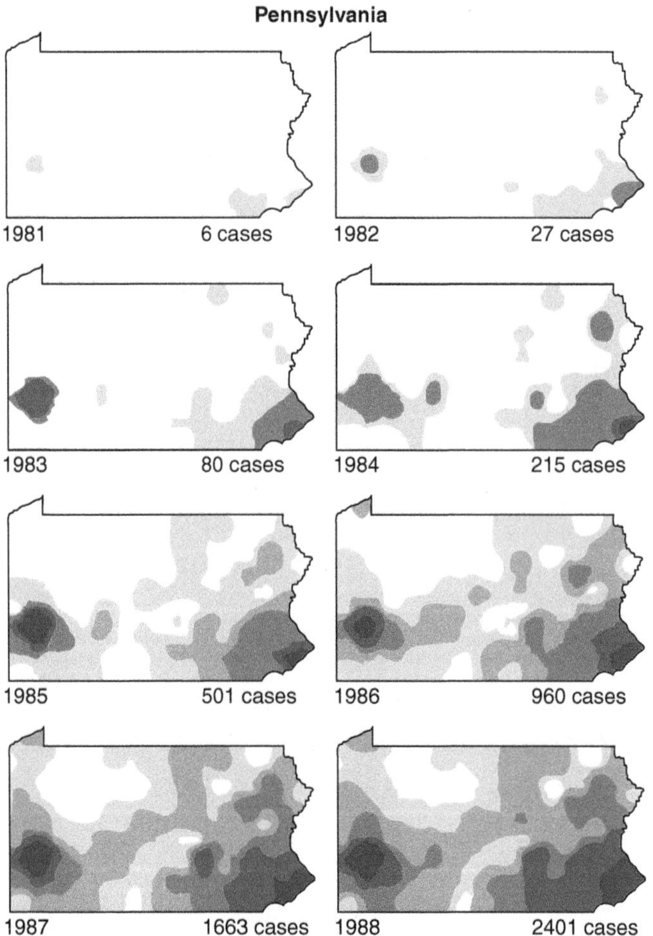

Figure 5–9 Diffusion of AIDS in Pennsylvania, Taken from a Colored and Animated Sequence Made for Television. Shading Increases Geometrically, 1, 3, 9, 27, 81, 243, etc.

Source: From "AIDS: Predicting the Next Map" by Peter Gould, in *Interfaces*, 21(3), May/June 1991. Reproduced with permission of Institute for Operations Research and the Management Sciences in the format republish in a book via Copyright Clearance Center.

also see the way in which the structure of the human landscape, the major turnpikes and roads, channels the epidemic." Similar maps have also been developed for the spread of AIDS in the western United States from 1981 to 1988 (Gould, 1991).

AIDS in the Less Developed Countries

Today it is nearly impossible for most of us to comprehend the terrible extent to which HIV/AIDS is affecting life in the less developed world, especially in Africa, which has more than 23 million people with HIV. As Carol Ezzell (2000, 96) noted, "AIDS is destined to alter history in Africa—and, in fact, the world—to a degree not seen in humanity's past since the Black Death."

In Sub-Saharan Africa, the most heavily affected region by the HIV/AIDS epidemic, an estimated 22.9 million people are living with HIV. This comprises about two thirds of the global total. In 2010 about 1.2 million people died from AIDS in sub-Saharan Africa and 1.9 million people became infected with HIV. Since the beginning of the epidemic 14.8 million children have lost one or both parents to HIV/AIDS (UNAIDS, 2012). Despite these statistics, AIDS mortality is decreasing in Sub-Saharan Africa. This is due mainly to the scaling up of treatment in the region.

"AIDS is destined to alter history in Africa—and, in fact, the world—to a degree not seen in humanity's past since the Black Death."

Approximately 95 percent of all HIV/AIDS cases today occur within the less developed nations, where poverty, poor health care, and limited public health systems are unable to cope (Figure 5–10). Gould (2005, 473) found for AIDS in Africa that "Vulnerability to the disease needs to consider not only exposure, but also each society's potentiality to intervene to mitigate its potentially devastating effects, whether through facilitating behavioural change or in more effective health provision." Kohler, Behrman, and Watkins (2007, 1) concluded in their study of HIV/AIDS risk perceptions, that "The most important empirical

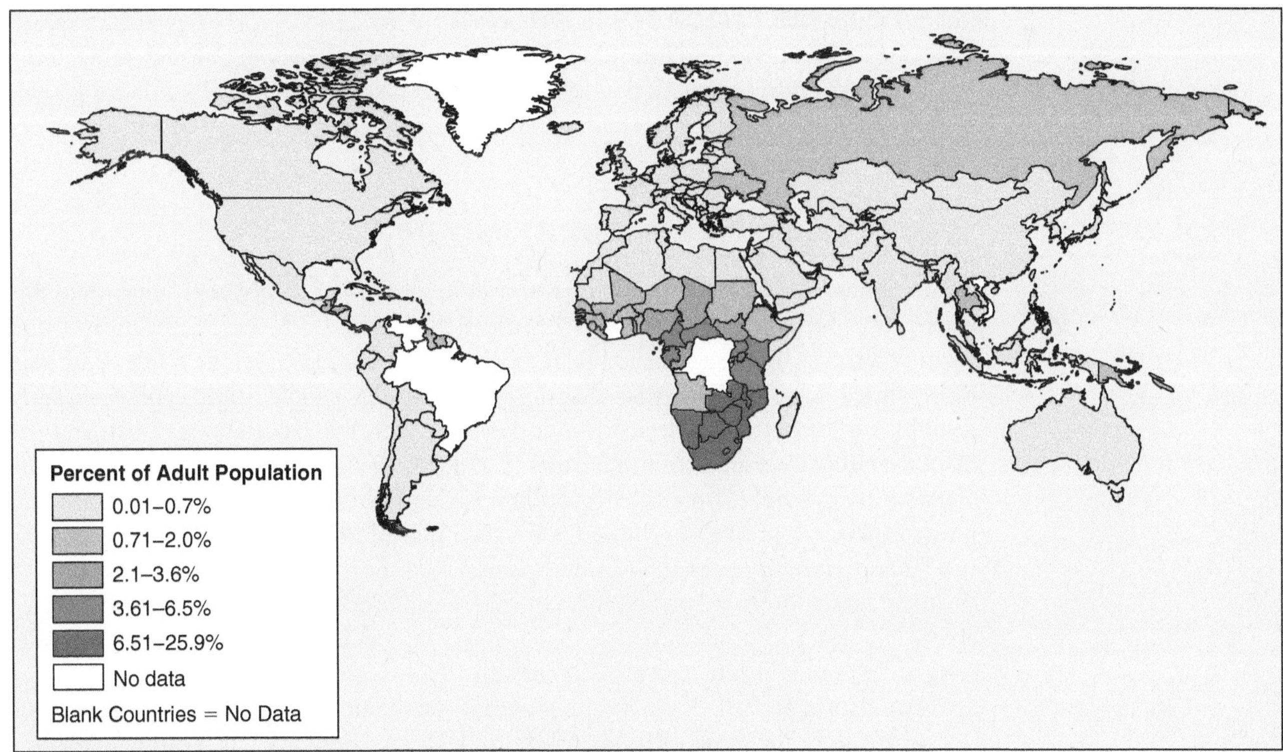

Figure 5–10 Percent of Population Age 15–49 with HIV/AIDS
Source: CIA World Factbook, 2009.

result is that social networks have significant and substantial effects on risk perceptions and the adoption of new behaviors even after we control for unobserved factors." Though HIV/AIDS is far more prevalent in Africa than anywhere else in the world today, the rate of infection has actually slowed in Sub-Saharan Africa because so many people already have the disease. On the other hand, areas of rapid growth of new HIV infections include North Africa, Eastern Europe, and Central and East Asia. For example in China, the cumulative number of AIDS cases through 2009 is 326,157. The majority of these cases were reported in Yunnan Province. Research indicates that the major route of transmission is through sexual contact and that rapid internal migration has contributed to the spread of the disease throughout the country (Jia et al., 2011).

Needless to say, HIV/AIDS has wreaked havoc on the demography and economy of many Third World nations. Unlike degenerative diseases such as heart attacks and various cancers, AIDS kills working-age adults and even children. In its wake we can expect to see millions of orphans as young parents succumb.

UNAIDS (2010) reports that treatment for AIDS is increasing worldwide, most notably in Africa. In 2009, in Sub-Saharan Africa, nearly 37% of people eligible for treatment were able to access life-saving drugs, as were 42% in Central and South America, 51% in Oceania, 48% in the Caribbean, and 19% in Eastern Europe and Central Asia.

Though treatment must be considered, prevention seems more likely to pay sizable dividends, and its cost is primarily for education. Roughly 75 percent of HIV infections worldwide are a result of unprotected sexual intercourse, so that is where prevention needs to be focused. Changing people's sexual behavior, however, is not easy, especially in poor countries. It is first a matter of education, then a matter of getting people to adopt new patterns, including the use of condoms, the limiting of sexual partners, and the controlling of sexually transmitted diseases. Another major cause of HIV transmission is shared needles by users of injected drugs. People must first make the connection between risky behaviors and HIV.

There are signs of hope. In Uganda, for example, the rate of new HIV infections declined during the 1990s, at least in part because of a successful campaign to halt spread of the disease. Reductions in infections have also been seen in Senegal, Kenya, and Zimbabwe. Concerted government action in Thailand has helped decrease new HIV infections by promoting safe behaviors, including condom usage. Cambodia has also had some success in curbing high-risk behavior.

The Pattern of World Mortality

The current world pattern of mortality is shown in Figure 5–11. Any interpretation of this pattern must be approached cautiously because the crude death rate shown on the map is greatly affected by the age structure of a population. For example, the crude death rate in the United States is higher than that of Mexico; yet few people would expect mortality conditions to be better in Mexico. Two other measures, life expectancy at birth and the infant mortality rate, are better indicators of actual mortality conditions. For Mexico the infant mortality rate is more than four times higher than that of the United States; and the life expectancy at birth in Mexico is less than that of the United States. Maps of both world infant mortality rates and world life expectancy at birth have been discussed already. The main reason for now considering a map of world crude death rates is that it provides us with a look at one of the two rates that are used to calculate the rate of natural increase. Areas of low crude death rates may, if they also have high crude birth rates, be areas of rapid population growth. Areas with high crude death and birth rates are potential areas for fast growth in the future, since typically crude death rates have fallen earlier and faster than crude birth rates.

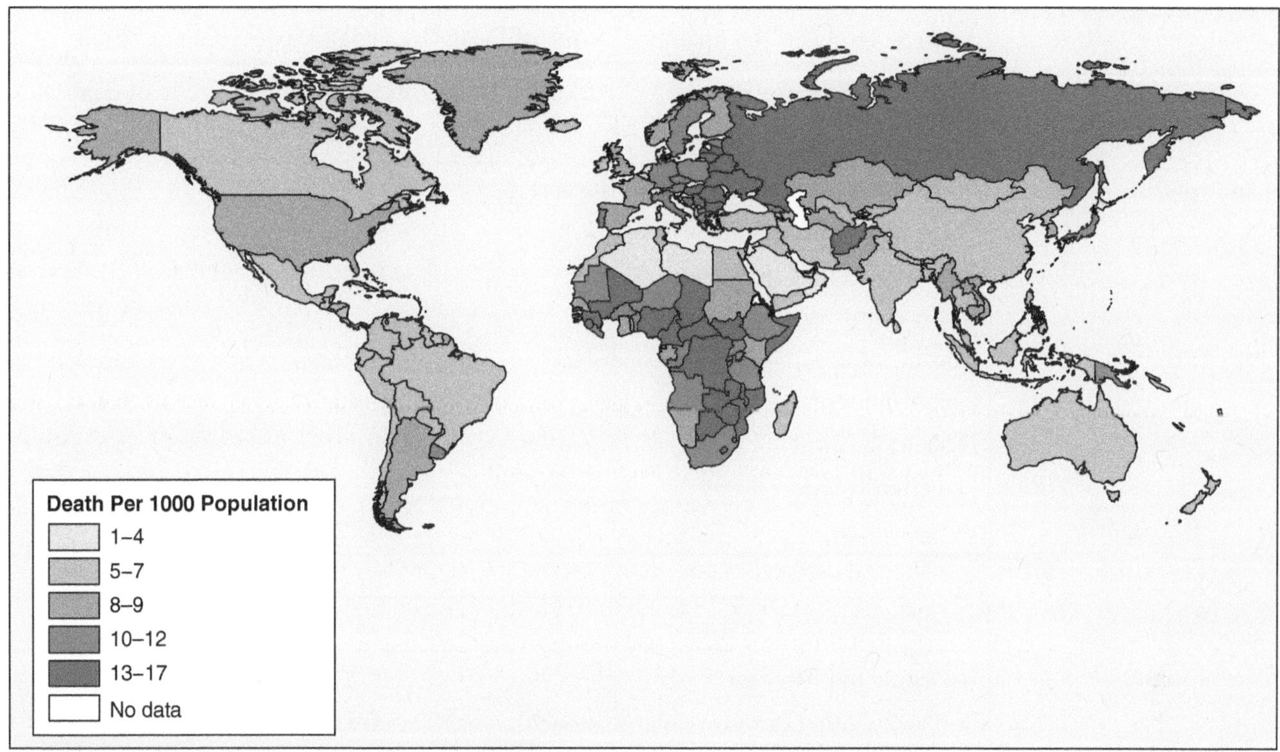

Figure 5–11 World Crude Death Rates, 2012
Source: 2012 World Population Data Sheet (Washington D.C.: The Population Reference Bureau, Inc.).

In 2012 the crude death rate for the world was 8 deaths per 1000. Death rates of 8 or less are found in many more developed countries, from the United States to Australia, as well as in some of the less developed countries of Latin America. Most of Western Europe and Asia, along with parts of North Africa, have death rates between 10 and 20. The highest death rates, those above 20, are found in Africa, as well as in Afghanistan. Barring major catastrophes, such as widespread famines, epidemics, and nuclear holocaust, it is likely that the epidemiologic transition will continue to reduce world mortality rates. However, as the AIDS pandemic has shown, there are no guarantees about the future course of mortality.

As Table 5–16 makes clear, in keeping with our earlier discussion of the epidemiological transition, causes of death differ for countries at different income levels. Within the heavily populated tropical areas of the world there are many diseases that currently are major killers even though we have ways to control them. According to Sachs (2007, 33A) these are NTDs or neglected tropical diseases, which he describes collectively as "... hellish infections whose combined impact on disease, disability and death rivals the impacts of AIDS, tuberculosis and malaria, yet they are far less known, partly because they are diseases that afflict only the poor in the tropics." He has encouraged the rich world to invest more in the health of the poor, but whether this happens on a large scale or not depends on many variables. One of the NTDs, guinea worm, is now on the decline and may be eradicated if current trends prevail. Sapolsky (2005) presented clear evidence of how much health and poverty are interrelated, finding that stress and other factors make the poor more vulnerable to diseases and less likely to live as long as those who are better off.

Though it is not one of the NTDs, malaria remains prevalent in the tropics, even though it was once much more widespread in higher latitudes, even in Scandinavia. According to Dunavan (2005, 78), "The events surrounding malaria's exit from temperate zones and, more recently, from large swaths of Asia and South America reveal as much about its perennial ties to poverty as about its biology." Malaria still claims more than

Table 5–16 The Ten Leading Causes of Death by Broad Income Group, 2008

High-income countries	Deaths in millions	% of deaths
Ischaemic heart disease	1.42	15.6
Stroke and other cerebrovascular diseases	0.79	8.7
Trachea, bronchus, lung cancers	0.54	5.9
Alzheimer and other dementias	0.37	4.1
Lower respiratory infections	0.35	3.8
Chronic obstructive pulmonary disease	0.32	3.5
Colon and rectum cancers	0.30	3.3
Diabetes mellitus	0.24	2.6
Hypertensive heart disease	0.21	2.3
Breast cancer	0.17	1.9

Middle-income countries	Deaths in millions	% of deaths
Ischaemic heart disease	5.27	13.7
Stroke and other cardiovascular disease	4.91	12.8
Chronic obstructive pulmonary disease	2.79	7.2
Lower respiratory infections	2.07	5.4
Diarrhoeal diseases	1.68	4.4
HIV/AIDS	1.03	2.7
Road traffic accidents	0.94	2.4
Trachea, bronchus, lung cancers	0.57	2.7
Diabetes mellitus	0.87	2.3
Hypertensive heart disease	0.83	2.2

Low-income countries	Deaths in millions	% of deaths
Lower respiratory infections	1.05	11.3
Diarrhoeal diseases	0.76	8.2
HIV/AIDS	0.72	7.8
Ischaemic heart disease	0.57	6.1
Malaria	0.48	5.2
Stroke and other cardiovascular disease	0.45	4.9
Tuberculosis	0.40	4.3
Prematurity and low birth weight	0.30	3.2
Birth asphyxia and birth trauma	0.27	2.9
Neonatal infections	0.24	2.6

Source: World Health Organization, 2008.

600,000 lives a year, though we have the tools to fight the disease, and thanks to donations by the United Nations, the United States, and the Bill and Melinda Gates Foundation, among others, the outlook for defeating malaria is looking better now than it ever has.

No one wishes to see death rates increase anywhere, but clearly the planet is becoming more strained in its ability to deal with population growth, most of which occurs in the less developed countries, those least able to deal with rapidly increasing numbers. Given that some balance between birth rates and death rates must be achieved in order to curtail world population growth, we continue to believe that lowering fertility rates is much preferable to raising death rates.

Climate and Health

Even a casual look at the geographic distribution of death rates (Figure 5–11) suggests that the pattern probably has some connections to global climates. Some of those connections have already been noted, especially those involving tropical diseases. Further connections can be made by noting that tropical nations tend to have lower incomes than temperate nations. Though the general distribution of mortality and morbidity is certainly not determined by climate, the pattern reflects some climatic influences. Furthermore, those influences may change as the world's climate changes.

Humans have adapted to a wide range of climates, but those climates offer a variety of different risks for health, from malnutrition to infectious diseases. For example, climates that offer little in the way of agricultural productivity restrict diets and create problems for humans because they struggle to get sufficient nutrients from limited diets. This is especially true in the polar areas, where short growing seasons inhibit most plant growth and people depend almost entirely on high-protein diets from fish and animals. Another example is the tropics, where plant growth may be rapid but the climate also favors mosquitoes and other disease-transmitting vectors. Malaria is a major killer in such environments, and diets are often deficient in proteins.

Weather affects human health as well. Rapid changes in temperatures and atmospheric pressure can have health effects, sometimes dangerous ones. The very young and the elderly are especially sensitive to extremes of heat and cold. Everything from heat stroke and hypothermia to heart and respiratory problems can be created by such extremes. Unusually high temperatures in Europe during the summer of 2003, for example, were associated with close to 30,000 more deaths that summer than in previous ones. In urban areas stagnant air masses can trap pollutants, which can then reach levels that affect our health as well. Smog episodes have long been correlated with health concerns. Heavy rains and floods can wreak havoc with our health, as India learned with a cyclone in Orissa in 1999 and as Americans learned when Katrina destroyed much of New Orleans in 2005.

As we will discuss in more detail in Chapter 10, Earth's climate right now is warming, so it is worth considering here some of the possible impacts of that warming on human health. Figure 5–12 provides a look at various ways that warming could impact us. Everything from supplies of drinking water to sufficient food production could be affected, though those and other impacts will vary geographically. Even as more rain may fall in some areas, it seems probable that considerable areas will have less rain, and that rainfall will become less predictable as well. Already, high latitudes are experiencing substantial changes in temperatures, and areas such as the Southwestern United States and Northern Africa seem to be drying out.

Among the many effects that global warming might have, the following seem likely:

- Heatwaves such as the one that affected Europe during the summer of 2003 may become more frequent.
- Water supplies are likely to become scarcer in places that are already marginal with respect to rainfall, especially if local populations continue to grow.

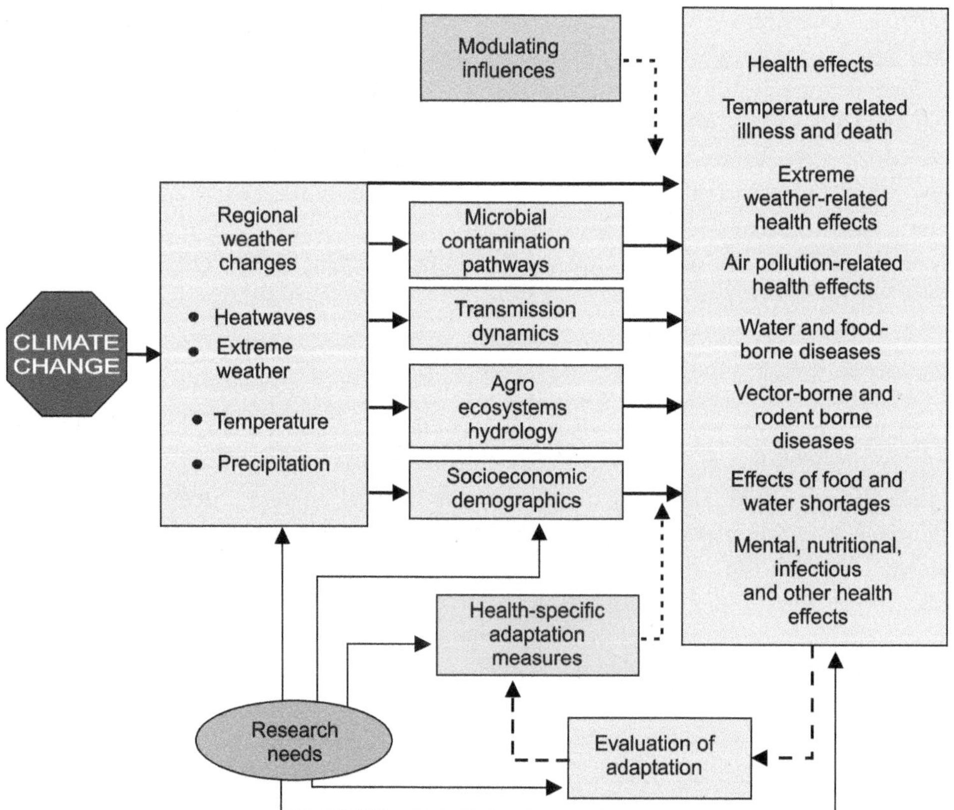

Figure 5–12 Health Effects Related to Global Climate Change
Source: World Health Organization.

- The seasons for many infectious diseases are likely to get longer.
- Coastal flooding may become more prevalent as sea levels rise. This could be especially problematic in areas that are low-lying and densely populated, including many delta regions such as the Nile and the Ganges-Brahmaputra. Low-lying islands will be threatened as well, and there are hundreds of them in the Indian and Pacific Oceans.
- Agricultural productivity may be affected positively in some places and negatively in others, and the latter may exacerbate malnutrition in large parts of the poor world.
- Tropical storms may increase in either frequency, intensity, or both, and this will be happening at the same time that increasing populations are living closer to oceans. Of the world's ten most crowded cities, eight of them are located on coasts: Tokyo, Mumbai, New York, Shanghai, Lagos, Los Angeles, Calcutta, and Buenos Aires.
- In Africa, where millions already suffer from a lack of food and essential nutrients, it is likely that wet areas bordering the rainforests will get wetter whereas sub-humid areas adjacent to deserts will get drier. Vegetative cover, animal habitats, and biodiversity will all be harmed, and crop yields in many areas are predicted to decline.

Weather, climate, and global warming all will affect human health far into the future. We can improve our ability to deal with each of these, but the most difficult problems may arise as we try to focus more attention on the possible health effects of global warming.

CHAPTER 6

Fertility: Patterns and Trends

Key Terms

fecundity
fertility
child-woman ratio
general fertility rate
total fertility rate
gross reproduction rate
net reproduction rate
direct maintenance costs
opportunity costs

The causes and consequences of fertility patterns are of major concern to governments everywhere and other interested groups. At one time high reproductive rates were characteristic of most societies, mainly as a response to prevailing high death rates—survival was at stake. However, death rates are either already low or falling in most regions of the world today, whereas fertility patterns differ considerably from place to place. Explanations of spatial variations in fertility are drawn from many sources and differ from place to place as well as from time to time. As Robertson (1991, 24) noted, ". . . the processes and values of reproduction are inextricably tied up with the processes and values of economic production and political control." Leete (1999) provides more on values and fertility change, as does Bryant (2007).

Measures of Fertility

Before proceeding, it is useful to differentiate between two terms that are sometimes confused, fertility and fecundity. **Fecundity** refers to the biological capacity for reproduction or the ability to reproduce, whereas **fertility** refers to actual reproductive behavior. One measure of fertility has already been introduced, the crude birth rate; others deserve consideration.

Fecundity refers to the biological capacity for reproduction.

Fertility refers to actual reproductive behavior.

Child-Woman Ratio

The **child-woman ratio** is an indirect measure of fertility that is used to estimate fertility in situations where birth records are deficient or nonexistent, mainly in the less developed countries.

The **child-woman ratio** is an indirect measure of fertility that is used to estimate fertility in situations where birth records are deficient or nonexistent, mainly in the less developed countries. It is "indirect" because it does not employ births in the measure at all. It is calculated as

$$CWR = \frac{P_{0-4}}{F_{15-44}} \times 1000,$$

where CWR = child-woman ratio,
P_{0-4} = the total number of children under five years of age, and
F_{15-44} = the total number of females between 15 and 44 years of age (This age group is considered to be the child-bearing age group. Sometimes the 15–49 age group is used instead.)

The child-woman ratio indicates the number of children under five years of age per 1,000 females that are in their child-bearing years. One disadvantage of using this ratio as a measure of fertility is that the number of surviving children is affected by prevailing mortality rates. Seldom, however, will you encounter this particular measure today.

General Fertility Rate

The **general fertility rate** is a more refined measure of fertility than either the crude birth rate or the child-woman ratio.

The **general fertility rate** is a more refined measure of fertility. As with the crude birth rate, actual numbers of births are used in the numerator. In the denominator, however, the total population is replaced by the total number of females in the child-bearing age group, usually 15–44 years of age. The general fertility rate is calculated as

$$GFR = \frac{B}{F_{15-44}} \times 1000,$$

where B = total number of births in a one-year period, and
F_{15-44} = total number of females between the ages of 15 and 45 at midyear

This measure gives us the number of live births per 1,000 women in the child-bearing age group.

Age-Specific Birth Rate

Age-specific birth rates are useful because child-bearing varies considerably with age. The age-specific birth rate is analogous to the general fertility rate, but instead of having the total number of females in the child-bearing age group as the denominator, it has the total number of women in a smaller age group, such as a one-year or five-year age group. The numerator, then, is the total number of children born in any given year to mothers in that specified age group. Normally, five-year age groups are used. There can be either six or seven age groups, depending upon the upper age limit that is used. In generalized form the calculation of age-specific birth rates is

The age-specific birth rate is analogous to the general fertility rate, but instead of having the total number of females in the child-bearing age group as the denominator, it has the total number of women in a smaller age group.

$$ABSR = \frac{B_a}{F_a} \times 1000,$$

where B_a = the number of births to females in the age group designated by "a" (e.g., a = a five-year age group 15–19 year olds), and
F_a = the total number of females in the age group "a" at midyear

If the child-bearing ages of 15–44 are used in five-year age groups, then there are six age-specific birth rates, one for each of the age groups 15–19, 20–24, . . . , 40–44. Often a seventh is added for the age group 45–49. For many countries sufficient data for calculating the age-specific birth rates are not available. Births occurring to women who are outside of the normal child-bearing age range are arbitrarily assigned to either the 15–19 year age group, if they occurred to females below age 15, or to the 45–49 year age group if they occurred to females above fifty years of age. For most populations there are relatively few births outside of the 15–49 age range.

Total Fertility Rate

The **total fertility rate** is the most useful way to summarize the age-specific birth rates for a population. It is the average number of children a woman would have *if* she were to have children at the prevailing age specific rates as she passed through her reproductive years. The total fertility rate is calculated as

Total fertility rate is the average number of children a woman would have if she were to have children at the prevailing age specific rates as she passed through her reproductive years.

$$TFR = 5\sum_{7}^{a=1} \frac{B_a}{F_a}$$

where B_a = the number of births to females in age group "a" in a one-year period,
F_a = the midyear number of females in age group "a," and
a = five-year age groups as follows: 15–19, 20–24, 25–29, 30–34, 35–39, 40–44 and 45–49, for a total of 7 different age groups

The five preceding the summation sign is there because we are using five-year age groups. If one-year age groups were used, the five would disappear from the equation.

In the United States in 2012 the total fertility rate was 2.1, near "replacement" level, but above the 1.8 that it had been only a few years earlier. In 2012 the world's total fertility rate was 2.5, continuing a gradual decline. Figure 6–1 provides some idea of the range of values that countries show for total fertility rates. Countries that maintain a total fertility rate of less than 2.1 for a sustained period of time will experience negative population growth, as we saw earlier in our discussion of different population projections for the United States. Both the causes and consequences of negative population growth

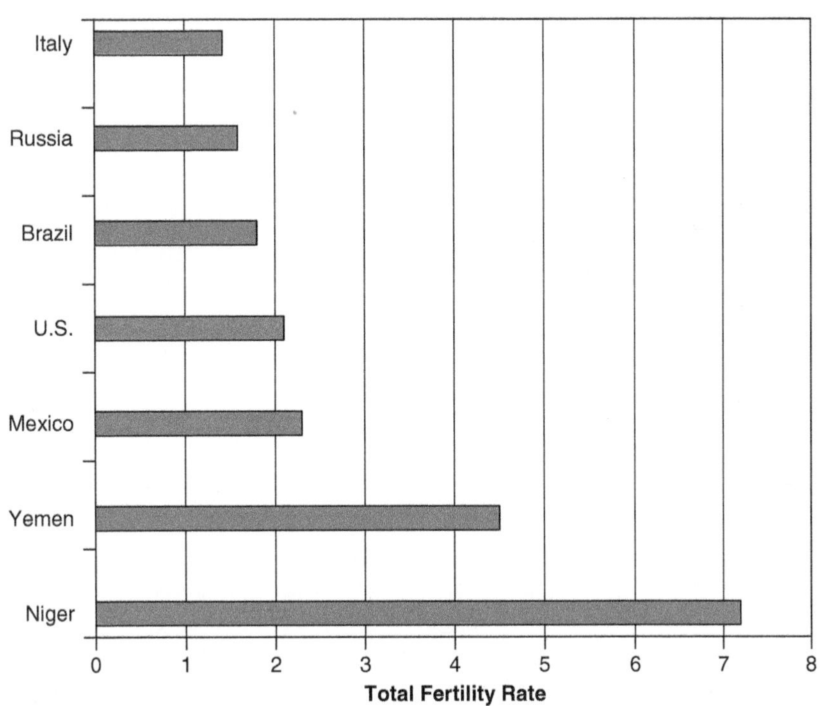

Figure 6–1 Total Fertility Rate: Selected Countries, 2012
Source: Data from the CIA World Factbook.

have received considerable attention (examples include Teitelbaum and Winter, 1985 and Davis, Bernstam, and Ricardo-Campbell, 1987). Bongaarts and Feeney (1998) suggested a way to modify the total fertility rate to include changes in the timing of childbearing. Longman (2004) is concerned that low birth rates threaten world prosperity; Morgan (2003) questions whether low fertility may actually be a crisis.

Gross Reproduction Rate

The **gross reproduction rate** is the same as the total fertility rate except that only female births are counted.

The **gross reproduction rate** is the same as the total fertility rate except that only female births are counted. Thus, it gives the average number of daughters that a woman would have if she passed through her entire reproductive life at the prevailing age-specific birth rates. The gross reproduction rate can be found by multiplying the total fertility rate by the proportion of births that is female. The gross reproduction rate is calculated as

$$GRR = 5\sum_{a=1}^{7} \frac{FB_a}{F_a}$$

where FB_a = the number of female births to females in age group "a" in a one-year period,
F_a = the midyear number of females in age group "a," and
a = five-year age groups.

The **net reproduction rate** is the same as the gross reproduction rate except that it is reduced somewhat to allow for the fact that not all women will live through their entire reproductive period.

Net Reproduction Rate

The **net reproduction rate** is the same as the gross reproduction rate except that it is reduced somewhat to allow for the fact that not all women will live through their entire reproductive period. A net reproduction rate of 1.0 would indicate that on the average females are exactly replacing themselves.

Major Determinants of Fertility

The reproductive behavior of a population results from a complex set of biological, social, and economic factors that operate differently from place to place as well as from time to time. In order to explain spatial and temporal variations in fertility, it is necessary to consider its determinants.

Biological Determinants of Fertility

As noted earlier, it is necessary to distinguish between fertility, actual reproduction performance, and fecundity, the biological capacity for reproduction. However, they are obviously related. Fecundity may be affected by a number of physical factors including age, health, nutritional status, and even the physical environment.

Age and Fecundity

In most societies reproduction is accomplished mainly by young adults. In calculating fertility rates, the female reproductive years are generally assumed to be ages 15–44 or 15–49. Puberty marks the onset of reproductive capacity. For females, menarche, or the beginning of menstruation, denotes puberty. The change from infecund to fecund is not a sudden change from one state to another, but rather a gradual increase through a period of adolescent subfecundity to a mature fecundity, reached at roughly ages 26–30 (Doring, 1969). The female's fecund period ends with menopause, when menstruation ceases. For males puberty also marks the beginning of reproductive capacity; however, the end of the male reproductive period is not so clearly demarcated.

Modern science is helping older women extend their childbearing years. For example, multiple births have increased significantly in the past two decades, mainly among women in their forties. These women find it difficult to become pregnant, so they turn to fertility drugs for help, increasing the likelihood of multiple births. Costs are high, the industry is poorly regulated, and outcomes are tentative at best because failure rates are high. Nonetheless, in an affluent nation some infertile couples are willing and able to spend whatever it takes to have a child. Researchers are working to reduce multiple births.

Health, Nutritional Status, and Fecundity

A person's health may affect fecundity for varying periods of time. In general, good health and fecundity go together. A variety of diseases may impair fecundity either temporarily or permanently. The most important diseases in this respect are venereal diseases, which may, if left untreated, cause permanent sterility. Nutritional status affects fecundity as well, especially in cases of severe malnutrition, which may even lead to temporary infecundity. In turn, a woman's health may be adversely affected by reproduction, especially if she was already experiencing health or nutritional problems. As the National Research Council (1989, 15) noted, "By avoiding pregnancy, women with health problems may substantially improve their own chances for survival and good health."

"By avoiding pregnancy, women with health problems may substantially improve their own chances for survival and good health."

Good health and fecundity go together.

Environment and Fecundity

The literature on environment and fertility suggests that fecundity is affected by environmental factors. Research has focused on sperm quality and how environmental factors may affect it (Swan, 2006; Jorgenson et al., 2006). Exposure to pesticides and other pollutants has been connected to poor sperm quality and these effects have also been shown to vary geographically (Swan et al., 2003). Research also suggests that exposure to polychlorinated biphenyls (PCBs) and dichlorodiphenyl trichlorethane (DDT) causes a longer time to pregnancy by their effects on both males and females (Axmon et al., 2006). One rather dated study suggests that fecundity is related to altitude, though evidence is inconclusive (James, 1966).

Social Determinants of Fertility

Given a normally fecund population, fertility will be determined mainly by variations in social variables, including the following: marriage patterns, contraceptive practices, and abortions. Demographers Kingsley Davis and Judith Blake (1956) provided a framework that identified "intermediate variables" through which various social factors operate to affect fertility. They identified three sets of variables under the following headings:

1. factors that affect exposure to intercourse
2. factors that affect exposure to conception
3. factors that affect gestation and successful parturition

Fertility variations among nations, then, can be explained by understanding how social differences are being translated into fertility behavior through these "intermediate variables." Major sources of such variations include those discussed subsequently: the following three subheads correspond closely with the three suggested originally by Blake and Davis. John Bongaarts (1986) noted that the four most important factors in fertility declines have been the use of effective contraceptives, the age of marriage, the length of time after child-birth that a woman cannot conceive (because of either sustained breast-feeding or abstinence), and abortion (Robey, Rutstein, and Morris, 1993).

Marriage and Fertility

In most societies today, marriage is the socially recognized union within which procreation occurs. However, marriage is obviously not a necessary prerequisite for reproduction, thus out-of-wedlock births and variations in cultural norms from place to place must also be considered when possible. The formation of any heterosexual union, however temporary or permanent, can potentially result in reproduction, though data for unions other than marriage are often difficult to obtain. As Robertson (1991, 6) commented, "The way we reproduce seems to make the family a logical necessity, but the enormous variation in family patterns around the world is a reminder that these elementary relations of reproduction may be organized in many different ways." Nonetheless, whatever the nature of heterosexual unions may be, entering into them is a prerequisite for reproduction for most females. There are now three million people alive who owe their lives not to sex but to *in vitro* fertilization.

In non-contraceptive populations, especially, the age at which females marry or enter into heterosexual unions is an important determinant of fertility. Assuming that ages 15–44 are the child-bearing years for most women, it is logical that those who marry later will have fewer children on the average, providing that there are few illegitimate births. Typically, in the less developed countries today, the average age at marriage is fairly low, especially in comparison with the pattern in most of the more developed countries. In some nations, such as Panama, Honduras, and Bolivia, the legal marriage age is as low as twelve for females and fourteen for males. Where women marry young and practice little or no contraception, fertility is likely to be high unless other factors operate to lower it. In preindustrial European countries the average age at marriage was also relatively low, though there was a rise in the age at marriage beginning as early as the eighteenth century (Hajnal, 1965).

In the United States the median age at first marriage did not change considerably between 1900 and 1990, however, since 1990, it has risen markedly as is illustrated in Figure 6–2. This rising age at first marriage has also been associated with an increase in

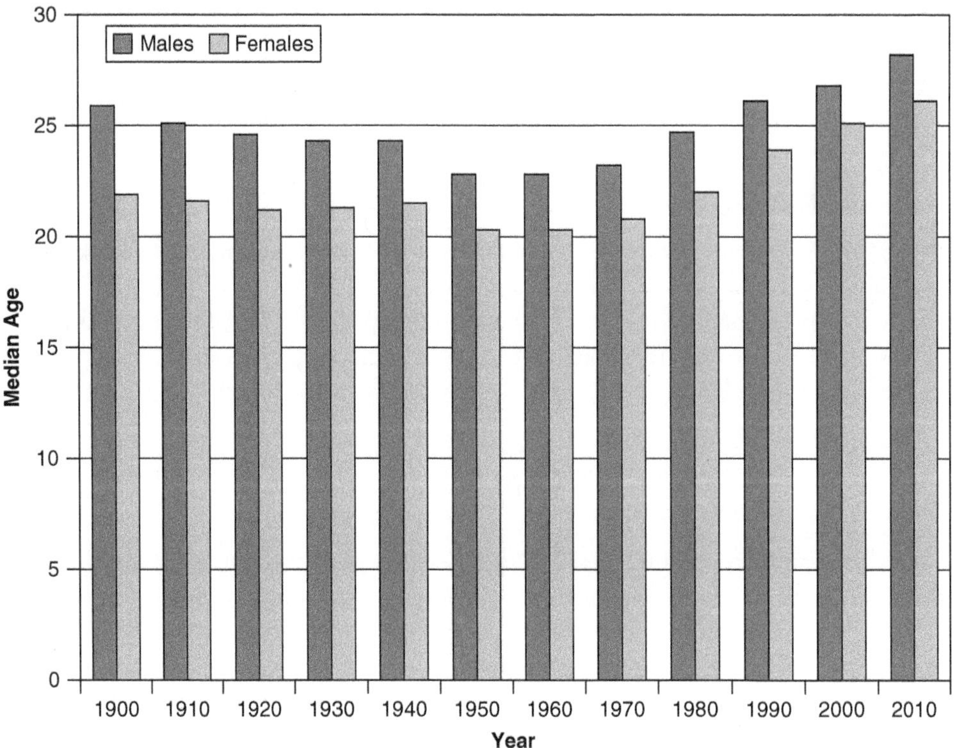

Figure 6–2 Median Age at First Marriage by Sex in the United States
Source: U.S. Census Bureau.

the percentage of those never married by given ages, as is shown in Figure 6–3. Mulder, Clark, and Wagner (2006) compare first union formation in the United States with the Netherlands and West Germany.

One feature that has changed considerably in the United States has been the percentage of children born to unwed mothers, a figure that has increased considerably in many western European nations as well. In 2010, about 41 percent of all births in the United States were to unwed mothers; many of these women are truly single, others are cohabiting but unmarried. About half of all marriages in the United States end in divorce, though that doesn't mean the end of sexual unions, obviously. Births to teenagers in the United States have dropped steadily in recent years. The rate of teen births fell to its lowest rate ever recorded in 2009, to 39.1 births per 1000 teens. Gray, Stockard, and Stone (2006) argued that the rising proportion of births to unwed mothers in the United States was in large part a matter of marriage behavior rather than fertility choice. In turn their study could have consequences for new policy choices with respect to fertility and family formation. Finally, as more couples have chosen cohabitation over marriage, researchers have found that such relationships are more violent than marriages (Brownridge, 2004). Kenney and McLanahan (2006) have recently questioned the cause and effect explanation of those observed results, arguing that a great deal of selectivity about who chooses which kind of relationship may be responsible rather than just the type of relationship involved.

Another change in America over the last few decades has been the rise in families with two working parents and lower levels of supervision for children, leading them to make choices that are not always best for their own welfare, as depicted, for example, in the film "Thirteen." In 1960, 70 percent of families in the United States had at least one parent who remained at home, and the "Ozzie and Harriet" family was common. By 2000, on

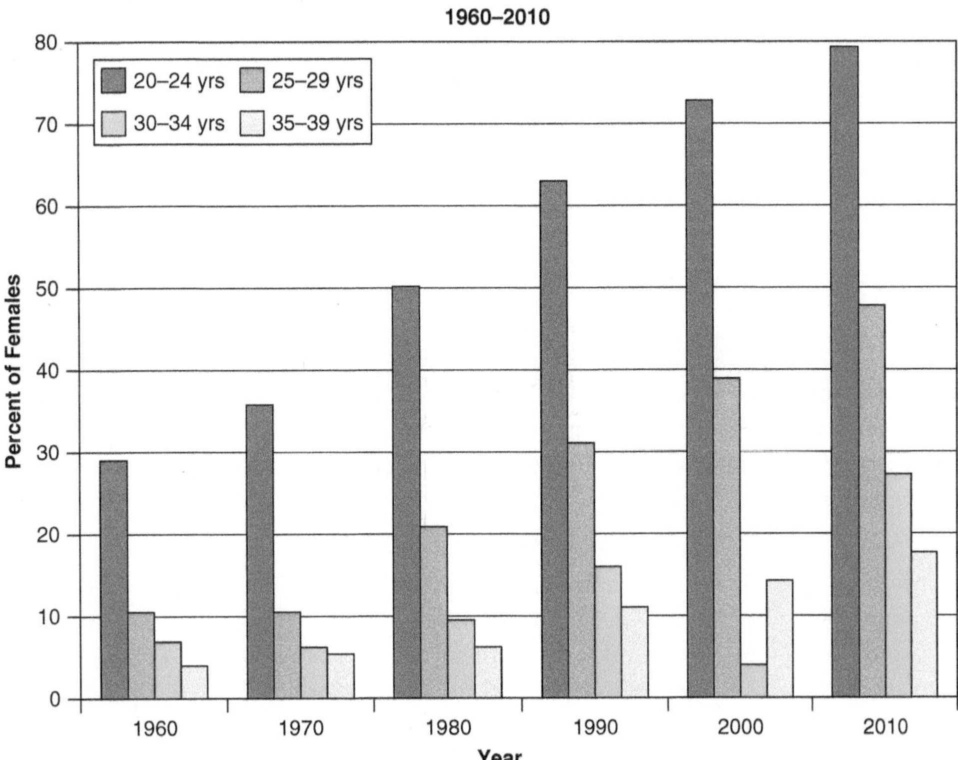

Figure 6–3 Percent of Females Never Married in the United States

Source: Data from United States Bureau of the Census, Marital Status and Living Arrangements: March 1990, Current Population Reports, Population Characteristics, Series P-20, No. 450; American Community Survey, 2000; 2010.

the other hand, 70 percent of families were headed either by single parents or by two parents who both worked outside the home. That proportion had risen to 73.5 percent by 2010. This has led to a decline in the average amount of time that parents spend with their children and a growing demand for changes, including longer school days, more (and cheaper) preschools, after school programs, and a host of other services. Problems arise when the talk turns to financing such changes, of course.

One of the steps that the People's Republic of China has taken to decrease fertility has been to encourage people to postpone marriage, ideally to around age 25 for females and 28 for males. Coupled with strong sanctions against premarital sex, there is little doubt that the higher age at marriage has been a factor in China's recent fertility decline. However, it is far from being the only factor. According to Zhang and Zhao (2006, 293) "China's fertility level has become a matter of considerable debate since the early 1990s." The 2000 census added more uncertainty because it found the total fertility rate was only 1.22, below what most other researchers felt it was at the time. The lowest estimates among other experts were 1.58, and they ranged up to about 1.87. Data problems and a huge population make such calculations difficult.

Another determinant of the fertility of a population is the percent of the population that marries. A high rate of celibacy is unusual in most populations, though there are a few exceptions, such as Ireland.

Related to marriage patterns, and also influencing fertility to varying degrees in different societies, is the amount of time spent between marriages after a divorce, separation, or death of a spouse. In general, the rate of dissolution of marriages is lower in preindustrial societies than in advanced ones, with such exceptions as the Islamic peoples. Societies have varying attitudes toward the remarriage of widows.

Given exposure to sexual intercourse within marriages or other types of heterosexual unions, fertility is influenced primarily by sexual behavior and contraceptive practice. Both the frequency of intercourse and abstinence for varying periods of time may affect fertility. Periodic abstinence may result from sickness or other reasons. Furthermore, some societies forbid intercourse during certain times, for example, while a woman is breast-feeding a child (Saxton and Serwada, 1969). Such practices could increase the spacing between children and reduce fertility. Geographic studies of variations in sexual behavior have bourgeoned in recent decades. Much of this work is related to the transmission of HIV, however, geographers have also studied other aspects of sexual behavior as well (Ford and Bowie, 1989; Uthman, 2008; Gavin et al., 2009; Messina et al., 2010; Hirsch et al., 2007).

Contraception and Fertility

A major determinant of fertility today is the degree to which contraception is practiced. Contraceptive practice is most common in the industrialized nations; however, with the increasing influence of family planning programs in many countries, more and more people are beginning to control their fertility, as noted some time ago by Nortman (1977, 3):

> Among the most far-reaching and visible developments going on in the world today is the change in contraceptive knowledge, attitudes, and practices—popularly known as the "KAP" of contraception. The past decade has seen a rapid acceleration in the historical trend to ever-increasing adoption and use of contraception. Contraception is now being practiced where it never was before, and people everywhere are turning to new and much more efficient methods.

The United Nations Department of Economic and Social Affairs (2011) reported that worldwide, in 2009, approximately 63 percent of married couples practiced some form of contraception. Contraceptive prevalence, however, varies by level of development. In more developed regions, the percentage of married couples using contraceptives in 2009, was 72.4 percent. In less developed regions, that percentage was 61.2 and in the least developed countries, it was 31.4.

Contraceptives may be classified into the following groups:

1. those that prevent the entry of sperm
2. those that avoid or suppress ovulation
3. those that prevent implantation

Included in the first group are coitus interruptus (withdrawal), condoms, spermicides, douches, and diaphragms, along with sterilization; the second group includes both "rhythm" methods and oral contraceptives; and the third category is the intrauterine device, or IUD. Abortions are in a separate category because they are used after conception has occurred, so they can serve as a backup to contraceptive failure.

Though contraceptive technology has lagged considerably since the innovations of the 1960s, there are some new contraceptive methods currently under investigation, including improved condoms (thinner and stronger), various injectables, and perhaps even a vaccine for men, along with improved implants, hormone-releasing vaginal rings, progestin-releasing IUDs, improved "morning after" pills, and perhaps even a vaccine for women.

One newer contraceptive device to have gained approval is Norplant, an implant for females that releases minute doses of progestin to inhibit ovulation for up to five years. Another is NuvaRing, a vaginal ring that is inserted for three weeks then removed for one. Other choices now include Lunelle, a monthly injection; Ortho Evra, a patch; and Mirena, a T-shaped, insertable device. All of these give females different systems for delivering the same hormones that are in most birth control pills.

The most controversial innovation has been RU 486, synthesized in 1980 and distributed in France during the 1980s (Ulmann, Teutsch, and Philibert, 1990). Given to females in tablet form along with a small dose of prostaglandin, RU 486 has the ability to terminate pregnancies, though it was not initially developed for that purpose. It is controversial not because of questions about its safety, but because it aborts rather than prevents pregnancies. Other "morning after" methods, such as Preven, use the same hormones that are used in oral contraceptives.

The usage of various contraceptives in the United States is shown in Figure 6–4. Among women between ages 15 and 44, the pill has become the most popular method for preventing pregnancies, followed by sterilization and condoms.

Abortion and Fertility

Sometimes considered a contraceptive, but actually a backup measure when pregnancy has already occurred, abortion has been, and continues to be, an important determinant of fertility. The role of abortion as a supplement to contraception is apparent in many places. Nortman (1977, 24) commented that:

> Studies in Japan, Latin America, and elsewhere have demonstrated that when societies are motivated to begin to control their fertility, the incidence of abortion (again, legal or illegal) tends to rise initially along with the use of contraception, but as contraception becomes more efficient, widespread, and available, abortion wanes, to be used only when contraception fails.

Sometimes considered a contraceptive, but actually a backup measure when pregnancy has already occurred, **abortion** has been, and continues to be, an important determinant of fertility.

In 2008, six million abortions were performed in more developed countries and 38 million in less developed countries (Guttmacher Institute, 2012). Figure 6–5 provides data on legal abortions in selected countries.

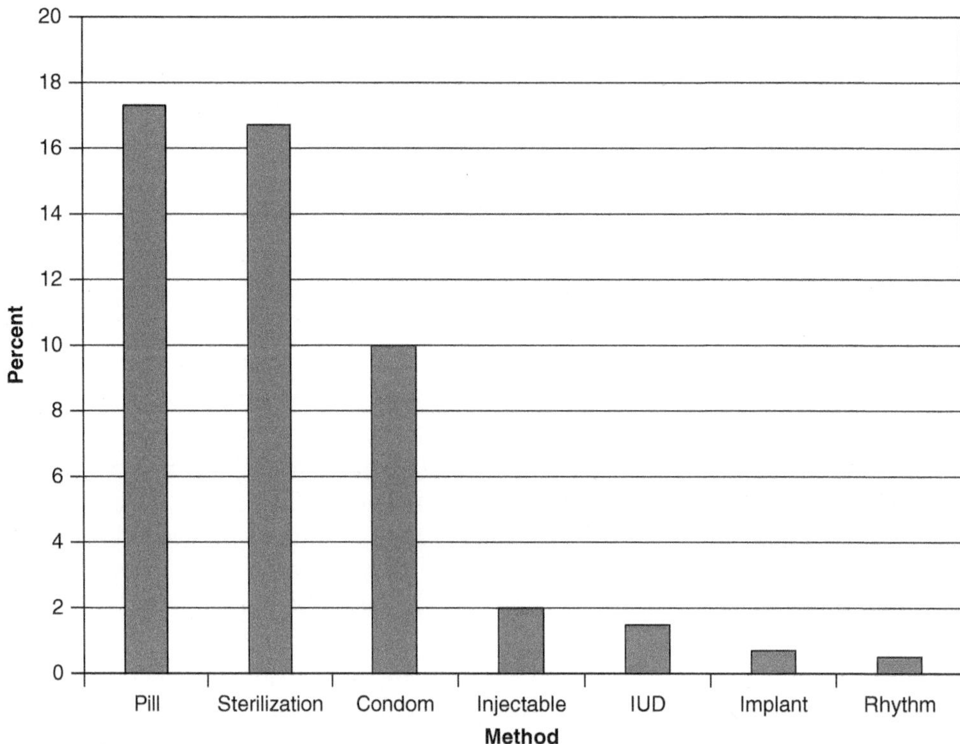

Figure 6–4 Contraceptive Usage in the United States, All Women Ages 15–44, 2006–2008

Source: Statistical Abstract of the United States, 2012. Table 98

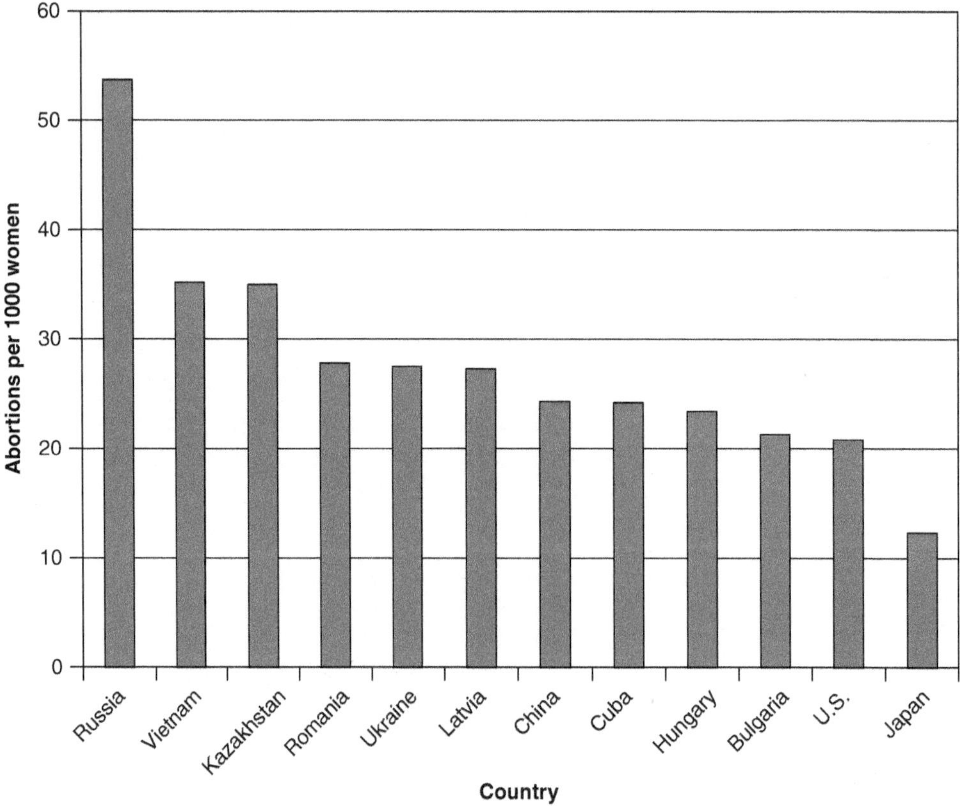

Figure 6–5 Number of Abortions per 1000 Women, Ages 15–44

Source: http://data.un.org/Data.aspx?d=GenderStat&f=inID%3A12

Note: All countries show data for 2004 with the exception of China (1998); Vietnam (2000); Bulgaria (2003); Hungary (2005); and Estonia (2005).

The impact of a sudden change in abortion laws is illustrated in some Eastern European countries. For example, abortion was legalized in Romania in 1956 and then severely restricted in 1967. Following the restriction, a rapid rise in fertility occurred. In some cases, easy access to abortions led to an increased dependence on abortion rather than contraception as a means of controlling fertility. In Poland, for example, Kulczycki (1995, 496) concluded that:

> The state's provision of easy access to safe abortion services in the late 1950s led to an excessive dependence on abortion. This situation was exacerbated by the state's failure to make modern contraceptives more readily available and more familiar, when it alone was able to do so.

In the United States, prior to 1970, it was relatively difficult for a woman to obtain a legal abortion. In the period 1972–1974 there were 2,229,070 legal abortions performed (Tietze, 1977). Despite two major Supreme Court decisions in 1973, Doe *vs.* Bolton and Roe *vs.* Wade, uniform abortion practices are not apparent in all states.

The spatial pattern of abortion rates in the United States may be seen in Figure 6–6, and it is clear that there remain substantial differences among the states with respect to abortion practices. Debates about legal abortions in the United States are on-going, however the original Supreme Court decision has remained in effect, though many have begun to argue that Roe *vs.* Wade is dead. Nonetheless, abortion remains an important political issue and anti-abortion sentiment still plays a role in the conservative agenda in the United States.

Geographer Patricia Gober studied abortion from the combined perspective of population geography, medical geography, and political geography, noting that previous studies had focused only on one of those segments at a time. Considering both historical factors and the current supply and demand for abortions by state, Gober (1994, 247) found that, "In the debate over the importance of supply versus demand, the outcome does not appear to be one of either/or. Both sets of variables give rise to state variations in abortion rates."

"The more acutely people are aware of economic constraints, the more likely they are to exert control over reproduction."

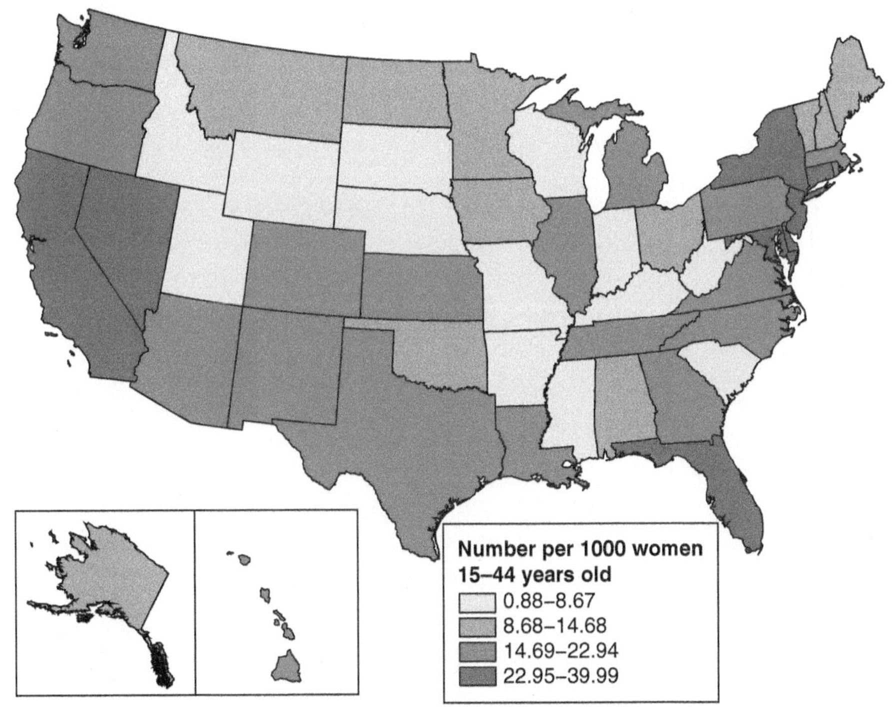

Figure 6–6 Number of Abortions per 1000 Women (aged 15–44) in the United States, 2008

Source: Statistical Abstracts of the United States, 2012.

Example

One unexpected side effect of abortions in some countries has been the selective abortion of female fetuses. In places such as China and India male children are still preferred. Ultrasound can be used to identify the sex of a fetus, making it easier for couples to decide whether to abort a pregnancy if it is a female. For that reason India outlawed sex-determination tests in 1994, but they are still readily available, and the number of girls per 1,000 boys has been dropping steadily since the 1980s. In Punjab in 2011 there were less than 895 girls for every 1,000 boys, well below what a normal sex ratio without selective abortion would have produced. Such distortions in sex ratios will have significant effects in the future—as females become scarcer, for example, marriage partners will be much more difficult to find. Already in China this has become a significant problem. In 2012, the sex ratio at birth in China was 113 males to 100 females.

Economic Determinants of Fertility

It has been suggested that fertility decline in the industrialized countries occurred as a response to economic development, with its concomitant changes in social mobility and population redistribution. Modernization remains the driving force behind the demographic transition model, and, as Robertson (1991, 70) noted, "The more acutely people are aware of economic constraints, the more likely they are to exert control over reproduction." As Figures 6–7 and 6–8 indicate, for the countries shown, there is a relationship between crude birth rates and both GNP per capita and the level of urbanization—variables that serve to some degree as measures of the level of development.

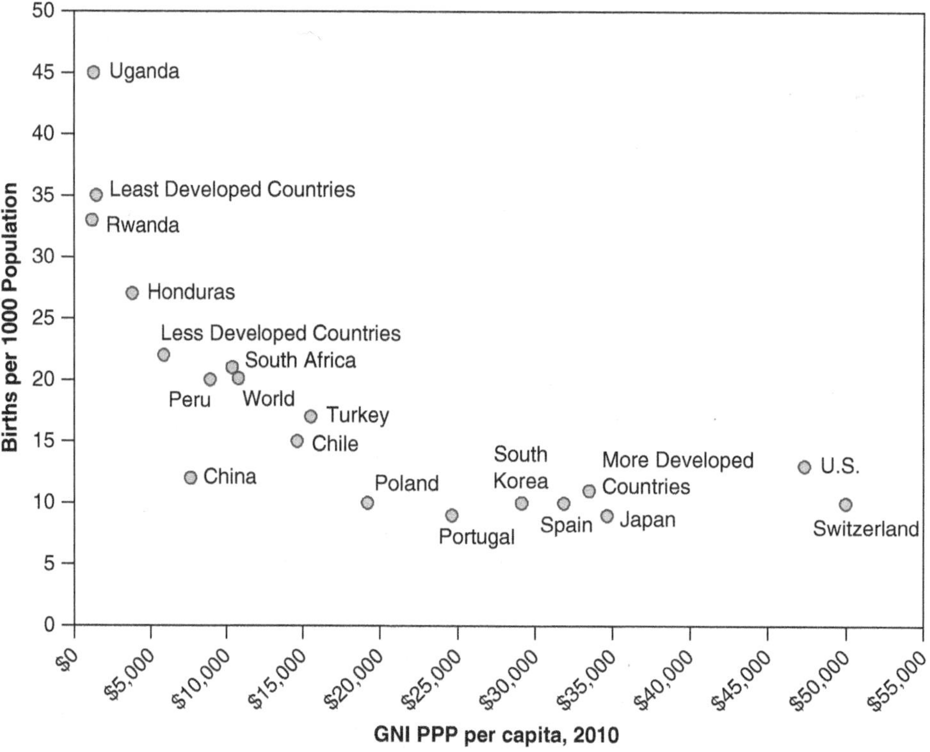

Figure 6–7 The Relationship between Per Capita Gross National Income Purchasing Power Parity (GNI PPP) and Crude Birth Rates

Source: Data from 2012 World Population Data Sheet. (Washington, D.C.: Population Reference Bureau, Inc.)

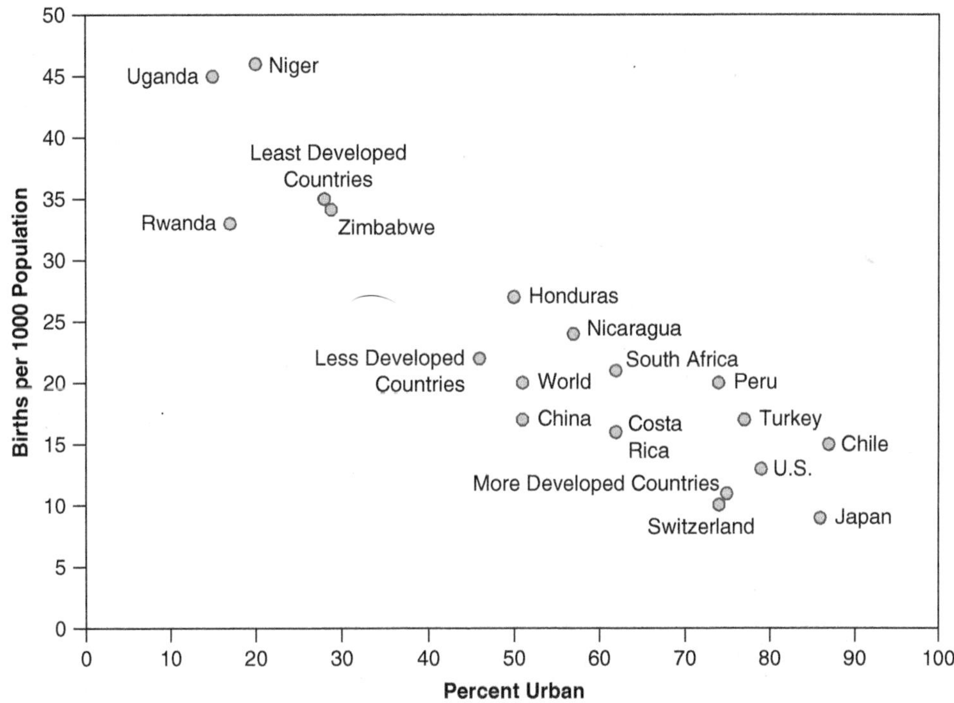

Figure 6–8 The Relationship between Level of Urbanization and Crude Birth Rates

Source: Data from 2012 World Population Data Sheet. (Washington, D.C.: Population Reference Bureau, Inc.)

One notable area of fertility research is concerned with developing an economic theory of fertility. Espenshade (1977, 3) pointed out that:

> Children are valuable. Most fundamentally, they provide for the continuation of the human species. If births were to cease, mankind would become extinct within the span of our lifetime. But this reason for wanting children is usually not uppermost in the minds of parents. To them, children are sources of joy and happiness, companionship, and pride. In some circumstances, children may also be prized because they are a potential means of support and security once parents are no longer able to provide for themselves. At the same time, children are costly. They put added pressure on family resources, and they can in other ways curtail the activities and opportunities of parents.

The essence of an economic theory of fertility is that a couple's decision to produce, or not to produce, a child is based on the costs and benefits of the child, as perceived by the couple. Such a theory, then, begins with a consideration of the costs and benefits of children. Costs and benefits have both economic and noneconomic components.

The Value of Children

Hoffman and Hoffman (1973, 20) suggested that the value of children may be thought of as "the functions they serve or the needs they fulfill for parents." A variety of terms for this concept have been used in the literature, including satisfaction, benefits, utilities, rewards, gains, gratifications, advantages, and positive general values. As Espenshade (1977, 4) commented, "The value of children is used to mean that collection of good things parents receive from having children." The role children play in caring for aging parents is important in many places (Boldrin and Jones, 2002).

Children may have both economic and noneconomic values. The noneconomic values are mainly psychological in nature and include the psychic satisfactions that parents derive from having and rearing children. Among these noneconomic values are the following: adult status and social identity; expansion of the self; primary group ties; stimulation, novelty, and fun; creativity and accomplishment; power and influence; and social comparison and competition.

Two types of economic benefits of children are the following:

1. children as a source of financial security in old age and in emergencies (Rendall and Bachieva, 1998)
2. the value of children as productive agents (Leibenstein, 1963)

Espenshade (1977, 5) stated that:

> There is a tendency to assume that the economic values of children are most salient in the less developed countries, especially in rural areas. In fact, it is widely believed that this is the major reason parents in such regions want large families. As a society modernizes and achieves higher levels of economic and social development, the economic value of children declines in importance. The extension of compulsory schooling and the enactment of child labor laws reduce the economic contribution from children. Similarly, to the extent that social security becomes institutionalized in such programs as public health and welfare measures, pension plans, and private annuity and life insurance programs, parents can relax their dependence on children as a source of old-age support.

More than three decades ago Mahmood Mamdani (1972, 113) made the importance of children in Third World countries clear in the following discussion of the failure of birth control in Manupur, a small village in India's Punjab:

> To Hakika Singh, the solution to his financial troubles is not to reduce the size of his family he has to support, but to *increase* it. It is the family that will support him, will even be his salvation. Admittably, he must take one chance: the next baby may be a girl instead of a boy. But since he is faced with utter financial disaster if he does anything else, Hakika Singh is willing to take that chance. Even with a number of sons, he may fail; yet they are his only hope. Even if the chances for success are low, his sons are his only route to success. As we concluded our conversation, his parting words were: "A rich man invests in his machines. We must invest in our children. It's that simple."

The Cost of Children

Direct maintenance costs are actual monetary outlays required for the support of children. These costs include food, clothing, housing, educational expenses, and medical expenses.

Opportunity costs measure opportunities that parents must sacrifice to have and raise children.

Costs may be considered as the disadvantages of children. Among the terms used for the costs of children are dissatisfactions, disadvantages, disvalues, penalties, and negative general values. These costs may be both economic and non-economic.

Among the non-economic costs are included the emotional and psychological problems that children impose on parents. Any parent is aware that raising children causes anxieties about such matters as the child's health and behavior. Among the economic costs are:

1. direct maintenance costs
2. opportunity costs

Direct maintenance costs are actual monetary outlays required for the support of children. These costs include food, clothing, housing, educational expenses, and medical expenses. **Opportunity costs** measure opportunities that parents must sacrifice to have and raise children.

According to Espenshade (1977) three types of opportunity costs may be recognized. First, a lower standard of living may result as certain consumption expenditures are foregone. Second, children may reduce a family's ability to save and invest. Third, and of increasing importance in the United States as well as elsewhere, the wife may sacrifice her career and earnings to raise children. Espenshade (1977, 6) noted that:

> The relative importance of the three kinds of economic opportunity cost is likely to vary according to the level of economic development. In the less developed countries, for example, there is a tendency to think that children scarcely affect consumption standards or the ability to save and invest. But whether or not this is actually the case depends on the level of aspirations among the population.

The costs and benefits of having children vary from time to time and place to place. In general terms, in preindustrial, primarily agrarian societies, the perceived costs of having and rearing children are low, whereas the perceived benefits of having and rearing children are high. Thus, in such societies couples will tend to have large families.

With economic development, industrialization, and urbanization underway in a society, the perceived costs and benefits of having and rearing children are altered. As people become urban dwellers, and more frequently pursue non-agricultural jobs, they begin to see that the costs of children are rising and the benefits of children diminishing. Families in urban areas may find housing, education, and medical costs all more expensive. At the same time they may find more goods available to them. Thus, they may find that children are not as attractive to them as they are to their rural counterparts.

In the more developed countries a large percentage of the population is urban, and most of the labor force is employed in the secondary, tertiary, and quaternary sectors of the economy. Females find more job opportunities outside the home in these more developed countries, and their participation in the labor force has increased greatly during the last several decades. The costs of having and rearing children are quite high. With the costs of children high and their perceived benefits low, the modern couple in an urban area of an industrialized country is likely to desire a small family. Figure 6–9 shows the direct costs of raising a child to age 18 in the United States. Add on top of that a substantial figure for sending a child to college and it is no wonder that most couples today in the United States are opting for small families, as they are in Europe and Japan (Foster, 2000).

Frejka and Sardon (2004) studied changes in cohort fertility among women in low-fertility countries during the last half of the last century. They were especially interested in changes in the ages of mothers over the time period, noting that there were considerable variations among the nations involved. They argue that low fertility is likely to continue among younger cohorts of women in those countries.

According to Peggy Orenstein (2001) young women in Japan are increasingly opting to remain single, live at home with their parents, and spend their entire income on themselves. These young women have become what sociologist Masahiro Yamada (1999) has termed "Parasite singles." Orenstein (2001, L1) noted that "More than half of Japanese women are still single by 30. . . . Although they earn, on average, just $27,000 a year, they are Japan's leading consumers, since their entire income is disposable." Though fertility is down, conception before marriage is rising (Kashiwase, 2002).

Though these young women help drive the Japanese economy today, their very low levels of reproduction are apt to hurt it tomorrow. Low fertility is leading to labor shortages, leaving the Japanese economy weaker than it might be otherwise, especially when Japan has little interest in promoting labor immigration. Parasite singles are thought to be both a remnant of the bubble economy that burst in the 1980s and a projection of the nation's pessimistic view of the Japanese economy.

More recent data shows that Japan has roughly 3 million people between the ages of 35 and 44 who continue to live with their parents, and 11.5% of those dubbed "parasite singles," are unemployed. The parasite singles are believed to be those who were in their 20's and 30's in the 1990s, lived with their parents, and never got married (Westlake, 2012).

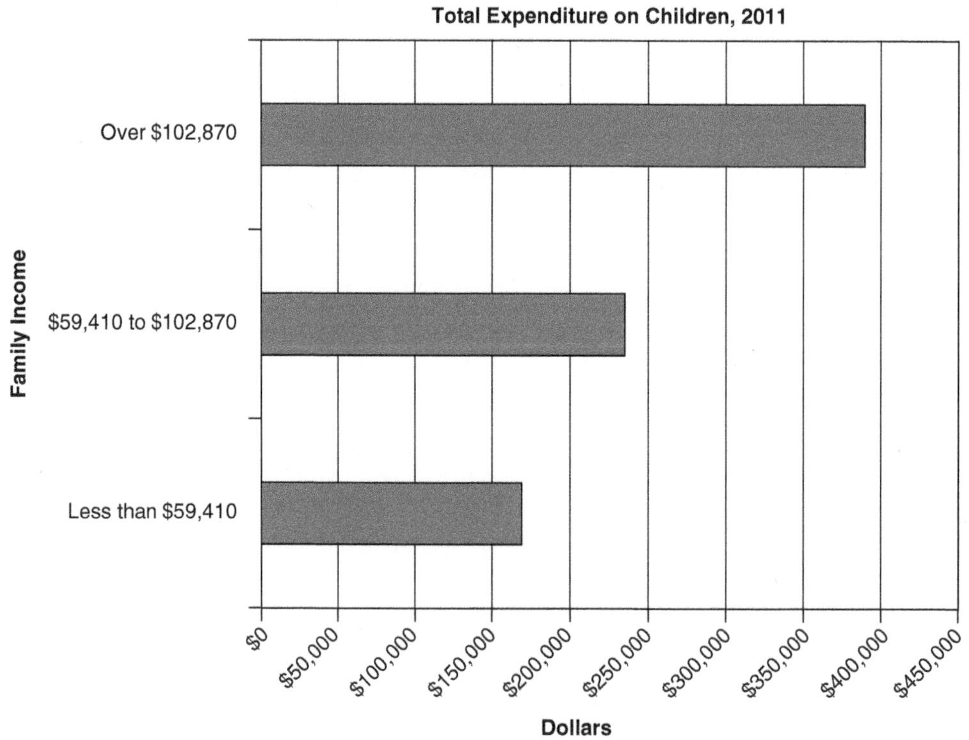

Figure 6–9 Estimated Expenditure to Raise a Child to Age 18 by Husband-Wife Families, United States, 2011

Source: United States Department of Agriculture, Expenditure on Children by Families, 2011.

Note: These are direct expenditures only; they do not include either childbirth and prenatal costs or opportunity costs. The latter could now run into the hundreds of thousands of dollars.

Elsewhere low fertility has been studied carefully as well. MacInnes and Díaz (2007) found that Scotland, contrary to popular belief, was not being affected as much by low fertility as many writers have suggested. Rather, they argued that Scottish fertility was not quite as low as it is elsewhere and that early deaths rather than a dearth of births was much of its demographic dilemma. Ogden and Schnoebelen (2005) give earlier low birth rates in France at least some credit for the decline in household size in Paris over the last two decades or more. By 1999 over half of the households in the city of Paris were one-person households, and 80 percent of them were either one- or two-person households. Andersson, et al. (2006) looked at sex preferences for children in some Nordic countries and found that modernization and equal opportunities for women did not always lead to parental gender indifference.

One final consideration of an economic view of fertility is what we might learn about population growth itself and the likelihood of controlling it. McKenzie and Tullock (1989, 97) argued that:

> Economists are in general agreement that the optimum quantity produced of anything is that quantity at which the marginal cost of the last unit is equal to the marginal benefits of it. This will be the case if all costs and benefits are actually considered by the individual making the decision on the output. And this rule of thumb holds for the production of children.

They go on to note, however, that seldom do couples have to assume the total costs of their fertility decisions—some of those costs are born by others, either directly in such forms as increased congestion or indirectly through taxes for such things as schools. At the same time, they note that population growth will be controlled in response to changes that are likely to occur in the child production process, including the following:

1. resource scarcity will force up the costs of children
2. less favorable tax structures and the availability of contraceptives and abortion
3. greater incentives for improved contraceptive technology with the rising costs of children

Kenya: A Look at High Fertility

With a total fertility rate of 8.0 in 1988, Kenya began the new decade with the dubious distinction of having the highest fertility level of all countries in the world. According to a study, Mott and Mott (1980, 7) noted that "Tribal loyalties and the high value placed on children by Kenya's vast population majority of rural families, and especially its women, combine to dampen national efforts to reduce population growth." Kenyans perceive that the relative place of a tribe in the political arena is primarily a function of tribal size, a situation that frustrates efforts to lower fertility.

"Tribal pressures affect Kenyan's behavior more than pronouncements arriving from the national seat of government," according to Mott and Mott (1980, 7), "but what ultimately counts is what an individual perceives in his or her own best interest." In a country in which the vast majority of the people work on the land, children are viewed as essential. They are the key to survival and are closely tied to status. They are a ready supply of labor, especially during peak times in the agrarian life cycle. Because much of the farm work is done by females, they perceive children as the major way in which the burden of farm chores can be relieved. Thus, to them the benefits of children are clear and tangible, whereas the costs are minimal. Of course, children also represent a major source of security for both males and females. Again according to Mott and Mott (1980, 9), "the vast majority of rural parents must still look to children for financial—and physical—support in old age."

To rural Kenyans children have a considerable prestige value as well. More children increase the chances that at least one of them will turn out to be a success or to marry well. This reinforces the tendency for women to have lots of children.

Because opportunities outside of the home are extremely limited for rural Kenyan women, the opportunity cost of having children remains low. Low actual costs and low opportunity costs, combined with high perceived benefits of children, lead to a perpetual cycle of high fertility. Nonetheless, in 2012, Kenya's total fertility rate had dropped to 4.4 as couples were responding to changing perceptions about the costs and benefits of children and to an increasing availability of information about family planning and contraceptives.

Shreffler and Dodoo (2009) found that land pressure in rural areas of Kenya influenced a rural fertility decline. Moreover, Kenya's fertility rate remains relatively high and research suggests that the HIV/AIDS epidemic in Kenya may be responsible (Doskoch, 2010). Elevated child mortality and reduced duration of breastfeeding (therefore shortening the length between pregnancies) are factors which might contribute to sustained high fertility.

Figure 6–10a illustrates the typical triangular-shaped population pyramid that is typical of countries with high birth rates and declining death rates. It portends more population growth in the future if nothing extraordinary happens. Figure 6.10b suggests that by 2050 projected low fertility will have begun to reshape Kenya's population pyramid considerably. Table 6–1 shows how rapidly Kenya has grown since 1950 and offers projections for future growth until 2050. If the projected population of 70.8 million is reached in 2050 then it would represent more than a ten-fold increase in that nation's population over one century.

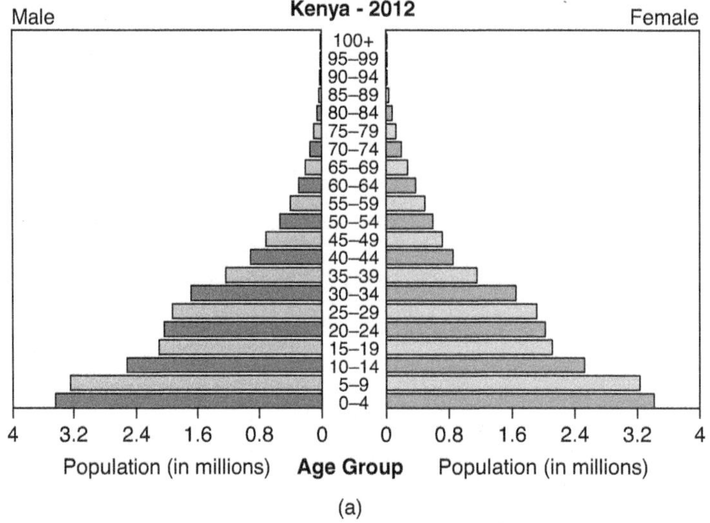

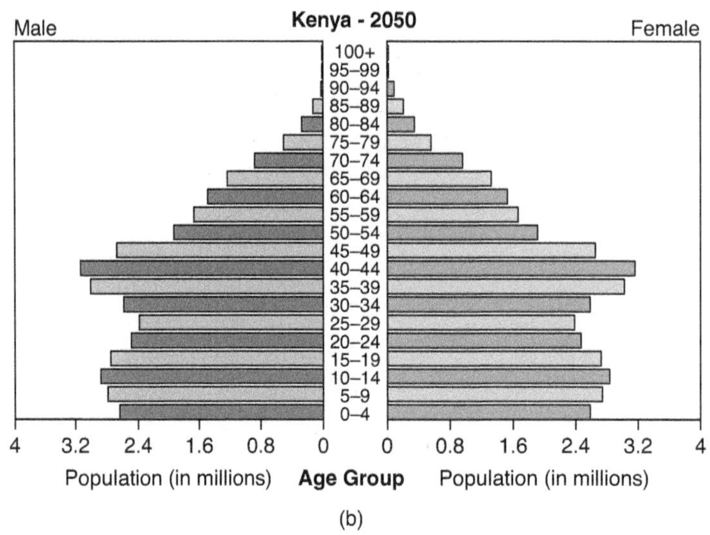

Figure 6–10 Population pyramids for Kenya in 2012 and 2050
Source: U.S. Census Bureau, International Data Base.

Table 6-1 Midyear Population Estimates and Average Annual Period Growth Rates for Kenya: 1950 to 2050 (Population in thousands, rate in percent)

Year	Population	Year	Population	Period	Growth Rate
1950	6,121	2005	35,246	1950–1960	2.9
1960	8,157	2006	36,316	1960–1970	3.2
1970	11,247	2007	37,409	1970–1980	3.7
1980	16,331	2008	38,541	1980–1990	3.6
1990	23,354	2009	39,708	1990–2000	2.7
2000	30,606	2010	40,843	2000–2010	2.9
2001	31,423	2020	49,858	2010–2020	2.0
2002	32,314	2030	56,522	2020–2030	1.3
2003	33,245	2040	64,059	2030–2040	1.3
2004	34,223	2050	70,756	2040–2050	1.0

Source: U.S. Census Bureau, International Data Base.

Italy: A Look at Low Fertility

At first glance Italy seems an unlikely candidate for an example of low fertility. It is a Catholic country in which family has been, and remains, an important institution, a land where the Vatican and the Pope still command considerable respect. Nonetheless, the total fertility rate in Italy has fallen to 1.4, substantially below replacement level; in some cities, such as Bologna, the total fertility rate is actually below 1.4. One description of the local impact of low fertility in Italy was this: "The most noticeable thing about Laviano is how quiet it is. Thirty years ago, the village, which now has a population of 1,500, saw as many as 70 babies born every year. In 2002, there were only four births" (Anonymous, 2004). Whereas in the mid-1960s there were about one million births annually in Italy, the number is now half that or less, substantially reducing the number of future parents and assuring that the total fertility rate is likely to remain low for some time to come.

As the Population Reference Bureau Staff (2004, 16) noted, "The struggle between traditional values and contemporary social reality in Italy helped push fertility to unprecedented lows in the past decade." Young Italians of both sexes live at home longer than they used to and remain single longer than they used to; cohabitation and out-of-wedlock pregnancies are still discouraged. Young Italians, of course, are lured by the same alternatives to marriage and childbirth that affect fertility in so many other industrialized countries—an array of relatively inexpensive consumer goods versus increasingly more expensive children. Success in the job market requires more education than it used to, which in turn requires more years of learning and provides less time for marriage and childrearing. Not only do young Italian women want rewarding careers, they want help at home from Italian men if they are going to become mothers. Such help so far has not been forthcoming. In 2003 the Italian government offered payments of 1,000 euros to mothers who had a second child, but that incentive was hardly enough to convince many Italian women that they should have a second child.

Despite proximity to the Vatican, and the social discouragement of cohabitation, there is no evidence that young Italians are less interested in sex than are their counterparts anywhere else in the world. Access to contraceptives and even abortion are much easier now in Italy than was the case a generation or two ago, so unwanted pregnancies are more easily avoided.

Figure 6–11a shows what low fertility and low mortality have already done to Italy's population pyramid. With a total fertility rate of only 1.4, Italy will face a shrinking demographic future unless its population is augmented by considerable immigration. Figure 6–11b shows that Italy's projected population pyramid in 2050 will still be strongly shaped by sustained low fertility even though it is projected to rise slightly by 2050 to a total fertility rate of 1.7. Table 6–2 shows a population that will have grown from 47.1 million in 1950 to a projected 61.41 million in 2050. Though on the surface that does not appear unusual, a look at the intervening data show that growth in Italy is projected to begin a decline in 2030 through 2050.

In Italy, as elsewhere, there are consequences, often unintended, of sustained low fertility levels. First, as more and more families opt for only one child, biological relatives will become scarcer. Second, the society will age rapidly and there will be fewer wage earners to support each retiree. Already about one-third of Italian wages are going to support pensioners. Italy has therefore recently taken steps to reform its state pension age requirements. Third, entrepreneurial activity seems to be inversely related to age structure, more vigorous in younger populations and less vigorous in older ones (Longman, 2004). Finally, the shrinking labor force in Italy may produce unwanted results as the nation's gross national product shrinks as well. Projections suggest that the Italian labor force will shrink by about 40 percent by 2050, a substantial change. Similar projections have been made for other European countries as well. Though less is said about it, immigration may end up filling the labor gaps that will increase throughout Europe in the decades ahead, but right now in most countries that solution is not a popular one.

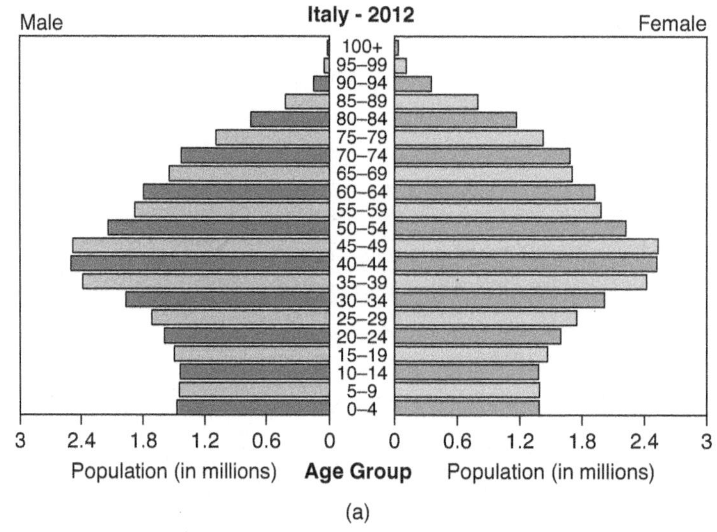

(a)

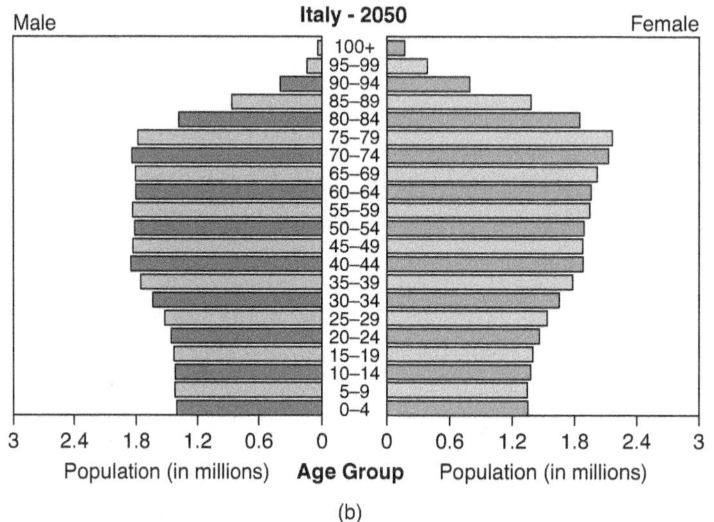

(b)

Figure 6–11 Population pyramids for Italy in 2012 and 2050
Source: U.S. Census Bureau, International Data Base.

Table 6-2 Midyear Population Estimates and Average Annual Period Growth Rates in Italy: 1950 to 2050 (Population in thousands, rate in percent)

Year	Population	Year	Population	Period	Growth Rate
1950	47,105	2005	59,037	1950–1960	0.6
1960	50,198	2006	59,272	1960–1970	0.7
1970	53,661	2007	59,626	1970–1980	0.5
1980	56,451	2008	60,091	1980–1990	0.1
1990	56,743	2009	60,461	1990–2000	0.2
2000	57,784	2010	60,748	2000–2010	0.5
2001	57,931	2020	62,402	2010–2020	0.2
2002	58,076	2030	62,622	2020–2030	0.3
2003	58,336	2040	62,318	2030–2040	0.0
2004	58,716	2050	61,415	2040–2050	−0.1

Source: U.S. Census Bureau, International Data Base.

Fertility Differentials

The fertility of a particular population results from a complex of different factors, and different groups within a population may respond to similar factors in different ways. Thus, different groups within a population are likely to display different fertility levels. Such differential fertility within a nation may in turn help explain regional fertility patterns. Important observed fertility differentials include rural-urban, income, education, and ethnic differentials.

Important observed fertility differentials include rural-urban, income, education, and ethnic differentials.

Rural-Urban Fertility Differentials

As noted already, fertility tends to be related to the level of urbanization when viewed at the world scale. Differential fertility between rural and urban areas within a nation is also common. Michielin (2004) studied the relationship between low fertility in Turin, Italy, and related it to migration. Weeks, et al. (2004) focused on fertility variations within Cairo, Egypt. In a study of fertility patterns in Guatemala, De Broe and Hinde (2006) found that the total fertility rate in rural areas at the end of the twentieth century was 5.8, compared to an urban total fertility rate of just over 4.0, a considerable difference. These authors also noted that fertility in Guatemala has declined much more slowly than it has in most other Latin American countries, though that decline has accelerated in recent years. Liu (2005, 411) focused on fertility levels in four Chinese cities and found that "Very low fertility and reproductive desire were found in these places, together with a high proportion of childless families and a high male/female sex ratio." Even in China it appears that urban areas discourage females from having children. Agyei-Mensah (2006) looked at the ongoing fertility transition in Ghana and found that it lagged behind in rural areas, especially in the northern region, but was most advanced in the Greater Accra region. In a more general but related look at fertility variations and declining fertility, Lutz, Testa, and Penn (2006) discovered that population density (rather than urban versus rural) was negatively related to fertility preferences, even after controlling for other social and economic variables.

Income Differentials

Previously we observed that fertility tended to be related to levels of economic development, as measured by GNP per capita for a sample of countries. Within a nation, fertility differences also tend to exist for various income groups. In general, fertility tends to be highest for the lowest income groups and to decrease with increasing income levels, though occasionally it will rise again at very high income levels. Part of the economic explanation for this rise is that those at high income levels can afford more of everything, including children. Data for the United States for 2006 appear in Figure 6–12. These data suggest a direct relationship between fertility and income, although fertility for the Less Than $10,000 income group is the lowest. This is explained by the fact that disproportionately larger numbers of women in the youngest age groups are in the lowest income group. These data are taken from the American Community Survey (ACS) and age of the women is shown at the time of the survey which could be up to one year after a child's birth. With these data, systemic underestimates of fertility are produced for the youngest age group, and systemic overestimates are produced for oldest age group. This data collection issue does not distort the data for the middle age groups (Dye, 2008).

Aassve, Billari, and Spéder (2006) found that in Hungary changes in fertility were being driven by reduced employment, as well as increased educational enrollment. From a different perspective, Martin (2006) studied the effect of family structure on income inequality in the United States. In a study of fertility in the Ecuadorian Amazon region Carr, Pan, and Bilsborrow (2006, 17) discovered that ". . . women from households with a legal land title had fewer than half as many children as those from households without a title."

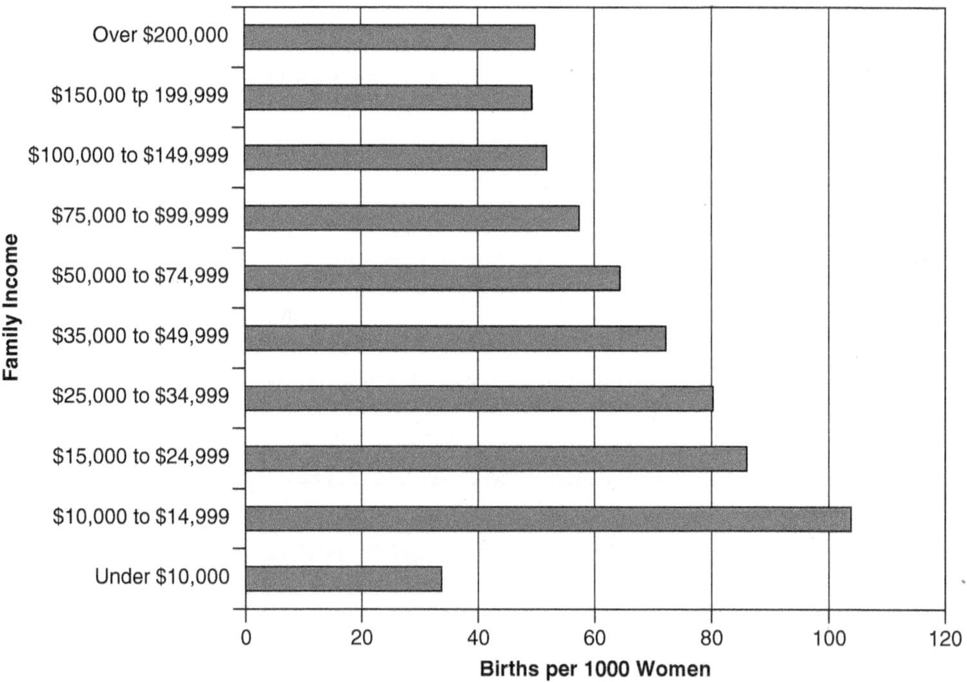

Figure 6–12 Births per 1,000 Women in the United States by Family Income, 2006
Source: U.S. Census Bureau, Fertility of American Women, June 2008 Current Population Reports, Series P-20–558.

Educational Differentials

Like income, education tends to be negatively related to fertility. As Hawthorn (1970, 92) commented, "It is true of all societies that awareness of methods of birth control varies directly with urban background or residence, a higher than average education and a higher than average income. . . ." However, beyond the knowledge of contraceptives, education may motivate couples to limit family size because of their greater awareness of the costs and benefits of children. According to the Population Reference Bureau Staff (2004, 18), "Education is the single most important determinant of the average age at marriage and age at first birth in the Middle East and North Africa because women in the region tend to give birth soon after marriage." Also, for more educated females the opportunity costs of having and rearing children are likely to be higher than for their less-educated counterparts and they are likely to marry at somewhat later ages. Figure 6–13 shows the relationship between educational attainment and fertility for 2006 in the United States, though these data do not clearly show the relationship described above. Paradoxically, the relationship between fertility and education has not shown up, at least not very clearly, in most of Sub-Saharan Africa (Eloundou-Enyegue and Williams, 2006). In a study of fertility among Palestinians in Israel, Nahmias and Stecklov (2007, 71) found that ". . . as educational levels increased among Israeli Muslim women, the strength and nature of the relationship between education and fertility has changed at both the individual and community levels."

Race and Ethnic Differentials

Different race and ethnic groups often have different fertility levels than the national populations of which they are a part. For example, in the United States, African-Americans, American Indians, and Mexican-Americans tend to have higher fertility rates than Whites. Such differentials tend to exist in most nations where there are significant numbers of the national population who belong to different race or ethnic groups. Figure 6–14 provides

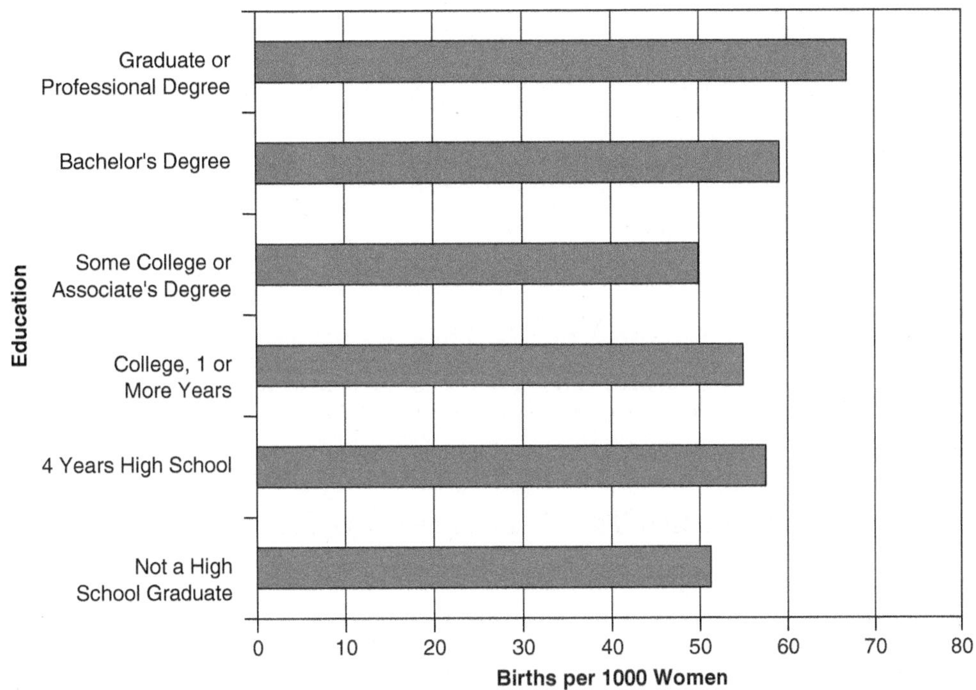

Figure 6–13 Births per 1,000 Women in the United States by Educational Attainment, 2006

Source: U.S. Census Bureau, Fertility of American Women, June 2008 Current Population Reports, Series P-20–558.

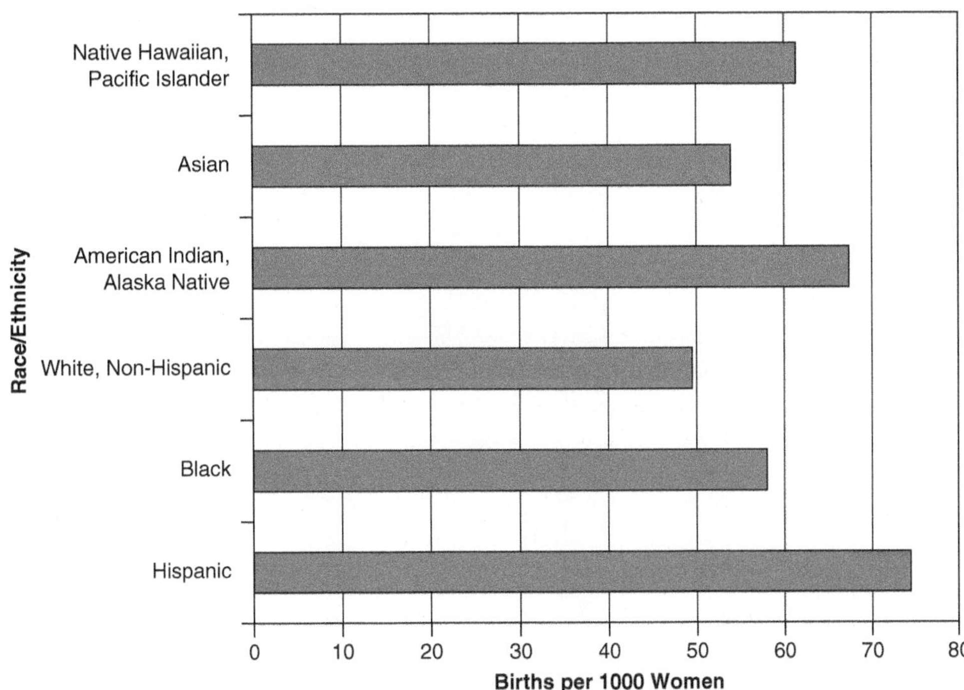

Figure 6–14 Births per 1,000 Women in the United States by Race and Hispanic Origin, 2006

Source: U.S. Census Bureau, Fertility of American Women, June 2008 Current Population Reports, Series P-20–558.

information for such differentials in the United States by race and ethnicity with Hispanics having the highest fertility rates.

Age Differentials

Women of child-bearing age in a society are not equally likely to have children at any given age. In most societies fertility is more concentrated in certain age groups, as is apparent for the United States in 2006 in Figure 6–15.

Fertility differentials are useful and important for explaining spatial and temporal differences in fertility. We must keep in mind, however, that fertility differentials of different types are probably not independent. Educational differences in fertility, for example, may also reflect income differences as well. Ethnic differentials also reflect income and educational differences. Furthermore, some patterns, once established, may be self-perpetuating. For example, Kahn and Anderson (1992, 54) found that ". . . it appears that both teen marriage and childbearing behaviors tend to be reproduced across generations." Finally, statements about the causes of fertility differentials should be made with care. Furthermore, many other fertility differentials can be noted, including religious and occupational differentials.

Spatial Fertility Patterns and Trends

Fertility varies in both space and time. Such variations result from similar variations in the determinants of fertility, as well as variations in the response to particular determinants. The map in Figure 6–16 shows the recent spatial pattern of crude birth rates in the world. The rates vary from a high of 46 in Niger and Zambia to a low of 6 in Monaco. In general, the world fertility pattern reflects the pattern of economic development, as shown in Figure 6–17.

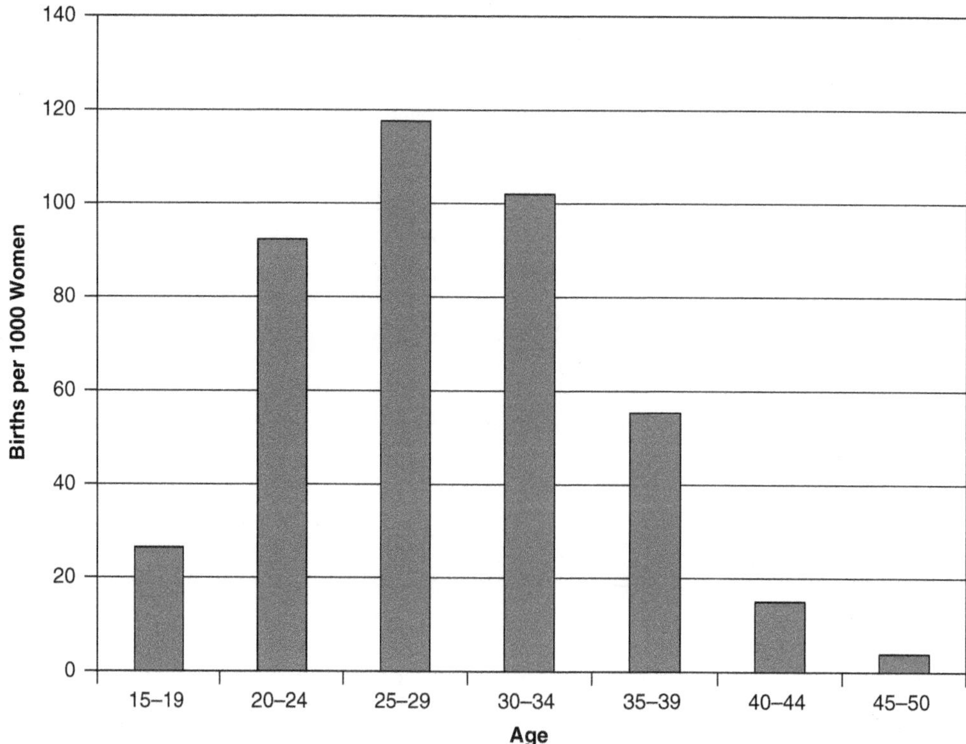

Figure 6–15 Births per 1,000 Women in the United States by Age Group, 2006
Source: U.S. Census Bureau, Fertility of American Women, June 2008 Current Population Reports, Series P-20–558.

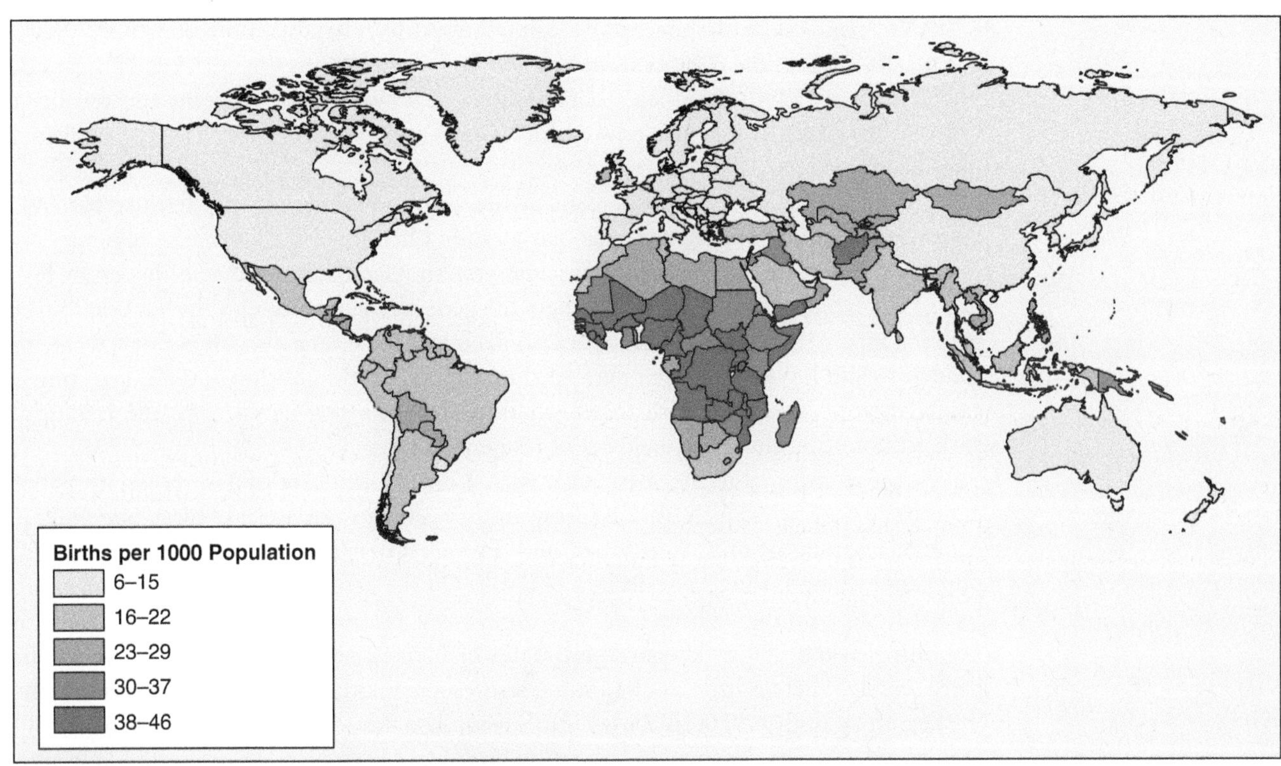

Figure 6–16 World Crude Birth Rates, 2012

Source: 2012 World Population Data Sheet. Washington, D.C.: The Population Reference Bureau.

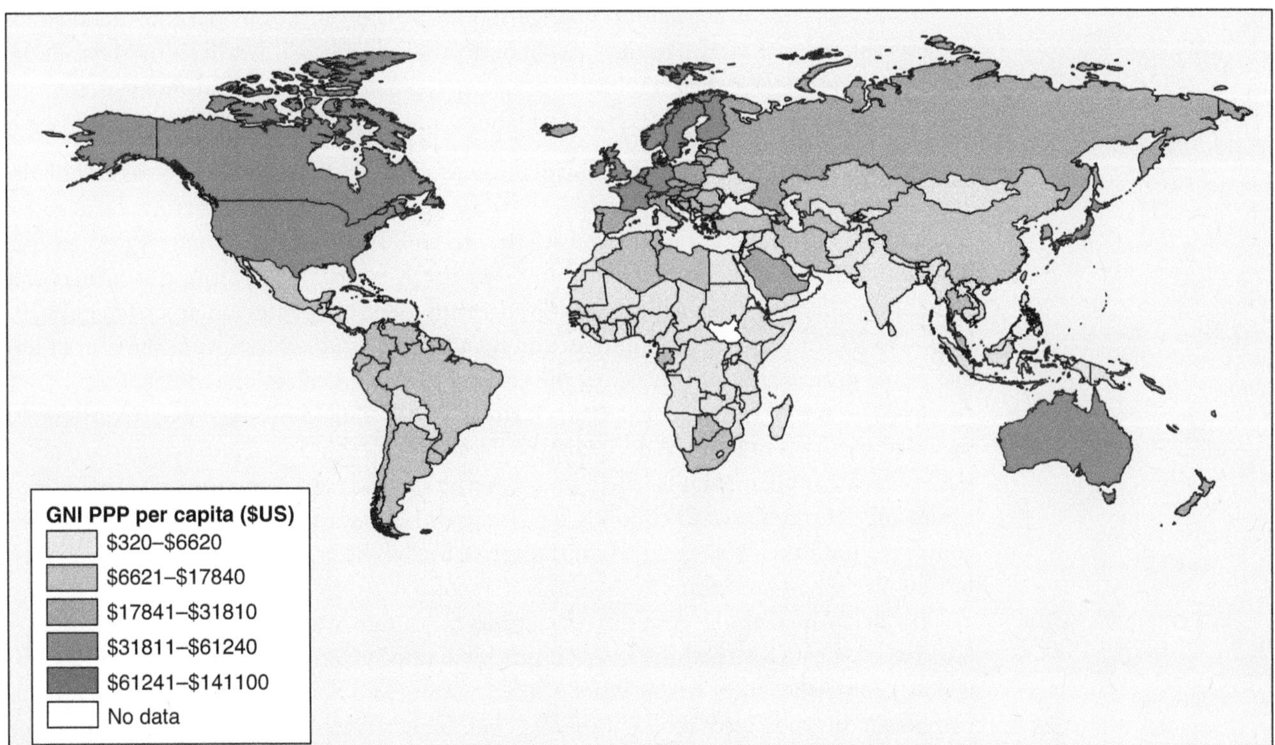

Figure 6–17 World Per Capita Gross National Income, Purchasing Power Parity, 2012

Source: 2012 World Population Data Sheet. Washington, D.C.: The Population Reference Bureau.

Africa has a crude birth rate of 36, and many African nations have crude birth rates of at least 40. Among the major exceptions are Tunisia and Morocco, with crude birth rates of 19. Bledsoe (1990) provided a detailed look at how marriage patterns and fertility in much of Africa are related and how education and other variables are changing traditional marital arrangements.

For Asia, with close to 60 percent of the world's population, the fertility pattern is more varied than it is in Africa. Asia has a crude birth rate of 18, and the highest Asian crude birth rates are found in southern and western Asia. Saudi Arabia, with a crude birth rate of 22, is one of many examples where the generalization about an inverse relationship between income and fertility fails to hold. This is true for several oil-rich nations, where the sudden wealth has not yet been paralleled by general modernization and a sharp contrast exists between extreme wealth (often concentrated in a few hands) and the traditional socioeconomic fabric of the majority of residents.

In broad regional terms, East Asia, with a crude birth rate of 12, stands in marked contrast to Southwest Asia. In East Asia, fertility varies from a crude birth rate of 23 in Mongolia to one of only 9 in Japan and Taiwan. Of major importance in this highly populated region is the recent decline of fertility in the People's Republic of China, which in 2012 had a crude birth rate of 12, a figure that represents a slight increase from the beginning of the 1980s. However, estimates of China's population and vital rates differ considerably. There is little doubt among researchers that Chinese fertility has fallen since the beginning of the 1970s, but with insufficient data the actual rate is unknown, probably even to the Chinese. Extremely low crude birth rates have been reported in many urban areas. When discussing Chinese population, despite its major demographic significance, we are still left at a point described by Zhao and Xiaomu (2010, 451) as follows:

> The study of China's fertility has been in a strange situation since the early 1990s. On the one hand, the growing number of censuses, fertility surveys and annual surveys of population change should provide sufficient data for detailed investigation of recent fertility decline; on the other hand, key fertility data are increasingly affected by problems of under-reporting and internal inconsistency that form an obstacle to such investigations.

In South Central Asia the crude birth rate was 23 in 2012. India, Asia's other population giant, reached a population of one billion in the year 2000. India has a crude birth rate of 22. With an annual natural increase of 1.5 percent in 2012, India continues to grow steadily, despite three decades of attempts to control fertility. Krishnan (1989) found that there were widespread differences in fertility within India, with crude birth rates ranging from only 20.4 in Goa to 39.7 in Rajasthan. Guilmoto and Rajan (2001) studied fertility at the district level and found decreases in fertility across the country with marked regional differences.

In 2012, Latin America had a crude birth rate of 19 and, like Asia, showed considerable regional variations in fertility. According to Brea (2003, 16), "Social and economic changes have played a fundamental role in Latin America's fertility decline, primarily by molding values and attitudes toward childbearing. Increased access to family planning has allowed couples to limit their family size. Contraceptive use has been an important determinant of fertility decline in the region."

In 2012, in Central America the crude birth rate was 21, with a high of 29 in Guatemala. Costa Rica had the lowest crude birth rate in Central America, 16, a figure that is down somewhat from recent years. Only in temperate South America and parts of the Caribbean do crude birth rates get much lower. Growth rates are high throughout most of Latin America because of the prevalence of low crude death rates, though declining birth rates should reduce growth rates and have already done so in many countries.

In North America, Canada had a crude birth rate of 11 and the United States had one of 13. The annual rate of natural increase for the two together was around 0.5 percent.

Though these growth rates are slow compared to those in the less developed countries, population growth in the United States and Canada contributes substantially to the growth in demand for energy and other resources. Immigrants add many births in the United States (Jonsson et al., 2004). Coleman (2006) has gone so far as to suggest that the fertility of immigrant populations in both the United States and a number of European countries is significant enough to be considered a "third demographic transition." Whether this term is broadly adopted in the demographic literature or not, there is little doubt that in the United States high rates of immigration and high fertility among those immigrants have rapidly altered the ethnic composition of the nation. In another study of migration and fertility, Kulu (2006, 147) noted that "Childbearing behaviour of migrants has thus not only emerged as an interesting research topic, but has also become critical to the understanding of current and future fertility trends in many societies."

The contribution of immigrant births to U.S. population growth has, however, become variable due to factors such as the recession that began in 2008. The Pew Hispanic Center reported that the U.S. birth rate decreased to its lowest in 2011 in large part due to the dramatic decrease in births among immigrant women. This report suggests that these declines are connected to the rising unemployment and poverty levels among Hispanics which occurred after the recession began (Livingston and Cohn, 2012).

Similarly, Europe had an overall crude birth rate of 10. Crude birth rates within Europe vary from a low of 6 (Monaco) to a high of 16 (Ireland). In general, crude birth rates are lower in Western Europe (10) and higher in Eastern Europe (11), though such differences are not large.

In Oceania the crude birth rate was 18. Within the region the rate varies from a low of 14 in Australia, New Zealand and Palau to a high of 32 in the Solomon Islands. Again, the rates tend to reflect differences in economic development.

Clearly, despite several exceptions, fertility patterns at the world scale reflect the spatial distribution of economic development. Within the realm of the more developed countries crude birth rates have fallen to quite low levels, sometimes even below crude death rates, as in Germany and several eastern and southern European countries. Total fertility rates are currently below replacement level in many more developed countries, including the United States, much of Western Europe, all of Eastern Europe and all but one southern European country.

Fertility has begun to decline in a number of less developed countries also, including the People's Republic of China, Taiwan, South Korea, Panama, Costa Rica, and Mexico (Murthi, 2002; Hussain, 2002). Using considerable information from area studies, Thornton and Lin (1994) provide an engaging account of fertility changes in Taiwan, noting the way that the Chinese population was able to maintain considerable continuity in a society that was undergoing rapid modernization; the family and fertility changes are considered within the broader context of social change.

Everyone interested in population keeps an eye on China because of its size, if for no other reason. In 1979 the Chinese government decided to promote the one-child family in a dramatic effort to bring population growth to a halt, even though Chinese fertility had dropped during the 1970s. In 2012, China's crude birth rate was 12, a very low figure considering that country's low income level, but up somewhat from what it had been in 1980. However, the rate of annual increase was 0.5 percent, which is considered problematic because of the base population of more than 1.3 billion people.

Fertility changes in the years ahead may be critical, especially as fertility in more and more countries drops below replacement level. We agree with geographer Paul Boyle (2003, 622), who noted that "Spatial variations (or the lack of them) can be a useful test of the comprehensiveness of grand theories of population change and it would be a shame if population geographers did not take up the challenge to investigate these theories a little more enthusiastically."

CHAPTER 7

Fertility: Family Planning Programs

Key Terms

unmet need for contraception
population policy
population program
pronatalist policies
antinatalist policies
KAP surveys
Comstock Laws
International Planned Parenthood Federation (IPPF)

In 2012, The United Nations Population Fund (UNFPA) declared family planning a basic universal human right. Family planning programs have been largely responsible for reducing global birth rates, but most notably those in less developed countries. The health and economic benefits of family planning are numerous, and the UNFPA cites many of these benefits in its annual report (2012). Research has shown that family planning efforts result in healthier children and mothers, and the evidence strongly suggests that family planning contributes to economic growth, though indirectly. Family planning is one of the most cost effective public health interventions. Through family planning, maternal mortality is reduced, deaths and illness among women are reduced, child mortality is reduced while health is improved. Family planning also reduces unwanted pregnancies and evidence suggests that this reduces the demand for abortions. More attention is being paid to addressing the reproductive health needs of women and couples through family planning now than ever before (Singh and Daroch, 2012).

Family planning programs and national policies regarding population issues began in the early 1960s. India was the only country in 1960 that had such a program. Since then we have seen a significant increase in support for family planning activities in both the more developed and less developed countries. Today, the vast majority of people live in nations that have formulated and adopted national policies to reduce population growth.

The global diffusion of family planning practices has been aided by both public and private organizations. This diffusion has resulted in dramatic decreases in fertility in less developed countries. In the early 1960s only about 18 percent of the women of childbearing age in the less developed nations practiced family planning. The total fertility rate was slightly over 6 children per woman in the less developed world compared with

2.7 children for the industrialized nations. Today, the total fertility rate for the less developed countries has dropped to 2.6 while that for the more developed nations is 1.6. Almost 60 percent of all couples of childbearing age in the less developed world and 72 percent in the more developed world use some means of birth control. In terms of modern methods of contraception, in the less developed countries, 54% of married women of childbearing age use these types of contraception while 63% of women in more developed countries use a modern method of contraception (Population Reference Bureau, 2012). In 1992, the World Health Organization estimated that as many as 100 million acts of human intercourse occur each day resulting in 910,000 conceptions (Khanna et al., 1992).

In 2012, contraceptives prevented 218 million unintended pregnancies in less developed countries. This in turn prevented 55 million unplanned births, 138 million abortions (of which 40 million are unsafe), 25 million miscarriages and 118,000 maternal deaths (Singh and Darroch, 2012).

As Potts (2000, 90) succinctly pointed out, "The quality of life of a large proportion of humanity during the coming century—and the future size of the global population—will depend critically on how quickly the world can satisfy the currently unmet demand for family planning" more so in the less developed world. The Population Reference Bureau (2012) broadly defines women who have an **unmet need for contraception** as those who want to delay or stop childbearing but are not using contraception.

Women who have an **unmet need for contraception** are those who want to delay or stop childbearing but are not using contraception.

According to Singh and Darroch (2012), as of 2012, the number of women who have an unmet need for modern contraception was 222 million. Between 2008 and 2012, this number declined slightly in the less developed world, but increased in some subregions, as well as in the 69 poorest countries. The unmet need is especially high in Sub-Saharan Africa, but pockets of unmet need are widely distributed throughout many poor countries, from Senegal to Cambodia to Haiti. Governments can save considerable sums on education, health, and other costs when women are able to avoid unwanted pregnancies, so investments in family planning and the provision of contraceptives produce benefits that far exceed their costs.

Two terms that are often, but not accurately, used interchangeably need to be introduced and defined—namely, the terms *policy* and *program*. A national **population policy** is an official government policy that is specifically designed to affect the size and growth rate of a population, the distribution of a population, or its composition. The important feature of such a policy is that the government *intentionally* plans to control one or more demographic variables. As noted in a United Nations (1969, 178) report:

A national **population policy** is an official government policy that is specifically designed to affect the size and growth rate of a population, the distribution of a population or its composition.

> All governments design policies, adopt administrative programs, and enact laws which intentionally or unintentionally directly influence the components of population growth—fertility, mortality, and international migration—as well as the internal redistribution of the nation's inhabitants. However, such measures represent national population policy only when implemented for the purpose of altering the natural course of population movements.

Once a government policy is adopted and a set of objectives has been specified, a **population program** may be initiated. The program includes the various means and measures that must be utilized in order to achieve the objectives of the population policy. For example, a government establishes a policy for the expressed purpose of lowering fertility. A family planning program may then be established to provide information and methods for controlling conception. Such programs may be official government programs or they may be private programs aimed at aiding the government in operationalizing its policy. According to Ross and Frankenberg (1993, 13):

A **population program** consists of the various means and measures that must be utilized in order to achieve the objectives of a population policy.

> The principal functions of the programs are to provide services, information, persuasion, and legitimation; countries vary greatly in the relative emphasis they give to these aspects. They also differ in coverage of the population for

each function. Some programs operate only in clinics in the larger cities; at the other extreme, some deliver a variety of services to the doorstep in the villages.

Explicit population policies are aimed at changing selected demographic variables. However, these demographic variables may also be influenced indirectly by other economic and social policies, including tax laws (income tax deductions for children, for example), public education, welfare, health, and various development programs. Furthermore, not all governments adopt national population policies, though they may, as in the case of the United States, support, or at least permit population-related programs.

Objectives of Population Policy

Most national population policies, especially in the less developed countries, have been conceived to decrease high rates of population growth. The focus of such **antinatalist** policies is on decreasing fertility. However, some earlier population policies in the more developed countries were designed to encourage higher birth rates. As Schroeder (1976, 6–7) noted:

Antinatalist policy is a population policy that aims to slow population growth by attempting to limit the number of births.

> Declining fertility led governments to proscribe abortion and contraception, and to adopt measures aimed at promoting large families, early marriages, and increased immigration. In most countries, the decision to promote more rapid population growth was reinforced by other factors—religious, military, or economic.

Examples of **pronatalist** attempts to encourage higher birth rates include egalitarian and welfare measures in Sweden, the pronatalist policy of Hitler's Germany, and family subsidies in France and cash awards for second births in Italy. Similarly, for the old Soviet Union, Thomlinson (1976, 604) commented that ". . . worries over the low birth rate and consequent slow rate of population increase were among the reasons that led Soviet leaders to revoke legalization of abortions in 1936. . . ." Though changes in some of these earlier policies have occurred in recent years, many more developed countries still have policies, direct or indirect, that could be described as pronatalist. Between 1976 and 1989 the number of governments that limited access to modern methods of birth control fell from 15 to 6. Very few nations currently restrict access to modern birth control methods. Among these are Iraq, Saudi Arabia, and the Vatican. The number of nations that do not directly support modern birth control fell from 28 to 20.

Pronatalist policy encourages higher birth rates.

Government policies designed to reduce mortality are virtually universal. The bases of such programs are mainly economic, but they are also humanitarian in nature. Programs for carrying out death control policies are many and varied. Unlike programs to decrease fertility, those designed to increase health and welfare meet with universal acceptance. Examples include disease control programs, work-safety programs and medical care programs.

Government immigration policies are also common in most countries, though they are designed for different purposes in different places; most have become quite restrictive. Their impact is mainly on the size and composition of the national population, though arguments that favor immigration are more generally made in the realm of labor market considerations.

The development of population policy is usually slow and is often fraught with difficulties; fertility decisions are personal and private, though their social consequences are easy to see. The Population Council suggested the following series of eight steps that might ordinarily characterize the sequential development of a policy to limit fertility (Nortman, 1973, 11):

1. Government interest in population as it impinges on development planning;
2. A pronouncement on population matters by some responsible public official;

Kap surveys determine the extent of knowledge, attitude, and practice of contraception.

3. The establishment of a commission to study the demographic situation;
4. Surveys to determine the extent of knowledge, attitude, and practice of contraception (**KAP surveys**); the establishment of pilot population projects; and course work in demography at the university level;
5. The allocation of a new or increased budget for family planning, usually through official health programs;
6. The elaboration of specific demographic targets and time frames;
7. Establishment of a family planning apparatus somewhere in the government to implement programs;
8. Integration of all previously existing population related programs and development of subsidiary activities such as public information and evaluation.

It should be apparent that the path to government adoption of a fertility control policy, as well as the development of programs designed to meet the goals of that policy, is a lengthy and controversial one. Still, the increasing awareness of such policy needs helps to create the necessary atmosphere in which to adopt and implement growth control policies.

Out of the adoption of fertility control policies has come the need for family planning programs; they are normally government supported and administered. Their stated primary purpose is to provide birth control information and services on a voluntary basis to people who desire such information. The difficulties of the task set out for family planning programs are many, but the perceived need seems to justify their existence.

Family Planning Programs

Though family planning programs seem relatively new, their history is not. Amidst the almost universal lowering of total fertility rates in the last 6 decades, the purposes, objectives and goals of family planning programs have been developed within different ethical, social, cultural and political contexts both in the more developed and less developed worlds.

A Brief Historical Sketch

Perhaps the earliest attempt to advocate the use of contraceptives to control family size occurred early in the Industrial Revolution in England with the publication of *Illustrations and Proofs of the Principle of Population* by Francis Place in 1822. Place's argument in favor of controlling family size was economic. He argued that workers could receive higher wages and better working conditions if only they could limit the supply of laborers available. Following Place, similar books were published in England and in the United States. However, during the 1870s legal attempts to suppress such literature were made in both countries.

In the United States the **Comstock Laws** (passed by Congress in 1873) forbade the dissemination by mail of information about birth control.

In the United States the "**Comstock Laws,**" which forbade the dissemination by mail of obscene, indecent or immoral materials, which included information about birth control, was passed by Congress in 1873. Similar laws against the distribution of birth control literature were also passed by many states and the importation of such literature was outlawed in 1890.

In the United States the chief voice in the early campaign for family planning was that of Margaret Sanger. Following the efforts of Sanger and others, most of the Comstock Laws were repealed in 1916 (the last of the Comstock Laws were repealed by the Massachusetts legislature in 1966).

The same year in which the Comstock Laws were repealed, Margaret Sanger opened America's first birth control clinic in Brooklyn, New York. A similar clinic, aimed at serving poor women, was opened five years later by Marie Stopes in England. By the mid-1930s the medical profession voiced its acceptance of family planning through a qualified endorsement of birth control by the conservative American Medical Association (AMA). The history of landmark events in the international family planning movement is outlined in Table 7–1.

Table 7–1 Landmark Events in International Family Planning	
Early Advocates	
1860	Malthusian League founded in England to spread information on birth control.
1873	Comstock Laws in the United States enacted to prohibit advertising or prescription of contraception (later repealed in 1916).
1916	First free birth control clinic opened in Brooklyn, New York, by Margaret Sanger. Clinic is a precursor to the Planned Parenthood Federation of America.
1921	First birth control clinic opened in England by Marie Stopes.
1937	American Medical Association gives qualified endorsement of birth control.
1942	Margaret Sanger incorporates her clinic and the American Birth Control League into the Planned Parenthood Foundation of America.
Post-World War II Developments	
1946	United Nations Economic and Social Council establishes a population commission representing member governments and a population division within the Secretariat.
1951	India adopts family planning as part of its economic program.
1952	The International Planned Parenthood Federation (IPPF) was founded in 1952 at the Third International Conference on Planned Parenthood in Bombay, India. The Planned Parenthood Federation of America is one of its major supporters.
1952	John D. Rockefeller, III establishes the Population Council as part of the eugenics movement.
Rapid Program Expansion	
1960	Oral contraceptives are introduced.
1961	Plastic IUDs become available.
1967	The trust fund for the United Nations Fund for Population Activities (UNFPA) is created.
1968	United Nations International Conference on Human Rights issues the Teheran Proclamation of which Article 16 states: "Parents have a basic human right to determine freely and responsibly the number and spacing of their children."
	Pope Paul VI issues ***Humanae vitae*** banning the use of artificial contraception.
	Paul Ehrlich publishes ***The Population Bomb***.
	U.S. Congress first allots foreign aid funds for family planning.
1972	President Nixon signs the Family Planning Services and Population Research Act, providing federal funding for family planning services in the U.S.
1973	U.S. Supreme Court upholds ***Roe v Wade*** decision, limiting the right of states to interfere in the private decision of a woman and her doctor to terminate a pregnancy.
1973	The Helms Amendment is passed prohibiting the use of foreign assistance funds to pay for abortions as a method of family planning.
1974	United Nations holds first World Population Conference in Bucharest. The United States urges countries to adopt policies aimed at slowing growth, primarily through family planning.
1979	The People's Republic of China begins campaign of "one child for one couple."
Political Realignments and New Paradigms	
1984	The Second UN World Population Conference is held in Mexico City. Most Third World countries favor slower population growth and family planning. The U.S. shifts its position, stating that population growth is a neutral phenomenon.

(*Continued*)

Table 7–1 Landmark Events in International Family Planning (Continued)	
Political Realignments and New Paradigms	
1985	The Mexico City Policy ("Global Gag Rule") is enacted that prohibits the U.S. government from supporting any organization that supports or participates in the management of a program of coercive abortion or involuntary sterilization. Support for the UNFPA and IPPF is suspended.
1994	The International Conference on Population and Development in Cairo leads to a paradigm shift (from simple demographic targets to the rights and needs of individuals) in in the focus of global population policies and strategies.

Source: Peter J. Donaldson and Amy Ong Tsui (1990) "The International Family Planning Movement," *Population Bulletin* 45:8. Courtesy of the Population Reference Bureau, Inc.; Robinson and Ross (2007); Sen (2010).

The **International Planned Parenthood Federation (IPPF)** is a global non-governmental organization with the broad aims of promoting sexual and reproductive health, and advocating the right of individuals to make their own choices in family planning.

After World War II there was a significant increase in both the stature and the number of private groups that promoted family planning activities. The **International Planned Parenthood Federation (IPPF)** was established in 1952. The IPPF fostered the gradual expansion of family planning services in less developed countries. Its nongovernmental family planning associations were able to demonstrate the high degree of demand in these countries as well as the acceptability of family planning among large shares of the public (Robinson and Ross, 2007; Claeys, 2010). According to Donaldson and Tsui (1990, 10):

> The IPPF is the most prominent and widely recognized private sector effort to support family planning internationally. Its objectives were (and are) to promote family planning and population education, to support family planning services, and to stimulate and disseminate research on fertility regulation. In 1990, IPPF had 107 family planning member affiliates representing 150 countries.

The United Nations established the Population Commission, and John D. Rockefeller III established the Population Council. The vision for the Population Council was one of producing research that could inform population policies and programs. The Population Council remains one of the most important agencies dealing with population issues, especially population control and family planning.

Recent Developments in Family Planning Programs

A surge of activity and rapid expansion of the family planning movement occurred during the 1960s and 1970s. This expansion came about as a result of the convergence of two major streams of thought and action about family planning. The first emanated from the birth control movements of Sanger and Stopes and their focus on the individual women and their wellbeing. The second stream emanated from the neo-Malthusian focus on unchecked population growth, its effects on food and other resources, the resultant impoverishment and deprivation from rapid population growth, as well as the potential for political instability which might stem from these effects (Robinson and Ross, 2007).

This expansion was supported in both the less and more developed nations. Funding for family planning services in less developed nations came from the United States and Sweden before 1968. In 1968 the United States Congress specifically allocated funds to support family planning activities. By the mid-1990s the number of donor countries had increased to 21, with the U.S., Germany, Japan and the United Kingdom providing over 75 percent of international population control funds (Seltzer, 2002). Private foundations have also been an important source of funding for these programs as well as the less developed countries themselves in more recent times.

Family planning became a central component of the United States Agency for International Development (USAID) program which was developed in 1965 and became the most influential of the population donors. In the following two decades, USAID contributions to population and family planning assistance throughout the less developed nations totaled $3.9 billion, making it the largest single donor of population aid (Gillespie and Seltzer, 1990).

Another important event during this period was the establishment of the United Nations Fund for Population Activities (UNFPA) in 1967. It was the most influential of the UN population programs. The goals of UNFPA were ". . . to provide an awareness of population problems and their relationship to social and economic development, the implementation of population policies, and the spread of family planning to better the health and well-being of the individual, family, and community" (Donaldson and Tsui, 1990, 11).

The World Bank became involved in population assistance in the late 1960s. Its contribution has been in the form of low interest loans with long payback periods to the least developed nations (Seltzer, 2002).

Americans became more aware of population issues through the publication of Paul Ehrlich's book, *The Population Bomb*, and a number of organizations were founded to publicize the problems associated with rapid population growth (examples include Zero Population Growth, the Population Crisis Committee, and the Population Institute). Population studies also increased at the nation's academic institutions including Brown, Chicago, Michigan, North Carolina, Pennsylvania, and Princeton.

Because of these population control efforts, a dramatic decline in fertility rates occurred in almost every region of the less developed world except sub-Saharan Africa. Overall, these fertility declines were accompanied by an increase in contraceptive use among women from 10 to 60 percent.

At the World Population Conference in Bucharest in 1974, less developed countries pushed back against the assumption generally held by the more developed countries that population control through use of family planning programs was necessary. Many less developed countries expressed discontent with the more developed countries' advocacy of global demographic goals and targets that were being used to frame population policy. Not enough attention had been given to the wider underlying causes of high fertility in less developed countries. A more balanced approach to population policy was desired not only by these less developed countries, but by many more developed countries as well (Robinson and Ross, 2007).

In 1984, at the second UN World Population Conference in Mexico City, the United States announced a shift in its commitment to population policies and family planning to one of neutrality on issues of population growth. Along with this shift was an unconditional opposition to abortion (Robinson and Ross, 2007). This change led the United States to withdraw its financial support for UNFPA and the IPPF. The central problem, as laid out by conservative political elements in the U.S., was the association that had been made between population policy and abortion. Interestingly, though abortion is legal in the United States, it was arguments that abortions were being performed as a part of some international family planning efforts that successfully curbed support for them among conservatives.

Fortunately for UNFPA and the IPPF, the cutting of funds by the United States was more than matched by additional funding support from the nations of Scandinavia, Western Europe, and Japan. Under the Clinton administration, funds for population activities expanded, though not without considerable debate in Congress. Support again diminished under the George W. Bush administration. President Obama expanded funds for population activities on the second day of his administration in 2009.

In 1994 an International Conference on Population and Development (ICPD) in Cairo was held and 179 countries agreed that population and development were inextricably linked. The empowerment of women and the meeting of people's needs for education and health, including reproductive health, were required for both individual advancement and balanced development. The 20-year Programme of Action adopted by the conference focused on individuals' needs and rights, rather than on achieving demographic targets. The ICPD Programme acknowledged the critical importance of place and space to gendered power relations, and recognized that girls and women are subject to subordination to men and that responsibility for change also lies with males and that their responsibility is essential for the advancement of sexual and reproductive health rights and family planning efforts. This paradigm shift produced a more complex but also a more promising approach with greater potential for change (Sen, 2010).

In 1995, the International Planned Parenthood Federation (IPPF) issued its Charter on Sexual and Reproductive Rights. The IPPF Charter connected sexual and reproductive health and rights issues with basic human rights concepts and conventions. It is based on 12 rights that are grounded in core international human rights instruments. This connection was based on research that demonstrated the human rights dimension of women's health. The Charter was developed with three necessary objectives: education, empowerment and equity. Educating people about the connection between their human rights and the way their lives are lived each day empowers them to act for change in their lives. Equity has long been seen as a goal of human rights education (Newman and Helzner, 1999).

Since the 1990s, major international agencies, such as Interact Worldwide, the World Health Organization (WHO), the UNPFA, Joint United Nations Programme on HIV/AIDS (UNAIDS), and the African Union have called for action against the HIV/AIDS epidemic in less developed countries (WHO, 2009). As a part of these efforts, the importance of linking reproductive health and HIV/AIDS policies, programs and services has been seen as an essential element in the provision of family planning services. Research has shown that integrating HIV/AIDS services and family planning services is an efficient method of reaching public health and development goals. Scientific evidence to support development of a set of defined best practices is sorely lacking however (Maynard-Tucker, 2009).

Implementing Family Planning Programs

The success of family planning programs depends on continuing financial support (RamaRao and Mohanam, 2003). In addition there must also be public support, strong political support, and good management and organization. The costs accrued in provision of family planning programs include direct costs (those involving the costs of commodities, supplies, and labor) and program and systems costs (expenses related to provision of contraceptives in general, such as staff supervision and training, family planning education and advocacy, construction of facilities, development of logistics systems and management) (Singh and Darroch, 2012).

> The success of family planning programs depends on continuing financial support.

The Guttmacher Institute estimates that in 2012, family planning programs for the 645 million users of modern contraceptives in the less developed world will cost about four million dollars per year. However, to address unmet need for contraception, 8.1 million dollars per year will be needed (Singh and Darroch, 2012). Donor countries and non-governmental organizations (NGOs) contribute much of the funding that is spent annually on family planning programs in the poor countries. In 2010, however, donor countries fell 500 million dollars short of their expected contribution to sexual and reproductive health services in less developed countries (United Nations, 2012). Reasons for this gap in funding may be due to budget cuts occasioned by the current global recession or changes in the ways that donor countries allocate resources. For instance, the United States, a major donor to international family planning for more than 50 years, began implementing a "graduating" process for countries who are recipients of its family planning aid. Countries that have reached benchmarks of a fertility rate of 3.0 or less and have a modern contraceptive prevalence rate of 55% are deemed by USAID to be ready for "graduation" from family planning aid from the U.S. (Bertrand, 2011).

In July of 2012, at the London Summit on Family Planning, donor countries and foundations pledged more than \$2.6 billion towards meeting the unmet need for family planning services in less developed countries by 2020. Less developed countries themselves also pledged to increase their support. More pledges have been received by donors since the

Table 7–2 Top 10 Donors to United Nations Population Fund (UNFPA), 2010

Top 10 Donors to UNFPA in 2010* (Contributions in millions US$)	
Donor	**Regular Funds[1] Contributions**
Netherlands	76.3
Sweden	60.6
Norway	54.1
United States	51.4
Denmark	37.1
Finland	33.7
United Kingdom	30.2
Japan	25.4
Spain	21.4
Germany	19.5

* Contributions valued in US$ at the time they were received using the United Nations Operational Rate of Exchange (arranged by descending order of regular resources). Provisional figures as of 25 April 2011.

[1] Contribution payments received in 2010.

Source: UNFPA, Report on Contribution by Contributions and Revenue Projections for 2011 and Future Years.

end of the summit (London Summit, 2012). A comparison of recent support of the United Nations Population Fund by its regular top ten donor countries is shown in Table 7–2.

Five types of activities have been particularly successful in the development of family planning activities, particularly in the less developed nations. These are:

1. Demonstration projects that were able to establish that there was a demand for family planning services and that these services could be delivered in a manner that was both medically safe and culturally acceptable;
2. The provision of contraceptives such as IUDs, condoms, spermicides, or the pill;
3. The support of training programs in population and family planning;
4. Assistance with surveys and national censuses; and
5. Other technical assistance, including that which would help the staff of family planning programs improve counseling on contraceptive side effects, expand the types of contraceptive methods offered, or conduct clinical trials to evaluate new contraceptive methods (Donaldson and Tsui, 1990, 18–20).

Though some combination of these and other activities may help meet the needs of growing numbers of potential parents, numerous obstacles remain. As Potts (2000, 90) noted, "An estimated 120 million couples in less developed countries do not want another child soon but have no access to family-planning methods or have insufficient information on the topic. Consequently, pregnancy too often brings despair instead of joy."

Family Planning in Less Developed Countries

Government support of family planning programs is widespread among the less developed nations. The number of men and women using contraception has grown more than sixfold since the 1960s. In 2012, the proportion of married women using modern contraceptives in the less developed world was around 57%. Prevalence of contraceptive use in less developed countries is strongly related to level of education, income, and rural or urban residence. In general, more educated and well-off urban women have higher levels of contraceptive use than their rural counterparts.

Recent estimates (Figure 7–1) suggest that roughly 60 percent of married women of childbearing age in the less developed nations use some form of modern or other family planning methods. There is considerable regional variation, however, ranging from a low of 31 percent in Africa to a high of 82 percent in East Asia. In Africa, more women are learning about family planning methods and searching for alternatives to repeated pregnancies and births than ever before. Fertility is declining in many African countries, even though economic development and wealth transfers would not predict such a result (Caldwell, 1994; Dow, et al, 1994; Sibanda, et al, 2003; Magadi and Curtis, 2003).

Sterilization of women is the number one method of birth control used in the world (30%). The intrauterine device or IUD is second (23%) followed by the pill (14%), the condom (12%) and other methods. Male sterilization accounts for 4% of methods of birth control used in less developed countries (United Nations, 2011).

Methods of contraceptive use vary geographically throughout the world. The prevalence rate of female sterilization is about 27% in Latin America and the Caribbean, about 23% in Asia, and 3.5% in Europe, and about 2% in Africa. In Africa, contraceptive prevalence is highest for the pill and for injectables (13%). In Asia contraceptive prevalence is highest for female sterilization and the IUD (40%) for the pill in Latin America and the Caribbean (16%). Male condom prevalence is highest in Latin America (10%) and Asia (7%) (United Nations, 2011).

In an effort to look at regional variations in family planning programs within the less developed world, Freedman (1990) divided these countries into three categories:

1. less developed countries with rapid development
2. countries with moderate development
3. countries with relatively little development

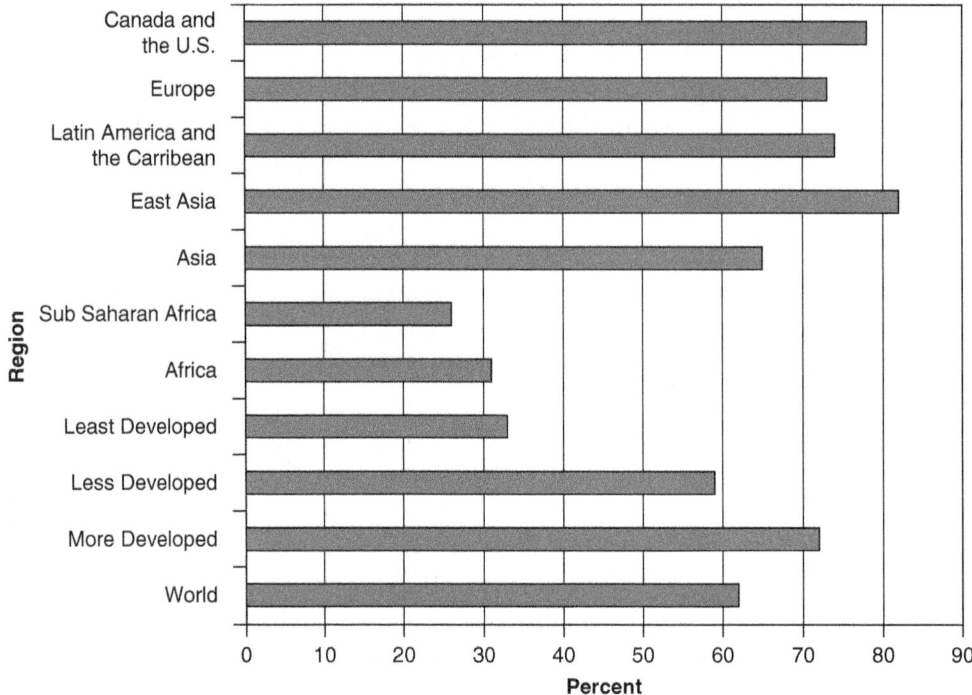

Figure 7–1 Contraceptive Use (All Methods) in World Regions, Married Women Ages 15–49

Source: Data from 2012 World Population Data Sheet (Washington D.C.: The Population Reference Bureau).

In countries with rapid development—such as Korea, Taiwan, Mexico, and Singapore—fertility rates have fallen rapidly. For example, in Taiwan in 1960, just before both family planning and development programs started, the total fertility rate was about 6 and contraceptive use was at low levels. By 1986, according to Freedman (1990, 37), ". . . only 26 years later, the fertility rate had fallen by 70 percent to 1.7—below replacement levels—and virtually all women were using contraception before the end of their childbearing years." Although Taiwan's family planning program coincided with rapid social and economic development, it is ". . . unlikely that the disadvantaged masses—the poor, rural, and illiterate—would have adopted family planning so rapidly without the family planning program" (Freedman, 1990, 37). In 2000, Taiwan's total fertility rate was 1.5, and in 2012, it stood at 1.1, well below replacement level.

Freedman's second category, those countries with moderate development, included China, Indonesia, and Thailand. In these countries, rapid fertility decline has taken place under initially rather unfavorable social and economic conditions. Freedman attributes this decline of fertility and increased use of contraception to a variety of factors, including the mobilization of the bureaucratic infrastructure and the ability of governments to incorporate the village masses into family planning programs.

China's one-child policy has received considerable support from the Chinese government, attracted much attention from demographers, and worked well in bringing Chinese fertility down to lower levels than could possibly have been predicted. Until recently, however, large families in China were badly stigmatized. That policy has been relaxed somewhat, though the one-child family model is still encouraged. Instead of outright fines for larger families, such families now must pay a "social compensation fee." Though the difference may not seem like much, for China it is a trend toward more gentle treatment of those who exceed the one-child limit. In part it is recognition that many Chinese, especially in rural areas, have more than one child. Ethnic minority groups in China have also been quite resistant to the one-child policy. At the same time, however, the Chinese are not abandoning their fight against overpopulation.

The third category includes those countries with little socioeconomic development and weak infrastructure. Most of the countries of sub-Saharan Africa, as well as India, Pakistan, Bangladesh, and the smaller countries of Southwest Asia, are in this category. While fertility declined in the less developed world more rapidly in the 1990s than in earlier decades, in countries such as Bangladesh and Egypt declines in fertility stalled more recently. Average fertility is higher in sub-Saharan African countries than in Asian, North African and Latin American countries. Fertility declines were evident in all four regions between 1992 and 1998 and 1998 and 2004. In Asia, North Africa and in Latin America the downward trend in fertility was steady in both periods, however, in sub-Saharan Africa, the decline was slower between 1998 and 2004 than it was between 1992 and 1998 (Bongaarts, 2008).

Though there are continuing debates about the efficacy of family planning programs, increasing evidence suggests that they are having a significant impact on fertility in many places (Bongaarts and Sindig, 2009; Le, et al., 2004; Gupta, Katende, and Bessinger, 2003). Most of the debate focuses on whether economic development and its concomitant social and economic transformations are a necessary precondition for fertility decline. Family planning advocates argue that an increase in the supply of birth control information and services can bring about significant fertility reductions without substantial economic development.

Family planning programs cannot take all the credit for fertility declines, but in many places they have had an impact. Other factors in recent fertility declines include improvements in literacy, improvements in the status of women (though they remain considerably disadvantaged relative to males in most less developed countries), rural development programs, income redistribution, and population education programs. Further improvements in the lives of women are necessary, however, not just for fertility control but also for the betterment of the societies in which they live.

A variety of factors are responsible for unmet need for contraceptives, including poor quality (and often poorly-financed) family planning services, limited access to these services, lack of contraceptive knowledge, weak motivation, and opposition from spouses (Blanc, Curtis, and Croft, 2002). Serving all women in less developed countries who currently have an unmet need for modern contraceptive methods would make a dramatic impact on development in terms of preventing an additional 54 million unintended pregnancies (including 21 million unplanned births), 26 million abortions (of which 16 million would be unsafe), and seven million miscarriages. In turn, 79,000 maternal deaths and 1.1 million infant deaths would be prevented (Singh and Darroch, 2012). Though family planning programs focus on providing contraceptives and knowledge about them, and though meeting unmet need for contraception seems crucial for most less developed countries, other considerations need to be made as well. For example, more attention needs to be given to unwanted sex, male sexual aggression, and male roles in families and households.

Before going on to family planning in the more developed countries, it is important to reiterate that funding for family planning is increasingly a problem. International aid has accounted for one-third of the funds, while the less developed countries themselves have supplied the remaining two-thirds of the funding. The less developed countries will be expected to increase their financial contributions in the future, though economic problems in many of them may make that difficult. The contribution by USAID in the 1990s was lower in real dollar terms than it was in the late 1970s. More significant, however, is the increasing demand and desire to practice family planning by people throughout the world. Less developed countries continue to increase their commitment and support for family planning efforts, while support from the historically largest international donor, the United States, is waning.

India

Among the less developed countries, India is generally believed to be the one whose government first favored lowering the rate of population growth. This antinatalist policy dates back to 1952, and it was incorporated into the first Indian Five-Year Plan. However, in the early years Indian family planning tended to be poorly financed. After a miserable start, followed by occasional lapses, the program was revitalized in 1965.

In 1965 the Indian Government established the Department of Family Planning, under the authority of the Ministry of Health and Family Planning. The Indian program is funded by the Central Government; however, most of the organization and operation of the program is left to the various states.

Until 1976, participation in the family planning program was voluntary. However, in 1976 compulsory sterilization was introduced in India, in the state of Maharashtra. From 1975–1977, then-Prime Minister Indira Gandhi suspended civil liberties and both the number of sterilizations and abuse of the program reached its peak. "Public sector employees needed to show a vasectomy certificate to get a transfer, promotion, food license, or an allocation of a house or medical care," noted Singh (1990, 127).

Until the introduction of involuntary sterilization, the mainstay of the voluntary family planning program had been sterilization, coupled with traditional contraceptives (especially IUDs and condoms). In the 1970s the Indian program attempted to improve communications and motivation. But by then need for convincing people of the advantages of smaller families was obvious to most observers.

A major element in Indian programs during the 1970s was a package of incentives. The major incentive was a financial award offered to men or women, with two or more children, who underwent a sterilization operation. The amount of the incentive was inversely related to the number of children. The maximum cash award was 150 rupees, more than a month's wages for the average agricultural worker. Beyond this cash incentive from the national government, the various states added other incentives, such as salary increases for sterilization, along with such disincentives as not providing maternity leave for children beyond the second child.

When the Janata Party assumed power in 1977, the family planning program was not dismantled. However, the program's emphasis was shifted considerably; only noncoercive methods were acceptable to the new government. Also, a new emphasis was placed on family welfare. Monetary incentives for sterilizations and IUD insertions were retained, with new rates of 100 rupees for a vasectomy and 120 rupees for a tubectomy, irrespective of the acceptor's current family size (Visaria and Visaria, 1981). Unfortunately, under the Janata Party people involved in the family planning program became demoralized and program achievements fell to low levels as well.

Between 1978 and 1981, however, the program gradually recovered. In 1979 the Working Group on Population Policy set a new long-range goal—India should reach replacement-level fertility by 1996 (a goal that has not yet been reached). In the same year the government planned for an increase in international family planning assistance for the 1980–85 period.

An analysis of data on contraception reveals some interesting changes. Approximately one-third (34.9 percent) of the total eligible couples in 1985–1986 were using effective contraceptive methods. Fifteen years earlier, in 1970–71, the comparable figure was 10.4 percent. The ineffectiveness of the program, however, becomes clear when birth rate statistics are analyzed. The decline in the birth rate during the same time period was very small, from 36.8 per thousand in 1970–71 to 33.9 per thousand in 1985. It was down to 24 by 2007 and was 22 in 2012.

The principal contraceptive method was sterilization, especially vasectomy, and it proved ineffective. According to an analysis of a village program by Singh (1990, 128–129):

> The implementation of the sterilization program left much to be desired for reducing fertility. The pressure of meeting quotas must have resulted in the sterilization of persons who were old or otherwise marginal to fertility reduction such as grandfathers, or unmarried older uncles. A father who voluntarily opts for sterilization usually does so after he has reached his desired family size, five or six children, with a sufficient number of sons.

According to Freedman (1990, 39), there was probably a consensus amongst most Indian and foreign observers that the program was poorly executed and ineffective for several reasons:

1. An obsession with unreasonably high targets for the program demoralized personnel;
2. There was too much of an emphasis on sterilization linked to a controversial incentive program;
3. There was little continuity of family planning program personnel;
4. Officials within the program were out-of-touch with the realities of Indian village life.

According to Roberts (1990, 88): "In India, where states function relatively independently and represent a range of religious and ethnic groups, the national coercive effort to reduce birth rates proved impossible to implement." Despite the partial success of family planning programs, India recently became the world's second "demographic billionaire," not a distinction that drew an ovation from the struggling Indian government. Taken in March of 2011, the latest census of India reported a population of nearly 1.2 billion, an increase of 17.7 percent since 1991. India's population has more than doubled since it began its first family planning efforts, and current projections suggest that it will surpass China in population by around mid-century. Today family planning in India focuses on the role and status of women. If women progress economically and educationally, then they will increasingly demand control over their own fertility, or at least that is the hope. Women are encouraged to think about child spacing, the advantages of smaller families, and the use of sterilization once they have reached their desired family size.

South Korea and Thailand: Two Success Stories

Two family planning programs that have achieved considerable success are those in South Korea and Thailand. The national program in South Korea was adopted in 1961 and has been one of the most successful in the less developed world. The achievements of the South Korean family planning program went hand-in-hand with impressive economic growth. Between 1962 and 1988, per capita income increased from $62 to $3,600, and the proportion of women who completed secondary school increased over 65 percent between 1965 and 1987 (World Bank, 1991).

Although these economic changes influenced people's views on family size, South Korea's family planning program also played a significant role in shaping people's perceptions of childbearing and family size. In the early 1960s family planning was practiced by less than 10 percent of the women at risk of pregnancy and the total fertility rate was 5.4. By 1991 the total fertility rate had fallen to 1.6 and the percent of married women using contraception had risen to 77 percent. In 2012 the total fertility rate was well below replacement level at 1.2.

The government family planning program that contributed to this impressive change in reproductive behavior in South Korea was, according to Donaldson and Tsui (1990, 40), ". . . based on a system of targets that established performance standards for the country's family planning outreach workers who were responsible for visiting women in their homes and encouraging them to begin to use contraception."

Despite the dramatic declines in fertility, South Korea did not change its family planning program. This has had ramifications for the dependency ratio in South Korea. In 2005, a pronatalist program was introduced, the Saero-Maji ("new beginning") Plan for the 2006–2010 period. In order to encourage childbearing, provisions were put in place such as tax incentives for childbearing, priority for the purchase of a new apartment, support for child care including a 30 percent increase in facilities, childcare facilities at work, support for education, and assistance to infertile couples. The Korean government announced in 2006 the Vision 2020 Plan to raise fertility and prepare for an aging society. The goal of the Vision Plan is to raise fertility to 1.6 children per woman by 2020 (Haub, 2010).

Thailand has also had a very successful family planning program. Between 1965 and 1991 the proportion of married women of reproductive age using a contraceptive increased from 15 to 68 percent. The total fertility rate also declined dramatically, from 5.7 in the mid-1960s to 1.7 in 2007. In 2012, Thailand's total fertility rate was 1.6.

The National Family Planning Program (NFFP), instituted in March of 1970, was aimed at providing information about contraceptive use to all citizens, especially those in rural areas. Non-physicians, mostly secondary school graduates and auxiliary midwives, were trained to distribute a wide variety of contraceptives. Thailand was one of the first nations to permit the use of Depro-Provera, an injectable contraceptive, and the contraceptive implant Norplant has also been introduced (Billingsley, 2005).

The Thai program also successfully linked national and international resources, as well as bringing together both public and private sector programs. These efforts were strongly supported by the country's economic development and by the cultural environment in which the family planning activities took place.

Throughout Asia fertility levels are falling. Aside from those already mentioned, Singapore, Hong Kong, and Sri Lanka have all experienced notable fertility declines; other countries in the region are likely to follow.

Family Planning in More Developed Countries

Because of the smaller populations, along with their lower rates of population growth, less emphasis is generally placed on discussions of family planning programs in more developed countries. Though the nature of population problems in more developed countries differs from that of such problems in less developed countries, it still deserves, and is receiving, increasing attention.

Earlier we pointed out that population policies in the more developed countries were generally pronatalist in the past. However, current policies in most more developed countries are more likely to favor lower fertility. Low fertility is characteristic of these countries, all of which have more-or-less completed their demographic transitions, and the use of contraceptives is widespread (Figure 7–2). In some more developed nations, especially Spain and Italy, total fertility rates are so low that demographers are no longer sure of what to say about them—nothing in the demographic transition model predicted total fertility rates that would drop as low as 1.2. Materialism, working mothers, and pessimistic views of the future are playing a role that no one foresaw.

Laws governing abortions have been liberalized as well, although in the U.S., recent changes supported by a powerful conservative agenda (spearheaded by the "religious right") have increasingly tried to restrict access to abortion, especially for the poor.

In Europe several nations have reached, or are close to reaching, zero population growth. Most of these countries have permissive policies with respect to contraceptives and abortions, but no official antinatalist policies. In some cases, for example France, even permissive policies are of recent origin.

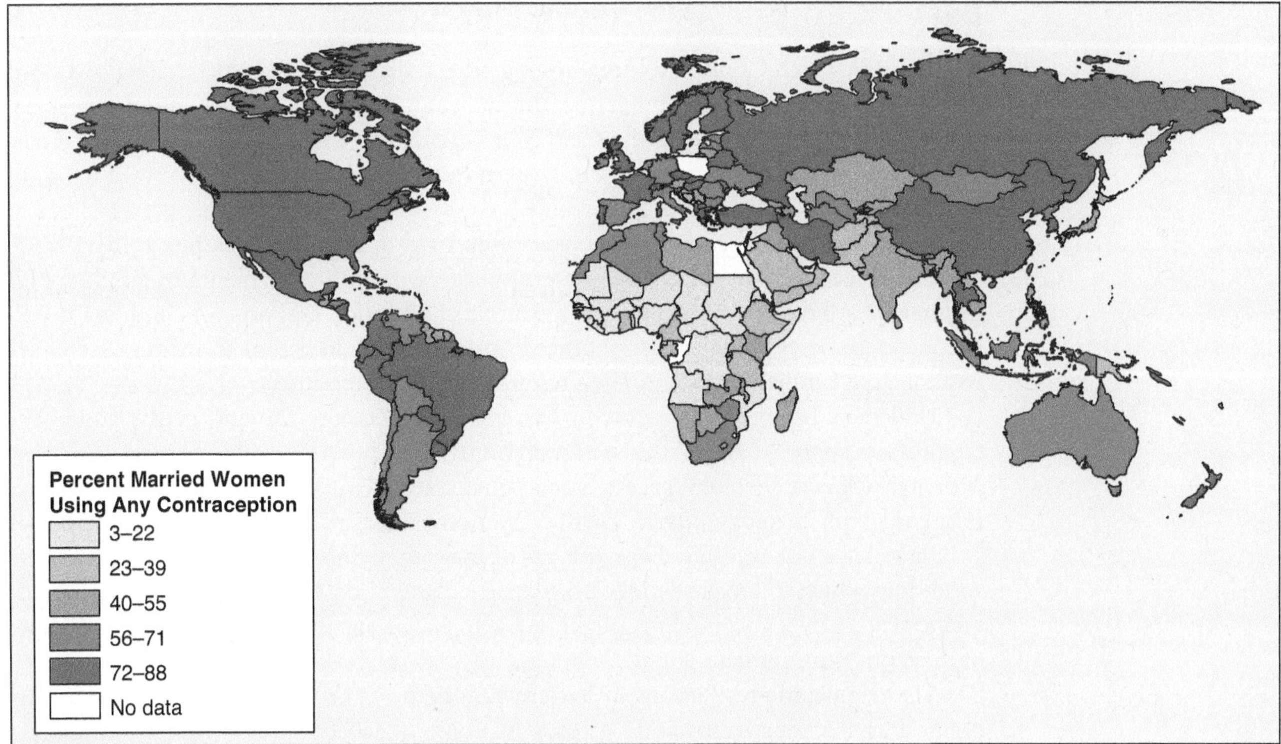

Figure 7–2 Percent of Married Women 15–49 Using Contraception
Source: 2012 World Population Data Sheet. (Washington, D.C.: The Population Reference Bureau, Inc.)

In the former Soviet Union population policy has been concerned mainly with population distribution. Higher fertility is being encouraged in selected areas, where growth rates are currently low. Both contraceptives and abortion, however, are readily available. Since the end of the Cold War, fertility in Russia has fallen considerably; the total fertility rate was 1.6 in 2012, well below replacement level. Furthermore, Russia's crude birth rate of 13 was below the crude death rate of 14, resulting in a rate of natural increase of -0.1 percent annually. Fertility has also declined in most of the former Soviet satellites; Hungary, for example, has a total fertility rate of only 1.2. In an intriguing study of low fertility in the former East Germany after the fall of the Berlin Wall, Witte and Wagner (1995, 394) concluded:

> Our finding that lower fertility is tied to individuals' own economic situation is an indication that some intimate aspects of behavior have already been changed through assimilation. Couples, knowing that unemployment is high, that their labor market value is relatively low, and that there are less generous maternity benefits, perhaps less flexible employers, and far fewer child care alternatives, respond rationally to socioeconomic change by limiting fertility.

In Asia, Japan, the largest more developed country in the region in terms of population, has a widely-accepted family planning program. Family planning services are available through more than 800 health centers throughout the country; abortions are readily available as well. The result is that Japan's total fertility rate has fallen to 1.4. This can be attributed to the increased entrance of females into the labor force, lack of support for working women with children, and a tendency for Japanese women to marry late (Oshio, 2008).

Throughout the industrialized world low fertility is common, even though explicit policies to encourage low fertility are not. A major element in low fertility has obviously been the effect of high and rising socioeconomic status, coupled with urbanization. Family planning programs often aid in the maintenance of low fertility levels.

Eastern Europe: A Modern Dilemma

In a world in which China and India, among others, struggle to alter birth rates downward through various policies and programs, the countries of Eastern Europe found that social and economic policies during the Cold War years led to undesirably low fertility levels. During the past thirty years fertility in Eastern Europe has followed a downward trend, in large part because of abortion.

In Eastern Europe governments have often tried to encourage couples to have more children. Incentives included paid maternity leaves, monthly payments to families with children, lump sum payments for the birth of a child, price supports for children's needs, tax breaks for parents, and a host of other features that would appear to encourage couples to have large families. But the desired results failed to materialize.

Understanding fertility trends in Eastern Europe before the end of the Cold War requires an appreciation of the status of women in these formerly socialist countries. Although Marxist ideology gave women equal status with men, reality was not always in accord with ideology. David (1982, 39) noted that in Eastern European countries ". . . laws have not eradicated the vestiges of male chauvinism or changed the economic conditions which still make childbearing a particularly heavy burden for women, whatever financial incentives governments might devise to encourage them to bear children." Many of these problems remain.

The democratic revolutions in Eastern Europe brought monumental changes to the political and economic structure of these nations. It will be interesting to follow these changes and to see if they have an impact on demographic behavior. So far, fertility has dropped in virtually all of these nations, as it has in Russia.

Law and Population Policy

Policies designed to affect demographic variables are often politically and emotionally sensitive issues. They often involve serious changes in preexisting laws, especially laws that have controlled the dissemination of knowledge concerning contraceptive devices.

Gazing into the future, it seems that family planning programs are likely to expand, to reach an increasing number of couples. Laws governing the sale of contraceptives will most likely change in favor of increasing their availability. Governments are likely to increase their support for family planning programs and population policies designed to slow rates of population growth. If all goes well, then the decline in the world's population growth rate will continue, and the present century should see a world population that is nearly stable in size, though probably not before we reach 9–10 billion.

CHAPTER 8

Migration and Mobility

Key Terms

migrate
migration
circulation
gross migration
net migration
migration stream
return migration
differential migration
primitive migration
group migration
free-individual migration
impelled migration
forced migration
"push-pull" models of migration
behavior
refugees
illegal immigrants
rural-urban migration
urbanization
megacities

One of the most distinguishing characteristics of humans is their propensity to **migrate**. The distance of these movements and their frequency are distinctive. This mobility is evidenced by the linguistic, social, and nationalistic mixing of much of the world's population. Though the human population has always been mobile, its mobility has accelerated with economic and technological progress, particularly in the fields of communications and transportation. Cultural adaptability has allowed humans to adjust to major ecological changes by employing mental abilities and technological skills. As Michael Parfit (1998, 11) observed, "In airports and seaports and railway stations, along forested borders, and even where steel and barbed wire make barriers that seem impenetrable, thousands upon thousands of people are on their way to somewhere new . . . they change the world." McNeill and McNeill (2003, 4) commented that "What drives history is the human ambition to alter one's condition to match one's hopes," and since humans first migrated out of Africa, movement has been one important means of finding a better life. Essentially, migration is the search for better opportunity (socioeconomic, religious, or political) at the local, regional, national, and global scale.

A change in residence intended to be permanent is to **migrate**.

One of the most distinguishing characteristics of humans is their propensity to **migrate**.

Migration is the movement of people from place to place, usually across some political boundary, for the purpose of changing their permanent place of residence.

Although **migration**, along with birth and death, are the primary processes of population change, migration differs from fertility and mortality in several ways. First, the biological processes of birth and death are uniform, discrete, and distinct events; for individuals

Migration differs from fertility and mortality in several ways.

each is a one-time-only event (save that growing number who at least claim that they crossed over to the other side, only to be brought back to life). Because migration is not a biological event, it does not have a uniform process. There is no upper limit to migration, but there is an upper limit to the number of children a woman can have. Goldscheider (1971, 49) remarked that "The total migration of communities or societies would not imply their demise, but their removal to another location." For example, the ghost towns of the American West were created when people left after the inducement that drew them there in the first place, typically minerals such as gold or silver, disappeared. In Maine today small towns are still disappearing from the map. Ghost towns are also found in other parts of the world more so in areas with harsh environmental conditions.

Second, because migration involves leaving one place and entering another, two populations must be considered: the population of the area of origin and the population at the destination. Changing mortality and fertility affect an area in a relatively simple way, whereas migration always simultaneously impacts two areas; in order to fully understand the migration process, both areas must be studied.

Finally, births and deaths are universals; that is, societies must have reproduction and as much control as possible over longevity in order to survive. On the other hand, migration is not a universal. Not everyone migrates, though in modern societies most people do. Furthermore, migration is a selective process that effectively has no upper limit. Some societies experience a great deal of migration, whereas others experience very little. In addition, migration can be repeated and even reversed.

Measures of Migration

At first glance it may seem that defining the term "migrant" would be relatively simple. However, several definitional problems can be identified. For example, if a migrant is simply defined as someone who "moves," then several questions must be answered. What constitutes a move? Should the move be permanent or do transients classify as migrants? Many people, such as commuters, shoppers, and tourists, frequently change their geographical position, but should they be considered migrants? There are not definitive answers to all these questions. According to demographer Ralph Thomlinson (1976, 267–268): "Demographers thus define persons as migrants if they change their place of normal habitation for a substantial period of time, crossing a political boundary." Geographers use the term **circulation** for activities such as shopping, commuting, and touring because each of these activities begins and ends at a person's place of residence; a considerable literature exists, for example, on commuting, the journey to work.

Geographers use the term **circulation** for activities such as shopping, commuting, and touring because each of these activities begins and ends at a person's place of residence.

Demographers distinguish between movers and migrants according to a single criterion. Movers are persons who change their place of residence; migrants are those whose change of residence takes them into a new political unit. All migrants are movers, but some movers are not migrants. Thomlinson also pointed out that getting information for all moves for all persons is almost impossible without a national registration system. However, it is usually possible to obtain a record of moves where people cross political boundaries (from a census, for example). Also, crossing a political boundary can be significant, particularly if a person moves from one country to another, even if the distance involved in such a move is short.

Geographer Curtis Roseman (1971) provided a conceptually useful way of describing migration. He distinguished between partial displacement migration and total displacement migration. His differentiation was made by considering a household's *activity space*, the set of places such as schools, work, stores, and recreational facilities with which the household interacts on a regular basis (circulation systems). Partial displacement moves are those that disturb only a portion of a household's activity space. For example, a family moves to a different part of town; the husband and wife keep the same workplaces, but the kids go to a different school and the family shops in a different shopping center.

In contrast, total displacement migration, which usually involves longer distances and is more disruptive, requires moving the entire activity space as well as the place of residence.

Basic Rates and Concepts

A standard set of migration rates and concepts has gradually evolved. They are useful in testing various migration hypotheses and facilitate the collection of data. In order to understand the migration literature, we need to understand the terminology and concepts.

Gross Migration

The sum of all the people who enter (in-migrants) and leave (out-migrants) an area is considered **gross migration**. Thus, it measures the total volume of population turnover in a community.

The sum of all the people who enter (in-migrants) and leave (out-migrants) an area is considered **gross migration.**

Net Migration

During any specific period of time a region may be receiving migrants from one region and losing migrants to another. The difference between arrivals and departures is **net migration**. When more people leave an area than move into it, net migration is negative. Positive net migration, on the other hand, occurs when more people enter than leave an area. Most countries on earth have negative net migration rates. The United States is one of the few countries with positive net migration rates.

The difference between the arrivals and departures is called **net migration.**

Out-and In-Migration

Each migratory event involves two actions: leaving one place and arriving at another. Leaving the place of origin is referred to as out-migration, whereas arriving at the place of destination is referred to as in-migration. If a move crosses an international boundary, then the term emigrant indicates an out-migrant and the term immigrant indicates an in-migrant. As we shall see, internal and international migration are very different processes in many ways, so it makes sense to use a different terminology for the two.

Leaving the place of origin is referred to as **out-migration**, whereas arriving at the place of destination is referred to as **in-migration.**

Areas of Origin and Destination

The area of origin is the place from which a migrant leaves, whereas the area of destination is the place at which the migrant arrives.

Migration Rates

The relative frequency of migration is called the migration rate. It is the number of migratory events divided by the population exposed to the chance of migrating. Donald Bogue (1969, 758) defined the following four such rates:

Because migration is a time-specific process, it is necessary to specify the time interval over which migration is studied or observed.

$$\text{Out-migration rate: } \frac{O}{P} \times k$$

$$\text{In-migration rate: } \frac{I}{P} \times k$$

$$\text{Net-migration rate: } \frac{I-O}{P} \times k$$

$$\text{Gross-migration rate: } \frac{I+O}{P} \times k$$

where O is the number of out-migrants from an area,
I is the number of in-migrants to an area,
p is the average or mid-interval population of the area, and
k is a constant, usually 100 or 1000.

These migration rates can be computed as specific rates if both the numerator and denominator refer to the same particular substratum of the population. There can be specific rates for age, occupation, income, race, sex, or any other particular subgroup for which data are available.

Migration Interval

Because migration is a time-specific process, it is necessary to specify the time interval over which migration is studied or observed. All other things being equal, the longer the time interval, the smaller the average size of the annual number of migrants, because a significant proportion of migrants return rather quickly to their places of origin. Even though migration data may have been computed on an average annual basis for two sample groups, if the migration intervals involved are not the same, then the data may not be comparable.

Streams and Counterstreams

Migrants who move from a particular origin to a particular destination over the same migration interval are considered part of a **migration stream**. Migrants who return to that origin during the same period are part of a counterstream.

Migrants who move from a particular origin to a particular destination over the same migration interval are considered part of a **migration stream**. Migrants who return to that origin during the same period are part of a counterstream. Keep in mind, however, that even if the stream of migrants between two places is exactly equal in numbers to the counterstream, the selectivity of migration is still likely to produce significant changes in both places. For example, the average age of persons in the stream may be younger than that of the counterstream; these age differences may in turn affect fertility, mortality, employment levels, incomes, and a host of other variables in both places.

Return Migration

To some extent **return migration** is an index of dissatisfaction with moves; it could even be construed as a "failure" rate.

While difficult to quantify because of scarcity of data, a certain number of people who make a move later return to their original location (though not necessarily to their original address). Because **return migration** is to some extent an index of dissatisfaction with moves; it could be construed as a "failure" rate. As we depend heavily on census data, however, we cannot differentiate between people who move into an area on the basis of whether or not they had previously lived in that area. Without doubt, some portion of most counterstreams is comprised of returnees, some of whom return because they have accomplished the objective for their original move.

Differential Migration

The study of the selectivity of migration and the differing rates between various social, demographic, and economic groups is referred to as **differential migration**.

The study of the selectivity of migration and the differing rates between various social, demographic, and economic groups is referred to as **differential migration**. This term is also used when studying the differences in the population composition of migration streams.

International and Internal Migration

International migration involves a change of residence from one country to another, whereas internal migration refers to a change of residence within a country. Because some territories have characteristics similar to those of nation-states, the distinction between international and internal migration is not always clear-cut. For example, difficulties arise when trying to designate migration between Puerto Rico and the United States. While migration can be broadly categorized into international versus internal; internal migration can be further broken into rural-urban, urban-rural, inter-urban, and even intra-urban. Other possibilities exist as well.

Types of Migration

Migration can take many forms. Though most people in the United States think of migration streams as comprised of individuals, or perhaps families, moving freely from place to place, this has not always been the typical migration pattern. Several types of migration have been defined, including

1. primitive migration
2. group or mass migration
3. free-individual migration
4. restricted migration
5. and impelled or forced migration

Most of these types were first recognized by William Petersen (1958).

Primitive Migration

This type of migration is associated with groups that are unable to cope with natural forces in their physical environments. You see, one method of coping with the deterioration of the physical environment of an area is to move from it. This is generally done by a group of people, and many times it is related to hunting or gathering food. When there is a disparity between the produce of the land (carrying capacity) and the number of people that must subsist from that land, a **primitive migration** takes place. According to Petersen (1975, 320): "... this can come about either suddenly, as by drought or an attack of locusts, or by the steady pressure of growing numbers on land of limited area and fertility."

When there is a disparity between the produce of the land (carrying capacity) and the number of people that must subsist from that land, a **primitive migration** takes place.

Group or Mass Migration

Most of the major population movements, certainly until the seventeenth century, consisted of the movement of groups of people. **Group migration** refers to the migration of a clan, ethnic group, or other social group that is larger than a family. Throughout history entire societies left their original domiciles (or residences) and laid claim to or invaded other areas. In some cases armies invaded other areas and some of the soldiers settled in the region of conquest. After an invasion, the native population was then either assimilated or displaced.

Another form of mass or group migration is colonization. The early stages of the colonization process involve group migration; but during later stages, most migration is by individuals or families (Simkins, 1970).

Group migration refers to the migration of a clan, ethnic group, or other social group that is larger than a family.

Free-Individual Migration

Free-individual migration was described by Fairchild (1925, 20) as follows:

> The movement of people, individual, or in families acting on their own individual initiative and responsibility without official support or compulsion, passing from one well-developed country (usually old and thickly settled) to another well-developed country (usually new and sparsely settled) with the intention of residing there permanently. A great deal of international migration since the seventeenth century has been characterized as this type, particularly migration to Australia, New Zealand and the Americas.

Free-individual migration refers to movement as a result of one's free will, choice or action.

Migration from Europe to the United States has been particularly important and, since the founding of the early colonies in the United States, it has involved some 30 to 40 million people. Much of the subsequent growth of the population of the United States is attributable to these immigrants. For example, an estimate by Gibson (1975, 158) was that "... about 98 million, or 48 percent, of the 1970 population (203 million) is attributable to the estimated net migration of 35.5 million in the 1790–1970 period." While similar estimates are unavailable for subsequent censuses, the proportion of Americans with a European ancestry in the 2000 and 2010 censuses was 61 and 62 percent, respectively. If current population trends continue, European ancestry will continue to dominate the U.S. population until the second half of the twenty-first century. Although the U.S. continues to draw European immigrants and 40 million of them gained legal permanent resident status between 1820 and 2011, the Europe-born population in the U.S. has declined steadily over time from 92 to 12 percent between 1850 and 2010.

Since the turn of the twentieth century numerous laws have been enacted to restrict the migration of people between countries.

Restricted Migration

Free migration has slowly been replaced by restricted migration. Since the turn of the twentieth century numerous laws have been enacted to restrict the migration of people between countries. In some cases the restrictions involve a complete ban on all movement of certain types of people, whereas in other countries migration quotas have been set up to curtail movements.

Restrictions on international movements to the United States in the twentieth century were primarily the result of increasing numbers of immigrants who desired to move to the United States, as well as the changing nature of the immigrants themselves—from Anglo-Saxon Protestants of northern and western Europe to Catholics of southern and eastern Europe. Immigration to the United States has been curtailed since 1921, when barriers were set up to restrict both the numbers and the types of migrants. A loosening of restrictions in movements to the United States began in the mid-1960s, resulting in a rapid increase in the annual number of immigrants entering the country since then. The loosening of immigration restrictions particularly started when the United States Congress passed the 1965 Immigration and Nationality Act (popularly known as the Hart-Cellar Act) that abolished the national origins quota system that had governed the country's immigration policy since the 1920s and replaced it with a preference system that privileged immigrant skills and family relationships with U.S. citizens and/or residents. Since then, the U.S. immigration stream has become global with significant inflows from Asia and Latin America. Like the United States, Australia and Canada initially opened up their doors to all immigrants but have since restricted immigration.

It is ironic that as the following words of Emma Lazarus' poem, "The New Colossus," were being inscribed and dedicated on the Statue of Liberty in the 1880s, the United States was about to impose racial barriers on migration and to establish a quota system:

> Not like the brazen giant of Greek fame,
> With conquering limbs astride from land to land;
> Here at our sea-washed, sunset gates shall stand
> A mighty woman with a torch, whose flame
> Is the imprisoned lightning, and her name
> Mother of Exiles. From her beacon-hand
> Glows world-wide welcome; her mild eyes command
> The air-bridged harbor that twin cities frame.
> "Keep ancient lands, your storied pomp!" cries she
> With silent lips. "Give me your tired, your poor,
> Your huddled masses yearning to breathe free,
> The wretched refuse of your teeming shore.
> Send these, the homeless, tempest-tost to me,
> I lift my lamp beside the golden door!"

Restrictive policies in most countries present difficulties for prospective international migrants. These people must either stay where they are or find illegal methods of overcoming restrictions. According to Thomlinson (1976, 288): ". . . two results of all these confining ordinances are that an enormous potential is being pent up in certain areas, and inequalities in the distribution of scarce goods and the standard of living are being aggravated." Pressures for more international migration continue to increase, especially as the more developed countries have low rates of population growth and the less developed countries continue to grow at medium to high rates. While the ratio of earnings in the richest countries to those in the poorest countries keeps increasing, the fraction of the world's population enjoying that wealth continues to decline. No one can believe right now that containing international movements will continue to be possible, at least not without committing an ever-expanding pool of resources toward that end. This is certainly true of the United States where after the 9/11 terrorist attacks, Congress and the Bush administration created the Department of Homeland Security in 2002 to, among other responsibilities, better protect the nation's borders. It combined 22 different federal departments and agencies into a unified and integrated homeland protection agency (United States Department of Homeland Security, 2012a). In 2010, the Obama administration heeded decades-long calls for the use of the military and the National Guard to aid in patrolling national borders and sent the National Guard to the U.S.-Mexico border (Greenhill, 2010). But despite all these measures, the U.S.-Mexico border continues to witness substantial amounts of illegal immigration and narcotics traffic. A key weak spot in the current United States' immigration system is its outdated immigration laws. However, these have proven difficult to reform since the Clinton administration. Hopefully, current President Obama and Congress will succeed in tackling the issue.

Impelled and Forced Migration

When the state or some other political or social institution is the activating agent in migration, then that migration is referred to as impelled or forced. There is a simple distinction between these two types: with **impelled migration** the migrant holds some degree of choice, whereas with **forced migration** the migrant has no power or control over the situation. For example, the Nazi policy between 1933 and 1938 that encouraged Jewish emigration by various anti-Semitic laws and acts would be considered impelled migration, whereas the policy that followed, with the actual forcing of Jews to leave their homes and in many cases actual extermination, would be considered a forced migration policy (Petersen, 1975, 321). Neither seems justifiable, of course, and differentiating between them makes little sense. Collyer (2006) wrote about the achievements and challenges of IASFM (The International Association for the Study of Forced Migration), which has held annual conferences for more than a decade now to discuss various aspects of forced migration.

When the state or some other political or social institution is the activating agent in migration, then that migration is referred to as **impelled** or forced.

With impelled migration the migrant holds some degree of choice, whereas with **forced migration** the migrant has no power or control over the situation.

Forced migration generally serves one of two purposes. First, it is a means whereby a potentially hostile group can be removed from a country, and second, it is a means of furnishing an unskilled labor force for certain areas. The transatlantic slave trade was a forced migration. According to Bouvier, Shryock, and Henderson (1977, 16):

> Forced migration, the most tragic of group movements, has meant flight and often enslavement for untold millions of human beings. Yet, despite the involuntariness of the act, this represents a major form of migration that has been prevalent throughout much of history. . . . The millions upon millions of 20th [and twenty-first] century refugees represent today's variation of "forced migration," which can be expected to continue as long as mankind insists on waging war or exercising total ethnic domination.

Tempting though it is to believe that slavery and the slave trade are things of the past, confined to the pages of history books, that is hardly the case. Tyler (1991), for example, pointed out that there are more slaves in the world today than at any time in the past, though most of today's slaves are either bonded laborers or child laborers, not chattel slaves. Even the latter are still with us, however, as children are sold in slave markets in Thailand, the Sudan, and a few other countries (Tyler, 1991). According to the National Underground Railroad Freedom Center (2004–2012, no pp):

> Slavery still exists today. Whether it is called human trafficking, bonded labor, forced labor, or sex trafficking, it is present worldwide, including within the United States and, increasingly, in your local community. An estimated 12–27 million people are caught in one or another form of slavery. Between 600,000 and 800,000 are trafficked internationally, with as many as 17,500 people trafficked into the United States. Nearly three out of every four victims are women. Half of modern-day slaves are children.

As the National Underground Railroad Freedom Center further states, there are four main differences between historical and modern forms of slavery (2012, no pp):

- There's no longer a need for legal ownership; people can be bought, sold and bartered among "owners" who take temporary possession;
- People caught up in slavery today can be purchased and sold for as little as $100 (compared to 10 times that much in the 1850s). As a result, people have become "disposable;" that is, easily replaceable.
- Slavery cuts across nationality, race, ethnicity, gender, age, class, education-level, and other demographic features
- Slavery's business side—human trafficking—is a global enterprise that can involve not just criminal gangs, but also corrupt law enforcement, drug dealers, and even families.

Most of today's child slaves are used for everything from sex slaves and child laborers to inexpensive fodder for "religious" sacrifices. A grim article in *Stern*, for example, suggested that as many as 10 million youngsters worldwide are involved in child-prostitution (Oberlander, 1992). Sachs (1994, 24) began an article about child prostitution with the following statement: "Once considered a universal crime, the trade in children's bodies is increasingly regarded simply as a business—with plenty of support from tour agencies, affluent travelers, and even governments." As Robert Scheer (1998, B7) bluntly noted, "Wherever foreign businessmen and tourists gather, they are readily serviced by sad-eyed economic refugees, often mere children, who have been lured to the sparkling centers of commerce in a desperate attempt to escape the poverty of their homelands."

Even the United States is not above reproach. In 2000 the Central Intelligence Agency estimated that about 50,000 women and children are brought illegally to the United States by "people-traffickers" annually to work as prostitutes, servants, and sweatshop laborers. Primary source regions for traffickers include Thailand, Vietnam, China, Mexico, Russia, and the Czech Republic. Most of these women and children are initially lured by the promise of good jobs in America. There is also a thriving domestic sex tracking industry in the U.S. that mostly uses children from poor families (Williamson et al., 2012).

Explaining Migration

Geography is among many disciplines that share an interest in understanding and explaining migration patterns. From the economists, with their focus on the role of labor and labor markets, to sociologists, with their eye for institutional changes, all are hampered somewhat in their search for explanations of migration by the lack of adequate data and other constraints. As Ritchey (1976, 399) noted, for example: "A major problem hampering theory development in migration research is a simple methodological problem that undermines the utility of many studies—the time to which measures of influences and of migration relate." These times, of course, are often different, depending on what data were available for the variables in use.

Migration, particularly that between nations, is among the most striking changes that can occur in a person's life. According to Bogue (1969, 801):

> In a high proportion of such moves the person not only changes his national loyalties, but also forsakes his native language, his cultural heritage and customs, his relatives and lifelong friends, and his occupation. So drastic is the change that many adults never make a complete adjustment in their lifetime, and it is only their children or grandchildren who are fully integrated into the receiving society.

Yet, often in the face of considerable odds, people have always made such moves; today more people than ever are on the move, and high rates of international migration over the coming decades seem assured. They are going to alter places considerably, as is already apparent in the United States and several western European countries. In a critical look at the impact of immigration on the United States in recent decades, Huntington (2004, 365) concluded that "America cannot become the world and still be America." Others disagree, and the literature on immigration and its impact on the United States and Europe is emotionally loaded.

Key Theories and Reasons for Migrating

The reasons why people migrate are varied. In a theoretical economic sense, migration is the mechanism by which human resources move to their highest-valued use; however, this ideal often goes unrealized. Among the major determinants of migration are distance, income differences, and information flows.

Repeatedly it has been shown that most migration takes place over short distances. The distance itself, however, is related to other factors that also inhibit migration. For example, distance is related to both the economic and psychic costs of moving. Furthermore, distance can determine the ease or difficulty with which one can collect reliable information about alternative destinations. Several studies have shown that the psychic costs of leaving friends and relatives may be significant, though they are difficult to measure. Diverse approaches to explaining migration are considered in subsequent sections; each is described only briefly, though each has generated a literature and a following of its own—none so far has received unanimous support.

a) Ravenstein, Lee, and the Push-Pull Model

The pioneering effort in migration theory was an early study by Ravenstein in the 1880s and was set down in his "Laws of Migration." He studied population movements in Great Britain and related migration to population size, density, and distance. He had a minimum of records for migrants in England, but in a fairly short time he extracted the essentials from those records and published a series of generalizations, most of which still hold true today. Some of Ravenstein's generalizations were the following:

1. Most moves cover only a short distance.
2. Females predominate among short-distance movers.
3. For every stream there is a counter stream.
4. Movement from the hinterland to the city is most often made in stages.
5. The major motive for migration is an economic motive.

With respect to the latter, Ravenstein (1889, 286) stated the following: "Bad or oppressive laws, heavy taxation, an unnatural climate, uncongenial social surroundings, and even compulsion . . . all have produced and are still producing currents of migration, but none of these currents can compare in volume with that which arises from the desire inherent in most men to 'better' themselves in material respects . . . . " It is somewhat surprising (but a tribute to him) that, from Ravenstein's day until now, relatively few attempts have been made to extend his generalizations, or to devise additional ones, or to gather his generalizations into a more theoretical framework. That is not to say, however, that migration researchers have been sitting still; recent efforts have been reaching along new, if still unproven, lines.

One important conceptual framework, accompanied by a set of hypotheses about the volume, streams, and characteristics of migrants, was broached by Lee (1966). He began by classifying the elements that influence migration into the following groups:

1. factors associated with a migrant's origin
2. factors associated with a migrant's destination
3. obstacles between the two that the migrant must overcome, which Lee calls intervening obstacles
4. personal factors

People move for a variety of reasons, including job changes, marriages, divorces, graduations, retirements, and trouble with the law. They may move because of conditions in their area of origin, conditions at their destination, or some combination of the two. Not all people in a particular area perceive conditions the same way, thus they may respond differently to the same stimuli. Lee suggested that a good climate was almost universally attractive, whereas a bad climate was undesirable. However, not everyone agrees on exactly what is or is not desirable. Recently some researchers have focused on the role of environmental preferences and migration patterns.

Lee summarized his ideas in the schematic diagram shown in Figure 8–1. The circles representing the places of origin and destination have pluses, zeros, and minuses. The pluses indicate elements to which potential migrants respond favorably, whereas the minuses are elements to which they react negatively; zeros stand for elements to which potential migrants are indifferent.

"Push-pull" models of migration derive from the observation that some elements of an origin "push" people to migrate, whereas some elements of a destination "pull" migrants toward it.

Conceptually, a potential migrant adds up the pluses and minuses for both the origin and one or more possible destinations and then decides whether the balance of pluses and minuses favors moving or staying. **"Push-pull" models of migration** derive from the observation that some elements of an origin "push" people to migrate, whereas some elements of a destination "pull" migrants toward it. However, before the migrant decides to move, another set of circumstances must be considered—Lee's intervening obstacles.

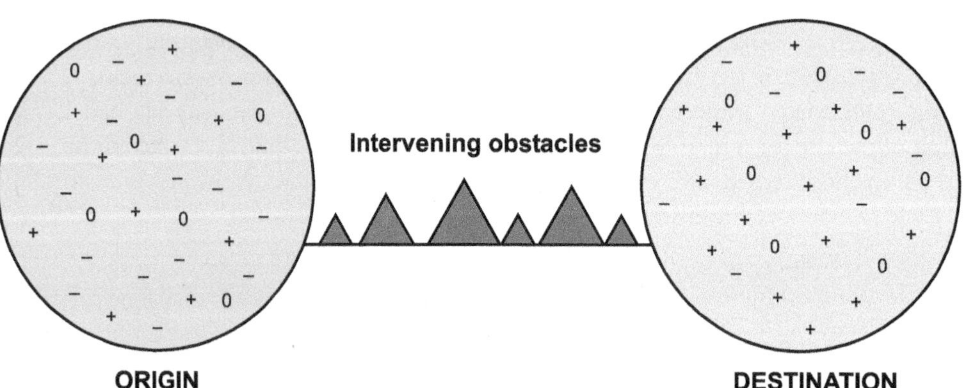

Figure 8–1 Lee's Migration Model

Source: "A Theory of Migration" by Everett S. Lee, in *Demography* 3(1), March 1, 1966. Reproduced with kind permission from Springer Science+Business Media.

These include such things as the actual cost of making the move, which is obviously related to distance, and psychic costs, such as the necessity of breaking ties with family, friends, and community, especially when long distances are involved. Furthermore, many personal factors may also influence a migration decision; for example, potential migrants may consider such things as health, age, marital status, and number of children. Clearly, a person's decision to migrate depends upon many considerations. Some of these considerations require further elaboration.

Operationalizing push-pull models is more difficult than conceptualizing them. Researchers must consider which variables to measure at origins and destinations, what time periods to measure them for, and how to include some measure of intervening obstacles. Most migration studies have been cross-sectional, using variables at origins and destinations to explain observed migration flows during a specific time interval (between two census periods, for example). Most economic and social models of migration are actually specific cases of the generalized push-pull model.

b) Two Micro Approaches from the Economists

The two economic models discussed in this section are micro-level in the sense that they focus on the rational behavior of individuals who are potential migrants rather than the macro-level system within which they are planning to migrate. Modernization theory and neoclassical economics have focused attention primarily on labor mobility; the basic argument is that migration flows are necessary and rational adjustments to geographical differences in the supply and demand for labor, as measured primarily by regional differences in real wage rates and unemployment rates.

In modernizing economies both of these models help explain the cityward movement of people as industrialization progresses. Higher wages in cities, coupled with diminishing opportunities for employment in agriculture, lead people, especially the young, to move steadily toward expanding urban areas. They also help explain why the globalization of the economy is attracting people in new directions as opportunities change. Large wage differentials between the more developed and the less developed nations do not go unnoticed, especially as global communications systems become increasingly common (think of the ubiquity of CNN, for example). The surplus of low-wage labor in the poor countries needs to come together with the large-scale capital accumulations in the rich ones. Two trends are then set in motion. One is a flow of capital outward from the more developed countries into the poor countries, where it can take advantage of low wage levels. The other is the flow of surplus labor from the poor to the rich countries, where demographic stagnation and affluence are creating a notable demand for unskilled labor. Once confined primarily to agriculture, nations such as the United States now need labor in many service sectors—hotels and restaurants—and in more industries as well—apparel and construction, for example. So long as large wage differentials persist, the economists will tell us that labor migration will follow (even if it means crossing borders that are otherwise marked "closed"). Migration is an integral part of the globalization of the economy.

The factor mobility model of migration is based on wage rate differentials, which in turn reflect geographic variations in the supply and demand for labor, one of the major factors of production.

(i) The Factor Mobility Model

The factor mobility model of migration is based on wage rate differentials, which in turn reflect geographic variations in the supply and demand for labor, one of the major factors of production. We could argue that migrants move up the wage rate gradient; that they move from low-wage areas to high-wage areas, in other words. In a *perfect* world, migration would then serve as an equilibrium mechanism. As people left low-wage areas, the smaller remaining labor supply would drive wages upward. At the same time, the growing supply of labor in the high-wage area would tend to depress wages there. Theoretically, migrants would move between the areas until wage rates were equal in both places; however, imperfections exist in the market mechanism and people are not perfectly mobile. For example, many Appalachian Americans have chosen to stay in Appalachia despite low wages and high unemployment rates, partly because they have had deep attachments to the region and partly because they have been unaware of opportunities elsewhere.

Some important earlier findings of a study by the United States Department of Labor (1977) are worth considering.

1. Household-level unemployment or job dissatisfaction does operate as a "push" factor in migration. People who are recent arrivals in an area, and who fail to immediately find employment, are especially likely to move again (often becoming return migrants).
2. Local economic conditions, such as unemployment rates, do affect outmigration, but only among those who are unemployed.
3. Unemployed individuals and others currently seeking work are more responsive to such economic determinants as family income, local wage rates, and expected earnings increases than are persons who are satisfied with their current jobs.
4. Families are much more likely to move in a given time period if they moved in the recent past, mainly because of a tendency for people to return to places that they recently left.
5. Families that have moved several times previously are more likely to move again than are families that have made only one or no moves recently, if the multiple moves were nonreturn moves. Such people are often referred to as *chronic movers*.
6. More important now than ever before, spouses are not passive, secondary migrants, but have a significant influence on the family's decision to move. Also, contrary to the results of some studies, families with working wives are not necessarily less likely to move than are families in which the wife does not work outside the home. These families will seek locations that maximize both incomes.
7. Surprisingly, the study found that age and education, typically found to be among the strongest correlates of the propensity to migrate, appear to be relatively unimportant in explaining the migration of married couples when other migration determinants, many of which vary with age and education, are held constant.

(ii) The Human Capital Model

Developed by Sjaastad (1962) and nicely summarized in Clark (1986), the human capital model is an alternative economic model. Because individuals seek to improve their incomes over the long run, movements take place when individuals or families decide that the benefits of a move outweigh the costs. The model is in many ways an improvement over the more simplistic factor mobility approach, though it is not unrelated to it conceptually. Clark (1986, 67) identified the following advantages of the human capital model over competing models:

1. Benefits occur over a period of time, which helps explain why migration rates drop with age.
2. Psychic as well as monetary costs and benefits can be included.

c) The "New Economics Approach"

The "new economics approach," an unfortunate name, perhaps, is a useful expansion on the micro models discussed above. The primary difference is this: the "new economics approach" argues that larger social units, primarily the family, rather than individuals, must be considered in the making of migration decisions. Rather than income maximization, proponents of this approach believe that risk minimization is the motivating factor in migration decision-making.

For example, consider a farming family in Mexico. If a son decides to migrate to the United States because of higher wages there, his decision affects the entire family's welfare. If he goes, his labor is lost at home; however, if the family can manage the farm without him, then remittances from him will increase the family's overall income and welfare. The remittances also act as a counterbalance for low food prices at home or poor crops. The family, then, decides as a unit whether or not the son should seek employment elsewhere.

> The "new economics approach" argues that larger social units, primarily the family, rather than individuals, must be considered in the making of migration decisions.

d) A Structuralist View

The essence of the structuralist view of migration, primarily international migration, is derived from neo-Marxist ideology or from world systems theory. It adds another dimension to the above economic arguments, maintaining that the wage rate differentials may stimulate migration but they are maintained by dependency relationships between countries of the periphery and those at the core. Briefly, the argument that structuralists make is that uneven development (an outcome of global capitalism) generates not just wage differentials but also dependency relationships. Migration results from differences in social class position, this view tells us, in a world divided into a few economic centers or cores (Japan, Western Europe, and North America) and a host of countries at the periphery, many of them former colonies of more developed countries. As Goss and Lindquist (1995, 322) noted, "The processes of underdevelopment create and sustain a dual labor market at the global level; . . . the Third World periphery provides for the reproduction of cheap labor that is selectively recruited by the core to counter falling rates of profit . . . ."

> The argument that structuralists make is that uneven development (an outcome of global capitalism) generates not just wage differentials but also dependency relationships.

e) A Structuration Approach

Another approach to the study of migration is the structuration approach, based primarily on the theory of structuration set forth by Anthony Giddens (1982, 1984), who concerned himself primarily with the "structure-agency problematic." Goss and Lindquist (1995) proposed this approach as a means of explaining international labor flows—we can only sketch for you here their major ideas.

According to Goss and Lindquist (1995, 335):

> . . . we will attempt to develop an alternative approach to international labor migration, drawing up Giddens's conception of agency and structure, the notion of institutions as "sedimented" social practices, the effects of the operation of rules and resources or "modalities of interaction" within institutions, the strategic actions of individuals, and time-space distanciation. More specifically, we suggest that what have previously been identified as migrant networks be conceived as migrant institutions that articulate, in a nonfunctionalist way, the individual migrant and the global economy, 'stretching' social relations across time and space to bring together the potential migrant and the overseas employer.

The essential idea here is that in the complex world of today's global economy there are intermediaries (migrant institutions) which function to bring together potential migrants in less developed nations with opportunities offered within the more developed or rapidly developing nations within the global economic system. As Goss and Lindquist (1995, 335) put it, "International labor migration can then be conceived as a process whereby individuals transcend the limits to presence-availability and negotiate their way across boundaries between locales in order to establish presence and control over resources in a distant place . . . ."

On their own, neither individuals in the Third World nor employers in wealthier countries are likely to succeed in matching their needs. Formal and informal organizations often, and increasingly, serve the function of "matchmaker" between the two. The authors use the Philippines as an example. Foreign employers are not allowed to directly recruit Philippine laborers; rather, they must go through licensed recruitment agencies—overseen by the Philippine Overseas Employment Agency—in that country. These recruitment agencies, then, are able to decide which potential migrants are going to be linked via migration to foreign employers in search of their labor. Regardless of where they live, these migrants ultimately depart from Manila.

Given the likely long-term persistence of structural differences in the global economy, it seems promising to look at the role of such migrant institutions in establishing and perpetuating streams of labor migration.

f) A Behavioral View

Behavior is the range of a person's observable actions and mannerisms within a given environment.

Among geographers, Julian Wolpert (1965) was one of the first to approach migration by concentrating on individual rather than on group **behavior**. He argued that understanding and explaining migration patterns was dependent on sorting out constants in migration behavior. Wolpert identified three central concepts of migration behavior:

1. place utility
2. field theory approach to search behavior
3. life-cycle approach to threshold formation

According to Wolpert (1965, 162), place utility ". . . refers to the net composite of utilities which are derived from the individual's integration at some position in space." Dissatisfaction with one's current location, then, is the major stimulus for beginning a search for another location. Place utility may be either positive or negative, depending upon how an individual perceives his or her location. Of considerable importance, then, is how the individual perceives the utility of his or her current place relative to the perceived utility of other places. According to Wolpert (1965, 162):

> The utility with respect to these alternative sites consists largely of anticipated utility and optimism which lacks the reinforcement of past rewards. This is precisely why the stream of information is so important in long-distance migration—information about prospects must somehow compensate for the absence of personal experience.

Subsequently, Wolpert (1966) suggested that the study of residential mobility should be viewed in terms of stresses in the current place of origin and the potential migrant's threshold for stress. Thus, the decision to move may be viewed as being dependent on both the alternatives available at a given time and a person's ability to cope with stress. Somewhat earlier, Rossi (1955) noted that residential complaints and dissatisfaction were important determinants of the decision to move. Wolpert proposed that a person's tolerance for stress could be measured by a "threshold function." Once a person's stress threshold is surpassed in a particular location, that person is likely to move.

The role of the life cycle in generating migration has attracted attention from other researchers as well, focusing on the related notion of "life course," a somewhat broader and more complex interpretation of changes that people undergo as they pass through their lives. Withers (1997), among others, has championed the importance of life course in understanding changing migration patterns. The life course approach has proved useful in studies of residential mobility. Billari and Liefbroer (2007) studied the impact of age norms on leaving home.

Despite the many reasons for which individuals choose to change their places of residence, it is generally agreed upon by researchers that the study of movement behavior should focus on two major decisions:

1. the decision to seek a new residence
2. the search for, and selection of, a new residence (Moore, 1972). Roseman (1977) also emphasized the need for considering separately the decision to move and the decision about where to move.

The Role of Information

Wage rate differentials alone fall short of fully explaining most migration patterns, though the prediction of movement up the economic gradient tends to hold true. Information is also important in the decision to migrate, especially information about alternative destinations. People are hesitant to move to areas about which they know little or nothing. Nelson (1959) found that friends and relatives who had previously moved from one place to another were especially good sources of information about their current location. Often this "friends and relatives effect" leads to chain migration, reflecting a directional bias on movers who leave a particular place. What we find, then, is that the distribution of past migrants tends to be a good predictor of future migration patterns (Greenwood, 1975). Migration streams, once established, are likely to be perpetuated for a long period of time. With respect to international migration, Goss and Lindquist (1995, 321) recently commented that ". . . it is as important to examine the flow of information and the role of previously successful migrants in recruitment networks as it is to identify objective income differences in the explanation of migration." It is also important to recognize that there is often a non-economic component in the decision to migrate (Halfacree, 2004).

Gender and Migration

At least until recently, most migration studies, especially those that focused on national and international labor flows, paid little or no attention to gender. However, as Castles and Miller (1993, 31) remind us, ". . . the migrant is also a gendered subject, embedded in a whole set of social relationships." Because of their traditional roles in patriarchal societies—primarily wives and mothers—female migrant workers have often been offered lower wages than males (Phizacklea, 1983). They have often been seen as easier to control than males as well.

Not only are female migrants sought for employment in many traditionally low-paying sectors of modern economies—including the garment industry, motel and hotel workers, maids, and nannies—they have also often been sexually exploited. In a study of Filipina migrant entertainers, Tyner (1996) considered the importance of perspective in migration studies. He deconstructed four controlling images of these Filipina migrants—the Other, the prostitute, the willing victim, and the heroine—and argued that these images in turn reflected the intentions, objections, and motives of the observers as they look at Filipina migrants. In concluding, Tyner (1996, 89) commented that:

> The discourse surrounding the causes and consequences of migration is equally informed by a set of images. These images reflect the observer's intention, objectives, and motives, as well as the political, social, economic, historical and geographical context of the encounter . . . . Only by transcending these representations, and uncovering the underlying sentiments associated with them, will meaningful dialogue—and hence policy formulation—be possible.

Another study of gendered migration focused on the importation of nannies into the United States (Hochschild, 2000), not just to care for children but also the elderly—in the process creating "global care chains." Paradoxically, many of the women who come to the United States to be nannies are themselves mothers, and they typically leave their own kids behind in the care of others. As the number of mothers in the work force in the United States has increased, so has the demand for people to take care of those children. Hochschild (2000, 35) comments that:

> At the first world end of care chains, working parents are grateful to find a good nanny or child care provider, and they are generally able to pay far more than the nanny would earn in her native country. This is not just a child care problem. Many American families are now relying on immigrant or out-of-home care for their *elderly* relatives . . . . But this often means that nannies cannot take care of their own ailing parents and therefore produce an elder-care version of a child care chain—caring for first world elderly persons while a paid worker cares for their aged mother back in the Philippines.

Though such migration streams may be rational responses to global capitalism's shifting opportunities, little is said of the "human costs" of such moves, when mothers leave their own children behind in poor countries to care instead for the children of working mothers in rich countries. Raising wages and respect for care-givers and encouraging males to contribute more to child care might improve the situation. Nonetheless, female migration from poor to rich countries is likely to continue to fill many care-giver niches, even though those women often leave behind children and elders that must themselves be cared for by others. As Hochschild (2000, 36) tells us, "Sadly, the value ascribed to the labor of raising a child has always been low relative to the value of other kinds of labor, and under the impact of globalization, it has sunk lower still . . . . And . . . the low market value of care keeps low the status of the women who do it." Critical and feminist geographers have further broadened our perspective on migration (Silvey, 2004).

Donato, et al. (2006) introduced a series of articles that updated and expanded work that geographers and others have been doing on gender and migration. Calavita (2006) discussed the gendered nature of immigration and various legal considerations, then focused on recent immigration to Spain and Italy. She concluded (2006, 104) that ". . . such exploration of immigrants' experiences in southern Europe reveals the surprising complexity of immigrants' multiple marginalities, and exposes the powerful contingencies of economic context, prevailing stereotypes, the particulars of state policy, and the agentive power of people struggling to survive." Piper (2006) looked at various aspects of migration as they relate to, and are shaped by, political factors. Sinke (2006) took a more historical approach, viewing the historiography on migration and gender in recent decades. She makes a good case for deepening the historical perspective that migration researchers use when studying gender and migration. Silvey (2006) provided an excellent overview of the contributions that geographers have made to the study of gender and migration. Feminist geographers have forced not only geographers but others to look at gendered migration in a new light. Mahler and Pessar (2006) illustrated ways in which gender has moved from the periphery closer to the center of migration studies, noting such factors as the role of gendered recruitment of immigrants in migration patterns. Clark and Huang (2006) focused on females and the problems of balancing migration and work, whereas Withers and Clark (2006) looked at housing costs and total family migration outcomes.

Gordon (2005) studied the changing sex ratio of immigrants to the United States, finding that (2005, 796) "The number of marriages between U.S. citizens or legal residents and non-citizens is highlighted as the dynamic factor determining the trend and proportion of male and female immigrants." From a different perspective, Taylor, Moran-Taylor, and Ruiz (2006) concentrated on how migration from Guatemala to the United States is impacting life in Guatemala. They found that (2006, 41) "Despite the advantages that migration brings to many families, especially in the face of a faltering national economy and state inactivity regarding national development, we conclude that migration and remittances do not result in community or nation-wide development." In fact communities and neighborhoods were little impacted, even though individuals obviously benefited from remittances. Conversely, other studies argue that "that remittances, even when not invested, can have an important multiplier effect. One remittance dollar spent on basic needs will stimulate retail sales, which stimulates further demand for goods and services, which then stimulates output and employment (Lowell and de la Garza, 2000 quoted in OECD 2006, 155). Finally, Cooke (2005) did an unusual study of what has become known as the "trailing wife" concept in migration, focusing only on same-sex couples. It has been well-established that when couples move because of a job change for the male (usually) the female who goes along with him to the new location has her employment opportunities harmed for perhaps several years. What he found was that (2005, 401) ". . . the migration of same-sex couples has no effect on the labour-market status and hours worked of men and women in gay and lesbian couples."

Residential Preferences and Migration

According to geographer Paul Schwind (1971, 150):

> Patterns of migration presumably are functions of, and should therefore contain information on, both the total magnitude of population systems and the aggregate preferences of migrants for relevant characteristics of regions. Migrants, however, probably respond not so much to factual statistical information about regions as to rather general perceptions of regional attractiveness.

Though geographers have accomplished a great deal in studies of place preferences, most of the work specifically relating to urban migration and city size preferences has been done by sociologists. Changing migration patterns in the United States since 1970 have stimulated researchers to seek explanations of migration patterns outside of the standard wage theory and its derivatives.

Studies of city size preferences are often concerned with the question of the need for a population distribution policy in the United States. Numerous proposals have been made, including those designed to aid depressed rural areas, such as Appalachia; to focus migrants on cities in the medium size range, thus decreasing in-migration to the most congested cities; and to guide the overall development of the American urban system. Fuguitt and Zuiches (1975, 491) noted that:

> An important element figuring in this discussion is concern about public preferences and attitudes on desirable places to live. A policy that provides community and housing options compatible with preferences should have a greater chance of success and could be expected to lessen any discrepancy between the actual and ideal distribution of the population.

Geographer Gundars Rudzitis (1991, 81) argued that: "Today, there is an increasing recognition of the need to incorporate noneconomic variables or surrogates into migration and regional growth models." He went on to note the importance of climate, trade-offs between climate and income, and the importance of what he called the "location-specific amenities model."

One criticism of many studies of city size preferences is that they failed to include an indication of the relative locations of cities, especially those in the smaller size categories. For many people a considerable difference in attractiveness exists between an isolated small town and a small town located within easy driving distance of a major city. Puzzled by the apparent contrast between survey results and actual migration patterns, Fuguitt and Zuiches (1975) added to a city size questionnaire a question asking those who preferred small towns whether they preferred those towns to be within thirty miles of a large city. What the distance-qualifying question showed was that a majority of those respondents who preferred small towns and rural areas said that they would like to live within thirty miles of a city of at least 50,000. Thus, we must be careful in interpreting the results of preference surveys that fail to include a distance-qualifying question or some relative location measure.

It appears that the perceived quality of life in small towns is their major attraction, however romantic the perception may be, whereas the lack of employment acts as a constraint on movement to them. As we have mentioned already, migration into small towns often destroys the very characteristics that attracted migration in the first place. Fuguitt and Zuiches (1975) concluded that if all people would move to the places they preferred, no mass exodus to remote areas would occur. Most people seem to prefer the best of both worlds, a quiet bucolic residential location and the opportunities and amenities that can only be provided by a metropolitan area.

Rudzitis (1991, 86) commented that there is a ". . . need for imagination rooted in regional historical reality if we are to better understand and promote the vitality of nonmetropolitan areas." He also suggested the need for geographers to consider such concepts as "sustainable development" in rural areas and small towns and the need for incorporating aspects of the relationship between local vitality and the natural environment in such areas. In concluding, Rudzitis (1991, 86) remarked that "A major shortcoming is the lack of studies of place, a comparative base from which to measure what might be expected . . . . A place, the world is what people take it to be, lying scattered openly on the surface, not as we social scientists often regard it, concealed beneath deceptive appearances." As geographer Peirce Lewis noted in his Presidential Address to the Association of American Geographers: "Good intellectual description provokes strong thought." (Lewis, 1985, 469) Many geographers are beginning to look more closely at the relationship between particular places and the migration streams that flow into and out of them. The relationship between mobility and attachment to place was explored by Bolan (1997).

Costello (2007) looked at urban-rural migration and some of its impacts on rural areas, a phenomena that has become widespread in the United States and parts of western Europe. Specifically, he focused on how such migration streams were driving house prices up in rural areas, decreasing local affordability and impacting the supply of rental housing adversely. Gallent (2007) focused on second homes and the concept of "dwelling." Gallent's approach is more philosophical than that of Costello's but both researchers call attention to the fact that rising affluence in recent decades has led to more people buying second homes. In turn this has stretched the long arm of many urban areas into the smaller towns and rural areas, where locals are often upset, priced out of local markets, and forced to compete for housing in markets that have been badly distorted by metropolitan affluence.

Selectivity of Migration

Migration is a selective process. Trewartha (1969, 137) described it thus:

> Assuming a sedentary population with an inducement to move, typically some individuals will leave and others remain where they are. But those who leave do not represent a random distribution of the biological and cultural characteristics of humanity in either the region of exit or entrance, for certain elements of the population tend to be more migratory than others. This is termed migratory selection.

Probably the most important and universally accepted migration differential is age. In movements within countries, as well as those between countries, it is older adolescents and young adults who predominate in most areas, as is apparent for the United States in

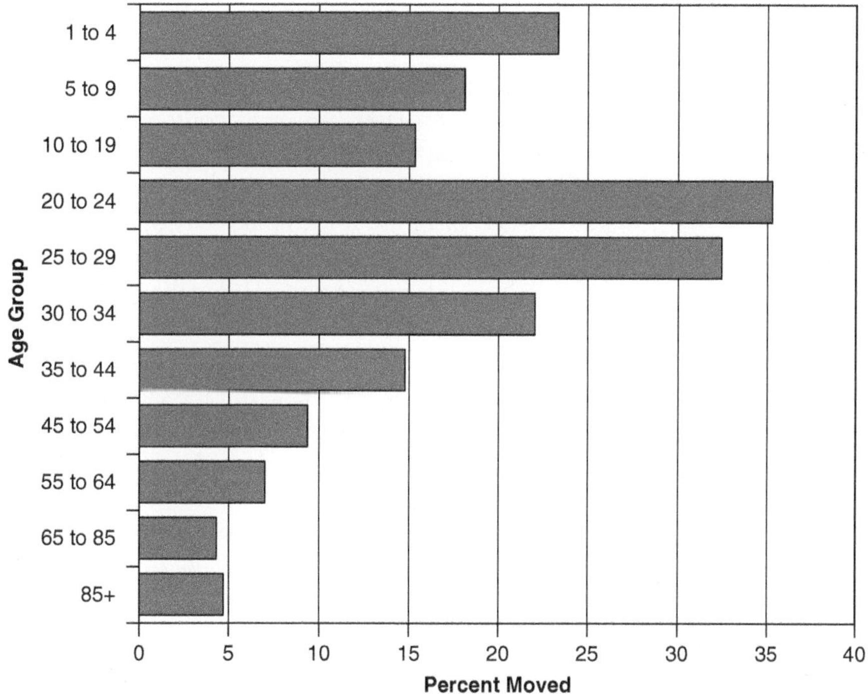

Figure 8–2 Total Mobility by Age of Movers, United States, 1999–2000
Source: United States Bureau of the Census, Geographical Mobility, March 1999–March 2000, Current Population Reports, P20-538.

Figure 8–2. Young people can generally adapt more easily to new conditions and, because they have only recently entered the labor force, they can change jobs more easily. Because of the age selectivity of migrants, we generally find that areas that have experienced a great deal of in-migration have a young age structure, whereas areas that have experienced a lot of out-migration are generally older in their population composition.

Marital status also affects migration selectivity. In the less developed areas of the world migration is usually undertaken by young, single adults. This was also probably true of the more developed world in past times. However, in the more developed countries today the married seem to be almost as mobile as the single.

Another facet of migration is sex selectivity, but whether the migration stream is male- or female-dominant depends upon a variety of factors. For example, in nineteenth-century Europe the movement from farm to city was primarily accomplished by young farm girls who went to work in the cities as domestic servants. On the other hand, the frontier towns in America were largely male-dominated. According to Trewartha (1969, 138):

> . . . there is some validity, no doubt, in the generalization that in less-developed countries migrants are predominantly male . . . . All the large commercial and industrial cities of India show a strong male predominance. The same is true in the newer settlements and mining towns of Negro Africa . . . .

Occupation and education are other variables that demonstrate migration selectivity. Unskilled workers are less likely to migrate than the skilled and semi-skilled; professional people are typically the most mobile occupational group. A considerable amount of research has been done on the relationship between education and migration selectivity. These studies have focused on the similarities and differences between the well-educated and the poorly-educated in regard to distances, rates, and direction of migration. Also, several studies dealing with education levels at the places of origin and destination of migrants have been undertaken.

In general, migrants have higher levels of education than nonmigrants, particularly if they have moved long distances. Folger and Nam (1967, 210) found that ". . . persons with higher levels of education are more migratory than persons with lower levels of education, and the difference in migration rates between poorly educated and well-educated persons increases for longer-distance moves." Similarly, data on migration in the United States between 1965 and 1970 pointed out the same pattern. For example, during this period, though the college-educated made up only 11 percent of the nation's total population aged 25 and over, they accounted for 25 percent of the people who migrated between noncontiguous states. Similar conclusions were reached with older age groups of the college-educated.

The above data and conclusions refer to relatively long-distance moves. An analysis of short-distance moves, such as from one house to another within the same county, reveals that in the United States the less educated are somewhat overrepresented; they tend to make more short-distance moves. They tend to change jobs more frequently, live in rental housing, and have more difficulty with the legal system.

With regard to changes in the educational status at the places of origin and destination, the data suggest relatively few significant changes at either place. In summarizing several studies, Folger and Nam (1967, 185) concluded that ". . . the effects of interregional migration on the educational attainment of the resident population are small." Though not directly related to education, migration is often very stressful (Mirdal, 2006).

Though there appears to have been relatively little effect on educational levels at places of origin and destination relative to internal migration in the United States, there has been a significant impact on the international scene. Educational self-selection, operating in the form of the so-called "brain drain," has become an international issue of concern. The less developed countries feel that the brain drain is hindering their cultural and economic development. Perhaps the most significant aspect is the migration of trained medical personnel to the more developed countries.

There are those, however, who argue that the brain drain does not necessarily have an adverse effect on the poorer countries. For example, they argue that discoveries made in one area will eventually benefit all areas, and that scientists and other highly educated persons should be in those areas where they can be the most productive. Though this may be true in the long run, there is little doubt that most less developed countries are in dire need of trained professionals if they are to proceed with economic and social development programs.

Interestingly, as the world economy becomes increasingly "globalized," we are likely to see increased movement among the more developed countries especially within Europe as well as from the poor nations of the world to the rich ones. In an interesting study of Americans who are seeking (and increasingly finding) employment abroad, Gray (1994) found M.B.A.s and others seeking employment throughout both hemispheres. Among reasons cited by interviewees were better opportunities for advancement and more responsibility at a younger age—most were finding the challenge of working abroad a positive experience. As Gray (1994, 46) concluded:

> Such optimism is comforting and, in a global perspective, almost certainly correct. National boundaries become ever less important in the world's economies; a job is a job, whether it be in Budapest, Buenos Aires or Birmingham, Alabama. Still, certain ancient human emotions have not yet adapted to the new realities.

On the other hand, keep in mind the growing number of Japanese and German auto workers who are now working in plants in America, building Hondas and Mercedes-Benzes. An escalation of movement across international boundaries seems inescapable in the decades ahead; it also will inevitably toss up new challenges for the survival of democratic societies everywhere. As Gat (2007, 66) pointed out, "All that can be said at the moment is that there is nothing in the historical record to suggest that a transition to democracy by today's authoritarian capitalist powers is inevitable, whereas there is a great deal to suggest that such powers have far greater economic and military potential than their communist predecessors did."

Consequences of Migration

Migration creates opportunities for geographic research. So far, we have focused primarily on the causes of migration, on the reasons why flows of migrants occur between places. However, migration can alter demographic, socioeconomic, and other conditions within societies. Because migration is a selective process, it produces changes in both the areas of origin and destination. The magnitude of these changes depends upon several variables. Because migrants are largely young, for example, areas of high out-migration will experience an aging of their populations. We see this in parts of the Great Plains area of the United States, as well as in parts of Appalachia (Werner, 2011, 12–15). In turn, as the age structure of an area becomes older, death rates increase and birth rates decrease, altering local age structures even more. In receiving regions, the opposite impact results. Thus, the effect of migration is not just a matter of numbers; selectivity by age, gender, race or ethnicity, education, income, and even ideology can be profitably considered.

Migration and its concomitant mixing of peoples have a variety of effects. There is a complex pattern of interaction between geographic, demographic, social, and economic factors, the consequences of which are therefore varied. Of particular significance to geographers is the impact of migration on space and numbers, an impact that takes two forms:

1. If destination areas absorb large numbers of people, their cities and towns will expand and, with the filling of the countryside, new lands open up, affecting the population distribution and density of the receiving area.
2. Conversely, the source areas for these migrants are likely to experience a decrease in population and a decline in density. Both population growth and decline produce far-reaching socioeconomic consequences.

Good examples of the impact of migration are the changes that took place in the United States and Ireland since the first third of the nineteenth century. Ireland has probably been more affected by emigration than any other country. Currently, the combined population of the Republic of Ireland and Northern Ireland is approximately 4.5 million people; in 1841 the Irish population was 8.2 million. Since that time, almost 5 million Irish have migrated to the United States and millions more have settled in Australia, Canada, and England.

An analysis of the Irish situation by Meenan (1958) pointed out that emigration has rarely been viewed as a beneficial force. Meenan (1958, 80) also noted some self-perpetuating features of the Irish situation and made the following statement:

> In each generation it is a movement of the younger sons and daughters. It is not an emigration of whole families and it never has been, except perhaps at the time of the famine. If it had been such, it might well have come to an end decades ago when the removal of families had led to a redistribution of farm holdings and the establishment of a new rural equilibrium. As things are, emigration enables those who are left to marry and rear families whose members will in their turn seek a living abroad. Certainly this is not all the story: A vicious circle is created whereby emigration causes underdevelopment, which in turn leads to further emigration.

Many believe that emigration can be advantageous to areas of dense populations. Studies of Italy's post-World War II emigration have shown that the economically disadvantaged regions, where most of the population was engaged in agricultural pursuits, provided most of the overseas emigrants. Population transfers relieved pressure on resources in these regions and made it possible for disadvantaged areas to increase their per capita incomes. Another factor noted in the study of Italy was that Italians living abroad sent money back to their homeland, thus reducing Italy's adverse trade balance. Similar circumstances have existed in other countries as well and may have helped to diminish the differences between the poor and the prosperous nations.

Since the 1990s, the role of migrant remittances in the economies of less developed countries has received significant global attention not least because these financial flows now dwarf the official development assistance of rich nations to these countries. In 2011, for instance, migrant remittance flows to less developed countries amounted to US$ 351 billion while official development assistance in the same year was about US$ 129 billion (African Economic Outlook, 2013; OECD, 2011). According to Plaza and Ratha (2011, 7) in 2010 "African migrants sent US$ 40 billion in remittances to African countries . . .". The remittances (i) help many Africans to weather negative economic shocks, (ii) support many African countries' creditworthiness and access to international capital markets, (iii) reduce the incidence and severity of poverty in many African countries, (iv) help households to diversify their sources of income and provide them with a source of savings and capital for investment, and (v) increase "household investments in education, entrepreneurship, and health, all of which have a high social return in most circumstances" (*ibid.*). In 2011, most (75 percent) of the Africans living outside the continent were from South Africa, the Arab Republic of Egypt, Morocco, Algeria, Kenya, Somalia, Tanzania, Ghana, Ethiopia, Uganda, Nigeria, and the Democratic Republic of Congo (*ibid.*, 5). Naturally, these same countries were some of the largest recipients of the continent's total annual remittances of about US$ 42 billion (African Economic Outlook, 2013). On average, remittances account for 5 percent of the GDP of most African countries and as a share of exports, represent 27 per cent (International Fund for Agricultural Development—IFAD, 2013).

Consequences of migration can be studied at many scales, from the above international example to that of the neighborhood or household. Changing metropolitan neighborhoods have long been studied by social scientists, though not always with a direct linkage to the migration processes that are altering them. Alba, et al (1995) is an excellent recent example of such studies; selected neighborhoods in the Greater New York metropolitan area are their focus. These neighborhoods, impacted considerably in the last two decades by immigration, are experiencing two different trends. On the one hand, the authors found that the entire area was becoming increasingly complex in its racial and ethnic characteristics; on the other hand, they found that segregated neighborhoods—primarily either African American or Latino, were becoming more numerous. The linkages between immigration and localized transitions in population composition deserve far more attention. Alba, et al (1995, 653–654) noted that:

> . . . the growing diversity of many neighborhoods may make it easier for their residents to ignore the problems besetting those who are left in all-minority neighborhoods. People, especially of the majority group, tend to judge diversity from their experiences in their own neighborhoods; seeing diversity among their neighbors, they may fail to realize that it leaves out a large part of a minority group. As our analysis has shown, this issue particularly affects blacks, a disproportionate number of whom do not reside in diverse neighborhoods.

At the household scale researchers have begun to look at such things as how households adjust when one or more members migrate—what happens to those left behind? Often, males migrate, leaving behind females who must then become household heads. Clearly, whether we focus on households in less developed countries, from which males have emigrated to take advantage of jobs in more developed countries, or households in less developed country metropolitan areas, from which males have moved, we are going to find changing roles for women and children. If remittances are not made to these households from those who left, then we can expect to find household incomes lower, and the fate of children within them more precarious. Everywhere we look—from the richest to the poorest nations—children are experiencing increasingly difficult roads to adulthood.

International Migration

There are many ways to classify the complex and diverse movements of human populations, but perhaps the most common division is to divide movements into two basic categories, international migration and internal migration. As Clark (1986, 74) has noted:

> International migration has a two-part structure: On one hand, it is composed of the labor streams seeking jobs in foreign countries; on the other, it is composed of the increasing flows of refugees from war and political disruption. However, neither of these flows takes place in a vacuum. The labor streams are inextricably woven together with changing urban development and there are increasingly dramatic social and economic impacts from global refugee migrations.

The Changing Nature of International Migration

Though international movements of people are as old as history, there was little useful information on the volume and nature of such movements before the nineteenth century. Archaeological evidence suggests that even the earliest ancestors of humans were wanderers. A progenitor of modern humans, *Homo erectus*, has been traced through fossil evidence to such widely separated places as China, Africa, Europe, and Java (Howell, 1965). The first "Americans" were most likely people who migrated from Siberia to Alaska over a land bridge that existed at various times during the Pleistocene. A Biblical example of migration was the movement of the Children of Israel out of Egypt to the Promised Land under the leadership of Moses. Many other examples of early migrations exist, and there is little doubt that all of human history has been dramatically affected by such movements.

One of the largest migrations in human history was the emigration from Europe that began in the sixteenth century and continued into the early twentieth century. Estimates are that over 60 million Europeans were involved in this migration. The primary places of origin for migrants were Germany, Russia, Poland, Spain, Austria, Hungary, Italy, Portugal, Sweden, the Netherlands, and the British Isles. The primary destinations for the emigrants were the United States, Canada, Australia, New Zealand, Brazil, Argentina, the British West Indies, and South Africa.

The flow of migrants out of Europe was relatively slow until the nineteenth century. Because of declining death rates in Europe during the early nineteenth century, the European population was growing rapidly. Also, Europeans began to feel that life was perhaps more rewarding abroad and that the overseas lands looked attractive. At the same time new technological improvements, such as the steamship, made migration easier. This is not to say that emigration from Europe was not fraught with dangers; geographer Beaujeu-Garnier (1966, 172) pointed out that

> . . . the exiles, often very poor, made lengthy journals on the deck or in the hold of a wretched vessel. Of the starving Irish who quitted their country in the famine of 1845, 6 percent died at sea, and counting also those who died soon after arrival, one in five of these unhappy creatures failed to make a home on the welcoming continent.

Early international migration from Asia was relatively small compared with the massive movement out of Europe, though millions of Chinese moved to nearby countries such as Thailand, Malaysia, and Indonesia. Migration from Africa was significantly different from movements out of other regions of the world before the twentieth century. The huge exodus of Africans was primarily involuntary. Slaves were taken from Africa and brought to the New World from the sixteenth century to the nineteenth century. Approximately 10 million slaves were imported into slave-using areas between 1451 and 1870. The number of slaves actually taken from Africa has been estimated at over 11 million, since many died en route to their destination. In terms of both the numbers and distance moved this was the greatest slave migration in history, a true diaspora, a concept that has received much attention in recent years (Carter, 2005).

International Migration: Patterns and Trends Since 1900

With more nation-states, more wars and regional conflicts, and more people during the twentieth century than ever before, it is little wonder that international migration has increased dramatically since 1900. Though much of this movement remains between poor countries, streams of migrants heading toward the rich countries have increased considerably as well. International migrants are driven by both "push" factors in places of origin, especially the lack of economic opportunities, and by "pull" factors in destination places, most often economic opportunities. Of course international movements of people take place for non-economic reasons as well, as we see in the continuing plight of the world's refugees, most recently those from Afghanistan.

Though it began much earlier, the modern global economy took shape during the twentieth century with the emergence of a tri-polar realm of rich places—Japan, North America, and the European Union—and a vast array of poor places—most of Africa, much of Asia (aside from Japan), and most of Latin America. During this same period of structural economic changes, demographic changes were occurring as well, shaped primarily by the demographic transition's course in selected places, mainly among the rich nations. Together, these economic and demographic changes have shaped, and continue to shape, modern patterns of international migration. Today, the rich nations are the world's major sources of capital, but the poor nations are the major source of people, hence labor. There seems little alternative in the future to movements that will bring capital and labor together, either by exporting capital from the rich nations to the poor ones or by importing labor from the poor nations to the rich ones. In reality both processes are going on, and both are likely to increase in the new millennium. Modern communications and transportation systems facilitate both the knowledge of differences among places and movements generated by that knowledge. Though the number of international immigrants has increased considerably over the last 100 years or so, their percentage of the total population has remained relatively small. In 2009, for example, there were an estimated 215 million people living in countries other than those of their birth, out of a total population of 7 billion. Most people still remain in the country in which they were born.

The twentieth century brought with it not only more nation-states but also more restrictions on international migration. Visas, passports, and other documents are evidence of changing attitudes toward movement between countries. As the current century began, only a few countries in the world actually welcomed international migrants—chief among them was the United States, but also included were Canada, Australia, New Zealand, and Israel. Even in these countries the welcome mat was not always out for just anyone, though on the other hand illegal entry into these countries is also a problem, especially in the United States.

In a world so divided into haves and have-nots, migration is one solution to improving lives and reallocating labor supplies. As Martin and Widgren (2002, 5) pointed out, "Another shift of population may occur in the next century as population growth and a lack of economic opportunities in one region foster large-scale migration to another, more prosperous, region." On the one hand, many rich nations are already experiencing labor shortages, and immigration can fill many of those niches (Goldstein, Lutz, and Scherbov, 2003; Bongaarts, 2004). However, this is not occurring anywhere without contention.

As Richburg (2004, 1) noted, "Across Europe, societies that were once solidly white and Christian are being recast in a multicultural light. The arrival of large numbers of people from the Middle East, East Asia and Africa . . . is pushing aside old concepts of what it means to be French or German or Swedish." For example, in Malmo, Sweden's third largest city, about 40 percent of the population is foreign-born. Immigration in Europe has led to the rise of political parties with anti-immigrant leanings, including the National Front in France, the Freedom Party in Austria, and the Pim Fortyn party in the Netherlands. Similar concerns have emerged in the United States. Concerned with the rise of immigration and notions of identity in America, but seeing such concerns emerging in other countries as well, Samuel P. Huntington (2004, 13) noted that "The more general causes of these quests and questionings include the emergence of a global economy, tremendous improvements in communications and transportation, rising levels of migration, the global expansion of democracy, and the end both of the Cold War and of Soviet communism as a viable economic and political system." Rosholm, Scott, and Husted (2006) found that in Scandinavia immigrant employment opportunities were actually declining. They suggested (2006, 318) that "A possible explanation is that the changing organizational structure–toward more flexible work organization—has resulted in a decrease in the attractiveness of immigrant employees due to the increasing importance of country-specific skills and informal human capital." Schou (2006, 671) added that ". . . increased immigration will generally worsen the Danish fiscal sustain-ability problem. Improved economic integration of immigrants and their descendants, however, may alleviate the problems of the public sector considerably."

Cultural differences emerge quickly, and many citizens don't prefer to use immigration to solve labor shortages, even as they grow older and seek to retire at younger ages. Nonetheless, low birth rates, aging populations, and mixed reactions to immigration are placing strains on the economies of most rich nations, especially Japan and most of the European Union (Johnson, 2004). Most of the nearly three billion people expected to arrive over the next 50 years will be born in the poor nations, and their impact on the rich nations may be overwhelming—they will not all remain in their homelands unless globalization leads to a much fairer distribution of wealth around the world.

At the beginning of the twenty-first century several major international migration patterns were discernible. Among them were significant flows of people from Asia and Latin America to the United States and Canada; flows from Asia, North Africa, and Eastern Europe into Western Europe; and flows from the Philippines and a few East Asian countries to the Arabian Peninsula, mainly Saudi Arabia. Eberstadt (2004) noted Asia's importance as a source of immigrants, though his focus was mainly on the relationship between demography and power. Smaller flows include those from Latin America and some Asian countries into Japan and from Southeast Asia into Australia. A majority, perhaps as much as two-thirds, of international immigrants today either enter countries illegally or enter them legally and then remain illegally after visas have expired or beyond planned visitations. Even in tiny Albania people are on the move as never before. According to Carletto, et al. (2006, 767) "Although migration, with the resulting remittances, has become an indispensable part of Albanian economic development, there is increasing consensus on the necessity to devise more appropriate, sustainable strategies to lift households out of poverty and promote the country's growth." Remittances are important elsewhere as well, from Mexico to Turkey. Cohen and Rodriguez (2005) looked at how remittances are reshaping rural Oaxaca and de Haas (2006) focused on how remittances were aiding regional development in southern Morocco. Without doubt, billions of dollars flow out of immigrant-receiving nations and back to places of origin.

Flows of immigrants and emigrants affect different places in different ways. Portes, Escobar, and Radford (2007) wrote about how transnational immigrant organizations pursue philanthropic projects that aid in local development. Williams (2006, 588) commented that "There are changing but increasingly important ways in which international migration contributes to knowledge creation and transfer." Asiedu (2005) looked at the benefits of return migrants to Ghana. Lucas, Amoateng, and Kalule-Sabiti (2006) studied migration streams between South Africa and destinations such as the United Kingdom, Australia, and New Zealand. They noted that at least some of this emigration was a "brain drain." Millar and Salt (2007, 41) focused on IT (information technology) and migration, arguing that ". . . the UK's broad-brush approach to managed migration is out of touch with the international sourcing policies of firms in the IT sector and has been weak in defending the competitive position of the resident workforce." Finally, Hugo (2006) looked at immigration responses to global change in Asia, noting what he called the three "Ds" and their importance: demography, development, and democracy. Other migration studies have focused on individual countries: Cavounidis (2006) on Greece, Gerber (2006) on Russia, Niedomysl (2005) on Sweden, and Raymer, Bonaguidi, and Valentini (2006) on Italy. Geographers are doing an abundance and variety of research in international migration, as these examples suggest.

Immigration policies vary considerably from nation to nation, and those of the United States are dealt with in a subsequent section. Here we mention some studies that touch upon the policy side of migration. DeVoretz (2006) outlined a series of economic criteria to assess immigration policy from three perspectives: the resident population, the immigrant, and the sending country viewpoints. Boswell (2007) added to the theorizing of migration policy. Epstein and Nitzan (2006) analyzed the endogenous determination of migration quotas, among other things, noting that the struggle over migration policies tends to come down to the strength of lobbyists among special interest groups. Hix and Noury (2007) argued that it was not so much economic interests as politics that ultimately shapes immigration policies.

Refugees: A Special Category

A **refugee** is defined as a person who is outside the country of his nationality . . . because he has or had well-founded fear of persecution . . . and is unable or, because of such fear, is unwilling to avail himself of the protection of the government of the country of his nationality.

The Protocol Relating to the Status of Refugees—agreed to by the United Nations in 1967—defines a **refugee** as a ". . . person who is outside the country of his nationality . . . because he has or had well-founded fear of persecution . . . and is unable or, because of such fear, is unwilling to avail himself of the protection of the government of the country of his nationality." According to the United Nations High Commissioner for Refugees–UNHCR (2011, 21) there were 10.5 million refugees in the world at the beginning of 2011. An additional 14.7 million people were "refugees" or internally displaced persons (IDPs) within their own countries (UNHCR 2011, 21). While both of these figures are estimates, what is known with more certainty, however, is that in recent decades, as a result of wars and famines (Bosnia and Rwanda are excellent examples), refugees and IDPs have increased in numbers. At the same time, geographers have shown an increased interest in refugees in recent years, and in one article Kliot (1987) noted that since the beginning of the twentieth century more than 100 million people have become refugees. As with other forms of migration, refugees affect both the region they leave and the one to which they move. Refugees often flee one form of conflict only to precipitate another often in neighboring areas or countries. For this reason, the "major refugee-generating regions [of the world] hosted on average between 76 and 92 per cent of refugees from within the same region [in 2010]. [The] UNHCR estimated that in 2010 some 1.7 million refugees (17% out of the total of 10.55 million) lived outside their region of origin" (UNHCR 2011, 24).

Kliot (1987) argued that among the major causal agents of current and future refugee movements are the following:

1. ethnic, racial, political, and religious conflicts within countries like Sudan and Afghanistan
2. local and regional wars, from South Africa and the Middle East to the breakup of Yugoslavia in the 1990s into Bosnia, Croatia, Kosovo, Macedonia, Montenegro, and Serbia
3. changing racial policies of many governments in Africa and Asia

Geographer Richard Black (1991, 281) noted that ". . . geographers must draw links between this field and their own longstanding interests in issues such as migration, 'natural' and other disasters, and the politics of conflicts which are often the immediate cause of refugee flows."

An excellent exposition of refugees is Zolberg, Suhrke, and Aguayo (1989). The authors begin by criticizing the accepted definition of refugees (cited above); instead ". . . we point out only that disagreement on these issues is unavoidable, because defining refugees for purposes of policy implementation requires a political choice and an ethical judgment" (Zolberg, Suhrke, and Aguayo 1989, 4). Their major fear is that acceptance of an earlier definition gives it credence and focuses research on ongoing practices. After thirty years, they suggest, perhaps the definition itself needs reconsideration because of the changing world system. The authors, after several detailed discussions, suggest that refugee studies need to focus more on the *root causes* of flight—included among them are poverty and social change—and the development of policies that might mitigate them. They note that the burden of asylum has fallen heavily on the less developed countries and argue the following (Zolberg, Suhrke, and Aguayo 1989, 282):

> . . . the richer states must, at a minimum, accept a greater financial obligation to assist the countries of first asylum in the South . . . . Refugee policy must be held up against the negative yardstick that at least it should not contribute to greater refugee flows in the future.

Though we have looked only briefly at refugees in this section, geographers are now studying refugees in many places. Rogge (1991), for example, has looked at the uncertain future for refugees in Southeast Asia and Everitt (1991) considered the refugee problem in Middle America since the 1970s. Luciuk (1991, 197) ended his study of refugees in Afghanistan by saying that "Since even the most optimistic estimates suggest that many years of economic, political, and social rehabilitation will be required before the land and its peoples recover from what has been described as a genocidal war, it seems evident that Afghanistan will remain a deeply troubled region of the world, and, as such, a source area of refugees and displaced persons, well into the foreseeable future." In surveying the refugee situation in Africa, Kenzer (1991, 200) concluded that the situation was ". . . horrible at best." This was said, prophetically, long before the 1994 debacle that almost completely destroyed Rwanda. Elsewhere on that continent other devastating events are probably not far off, perhaps in Burundi, or Nigeria, or Mozambique. Gao (2006) studied the international networks of Chinese asylum-seekers and their role in asylum-seeking processes.

Immigration: The United States

"They're coming to America," sang Neil Diamond some years ago in a hit tune, and so they are still. The United States has long been recognized as a nation of immigrants, even if its immigration history has been a checkered one. Today some 30 million foreigners enter the United States each year, mostly as visitors, but also as students and temporary workers. Globalization is likely to increase, not decrease, the exchange of people across international borders, both temporarily and permanently.

Immigration has played a key role in the social and economic development of the United States. If there had been no immigration to the United States during the nineteenth and early twentieth centuries, then population growth would have been much slower and the ethnic composition of the United States' population would be substantially different.

Immigration history in the United States can be divided into four distinct periods: colonial, old immigration, new immigration, and restricted immigration. The first three periods could be considered times of free-individual moves, whereas the last period involved significant restrictions on immigration.

The immigrants who came to the United States during the colonial period were confronted with a situation that was significantly different from that of later arrivals. Most were of British ancestry, though other ethnic groups such as the Dutch, Swedes, Germans, French, and Spanish were also represented.

The period of old immigration was approximately 1800 to 1880 and involved a considerable increase in immigration. The overwhelming majority, about 95 percent, of these immigrants came from western and northern Europe. Though the largest proportion of these immigrants came from England, large numbers of them were from France, Germany, Scotland, and Ireland.

The period of new immigration started around 1880 and was marked by a shift in the source of immigrants from western and northern Europe to eastern and southern Europe. This new wave of immigration was much larger than the old, and in the first decade of the twentieth century approximately 9 million immigrants entered the United States.

Since 1921 immigration to the United States has no longer been free. Restrictions on both the number and types of migrants are in effect. The origin of immigrants has shifted dramatically, however. Before 1900 immigrants came almost exclusively from Europe. Today they come mostly from Asia and Latin America. Martin and Midgley (2006, 4) note one of the many dilemmas associated with American immigration policy, "Foreigners enter the United States through a front door for legal immigrants, a side door for legal temporary migrants, and a back door for the unauthorized." As Massey (2006, 584) reminded us, "Over the course of U.S. history, attacks on immigrants have waxed and waned, yet in the long run American society has incorporated an ever-widening array of peoples and nationalities into the national franchise."

Key Contemporary U.S. Immigration Laws

Since the 1950s, U.S. immigration has been governed by the following incremental pieces of immigration laws.

Immigration and Nationality Act (INA) of 1952

This is the basis of U.S. immigration law. Prior to INA, U.S. immigration was governed through a variety of statutes that the McCarran-Walter bill of 1952 collected, codified and structured into the INA (U.S. Citizenship and Immigration Services—USCIS, 2013a). While Congress has since amended INA many times (e.g., 143 times between 1986 and January 2013) with other pieces of legislation such as those discussed here (U.S. Citizenship and Immigration Services—USCIS, 2013b), INA continues to serve as the basic body or corpus of U.S. immigration law. This incremental approach to immigration law exists because whenever "[the U.S.] Congress enacts a law, it generally does not re-write the entire body of law, or even entire sections of a law, but instead [it] adds to or changes specific words within a section. These changes are then reflected within the larger body of law" (ibid. no pp).

The Immigration and Naturalization Act of 1965 (popularly known as the Hart-Cellar Act)

Up to the 1960s, U.S. immigration policy was based on a national-origins quota system that had been in place since the 1920s. This system, which based its quotas on the representation of each nationality in past U.S. censuses, largely favored European immigration. But as the U.S. civil rights movement gained strength in the early 1960s, the quota system of unequal treatment on the basis of race and national origin became untenable. Before the passage of the Immigration and Naturalization Act of 1965, the quota system had, for instance, discriminated against groups likes Greeks, Poles, Portuguese and Italians in favor of Northern Europeans (History Channel, 2013). Thus the Hart-Cellar Act "replaced the national origins [quota] system with a preference system designed to reunite immigrant families and attract skilled immigrants to the United States" (Smith, 2009, no pp). Aside from the civil rights challenges to the pre-1965 immigration quota system, this law was a response to "changes in the sources of U.S. immigration since 1924. The majority of applicants for immigration visas now came from Asia and Central and South America rather than Europe" (*ibid.*). Nevertheless, the Hart-Cellar Act's immigration preference system continued to restrict the annual number of immigration visas (*ibid.*). This practice still stands.

Much more than any immigration law, the Hart-Cellar Act has, since its passage, fundamentally changed the demographic composition of American society by shifting the center of U.S. immigration to the non-European regions of Latin America, Asia and Africa. Dubbed the "law [that] changed [the] face of America" (Ludden, 2006, no pp.), the diverse U.S. immigration stream that this law unleashed in the 1960s not only partly contributed to the election of the first black president of the U.S., Barack Obama, in 2008; but is also likely to make the U.S. a non-White majority country by the middle of the twenty-first century (History Channel, 2013).

Refugee Act of 1980

Before the passage of this law, the United States did not have a general policy governing the admission of refugees. Thus, the U.S. used special legislation to admit refugees, such as those from Indochina in the 1970s (Smith, 2009). Through the Act, Congress affirmed the historic policy of the U.S. to respond to the urgent needs of people facing persecution in their homelands. Moreover, according to Kennedy (1981, 143) the law: (1) ended previous discriminatory treatment of refugees by offering protection to all persons meeting the test of the United Nations Convention and Protocol on the Status of Refugees of 1951, (2) capped annual refugee admissions at 50,000, (3) provided for an orderly but flexible procedure to deal with refugee emergencies that exceeded the annual quota, (4) gave Congress control over the entire refugee admission process, (5) created an explicit asylum provision in U.S. immigration law for the first time, and (6) provided a full range of federal programs to assist in the refugees' resettlement process.

Since 1980, the law has facilitated the admission and settlement of 2.6 million refugees in the United States (United States Department of Homeland Security, 2012b, 39). Between 2002 and 2011, most of these refugees were from Burma, Iraq, Somalia, Bhutan, Cuba, Iran, Ukraine, Liberia, Russia, Laos, Vietnam, Sudan, Ethiopia, Burundi, Democratic Republic of Congo, Eritrea, Afghanistan, and Moldova (*ibid.* 40). The vast majority of these countries have experienced disruptive events such as civil wars in that period. Similarly, the U.S. has affirmatively or defensively granted asylum to 178,591 individuals between 1990 and 2011 (*ibid.* 43). "To obtain asylum through the affirmative asylum process [one] must be physically present in the United States [while an immigration judge considers eligibility for asylum] . . . [Conversely, a] defensive application for asylum occurs when an individual requests asylum as a defense against removal [or deportation] from the U.S." (Refugee Council USA, 2013, no pp).

The Immigration Reform and Control Act of 1986 (IRCA)

This law "was passed . . . to control and deter illegal immigration to the United States. Its major provisions [stipulated] legalization of undocumented aliens who had been continuously unlawfully present [in the U.S.] since [January 1] 1982, legalization of certain agricultural workers, sanctions for employers who knowingly [hired] undocumented workers, and increased enforcement at U.S. borders" (U.S. Citizenship and Immigration Services—USCIS, 2013c).

The law came after years of political bickering on how to deal realistically with problems resulting from illegal immigration. Among other things, the law required employers to ensure that their workers were eligible to work in the U.S. by documenting their status through a U.S. passport, certification of U.S. citizenship, an alien registration receipt card, a foreign passport with a work authorization stamp, a U.S. birth certificate, a social security card and, in addition to one of the above, proof of identification in the form of a valid driver's license or a state identification card. However, because the law did not hold employers responsible for the authenticity of each of the above items, though they had to keep them on file for a period of at least three years from the date of their submission, it inadvertently led to the development of a cottage industry in fake document production in U.S. border states. Moreover, the law caused the traditional pattern of circular and repeat migration from Mexico to the United States to become more permanent. As Vernez and Ronfeldt (1991) noted, "Mexican immigration to the United States can no longer be characterized by the persistent image of the Mexican immigrant as a temporary worker staying here for a short period of time and leaving his or her family behind, if it ever could be so characterized." Most of the permanent "temporary" Mexican migrants have since significantly impacted the socioeconomic conditions of states in the southwest, especially California and Texas (Bean, Edmonston, and Passel, 1990).

Since the 1990s, the U.S. has been trying to make illegal border crossings from Mexico more difficult. Remote video surveillance sites, thermal infrared imagers, night vision goggles, long-range infrared systems, or fiber optic borescopes, and intelligent computer-aided detection systems have all been employed to help Border Patrol officers detect illegal immigrants. In response, immigrant routes have shifted to less-guarded interior sites. Close to three decades after IRCA, America now has about twelve million illegal immigrants. So IRCA has failed largely because the economic fundamentals that drive Mexican immigration to the U.S. have not changed.

The Immigration Act of 1990

Because of significant public concerns about the level of immigration to the U.S., Congress passed the Immigration Act of 1990 which instituted "the most sweeping reform in U.S. legal immigration law in the [previous] sixty six years" (Leiden and Neal, 1990, 328). While "the Act [introduced] for the first time an overall cap on worldwide immigration that includes the immediate relatives (spouses, minor children, and parents) of U.S. citizens," (*ibid.* 329), the new law "increased the limits on legal immigration to the United States, revised all grounds for exclusion and deportation, authorized temporary protected status to aliens of designated countries, revised and established new nonimmigrant admission categories, revised and extended the Visa Waiver Pilot Program, and revised naturalization authority and requirements" (U.S. Citizenship and Immigration Services-USCIS, 2013d). "The [law's] division of immigrant visas into three areas—family-based, employment-based, and diversity—reflects the different interests behind U.S. immigration policy" (Leiden and Neal, 1990, 329). Overall the law raised the annual legal immigration limit from 500,000 to 700,000 with some adjustments to be made after 1993—when a required review of immigration was to be made. In reality, between 1990 and 1993, the U.S., on average, admitted 1.3 million legal immigrants a year and has since maintained an annual average of 1 million (United States Department of Homeland Security, 2012b, 5). In sum, 22.2 million legal immigrants entered the United States between 1990 and 2011 (*ibid.*).

A key aspect of the 1990 law is that it also encouraged the immigration of skilled, educated, and wealthier people to the U.S. Thus, up to 10,000 slots per year were made available to investors who were willing to put at least $1,000,000 into an American business that employed ten or more workers; with only half that amount being required if the investment was made in a rural or depressed area. The law also sought to further diversify the U.S. immigration stream by creating a Diversity Visa (DV) program in which 55,000 immigrant visas were to be availed to applicants from countries with low rates of immigration to the United States starting in fiscal year 1995. Qualifying countries have since been determined annually as those that have sent less than 55,000 (reduced to 50,000 in the 1999 fiscal year) immigrants to the U.S. in the previous 5 years. While countries like Canada, mainland China, and the United Kingdom were ineligible from the start, others have since joined the list as the number of their migrants to the U.S. reached the stated threshold. At the continental level, Africa has benefited most from the DV program because of its historically low levels of immigration to the U.S. Therefore, while Africa sent 23,812 DV immigrants to the U.S. in 2011, Asia, the region with the next highest total, sent 15, 573 (United States Department of Homeland Security, 2012b, 27).

The Illegal Immigration Reform and Immigrant Responsibility Act of 1996 (IIRIRA)

IIRAIRA partly came into being to fix the weaknesses of previous immigration laws especially The Immigration Reform and Control Act of 1986 (IRCA). Thus, it was:

> "an effort by Congress to strengthen and streamline U.S. immigration laws. [Specifically] the Act was designed to improve border control by imposing criminal penalties for racketeering, alien smuggling and the use or creation of fraudulent immigration-related documents and increasing interior enforcement by agencies charged with monitoring visa applications and visa abusers. Employment eligibility verification guidelines [were] also incorporated into the Act, including sanctions for employers who [failed] to comply with the regulations and restrictions on unfair immigration-related employment practices, as well as provisions governing the disbursement of government aid to aliens" (Legal Information Institute, 2010, no pp).

As part of its attempts to improve border control, IIRIRA approved an expansion of border guards by 1,000 per year for five years, bringing the total number of border patrol agents to around 10,000 in 2000. It also introduced a pilot telephone verification program so that employers could more easily check on the immigrant status of job applicants and social service agencies could do the same for applicants for various government benefits, and altered several requirements for residents who sponsor immigrants (Germain and Stevens, 1996). IIRIRA's pilot telephone verification program has since morphed into E-Verify, "an Internet-based system that allows businesses to determine the eligibility of their employees to work in the United States" (U.S. Citizenship and Immigration Services—USCIS, 2012, no pp).

USA PATRIOT Act of 2001

On October 26, 2001, less than two months after the incredible events of 9/11, President George W. Bush signed into law the USA PATRIOT Act, formally known as the Uniting and Strengthening America by Providing Appropriate Tools Required to Intercept and Obstruct Terrorism Act of 2001. Title IV of the act, Protecting the Border, contains three parts that relate to immigration—Subtitles A, B, and C.

Subtitle A, entitled Protecting the Northern Border, contained a collection of stopgap measures that were designed to improve security at the border between Canada and the United States. Among its provisions were the deployment of adequate personnel on the U.S.-Canadian border and the limited authority to pay any necessary overtime.

Subtitle B, entitled Enhanced Immigration Provisions, was much broader, and of more concern to many supporters of immigration. Included in this section were new definitions relating to terrorism, mandatory detention of suspected terrorists, a foreign student monitoring program, future requirements for machine-readable passports, and the prevention of consulate shopping by foreigners seeking visas to the U.S.

Subtitle C, entitled Preservation of Immigration Benefits for Victims of Terrorism, contained such provisions as those dealing with special immigrant status, extension of filing and reentry deadlines, temporary administrative relief, and a notice of no benefits to terrorists or family members of terrorists. Whether these, or other immigration laws, will stop terrorists from entering the country remains to be seen. A critical element here is not the laws themselves, but the rigidity with which they are, or are not, enforced. Enforcement of previous immigration laws would probably have stopped most of the nineteen terrorists who participated in the 9/11 attacks from entering the United States. Lax enforcement continues, employers across the nation hire and exploit illegal immigrants, and politicians pretend to be concerned.

Since 2001, Congress has enacted many other piecemeal immigration laws (U.S. Citizenship and Immigration Services—USCIS, 2013b) but has so far failed to pass comprehensive immigration reform even though calls for the same have been voiced for years. During the Presidency of George W. Bush from 2001 to 2009, the issue of comprehensive immigration reform continued to get some popular but ineffective congressional and presidential attention. When President Obama was elected in 2008 with the support of many minority groups, including the offspring of recent immigrants, many hoped for comprehensive immigration reform in his first term of office. However, these hopes were soon derailed by congressional conservatives and by the President's own focus on the country's economic recovery from the 2007–2008 financial crisis. As President Obama's second term gets underway, prospects for comprehensive immigration reform look much more promising because of weaker congressional conservative opposition occasioned by, among other things, widespread belief that Republicans' hardline opposition to immigration reform cost them the 2012 presidential race. Only time will tell however if Congress and the President will this time around capitalize on the emerging national bipartisan consensus to repair "America's 'broken' immigration system" (Simendinger, 2013, no pp).

Illegal Immigration

Since the establishment of migration quotas, there has been relatively little concern about the origin of present-day legal immigrants to the United States. There is, however, increasing concern over the presence of immigrants who have entered the United States illegally. Many of the factors that help explain legal migration also apply to illegal migration. The attributes of places of origin and destination hold true for **illegal immigrants** as well as for legal immigrants. The undocumented immigrant is confronted by unfavorable conditions in his place of origin and tries to overcome these difficulties by emigrating. According to Johnson (2006, 1) "... for the first time in the last 10 years—if not the first time ever—the flow of illegal immigrants into the country is larger than the flow of legal immigrants." It is no wonder that illegal immigration has become a hotly debated topic.

The attributes of places of origin and destination hold true for **illegal immigrants** as well as for legal immigrants.

The problem of undocumented immigration is not just associated with the United States. Similar problems exist in Latin America, Asia, and Europe. As Black, et al. (2006, 552) noted, "Interest in illegal or undocumented migration in European states in recent years reflects an apparent growth in international mobility that falls outside the law." Apparent? That is academic-speak for a lack of confidence in statistics, which are scarce, as opposed to anecdotal evidence, which is abundant. Glytsos (2005, 819) found in Greece that "Legalization can hardly solve the problem of immigrant employment, nor can it pull all immigrants out of the underground labor market and integrate them into the Greek economy and society." DeBardeleben (2005) presents a broad and thoughtful discussion of most aspects of European immigration, including a series of case studies, and Alba, Schmidt, and Wasmer (2003) look specifically at how migration has helped reshape Germany since reunification. Two other books on European immigration are worthy of attention as well: Zimmermann (2005) and Geddes (2003). Whereas the chapters in Zimmerman focus mainly on individual countries, Geddes systematically views migrant and immigrant politics throughout Europe.

More recently France, Germany, and Austria have experienced rightward political shifts as voters have become increasingly concerned with the presence of immigrants, especially those who have, or are at least perceived to have, entered illegally. Garcia (2006) studied the effect of anti-immigrant sentiment on elections in an attempt to better understand how rightist parties have fared.

Presently the United States attracts the largest number of illegal migrants. There is no actual count of the number of illegal migrants in the United States, but estimates range from 12 to 20 million. Nevertheless, there are data on the number of illegal aliens who have been apprehended. Data from 1970 to 2006, shown in Figure 8–3, suggest approximately a 500 percent increase in the apprehension of illegals during that period. It has been estimated that only one out of every three or four persons attempting to enter illegally was caught. As shown in Figure 8–3, the largest proportion of undocumented immigrants to the United States comes from Mexico. For instance, of the 1,199,000 illegal aliens located

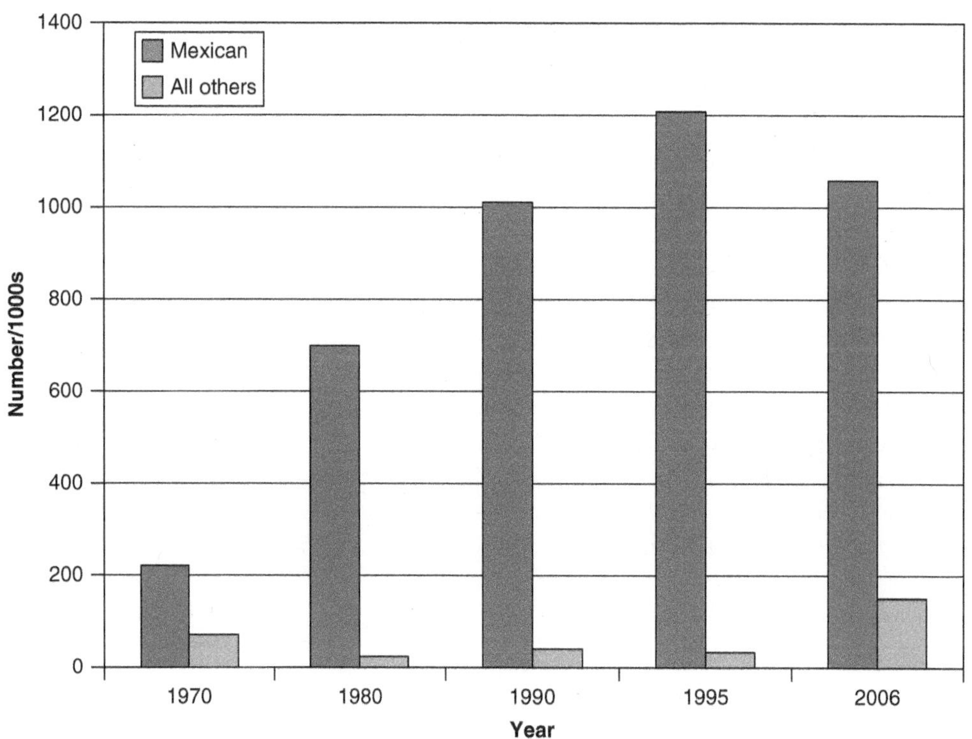

Figure 8–3 Deportable Aliens Located in the United States

Source: Statistical Abstract of the United States, 1995, Table 331, and yearbook of Immigration Statistics, 2006.

in the United States in fiscal year 1993, about 97 percent or 1,169,000, were from Mexico. Similarly, of the 10.1 million illegal aliens apprehended in the U.S. between 2002 and 2011, eighty-seven percent (or 8.8 million) were Mexicans (United States Department of Homeland Security, 2012b, 92). Mexican dominance of illegal immigration to the U.S. is because economic conditions in Mexico provide a "push" to migrate to the United States for employment. Moreover, Mexico's proximity to the U.S. greatly eases many Mexicans' legal and illegal migration to the U.S.

Guest worker programs are another source of illegal immigration to the U.S. and other more developed countries. While these temporary worker programs seem to be an ideal solution to the labor scarcity problem in many rich countries, these programs often fail because many guest workers end up staying. This is well-illustrated by the United States' *bracero* program and by Germany's *Gastarbeiter* program. In this global economic era when goods and financial capital move easily across borders, the "stickiness" of labor mobility has become a problem, mainly because all nations virtually restrict the free movement of people across their borders. This reality coupled with the near constant demand for cheap labor in the U.S. and other rich countries has created an active people smuggling industry. Thus, people-smugglers, whether "coyotes" in Mexico or "snakeheads" in China, continue to help meet this need through illegal labor supplies even as guest worker programs are being reconsidered in both the U.S. and Western Europe (Castles, 2006). Perhaps, as Jacoby (2006, 60), has argued, "The best way to regain control [of immigration in the United States] is not to crack down but to liberalize—to expand quotas, with a guest-worker program or some other method, until they line up with labor needs."

After the attacks of 9/11, after passage of the USA PATRIOT Act, and after repeated polls that show how dissatisfied Americans are with immigration in general, and with illegal immigration in particular, illegal immigrants are coming in ever larger numbers in the new millennium. According to Bartlett and Steele (2004, 51): "In a single day, more than 4,000 illegal aliens will walk across the busiest unlawful gateway into the U.S., the 375-mile long border between Arizona and Mexico. No shoe removal. No photo-ID checks. Before long, many will obtain phony identification papers, including bogus Social Security numbers, to conceal their true identities and mask their unlawful presence." Even before 9/11 President Bush had spoken about an amnesty program for illegal immigrants, and that talk, put off by the events of 9/11, but reemerging in 2004, is part of what drives further illegal immigration, along with the prospect of jobs. As Bartlett and Steele (2004, 58) pointed out, "Nonenforcement of employer sanctions, which is in keeping with the Federal Government's nonenforcement of immigration laws across the board, has been the equivalent of hanging out a HELP WANTED sign for illegals."

In the past many illegal migrants stayed close to the border after entry. However, recent data show that large numbers are now scattering throughout the nation, to New York, Houston, Chicago, Miami, San Antonio, and Los Angeles. The problem of illegal aliens is complex. Many interest groups are involved with both sides of the issue. Labor unions are fearful that jobs that could be held by United States citizens are being taken by illegal aliens. Those who favor limited or zero population growth see the influx of illegal aliens as a stumbling block in achieving a stable population. On the other hand, there are those who are against efforts to curb immigration, including agricultural interest groups that benefit economically from the cheap labor supply of illegals. Also, some Hispanic rights groups, as well as the United States Catholic Bishops, fear increased discrimination against Mexican Americans.

In the summer of 2007, Congress, under pressure from President Bush, sought to find some solution to the problem of illegal immigration, but its attempt ended in failure. Different sides of the issue are so divided, and so set in their views, that compromise has become all but impossible. From the building of hundreds of miles of new fences (probably with illegal immigrant labor for lack of any other) to amnesty for all those here illegally, from utter disdain for illegal immigrants to opening the border to everyone, politicians and the electorate are miles apart. The degree to which American immigration policy has failed was aptly summed up by Massey (2006, 582): "Despite spending in excess of $6 billion per year on immigration enforcement, net undocumented migration into the United States is currently running at a record 600,000 per year, the total undocumented population has reached an unprecedented 11.1 million persons . . . and border deaths are nearing 500 per year—all in addition to the roughly one million people who enter through legal channels each year." Broken is hardly the right word for America's immigration policy.

Even as the President and Congress continue to grapple with a "fix" to the country's immigration problem, illegal immigrants are changing communities across the country. In California's San Joaquin Valley, there are whole communities today that look like transplanted Mexican villages, not so much architecturally as culturally. In towns such as Huron, 97 percent of the population is Hispanic or Latino and Spanish is the primary language spoken by 98 percent of the population (U.S. Census Bureau, 2013), most signs are in Spanish, and the streets are lined with *taquerias* (taco shops), *carnicerias* (meat markets or butcher shops), and *panederias* (bakeries). Absentee landowners use this labor supply, exploit it, and leave communities stuck with high unemployment rates, low tax bases, poor schools, little medical care, and a host of other disadvantages. Across the nation, immigration, legal and otherwise, is changing places at accelerating rates. Even the Heartland is feeling such changes. As Bloom (2006, 62) noted, "By 2030, half of Iowa's population of three million is expected to belong to minority groups. By far the greatest number will be Hispanics working in low-level jobs." The same could be said of turkey and chicken processing plants in North Carolina, or in scores of other places from coast to coast. Bloom (2006, 68) was certainly correct in pointing out that "Unless something wholly unexpected happens, more and more immigrants will stream into rural America." Most likely the "unexpected" will not include any progress by a congress incapable of resolving any major issues at all anymore.

Legal Immigration

Aside from its illegal immigration challenge, the United States continues to attract many legal immigrants. In the 1820–2010 period, 76.4 million people obtained legal permanent resident status in the U.S. However, the admission of these immigrants has historically varied widely from decade to decade (Figure 8–4) depending on existing legal, political and socioeconomic conditions in the U.S. and abroad. Nevertheless, immigration to the U.S. has accelerated since the 1930s since the number of people who gained permanent residency in the U.S. in the 110-year period between 1820 and 1930 and in the 79-year period between 1931 and 2010 is roughly the same at 37.8 million and 38.6 million, respectively (Figure 8–4). It is also noteworthy that while most of the people who immigrated to the U.S. between 1820 and the 1960s were from Europe, most of those that have come since are from Latin America, Asia and the Caribbean (Figure 8–5). One illustration of the dramatic shift in the continental origin of U.S. immigrants is that there was no European country in the top 12 or even 16 national sources of immigrants to the U.S in the 2002 to 2011 period (Figure 8–5). Thus, while 185,055 Jamaicans immigrated to the U.S. between 2002 and 2011 (Figure 8–5), the United Kingdom which was the leading European source of U.S. immigrants in the same period had 146,751 individuals. The dramatic growth in U.S. legal and illegal immigration is recent decades is fueling debates in the United States about its immigration policy.

Reimers (2005) offered a comprehensive analysis of non-European immigration to the U.S. and showed that the now dominant U.S immigration stream from Latin America, Asia and the Caribbean has much in common with the mostly European one that came before it and that these new immigrant streams have also benefitted U.S. society in many ways. Zolberg (2006) also provided a thorough view of how immigration has shaped America and how Americans have argued about immigration policies as well. Fitzgerald

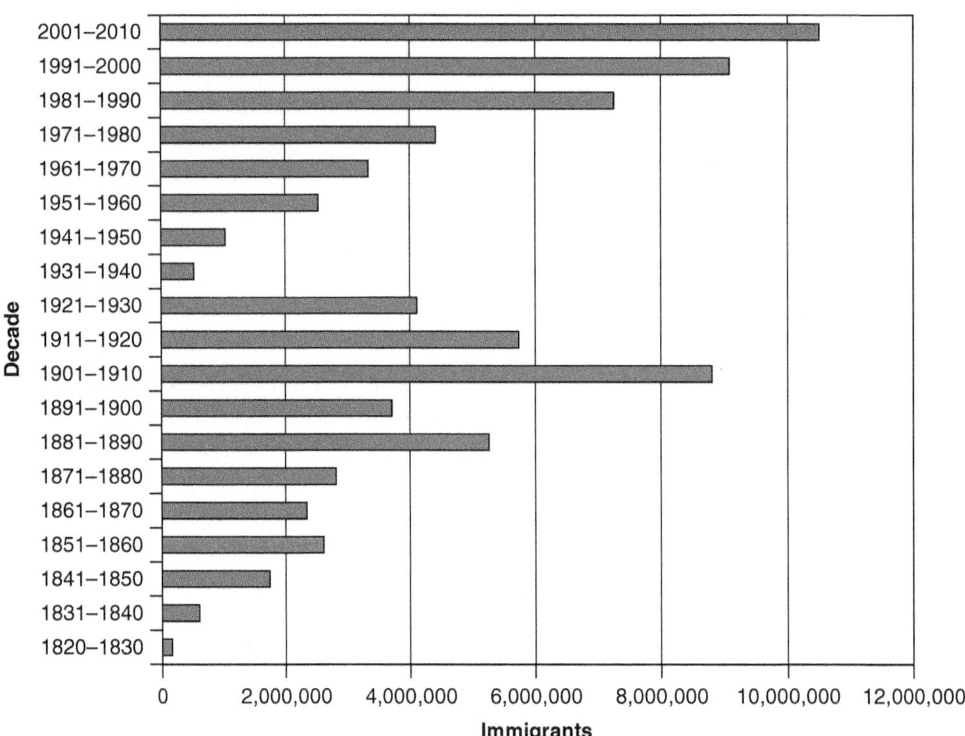

Figure 8–4 Legal Immigration to the United States, 1820–2010

Source: United States Department of Homeland Security (2012b). Yearbook of Immigration Statistics: 2011. Washington, D.C.: U.S. Department of Homeland Security, Office of Immigration Statistics, 2012, Table 1.

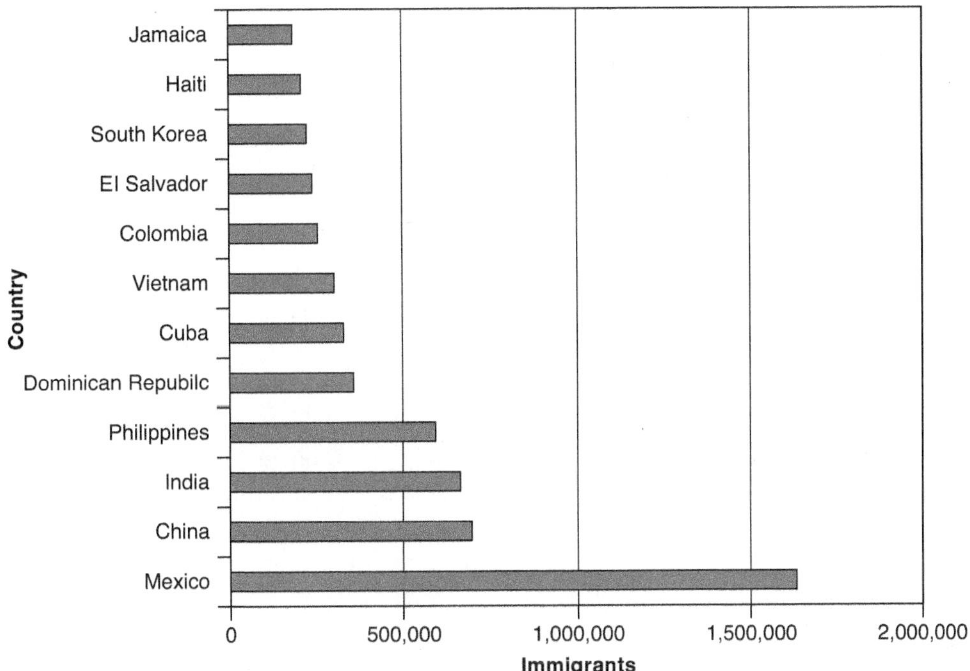

Figure 8–5 Immigrants Admitted by Top Twelve Countries of Birth, 2002–2011
Source: United States Department of Homeland Security (2012b). Yearbook of Immigration Statistics: 2011. Washington, D.C.: U.S. Department of Homeland Security, Office of Immigration Statistics, 2012, Table 3d.

(2006) looked on the south side of the U.S.-Mexico border and at Mexico's emigration policy. Ellis and Goodwin-White (2006:920) studied the movement of adult children of immigrants in the United States and found that "regardless of education, members of the 1.5 generation [children of first generation immigrants] of all race groups appear to be least likely to undertake an interstate move when resident in a state with high concentrations of immigrants." Finally, Ghandnoosh and Waldinger (2006) wrote about the politics of immigration in the United States, starting with the *Wall Street Journal*'s July 4, 1984, editorial page call for an open border and no controls on immigration. The journal's editors even proposed a constitutional amendment that "There shall be open borders." At the other end of the spectrum, however, are organizations such as FAIR (Federation for American Immigration Reform), which would all but close the borders entirely, at least for a while.

Haubert and Fussell (2006, 489) focused on pro-immigration sentiment in the United States and concluded that "cosmopolitans—people who are highly educated, in white-collar occupations, who have lived abroad, and who reject ethnocentrism—are significantly more pro-immigrant than people without these characteristics." One of the obvious reasons for this is that immigrants, especially illegal immigrants are no threat to those people. In a somewhat different study, Mora (2006, 885) found that ". . . policies which reduce trade and labor flows across the U.S.-Mexico border may inadvertently dampen the entrepreneurial activities of foreign-born residents in U.S. border cities."

The dramatic growth of U.S. immigration (Figure 8–4), whether legal or illegal, is also mirrored in many U.S. border states like California, Texas, Florida and New York. Figure 8–6 shows that California's population grew 37-fold (or 2,438 percent) between 1900 and 2012. By 2010, slightly more than 25 percent (one-fourth) of California's residents were foreign-born and in cities such as Los Angeles that percentage was slightly over 41. California also had most (25 percent) of the United States' foreign born population in 2010 followed by New York (10.8 percent), Texas (10.4 per cent) and Florida

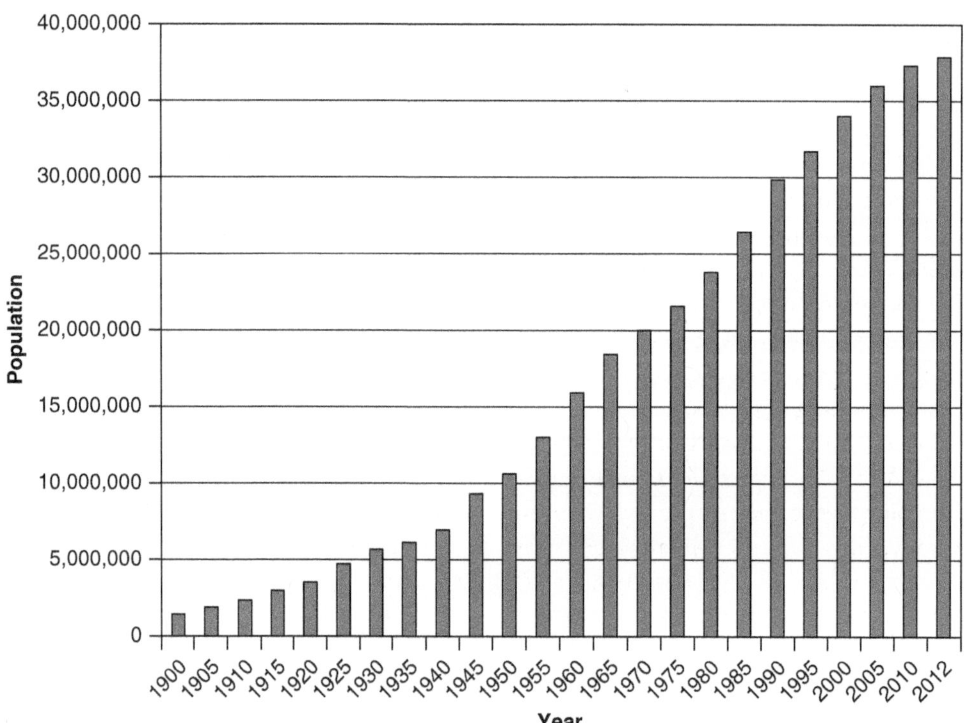

Figure 8–6 Population Growth in California, 1900–2012

Source: California Department of Finance, E-7. California Population Estimates, with Components of Change and Crude Rates, July 1, 1900–2012.

(9.2 per cent) (Grieco et al., 2012). Between 1995 and 2005, an estimated 98 percent of California's population growth was accounted for by immigrants and their children. Since 2004, the state has added 500,000 residents annually, mostly from immigrants and births to immigrants even though the state was simultaneously losing more than two million people through out-migration to other states especially Nevada, Arizona, and Texas. California's demographic picture is even further complicated by the fact that an estimated 2.6 million of its residents are there illegally.

Like its illegal counterpart, increased legal immigration to the U.S. is also having a significant impact on the country's culture, economy, and society and has helped to rekindle old emotive debates on the costs and benefits of immigration. According to West (2010, 20), immigrants contribute to the economic and cultural vibrancy of the U.S. by creating new businesses and jobs and by bringing or creating new ideas that enrich the country's intellectual property. On the negative side, immigrants take American jobs, depress American wages, and put pressure on local public resources though, on the whole, immigration appears to have a net positive impact on the United States.

Refugees

The U.S. has long played a central role in the global refugee crisis. This commitment to helping refugees was enshrined into U.S. law with the passage of the Refugee Act of 1980 as discussed on page 227 under the "**Key Contemporary U.S. Immigration Laws.**" Prior to that the U.S. had, since the end of its involvement in the Vietnam War in the mid-1970s, admitted numerous refugees from Southeast Asian countries under a special refugee act. Originally the government of the United States encouraged a refugee distribution policy that spread Southeast Asian refugees around the United States. However, subsequent migration patterns for these groups have redistributed their numbers considerably, and, like other Americans, they have been moving to the Sunbelt. As a result, concentrations of Asian refugees have developed in many areas, especially in Southern California, where Long Beach has the largest settlement of Cambodians outside of Cambodia and where nearby Westminster has gained the nickname "Little Saigon" as a reflection of the concentration of Vietnamese in that city. The redistribution of Indochinese refugees within the United States was carefully studied by Desbarats (1985), who found that refugees were concentrating in states that offered some combination of high public assistance benefits, lenient eligibility requirements for receiving those benefits, mild climates, reasonable employment opportunities, and large Asian communities.

Growing anti-immigrant rhetoric in the United States during the 1990s (exemplified by the passage of the anti-immigrant Proposition 187 in California) has resulted in some rethinking of U.S. refugee admission policies which are, however, far from uniformly enforced. Thus, refugees from Haiti are often quickly repatriated while the long-standing U.S. policy of accepting all Cuban refugees has increasingly given way to a more restrictive policy. Nevertheless, the U.S. continues to accept thousands of diverse refugees from all over the world on humanitarian grounds (United States Department of Homeland Security, 2012b, 39). For example, between 2002 and 2011, the U.S. welcomed 515,301 refugees; 85 per cent (or 440,940) of whom were from Burma, Iraq, Somalia, Bhutan, Cuba, Iran, Ukraine, Liberia, Russia, Laos, Vietnam, Sudan and Ethiopia (United States Department of Homeland Security, 2012b, 39). Geographers continue to explore the fascinating patterns of these refugees' distribution and settlement in the U.S.

Internal Migration

A large proportion of migrants never cross national borders; they move within their own countries. Such internal migration is often more difficult to assess, and only those countries that keep a complete dossier on each citizen can accurately reconstruct the nature of internal migration. However, in countries such as the United States, census information—supplemented by periodic surveys, Social Security data, and information on driver's license exchanges—can be used to trace internal migration patterns.

Though they are treated as independent phenomena, in today's globalized world economy internal and international movements are often linked. Skeldon (2006, 15) studied linkages between internal and international migration patterns in Asia and found that "The concentration of populations in urban areas can give rise to later international movements, but these international migrations themselves, by creating vacuums in areas of origin, can in turn generate internal migration." Similarly, it is easy to see how large immigrant destinations such as Los Angeles may well generate streams of out-migration to surrounding cities and even adjacent states. Anecdotal information suggests that even illegal immigrants may now be leaving California's largest city.

General Patterns of Internal Migration

Some generalizations can be made about the geographic patterns of internal migration. For example, one of the major movements in modern times has been the movement of people from rural to urban areas. This movement has been the result of many complex interactions, especially the shift of employment away from rural agricultural pursuits and toward urban industrial and service occupations. As geographer Wilbur Zelinsky (1971, 236) once noted, "The onset of modernization . . . brings with it a great shaking loose of migrants from the countryside."

The conquest of new territory has been an important force in the settlement of many countries. The movement of the frontier in the United States, from the original thirteen colonies to the Pacific Ocean, has its parallels elsewhere. Examples include the westward push of the pioneer fringe in Brazil, the northwestward expansion across Canada, the northward movement in Australia, and the movements of peoples from the Andes to the coastal plain and interior forests of South America.

Nomadism is another type of internal (and sometimes international) movement. Even at the start of the twenty-first century there are still people who are constantly moving about, often with little regard for international boundaries. Many nomads are still found today in the Sahel region of Africa, for example, and in the high latitudes of both Europe and North America.

Migration within the United States

Approximately four million Americans lived in a narrow belt along the Atlantic Ocean in 1790; only a hundred years later the United States occupied an area that extended to the Pacific Ocean. Movements during the twentieth century have been primarily westward and coastward, with a loss of population in rural areas and a concomitant concentration of people first in cities and then in suburbs and exurbs. Because of the large gains in population in the western states, as well as in Florida, Texas, and other states in the southwest, the mean center of the U.S. population for 2010 was farther west and somewhat south of its 2000 location, as is apparent in Figure 8–7. You can expect this trend to continue. The

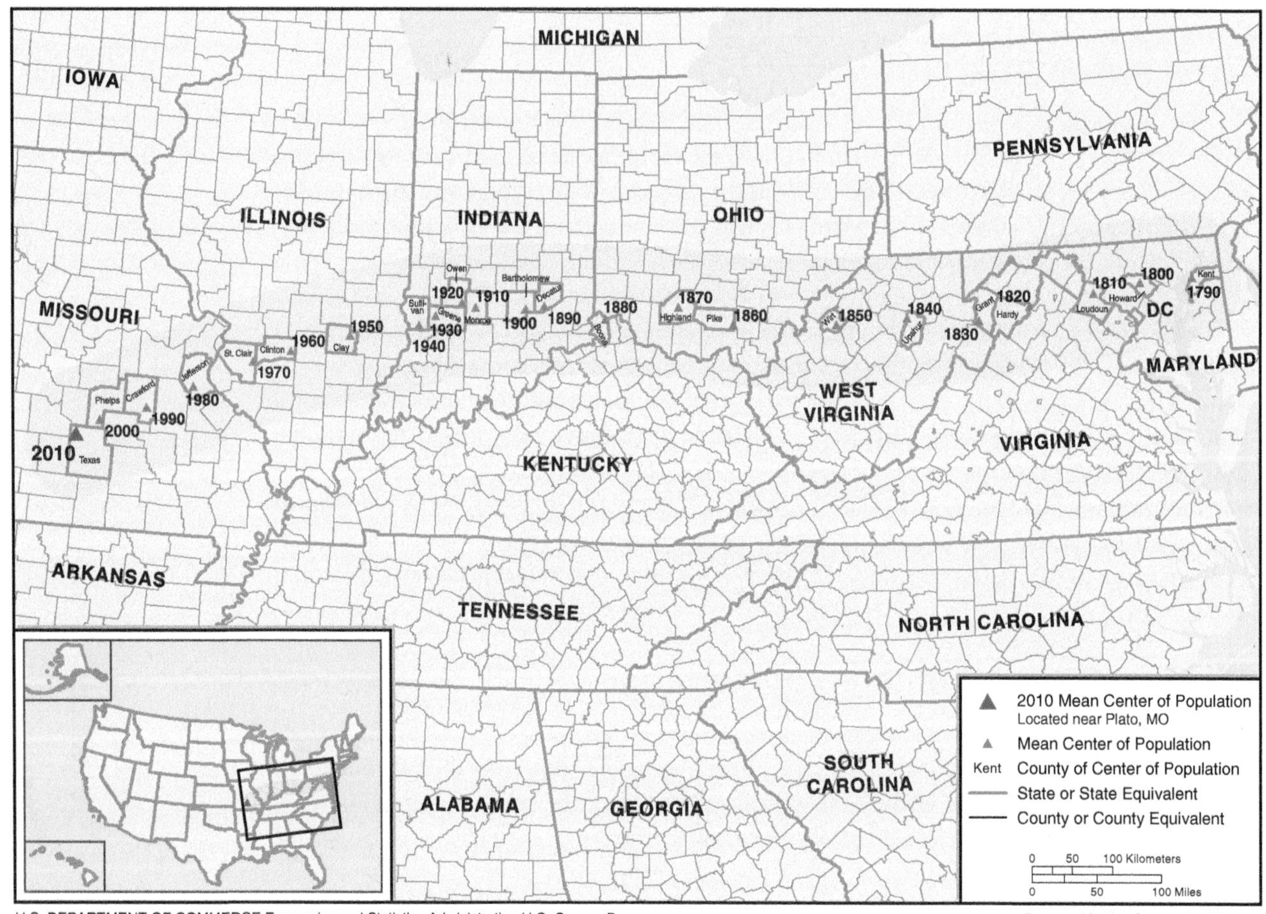

Figure 8–7 Mean Center of Population for the United States: 1790 to 2010 Show state boundaries as shown in this original: http://www.census.gov/geo/reference/pdfs/cenpop2010/centerpop_mean2010.pdf
Source: United States Bureau of the Census.

United States is indeed a nation of migrants, as well as a nation of immigrants. Since the annual mobility sample studies were first begun in 1947, about one-fifth of the American population has changed residence each year (about 17 percent in recent years).

The United States has entered a new era with respect to both the distribution of population and economic activities. Capital and people are increasingly being attracted away from the old industrial heartland and toward the South, Southwest, and West. Forces not only in the national economy, but increasingly in the global economy as well, are constantly at work changing locational advantages, altering economic landscapes, and encouraging consequent patterns of internal migration. The essence of American mobility was captured in the following comment by geographer Paul Simkins (1978, 204): "Although it does not appear as such in any of the state seals or flags, the wheel in many ways may be considered a symbol of America; the wagon or train wheel which early carried the nation westward, or the automobile wheel that moves it now increasingly to and from metropolitan centers." Americans have always sought to better their lot in life, whether they were pioneers seeking "elbow room" or folks in twentieth-century rural Iowa seeking the perpetual sunshine of Southern California.

Some salient features of recent movers give us a better sense of American mobility. For example, long distance moves have been increasing, even though the overall frequency of moves has slowed somewhat since 1990. Also, the most mobile age groups were 20–24 and 25–29-year-olds, and the least mobile race/ethnic group was non-Hispanic whites. Other observations include the following: married people move less than single and divorced people; one-third of renters move, and lower income people move more often than those in higher-income groups. People of different educational levels had similar mobility rates, though movers with a college degree tended to move longer distances than those without.

At the forefront of motives for migration, has been the hope of material gain, though the rise of the Sunbelt as a center of attraction for migrants suggests that other motives are also important, especially mild climates. Many seek to escape the long cold winters of the cities of the Northeast, as well as their smog, traffic congestion, and high crime rates. Though an overview of migration patterns can deal only with aggregates of people, remember that the motives for migration differ for various age groups and for other characteristics of people as well.

Despite the rural-urban migration that characterized the redistribution of people throughout the United States during the twentieth century, Americans have never been in unanimous agreement that cities, especially large ones, were desirable places. Industrialization concentrated people in cities, primarily for economic reasons. Urban locations often meant lower costs for industries, as well as better access to service needs and to local markets for goods. However, from the worker's viewpoint, industrial concentration in urban areas meant more jobs and higher wages, so to the cities they moved.

Once in the cities, however, most groups have tried to work their way outward again, giving rise first to suburbanization, then increasingly to exurbanization. Many problems in the cities today are at least in part related to suburbanization because it, like other migration processes, is selective—those economically able to move outward tend to do so. Though suburbanization was long considered an almost exclusively white process, Latinos and African Americans are increasingly moving toward the suburbs as well. In Los Angeles, for example, newly-arrived immigrants in formerly black communities such as Watts and Compton have accelerated the movement of African Americans toward such suburban locations as Bellflower, Lancaster, and Palmdale within Los Angeles County and to suburban parts of neighboring Riverside and San Bernardino counties.

Movement to the Sunbelt

The Sunbelt states are those of the South and Southwest, along with California, though there is not complete agreement from one list of Sunbelt states to the next. Colorado, for example, is sometimes included because of its growth pattern rather than its climatic claims. Furthermore, within the Sunbelt itself there is considerable variation in the growth and attractiveness of different states. Mississippi, for example, has hardly had the same experience that Texas or Florida has had. Since 1970 a dramatic shift of population away from states in the North Central and Northeast and toward the Sunbelt states has occurred. A disproportionate share of recent population growth in the United States has occurred in only three states: California, Florida, and Texas. Not only people, but also industry, are increasingly attracted to the Sunbelt.

People are increasingly moving away from the older areas of the North Central and Northeastern states, partly in response to environmental preferences and partly as a result of aggressive economic development in many of the Sunbelt states. Many people are also being attracted to the Rocky Mountain states and to the Pacific Northwest, especially to Portland and Seattle.

The Rural Renaissance

During the 1970s more people were leaving metropolitan areas than entering them. The long-established pattern of rural-urban migration underwent a reversal during the 1970s. As you might imagine, not all nonmetropolitan areas were equally pleased with the prospect of increased population growth and "rural urbanization." Conflicts often arise between growth advocates and those who seek to limit growth. Land use conflicts are also common. Many small-town residents fear that growth will destroy the very features that attract people to small towns. For example, they point to increasing crime rates, traffic congestion problems, and even smog as inevitable results of small-town growth. It is interesting to note that quite often it is the newcomer who is most vocal about limiting further growth and change; visit with ex-Californians in Oregon or Idaho, for example, and see if they invite you to stay.

The sudden reversal of long-term urbanization in America caught many people, including demographers and geographers, by surprise. Many had thought that most of the nonmetropolitan growth was just spillover from expanding metropolitan areas; however, even nonmetropolitan areas far from urban centers were participating in the new trend. Spillover, of course, had been commonplace, but a new pattern seemed to be at hand. Among its major causes were the following:

1. changes in communication and transportation technology that have taken away, or at least mitigated, the necessity of urban concentration
2. the expansion of highways that allow easy access to urban areas

However, regional variations in the rates of both metropolitan and nonmetropolitan growth rates may be observed.

The growth of retirement and recreational communities has also been a factor in nonmetropolitan growth. Early retirement, coupled with better benefits for retirees, frees an increasing number of people to seek locations based on personal preferences rather than economic necessities. Wherever the elderly choose to go, of course, services are required; thus jobs are created and further growth is stimulated. Not only Florida and the Southwest, but also the Ozarks, the Texas hill country, and California's "Gold Country," in the western foothills of the Sierra Nevada, are experiencing growth from substantial influxes of retirees. Retirees may be looking in new directions as well. From Hanover, New Hampshire, to Sequim, Washington, retirees are seeking smaller, less expensive, less crime-ridden areas, and they may be willing to trade off some sunshine for more peace and quiet.

The Canadian experience has been similar, as noted by Keddie and Joseph (1991). Though they found less concrete evidence of a rural turnaround during the 1970s in Canada, they did find that ". . . the data for the early 1980s point toward a return to the urban-dominated growth scenario prevalent in the 1960s" (Keddie and Joseph, 1991, 378). They also noted that changes in rural population trends generally were a good indicator of the economic fortunes of different regions. Lindgren (2003) studied counter-urban moves and the nature of those movements in the Swedish urban system. Stockdale (2006) discussed the concept of a "retirement transition" and its role in repopulating rural areas in England and Scotland, whereas Potts (2006) found that rural-rural migration was important in Malawi. Hall and Müller (2005) looked at the relationship between second homes and mobility.

It is worth noting that national and regional trends often obscure underlying patterns that may be seen for particular subgroups of the population, from retirees to ethnic groups. These various patterns are important as well, and receive considerable attention in the migration literature. Examples of such studies include Robinson (1986), who wrote about the reversal in the 1980s of the previous pattern of African American net migration out of the South, and McHugh (1989), who studied the internal redistribution of Hispanics within the United States. In August of 2005 Hurricane Katrina reminded us that natural disasters can generate notable migration streams and reshape cultural landscapes.

Places Left Behind

Like economic development and globalization, migration creates winners and losers, reshaping America's landscape impassionately and inexorably. Of the places left behind in this vast land, none stands out so clearly as the Great Plains, where a broad swath of population loss stretches northward from Texas and Oklahoma to the Dakotas, eastern Montana, and western Minnesota. As Jeff Glasser (2001, 19) remarked, "Up and down the Great Plains, the country's spine, from the Sandhills of western Nebraska to the sea of prairie grass in eastern Montana, small towns are decaying, and in some cases, literally dying out. The remarkable prosperity of the last decade never reached this far."

Many counties in this vast interior landscape have reverted back to "frontier" status, as their population densities have declined to below six persons per square mile. The population of the Great Plains actually peaked during the 1930s, and has declined since then, albeit unevenly in both space and time. Most of the declines have come in places that specialized in some combination of wheat farming, cattle ranching, and mining (including oil production). Without subsidies for farming and energy production, declines probably would have been even faster.

Farming in much of the Great Plains is in a depressed state, stung by low grain prices, rising production costs, and competition from imports. Additionally, young people are less interested than ever in continuing to live and work on farms—they go off to college then on to careers in places that offer them jobs. The Great Plains remains a harsh land of great vistas but declining opportunities, a place where some will always remain transfixed but others are more than willing to give it back to the buffalo. Poet and resident Kathleen Norris (1993, 110) described it best:

> Where I am is a place where the human fabric is worn thin, farms and ranches and little towns scattered over miles of seemingly endless, empty grassland . . . . But some have come to love living under its winds and storms. Some have come to prefer the treelessness and isolation, becoming monks of the land, knowing that its loneliness is an honest reflection of the essential human loneliness. The willingly embraced desert fosters realism, not despair.

In a nation where most people embrace "reality TV," cell phones, iPods and instant gratification, however, the Great Plains is likely to continue its decline as its young leave for cities and its elderly gradually die. Contraction feeds on itself—as small towns decline, property values shrink, schools and businesses close, and people have even fewer incentives to stay.

Rural-Urban Migration and Urbanization: Regional Comparisons and Contrasts

Rural-urban migration drives the urbanization process.

Urbanization is the percentage of a population that lives in urban areas and the process by which this number increases over time.

Rural-urban migration drives the urbanization process. The population of the world as a whole has been increasing rapidly, but the urban population has increased even more rapidly than the overall rate in recent decades. According to Berry (1993, 399):

> . . . transnational urbanward migration is an important force that redistributes urban growth across the globe. The sharply-rising rate of this migration reflects radical increases in the underlying migration propensities: a sevenfold increase 1900–80; and fourteenfold 1860–1980 . . . . Significant differences have emerged in the relationship across levels of development since World War II. Transnational migration remains a potent source of urban growth in a number of low-and middle-income countries, at the expense of other countries at these levels of development.

An analysis of **urbanization** by geographer Ray Northam (1975) suggested that it could be viewed chronologically as a three-stage process, as shown in Figure 8–8. In the

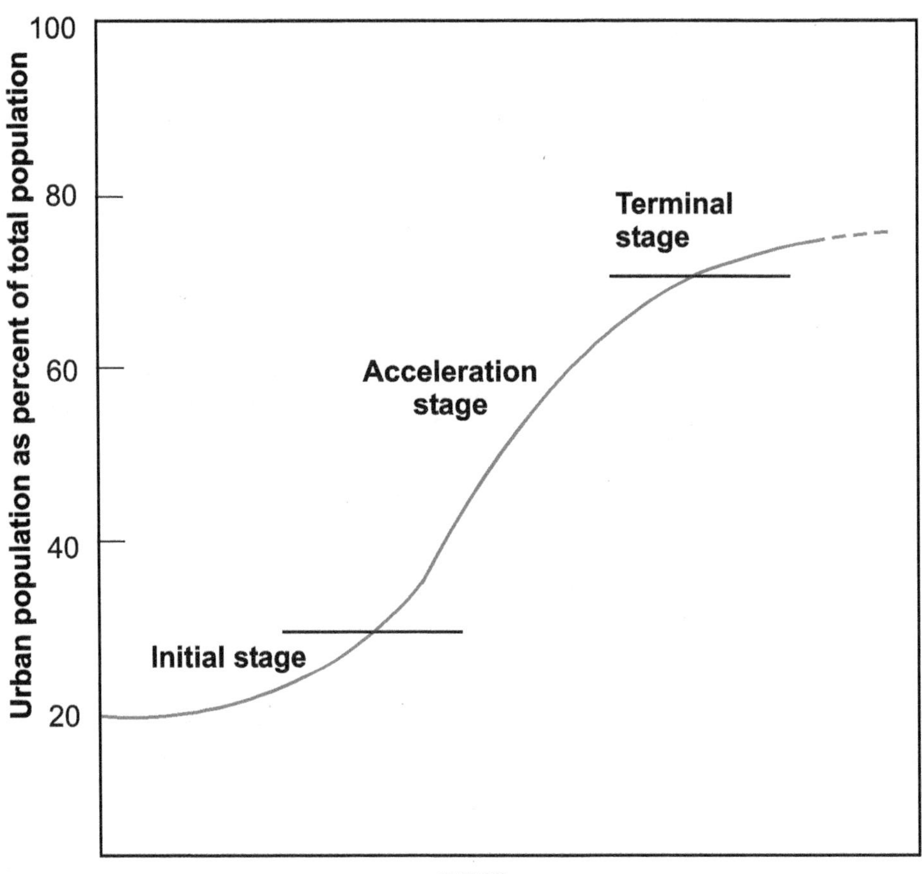

Figure 8–8 The Urbanization Curve and Stages of Urbanization

Source: From *Urban Geography*, Ray M. Northam. Copyright © 1975. Reproduced with permission of John Wiley & Sons, Inc.

Initial Stage the population is primarily rural, engaged in agricultural pursuits, and of a dispersed nature. The second stage, referred to by Northam as the *Acceleration Stage*, is a period in which an increasingly large share of the population lives in urban centers. During this stage there is a marked redistribution of the population, with the urban component rising from less than 25 percent to over 70 percent of the total population. The Acceleration Stage also involves a basic restructuring of the economy; there is a rapid shift from agriculture to industry, with a notable concentration of economic activity in the city. Large numbers of people are employed in manufacturing industries as well as in service and trade activities.

An analysis of **urbanization** by geographer Ray Northam suggested that it could be viewed chronologically as a three stage process, as shown in Figure 8–8.

The final stage in the urbanization process is referred to by Northam as the *Terminal Stage*. During this stage the urban population is approximately 60 to 70 percent of the total population (occasionally even reaching 80 percent). Once the urbanization curve goes above 70 percent, it tends to flatten out. For example, the urbanization curve for England and Wales since 1900 has tended to flatten out after it reached the 80 percent level.

Today we might add another stage to Northam's model; a stage in which deconcentration begins in highly advanced "post industrial" economies. Some time ago, Berry (1980, 13) noted that, "Urbanization, the process of population concentration, has been succeeded in the United States by counterurbanization, a process of population deconcentration characterized by smaller sizes, decreasing densities, and increasing local homogeneity, set within widening radii of national interdependence."

A worldwide regional analysis shows that the world's nations are in various stages of urbanization. As a general rule, the more developed nations might be considered as being in the Terminal Stage of urbanization. On the other hand, the less developed nations are primarily in the Acceleration Stage, though some nations that are largely agrarian in nature could still be considered in the Initial Stage. Figure 8–9 shows the world pattern of urbanization, with many African and Asian countries still being predominantly rural.

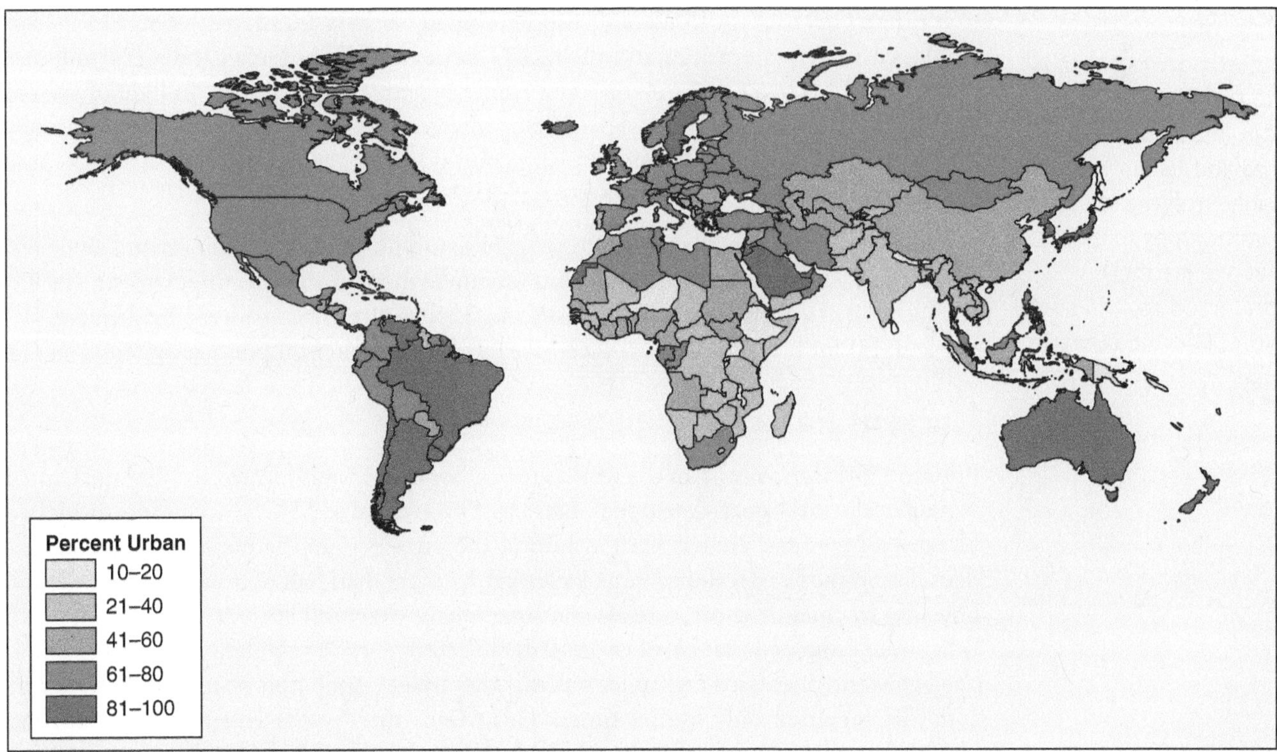

Figure 8–9 World Urbanization in 2012

Source: 2012 World Population Data Sheet. (Washington, D.C.: The Population Reference Bureau, Inc.)

Urbanization in Less Developed Countries

The growth of large cities is a well-recognized characteristic of more developed nations, but urbanization is equally evident in less developed countries. The twentieth century was the century of the urban transition; by the end of that century nearly three billion people, almost half of the world's projected population at that time, lived in cities. Two-thirds of those people lived in the less developed countries of Africa, Asia, and Latin America (United Nations, 1991). From crowded Cairo to Mexico City, from Rio to Mumbai and Beijing, poor cities, teeming with residents, are grappling with problems that seem hard even to imagine. Already, in places such as Calcutta, people go through the streets early each morning removing those who failed to survive the night.

In the first decade of the twenty first century, another very important urbanization milestone occurred when, for the first time in human history, the world's population became predominantly urban in 2008 (Population Reference Bureau, 2013). By 2011, fifty-one percent of the world's population (that is 3.6 billion people) lived in cities (United Nations 2012). While many more developed countries became urban in the first half of the 1900s, many less developed ones (more so those of Asia and Africa), are still urbanizing through rural-urban migration and natural increase.

Because of overcrowding and unemployment in the rural areas of many less developed countries, families leave the poverty-stricken countryside with the hope of making a livelihood in the city. Unfortunately, jobs are often equally difficult to find in the cities because the rate of urban population growth, which is often twice the rate of national population growth, exceeds the level of economic growth. This results in massive unemployment, underemployment (too many people for the task at hand), misemployment (production/provision of goods and services that contribute little to social welfare), crime, and the growth of urban slums (Sijuwade 2010). Another key aspect of current world urbanization is that less developed countries now host most of the world's largest **megacities** or urban agglomerations with 10 million or more residents. In 2011, seventeen (74 percent) of the world's 23 largest megacities were in less developed countries and this trend is set to continue well into the foreseeable future (United Nations, 2012, 6–7). Some of the less developed world's megacities and their 2011 populations are: Delhi, India (21 million), Mexico City, Mexico (20.1 million), and Lagos, Nigeria (10.7 million). In light of current and future urbanization trends, less developed countries are going to find it increasingly difficult to provide decent jobs and living conditions for their urban citizens. According to Gugler (1988, 1):

A **megacity** is a metropolitan area with a population of 10 or more million people. A metropolitan area is the totality of a densely populated core urban area and its surrounding highly integrated communities or those that are strongly connected to it in social and economic terms.

> Rapidly growing urban populations [1] have to find employment in urban labour markets characterized by widespread unemployment and underemployment, [2] increase demands for urban housing and services already considered inadequate, and, [3] for better or worse, exacerbate popular pressures on political systems.

Current Patterns

Prior to 1975 the majority of the world's urban population was located in what the United Nations calls the "more developed" nations. However, since 1975, according to United Nations estimates, there has been a shift; at the present time the majority of urban residents live in the less developed countries. Slightly more than half of the world's population now lives in cities. If current trends continue, this pattern will become more accentuated. It has been estimated that between 1950 and 2000 the urban population in the less developed countries increased eightfold, whereas the urban population of the more developed countries increased only by 2.4 times. This means that by the end of the last century approximately two-thirds of the world's 3.1 billion urban residents lived in the less developed countries, compared with the 1950 population, where only one-third of the urban population resided in these countries.

In the countries that are now highly urbanized—mainly the leading industrial countries of the more developed regions—urbanization advanced slowly at the beginning of the development process, then rose sharply in the beginning stages of industrialization, and finally tapered off when a saturation point was reached. Because of the relatively low rate of urbanization in Europe, the emergence of new political, social, and economic institutions kept pace with the urban influx. Urban growth in the less developed countries is taking place under much more difficult conditions. Several important factors that have distinguished the rapid and dramatic urban growth in the less developed countries were discussed by Beier (1976) and they are summarized below.

First, the increase in population growth since the last century (i.e., 1900s) is the single most important factor distinguishing present from past urbanization. Population growth rates in Europe during its period of urbanization were around 0.5 percent a year, whereas in the less developed countries today annual population growth rates are sometimes as high as 3.0 percent or more. In 2000 Africa had a rate of natural increase of 2.4 percent. Significantly higher growth rates mean that the less developed countries have both larger natural increases within their cities and larger population movements to their cities. This combination is straining the very fiber of many less developed country cities (Otiso, 2003, 221–222).

Second, because of widespread communication facilities, populations in the less developed countries are provided with more information about urban amenities and opportunities, thus increasing the "pull" of the city on rural populations. At the same time the cost of migration is lower because of better transportation facilities.

Third, unlike the urbanization experience of most of the more developed countries, urbanization in the less developed countries is generally confined within fixed territorial boundaries. Little opportunity exists for free migration of the surplus population to other countries.

It appears that the less developed countries will find it more difficult to urbanize than did the more developed countries, and that they will have to accomplish that task in a shorter period of time. Though there will be country-by-country differences, approximately one-half of the new residents of cities will be native born and the remainder will be newcomers. These new urbanites will most likely be unskilled laborers; they are also liable to be relatively poor, undereducated, and probably illiterate—women and children will suffer the most. Only if they benefit materially from their urban experiences can we expect significant declines in their fertility, and such declines are essential in most less developed countries if those countries are to avoid Malthusian controls on their populations.

India is a good example for comparing urbanization in the less developed countries with the earlier European experience. In 1951 India was only 11 percent urbanized, a level reached by many European countries between 1850 and 1900; at this level of urbanization slightly over 50 percent of the population of European countries derived their livelihood from agriculture, whereas 65 to 70 percent of India's population were agriculturalists at that same level of urbanization. Approximately 10 percent of India's population was employed in manufacturing, whereas about 25 percent of Europe's labor force had been so employed. Though a common thread of rapid urban growth runs through the less developed countries, widespread differences exist among them in long-term urban growth prospects and in the necessary resources to support such prospects.

What remains to be seen, of course, is the degree to which urbanization and the growth of large cities everywhere is actually sustainable. The litany of needs is considerable: health services, jobs, shelter, better schools, and incentives to keep urban sprawl under control are examples. Beyond that, as Brockerhoff (2000, 39) wonders out loud for us:

> What is much less certain is whether the horrific scenarios envisioned by some scholars will come to pass. Will earthquakes and hurricanes kill millions of people in big cities that are unable to prepare for or cope with such disasters? Will large and dense populations become breeding grounds for devastating new infectious

> diseases? Are ghettos a permanent and worsening aspect of the urban landscape in even the richest of countries? If cities swell with youth, but not with jobs, will violence erupt? Do increasingly volatile global financial movements impose an insurmountable barrier to informed urban planning?

Brockerhoff argues that just talking about these and other issues of large-scale urbanization is probably healthy. He ends his discussion on a positive note, writing that "These factors and others suggest that sustainable urban development, even under conditions of extreme population growth, is an attainable goal" (Brockerhoff, 2000, 39). We hope he is right.

Cities in Less Developed Countries

In 2011, about 2.7 billion people lived in the cities and towns of less developed countries and the urban population of most of these countries was growing rapidly (United Nations, 2012, 4). While 60 percent (3 in 5) of the world's urban residents now live in cities with fewer than 500,000 residents, this proportion will decrease to 50 percent (1 in 2) by 2025. On the other hand,

> cities of 1 million and more inhabitants, accounting for about 40 per cent of the world urban population in 2011, are expected to account for 47 per cent of the world urban population by 2025. Indeed, the future urban population will be increasingly concentrated in large cities of one million or more inhabitants. In fact, among the million plus cities, the megacities of at least 10 million inhabitants will experience the largest percentage increase. This increasing urban concentration in very large cities is a new trend which contradicts previous observations (United Nations, 2012, 5).

To illustrate, while New York and Tokyo were the only megacities (10 million or more residents) in the world in 1950, there were 10 by 1990 (50 percent of them in less developed countries), 23 by 2010 (74 percent of them in less developed countries), and an expected 35 by 2020 (80 percent of them in less developed countries) (Table 8–1). Since 1990, most of the world's megacities have been in Asia and will continue to be most common there well into the future. While western cities will continue to fade from the top ranks of the world's largest megacities, they will continue to do better than most of their non-western counterparts in quality of life measures such as access to basic urban services. In contrast, most Asian, Latin American and African megacities will continue to lack basic urban services such as sewage disposal systems and clean water for drinking.

It is hard for Americans to imagine how bad life can be in some of the largest urban agglomerations less developed countries. Writing about life in Lagos, Nigeria, Packer (2006, 74) said that "The really disturbing thing about Lagos's pickers and venders is that their lives have essentially nothing to do with ours. They scavenge an existence beyond the margins of macro-economics. They are, in the harsh terms of globalization, superfluous." Similarly, Kotkin (2006, 132) remarked that "in the early twenty-first century, at least 600 million urbanites in less developed countries survive in squatter settlements—called variously *barriadas, bidonvilles, katchi adabis, favelas,* and shantytowns." Despite miserable conditions, their inhabitants are better off there than in the countrysides, and the growth of these downtrodden areas accounts today for much of the overall urban growth in less developed countries. Even in American cities millions struggle to live decent lives. For example, in 2006, Los Angeles had about "12,000 homeless people living in the shelters, tent cities, cardboard condos, and flophouses that give Skid Row the dubious distinction of having the nation's largest concentration of homeless" (Streisand (2006, 50). By 2011,

Table 8–1 Number of World Megacities in 1950, 2000, 2010 and 2020: Population in millions

1950 (2)	1990 (10)	2000 (17)	2010 (23)	Projected 2020 (35)
New York (12.3)	Tokyo (32.5)	Tokyo (34.4)	Tokyo (36.9)	Tokyo (38.7)
Tokyo (11.3)	New York (16.1)	Mexico City (18.0)	Delhi (21.0)	Delhi (29.2)
	Mexico City (15.3)	New York (17.8)	Mexico City (20.1)	Shanghai (26.1)
	São Paulo (14.8)	São Paulo (17.1)	New York (20.1)	Mumbai (23.6)
	Mumbai (12.4)	Mumbai (16.4)	São Paulo (19.6)	Mexico City (23.2)
	Osaka (11.0)	Delhi (15.7)	Shanghai (19.5)	New York (22.4)
	Los Angeles (10.9)	Shanghai (13.9)	Mumbai (19.4)	São Paulo (22.2)
	Kolkata (10.9)	Kolkata (13.1)	Beijing (14.9)	Beijing (20.7)
	Seoul (10.5)	Buenos Aires (11.8)	Dhaka (14.9)	Dhaka (20.1)
	Buenos Aires (10.5)	Los Angeles (11.8)	Kolkata (14.2)	Karachi (17.7)
		Osaka (11.2)	Karachi (13.4)	Kolkata (16.6)
		Rio de Janeiro (10.8)	Buenos Aires (13.4)	Lagos (15.8)
		Dhaka (10.3)	Los Angeles (13.2)	Los Angeles (14.9)
		Cairo (10.2)	Rio de Janeiro (11.8)	Buenos Aires (14.8)
		Beijing (10.2)	Manila (11.6)	Manila (14.4)
		Karachi (10.0)	Moscow (11.4)	Shenzhen (14.2)
		Moscow (10.0)	Osaka (11.4)	Guangzhou (14.2)
			Cairo (11.0)	Istanbul (13.7)
			Istanbul (10.9)	Cairo (13.2)
			Lagos (10.7)	Rio de Janeiro (13.0)
			Paris (10.5)	Chongqing (12.4)
			Guangzhou (10.4)	Moscow (12.4)
			Shenzhen (10.2)	Kinshasa (12.3)
				Osaka (12.0)
				Paris (11.7)
				Bangalore (11.6)
				Wuhan (11.6)
				Jakarta (11.6)
				Chennai (11.2)
				Tianjin (10.9)
				Chicago (10.8)
				Lima (10.6)
				Bogotá (10.5)
				Hyderabad (10.2)
				Bangkok (10.2)

Source: http://www.guardian.co.uk/global-development/datablog/2012/oct/04/rise-megacities-get-data#data (based on United Nations *World Urbanization Prospects, the 2011 Revision*)

the number had grown to 23,539. While 63 percent of these were in the 25–54 age group, 34 percent were aged 55 and over. Nearly one-third of the homeless persons in Los Angeles suffer from physical disabilities, mental illness and substance abuse (Los Angeles Homeless Services Authority, 2011, 3–4).

Most of the future increases in urban populations in less developed countries will occur through the expansion of existing cities. There will be few new cities created, and those are probable only where new resources are discovered. Because most of the choice urban locations are already usurped by urban centers, and because it is less costly to expand these areas than to develop new cities, it is unlikely that many new urban centers will appear.

Migration Impact

As we noted previously, rural-urban migration is a major aspect of urban growth; in most countries today it is the driving force, as high fertility and high rates of unemployment make rural areas less and less desirable. Its significance varies over time and from country to country. In Latin America and other areas that already have a significant proportion of their populations living in urban areas, there will most likely be a decline in rural to urban migration in the next few decades. However, in nations that are largely rural, like many in sub-Saharan Africa, migration will continue to play a key role in urban growth and distribution patterns for a long time.

The reasons why people migrate are many and varied, but the economic motive, the desire to better one's self and family, still seems to predominate. In most cases the migrants' expectations of economic improvement are met.

Migrants come from varying cultural and socioeconomic environments and, according to Berry (1973, 82), they ". . . comprise a large and disparate array of social types both before and after migration."

Migration flows are important in explaining why large cities in the less developed countries are growing relatively faster than the medium-size and smaller cities. Migration has often occurred in steps—from rural areas to small local cities, and then to the larger urban areas—especially during the early stages of urbanization (Shaw, 1975, 45–46). The evidence suggests, however, that migrants are increasingly bypassing the smaller cities and moving directly to the larger urban areas. There are several reasons for this changing pattern, including the concentration of important economic opportunities in the larger cities and improved communication and transportation facilities.

Urbanization in the United States

Urbanization in the more developed countries has already reached Northam's Terminal Stage and, as we have suggested, it has even begun to go beyond that stage in some places to a new counter-urbanization stage. Because the latter is best recognized in the United States, and because the pattern of urbanization is quite typical of the more developed countries, we consider it in some detail rather than providing a more general overview of urbanization in the more developed countries.

Since the first census was taken in the United States in 1790 there has been an increasing concentration of Americans in urban areas (Figure 8–10). As the figure shows, while the United States was 5 percent urbanized in 1790, it became predominantly urban in 1920. By 2010, two hundred and twenty years later, 81 percent of the United States' population resided in urban areas. Because of the United States' low population growth rate, the country's current percent urban population is about the same as it was 2010. Many of the original urban centers in the United States that were forerunners of present-day urban agglomerations were located on the eastern seaboard and in peripheral areas and included Boston, New York, and Philadelphia. During the eighteenth century more cities

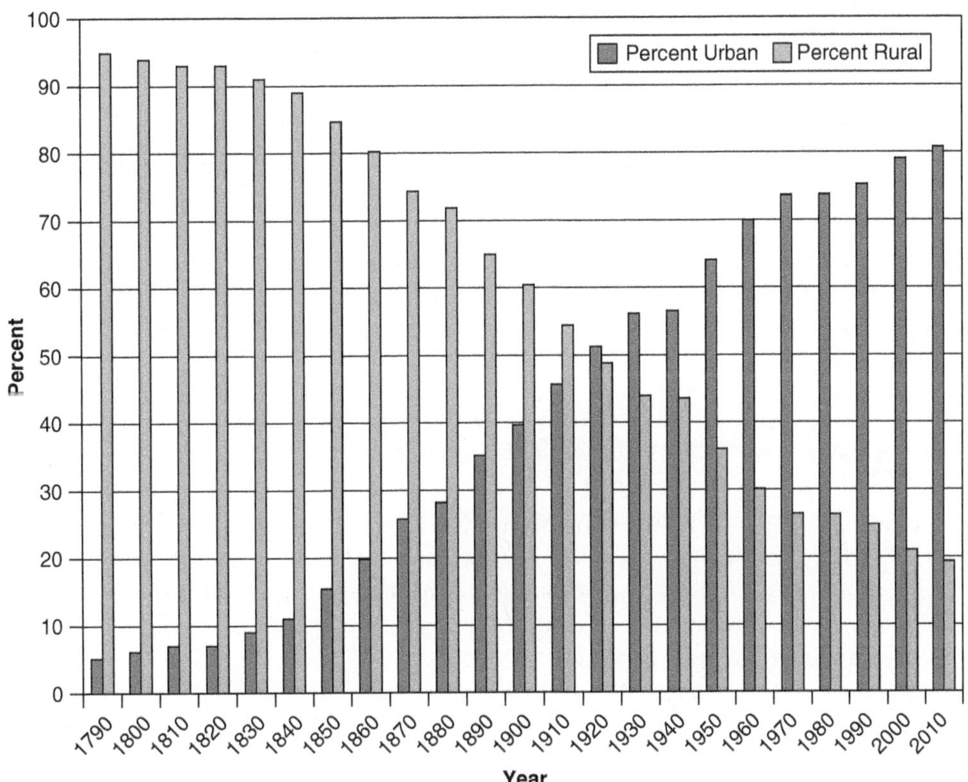

Figure 8–10 United States: Urban and Rural Population 1790–2010
Sources: U.S. Census Bureau (1993) "Table 4. Population: 1790 to 1990" http://www.census.gov/population/censusdata/table-4.pdf and U.S. Census Bureau (1995) "Table 1. Urban and Rural Population: 1900 to 1990" http://www.census.gov/population/censusdata/urpop0090.txt (Accessed January 28, 2013).

were formed on the eastern seaboard, but there were also large numbers of cities founded in the Mississippi and Ohio River valleys as well as other riverine sites. This was also the time when the first major western seaboard cities were founded.

All sections of the country had some sort of urban agglomeration by the end of the nineteenth century (1800s), and most of the previously founded cities continued to grow. Cities formed during the nineteenth century were concentrated in the Midwest, primarily on navigable rivers and the Great Lakes. Forty percent of the United States' population lived in urban centers by the end of the nineteenth century.

Rural-urban migration was a major component of urban growth in the United States in years past, but this pattern ended in the 1980s if not earlier. According to Berry (1990, 99), "We are a completely urbanized society, with very few true rural residents left amidst large-scale factory style agriculture. Urban growth now is a product of natural increase and of foreign immigration." The latter, however, has certainly increased in importance since 1980, especially in cities such as Miami, New York, and Los Angeles. Changing population compositions within these cities are altering everything from local architecture to local politics. Indeed by 2010, the "majority of the residents of 106 metropolitan and micropolitan areas [were] members of minority groups . . . [among] the leaders [were] Honolulu (80.7 percent minority), Los Angeles (67.6 percent) and Miami (63.8 percent)" (Thomas, 2012, no pp).

The nation's largest cities, and the suburban areas surrounding them, have grown continuously since the nation's inception. Between 1910 and 1988 the national population grew by 167 percent, whereas the metropolitan population grew by 449 percent.

Metropolitan areas with at least one million residents grew by 630 percent. In contrast, throughout this 78 year period the nonmetropolitan population stayed within a narrow range between 56 and 67 million (Frey, 1990, 5).

The "look" of American cities is also changing. According to journalist Joel Garreau (1991), the majority of metropolitan Americans live and work in areas that look nothing like our old downtowns. Garreau calls these new multiple urban cores "edge cities," and he argues, quite convincingly, that they represent an important change in metropolitan development since 1950. Beyond the edge cities, around the peripheries of most metropolitan areas, continued expansion converts rural landscapes to urban land uses, a process geographer John Fraser Hart (1991) calls "the metropolitan bow wave," a process that is likely to continue inexorably in most regions of the nation. As Hart (1991, 50) commented:

> As population increases, the built-up areas of urban centers will continue to expand outward. In the vanguard of the expansion will be a bow wave of competitive and contradictory activities where the least intensive urban uses of land are steadily displacing the most intensive rural ones. This urban-rural fringe of greenhouses, nurseries, truck farms, and dairy farms will shift outward before the built-up area. The urban-rural fringe for one generation becomes a suburb for the next, and new fringes arise farther out.

The outward extension of cities is often referred to with the catchall phrase "urban sprawl," even though it means different things to different people. According to Bruegmann's (2005) compact history of urban sprawl, sprawl is everywhere a product of the combined influence of affluence and democracy. The outward sprawl of cities is hardly new, though in its current form it owes its genesis in the United States to the middle of the previous century. Instrumental in sprawl have been not only affluence and democracy but also the automobile and highway systems, especially the Interstate system that was begun under President Eisenhower in the 1950s and by now has affected virtually every place in the nation.

Figure 8–11 shows the level of urbanization in the United States by state and county. According to the US Census Bureau (2012, no pp):

> "Of the nation's four census regions, the West continued to be the most urban [in 2010], with 89.8 percent of its population residing within urban areas, followed by the Northeast, at 85.0 percent. The Midwest and South continue to have lower percentages of urban population than the nation as a whole, with rates of 75.9 and 75.8, respectively. Of the 50 states, California was the most urban [in 2010 despite its vast agricultural production], with nearly 95 percent of its population residing within urban areas. New Jersey followed closely with 94.7 percent of its population residing in urban areas. New Jersey is the most heavily urbanized state, with 92.2 percent of its population residing within urbanized areas of 50,000 or more population. The states with the largest urban populations were California (35,373,606), Texas (21,298,039) and Florida (17,139,844). Maine and Vermont were the most rural states, with 61.3 and 61.1 percent of their populations, respectively, residing in rural areas. States with the largest rural populations were Texas (3,847,522), North Carolina (3,233,727) and Pennsylvania (2,711,092)."

Future Migration Trends

What will happen to migration patterns in the United States in the future is a matter for speculation; however, some current trends will in all likelihood continue. The volume of migration will continue to increase, and the westward and southwestward movements

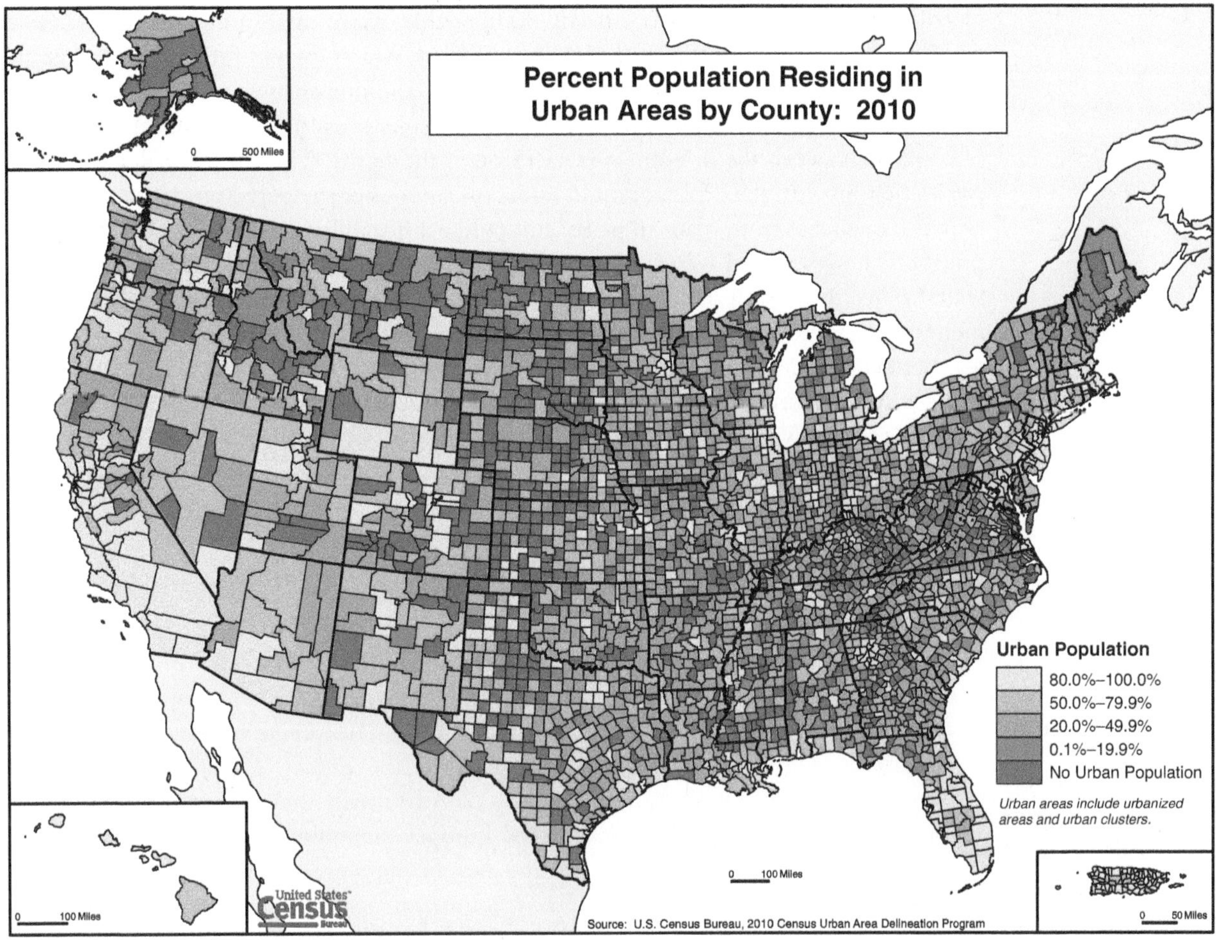

Figure 8–11 Percentage of Urban Population, 2010
Source: U.S. Census Bureau, 2010 Census Urban Area Delineation Program.

will continue as well. Movement between cities will continue, as will the movement from central cities to suburbs. Cities that are service centers, or that perform political or commercial functions, will grow more rapidly than manufacturing cities. Also, amenities such as climate and scenery will become increasingly important factors in people's decisions to migrate. Not all of these trends are good, however. For example, in 1950 about 10.2 million people lived in the states bordering the Gulf Coast of the United States from Texas to North Carolina. By 2006 that number had grown to 24.8 million, far more people to get in the way of devastating hurricanes such as Katrina which devastated New Orleans in 2005 and reduced its population from an estimated 452,170 on July 1, 2005, to 223,388 on July 1, 2006. It is still far from recovering its pre-Katrina population. Included in the 24.8 million Gulf Coast residents that are highly vulnerable to hurricanes are three of the nation's twenty most populous metropolitan areas: Houston-Baytown, Sugar Land (Texas), Miami-Fort Lauderdale-Miami Beach (Florida), and Tampa-St. Petersburg-Clearwater (Florida).

Elsewhere, the nature and extent of future migration will be primarily a function of economic trends, environmental quality and population growth. There are no longer any large open areas or "safety valves" available. With a combination of scarce resources and increasing population pressure, there most likely will be continued migration in even larger numbers in search of shelter, food, and a better lifestyle. As Martin (2007, 66) noted, "A key factor in this heightened mobility is the globalization of motorization. The vehicles and infrastructures of this motorization are substantial producers of ambient pollution and global warming." As we shall soon see, more people, more cars, more energy use, more mobility, and a changing climate are facts of life with which we and future generations are going to have to contend as we shape Earth's future and our own.

At the international level, free migration is no longer possible in most instances because of conflicts between the interests of a society and the rights of the individual. Continued population growth, the exploitation of resources at an unprecedented rate, vast differences in living standards from place to place, and political instability are all related to migration processes, however, and we can rest assured that the restless movements of people at all scales will continue as individuals and groups seek ways to better their lives in this increasingly complex and interrelated global society in which we find ourselves.

Better transportation and communication systems have always impacted the volume and direction of migration streams. Among the newest communication technologies the Internet, via the World Wide Web, is becoming influential. As Tyner (1998, 340) recently concluded, "Web-related recruitment has the potential for revolutionizing patterns and processes of international labor migration."

Broader questions about the survival of nations, democracy in multicultural societies, and the meaning of international borders are going to be asked ever more frequently. Castles and Miller (1993, 275) observed the following:

> National states, for better or worse, are likely to endure. But global economic and cultural integration and the establishment of regional agreements on economic and political co-operation are undermining the exclusiveness of national loyalties. The age of migration could be marked by the erosion of nationalism and the weakening of divisions between peoples. Admittedly there are countervailing tendencies, such as racism, the "fortress Europe" mentality, or the resurgence of nationalism in certain areas . . . . But the inescapable central trends are the increasing ethnic and cultural diversity of most countries, the emergence of transnational networks which link the societies of emigration and immigration countries and the growth of cultural interchange.

Finally, at a time when more than half of the world's people live in cities, for better or worse, we would do well to consider Kotkin (2005, 160), who wrote that "It is in the city, this ancient confluence of the sacred, safe, and busy, where humanity's future will be shaped for centuries to come."

CHAPTER 9

Population and the Environment

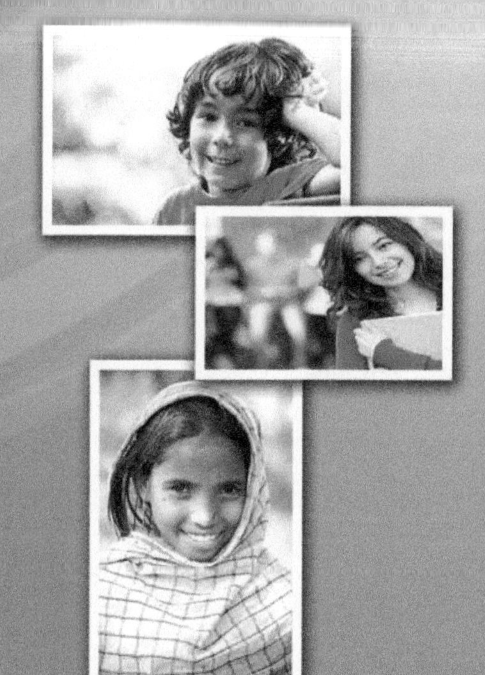

Key Terms

environmental degradation	deforestation
pollution	ecological diversity
crowding	overgrazing

So far we have considered the demographic side of population geography; here and in the next chapter we turn to linkages between the demographic system, especially growth, and two other systems, the environment and food supply systems, which are interrelated to a considerable extent. The health of both of these systems affects us all; they are essential if the human species is to sustain itself. Each and every one of us places demands on the natural world. Unfortunately, as Abernethy (1992, 235) pointed out, "In the short run, both high fertility and an extractive, exploitative approach to the environment work well."

The interrelationships among population growth, resource depletion, and the environment have been topics of concern for many scholars, including geographers (Park, 1993). As with many other complicated issues, dealing with the relationship between population and other variables engenders considerable disagreement among experts. As Repetto (1987, 3) noted, "Both with regard to the past and outlook for the future, the connections between population growth, resource use, and environmental quality are too complex to permit straightforward generalizations about direct causal relationships. Further, many of these connections are manifested only at the local level and may be missed when the perspective is regional or global, where much of the debate has focused." Similarly, de Sherbinin and Kalish (1994, 1) noted:

> It seems self-evident that human numbers and human demands on the environment are driving ecological change in such areas as global warming, ozone depletion, deforestation, biodiversity loss, land degradation, and pollution of land, water, and air. But it is a trickier matter to figure out how much of the resulting environmental impact is from increasing numbers of people, and how much from human behavior, including consumption habits.

On the one hand, there are those who feel that the root cause of most of our ecological problems is population growth—the Neo-Malthusians are perhaps the best example, led by biologist Paul Ehrlich. They argue that population growth is detrimental to the environment, especially when it is accompanied by affluence (so that a birth in India, for example, is going to do far less environmental harm over that child's lifetime than will a birth in the United States). Another view (more complementary than contrary) is that many variables other than population are critical elements in environmental deterioration, and the use or abuse of technology plays an especially significant role in environmental degradation—Barry Commoner has long supported this view. People's impact on the environment can be traced far back into history, and their attitudes toward the natural environment have varied considerably, both from time to time and from place to place. A third view (contrary to either of the above, especially to the former) is that of the most optimistic economists, led by Julian Simon. In their view population growth is hardly harmful; in fact it can be good because it provides more brains, increases the ability of humanity to innovate, and creates markets for goods and services.

The two extremes in the debate about population and the environment are represented by ecology (Ehrlich) and economics (Simon). As demographer Nathan Keyfitz (1994, 24) wryly noted, "Worldwide communication has never been easier, yet biologists and economists who eat in the same faculty club do not seem to communicate very effectively." He notes that:

> The central focus of modern economics is on growth: the idea that increasing productivity brings employment and prosperity. With sufficient economic growth, population growth is not to be feared. The resulting wealth, economists maintain, will in turn counteract the incidental ill effects of growth. Biologists, on the other hand, see the economy as embedded in a fragile ecosphere, upon which growth acts in dangerous, inscrutable ways (Keyfitz, 1994, 24).

One geographer, Crispin Tickell (1993, 224), has gone so far as to suggest that ". . . demography has much in common with economics: it is a combination of social studies, algebra and necromancy." Whether you feel that is too harsh or not, geographers can appreciate the nature of this debate better than most others because we have long grappled with our own disciplinary split personality—cultural/human and physical. Geographers are likely to have had training both in the social and natural sciences, as well as the humanities. Out of this historic development, geography has become an integrative discipline. This ability to see the "big picture," or think wholistically, often criticized by those in other disciplines, may give geography considerable importance as we struggle with growing global problems of every sort in the decades immediately ahead. For example, as geographer Dean Hanink (1995, 383) noted, "Location, it seems, matters in the analysis and solution of many environmental problems." Furthermore, geographer David Sauri-Pujol (1993, 9) noted that, "The environment is constantly transformed, degraded or improved by human action and so an understanding of human agency upon the Earth, as provided by human geography, remains essential if we are to understand and overcome our current environmental problems."

Impact of Early Civilizations

In order to understand the relationship of people to their physical environment in the modern world, it is helpful to trace the evolution of the people-environment interface. Several questions concerning this interface may be asked. What was the impact of early people on their environment? Have humans ever been part of an ecological community subject to control by ecological factors in the same way as other animals, or have humans always been able to influence their environment to a greater degree than any other animal?

Archaeological evidence supports the thesis that early humans had at their disposal potent ecological devices with which they could alter their environment, though as Hern (1990, 10) has noted, "Small scale human assaults on the environment had little or no lasting impact during the early Pleistocene, although local and regional impacts began to be seen in the late Pleistocene and Neolithic. . . ." Humans have utilized fire for at least half a million years, for example, and it is a significant tool for modifying the environment, particularly in semi-arid and arid areas. Early humans also were able to modify natural vegetation and capture or kill animals with the aid of implements that they had made.

Perhaps even more significant was the development of farming techniques, which provided humans with the ability to alter the natural environment considerably, and the concomitant diffusion of agriculture. The relationship between population and land was discussed by Chinese writers at the time of Confucius, and Plato estimated that the ideal population for a Greek state would be 5,040 landholding households (Keyfitz, 1972, 12). As Eckholm (1976, 17) noted:

> Over the course of ten thousand years humans have successfully learned to exploit ecological systems for sustenance. Nature has been shaped and contorted to channel a higher than usual share of its energies into manufacturing the few products humans find useful. But while ecological systems are supple, they can snap viciously when bent too far. The land's ability to serve human ends can be markedly, and sometimes permanently, sapped.

Population growth was one response to improved agriculture; it further increased the need for food and made it impossible for agricultural people to avoid altering the ecosystem. For example, the development of agriculture in the wooded or parkland environment of the Near East and its eventual spread across Europe eventually brought about the decimation of forests in order to open up more land for food production. Deforestation in turn frequently led to ecological disasters.

The collapse of numerous ancient agricultural systems has been traced to ecological imbalances. For example, in the semiarid regions of North Africa and the Near East this imbalance took the form of increased desiccation. It was not climatically induced; rather, it was a result of continued overgrazing and misuse of the land; some of these areas remain derelict today. Another example of ecological imbalance brought about by population pressure can be seen in Mesopotamia, where efforts to extend agriculture into drier areas with the use of irrigation brought about disastrous results, including salinization and wind erosion. Numerous examples of environmental problems and possible solutions are included in Diamond (2005). Despite its many benefits, as Wallach (2005, 45) reminds us, "One clear consequence of agriculture was the deterioration of diet." We will return to this theme in the next chapter, and it was discussed in Chapter 5 as well.

The pollution of air and water is not just a product of modern society either; evidence suggests that early cultures also were confronted with it. When Juan Cabrillo visited California in 1542, he noted that while anchored in San Pedro Bay he could see the mountain peaks in the distance but not their bases. A thermal inversion in the area trapped the smoke from Amerindian fires, bringing about air pollution as well as human health hazards.

The increased concentration of people in urban areas is also associated with water pollution problems. There was, at least in some urban cultures, a need to avoid water pollution problems in order to ensure safe drinking water. The contamination of surface water was sometimes recognized as a problem and led to the development of wells and water storage tanks. Canals were built in many areas in an effort to bring safe drinking water into urban areas. For example, the Aztecs under King Ahuitzotl brought in spring water through a stone pipeline.

In the previous chapter we noted Brockerhoff's optimism about the future of cities. A less sanguine view of urbanization, related to its environmental consequences, was suggested by McNeill (2000, 281–282):

> Twentieth-century urbanization affected almost everything in human affairs and constituted a vast break with past centuries. Nowhere had humankind altered the environment more than in cities, but their impact reached far beyond their boundaries. The growth of cities was a crucial source of environmental change. . . . For 8,000 years cities had been demographic black holes. In the span of one human generation they stopped checking population growth and started adding to it: a great turning point in the human condition.

McNeill recognizes that cities are complex, and that their support draws from a considerable area (often the entire world). He and others wonder if we can sustain urban behemoths with 20 or 25 or 30 million residents—cities that the United Nations projects will soon exist in several locations. As McNeill (2000, 287) commented, "Urban impacts extended beyond city limits to hinterlands, to downwind and downstream communities, and in some respects to the whole globe." Even as greater quantities of food, water, and materials moved into cities, greater volumes of garbage, human waste, and an assortment of pollutants moved out.

Environmental Degradation and Population Growth

Abundant evidence suggests that population growth and increased pollution went hand-in-hand in early civilizations. As the human population has increased in numbers and become more geographically concentrated, there has been an increased potential for disrupting the earth's ecosystems.

Some Dimensions of the Problem

The problem of **environmental degradation** can be broken down into several key factors, including the following: pollution, crowding and violence, global warming and ozone depletion, deforestation, decreasing ecological diversity, and overgrazing.

The problem of **environmental degradation** can be broken down into several key factors, including the following: pollution, crowding and violence, global warming and ozone depletion, deforestation, decreasing ecological diversity, and overgrazing. As discussed in this section, these factors have been among the consequences of human activity for several thousand years; many are becoming more critical as population and industrialization continue. We are inclined to agree with the following comment by Tickell (1993, 220): "We have the misfortune to be perhaps the first generation in which the magnitude of the global price to be paid is becoming manifest." At the same time, we are aware of success stories and know that with considerable cooperation many of the following problems may be slowed or even reversed in the future. Nonetheless, with more than 7 billion people sharing the planet and a globalizing economy that remains dependent on fossil fuels for its basic energy source, we are, at best, cautiously optimistic.

Pollution

When waste increases to the point that it can no longer be accommodated by the ecosystem, it becomes **pollution**.

One important natural function of the earth's ecosystem is the absorption of waste material. Although the waste from one organism can be an important input to other organisms, when waste increases to the point that it can no longer be accommodated by the ecosystem, it becomes **pollution**.

We are currently polluting our planet at an unprecedented rate. The reasons for this are many and complex but, as economist Kenneth Boulding summarized it, "Our desire to conquer nature often means simply that we diminish the probability of small inconveniences at the cost of increasing the probability of very large disasters" (Boulding, 1966, 14). More than five decades have passed since Boulding wrote that, and so far no serious calamity has befallen planet Earth. Moreover, our population has about doubled during that time, a reminder that Malthusians so far have not fared well in their predictions of the disasters that population growth would bring us. However, economists and others who take pride in how much population growth the world has been able to absorb never answer the following question, which is seldom even asked: Would we have been even better off if population had stabilized at say two or three billion?

Pollution's primary forms are biological or chemical (though filth, noise, and other irritants may be considered pollutants to many). Human population density often leads to increased biological pollution. Today, the crowding of large numbers of people into small places has brought about increased pollution almost everywhere. As a population increases, so does the accumulation of its human organic waste, for example. City water supplies may be contaminated as it becomes increasingly difficult to dispose of large volumes of waste.

Chemical pollution is another by-product of a growing population, coupled with modern technology. Many streams and lakes have been polluted by the addition of toxic chemicals. One of the most notable—and most tragic—cases occurred in Minamata Bay, Japan, where industrial waste containing mercury was dumped into the fishing waters. Local fishermen continued to eat their catch; the result was several thousand cases of a debilitating disease now referred to as Minamata disease, a form of mercury poisoning. Even today in the United States mercury is distributed throughout many water systems and Americans have been warned not to eat fish from numerous streams and lakes. Most of that mercury enters the atmosphere first, primarily from such industrial emissions as those of coal-fired power plants, then subsequently enters the ground and water systems. While the Environmental Protection Agency was reluctant to develop strong controls on such emissions under the leadership of President George W. Bush (2001–2009), current President Barack Obama has used his executive powers and those of the 1970 Clean Air Act to:

> . . . press the most sweeping attack on air pollution in U.S. history. He has imposed the first carbon-dioxide limits on new power plants, tightened fuel-efficiency rules as part of the auto bailout and steered billions of federal dollars to clean-energy projects. He also has proposed slashing mercury emissions from utilities by 91 percent by 2016 (Eilperin, 2012b, A07).

Pollution problems that were local in scope within earlier civilizations are now becoming global concerns, crossing borders indiscriminately and threatening international relationships in the process. Chemical wastes that break down slowly can ultimately reach the oceans, which are constantly used as dumping grounds. As the population increases on the continental margins, the oceans become common sinks for industrial wastes and garbage. Compounds such as DDT have been carried to virtually all parts of the ocean (don't forget, no matter what we might name them, there is only one ocean), and they continue to be used.

Air pollution is also affected by population growth and the geographic concentration of population, as geographers Charles Collins and Steven Scott (1993) noted in their study of air pollution in Mexico. Climatic changes are being induced by concentrations of people in urban areas, frequently with a concomitant increase in airborne pollutants. As Repetto (1987, 28) pointed out:

> Estimated global emissions of carbon dioxide from the use of fossil fuels and burning of biomass have nearly tripled since 1950. Emissions of the most active chloroflourocarbons have increased from negligible amounts to almost 700,000 metric tons per year in 1985. . . . Samples of air trapped in ice cores suggest that the methane concentration has doubled.

In addition to lead, the urban atmosphere often contains high counts of microorganisms; gases, such as carbon monoxide and oxides of nitrogen; and a variety of other chemical compounds, including sulfur dioxide. Particulates and smog in the atmosphere cut down on visibility and produce a number of effects on people, from eye-irritations to much more serious conditions.

The United States Office of Technology Assessment (1984) estimated that the effects of air pollution caused 50,000 premature deaths in the United States each year—about 2 percent of annual mortality. Especially vulnerable are the millions of people who are already suffering from asthma, emphysema, and other chronic respiratory disorders (National Research Council, 1985). The Environmental Protection Agency (EPA) and the American Lung Association estimate that 131 million Americans (42% of the total population) live in areas where air is unhealthy. This may lead to as many as 120,000 deaths each year (American Lung Association, 2013).

Although the Clean Air Act of 1970 has been successful in reducing air pollution, Ehrlich and Ehrlich (1990, 138) estimated that if population had not grown, air pollution would now be only a little more than half the 1970 level. Even Los Angeles (the butt of many a late-night comedian's jokes about smog) has cleaned up its act during the past several decades, though it remains the smoggiest city in the United States. Cleaner industries and cars have become the rule, the establishment of an Air Quality Management District (AQMD) in the region has helped fight major polluters, and new rules promise even stricter controls ahead. Nonetheless, the Los Angeles Basin continues to fill up with more and more people, driving hither and yon to work and play. Though, in speaking about Los Angeles, Lents and Kelly (1993, 39) point out that ". . . residents and business people seem to recognize the need to solve the serious air-pollution problems." Neither these authors nor the AQMD, however, suggest the contribution that sustained high population growth has made to the problem.

With increases in the human population and industrial output that have occurred over the past century, persistent contamination of the biosphere has become a global problem. Continued population growth and the concomitant new demand for food, shelter, and other goods and services will make it ever more difficult to bring pollution under control. China, as it rapidly develops, has run into numerous environmental problems, including pollution.

Now that more than half of us live in cities, it is more critical than ever to understand them and their effects on the environment. As Weatherford (1994, 175) noted, "Cities destroy. They consume the area around themselves, and if they cannot find new materials, they die. Historically, the destruction from urban areas has followed a pattern of destroying or severely disrupting first the animals and plants of the surrounding area, then, inevitably, the soil, the water, and finally the air."

Water alone, especially clean water, is becoming scarcer on a per capita basis as world population growth continues. Remember that 97 percent of the world's water is salt-water, and much of the remainder is locked up in ice sheets and glaciers. In the United States, water consumption per capita is about three times what it is in the Third World. Furthermore, as Americans continue to move southward and westward, they put ever greater demands on water supplies. Places such as Las Vegas and Phoenix must find water somewhere, or, heaven forbid, they might have to curtail their rapid growth. Worldwide, approximately 1.2 billion people lack safe drinking water, and over 2.6 billion lack adequate sanitation (World Health Organization, 2010).

Within most urban areas landscaping is the major consumer of water, but beyond those areas it is agriculture that is the world's primary consumer of fresh water, consuming perhaps 95 percent of all the water that humans use. During the twentieth century water consumption increased faster than population growth, mainly because more irrigation was required to meet the world's food needs. While Americans sit in air-conditioned houses in Las Vegas and drink bottled water (often no more than bottled tap water at an inflated price), many Africans struggle to find enough water to get their crops to grow and have no source of safe drinking water. At some point water will be worth fighting for, so it is one likely source of future conflicts. Shifts in rainfall patterns will further increase the likelihood of such conflicts, especially in such places as the Sahel region of northern Africa.

Crowding and Violence

Though we have placed crowding and violence in the same category, we are not arguing that there is a clear-cut cause-and-effect relationship between the two, nor between them and population growth. However, it seems clear that population growth at least exacerbates such problems.

The pioneering work of anthropologist Edward Hall, dealing with how different cultures perceive and evaluate personal space, has been of considerable interest to geographers and other social scientists (Hall, 1969). As the population of an area grows, there will be more people per square mile; nothing can change this. Similarly, personal space becomes smaller and smaller. Evidence suggests that individuals and cultures can tolerate different levels of crowding. What, then, is the effect of crowding on specific populations?

Studies with animal populations (including rats, mice, and deer) suggest what happens when overcrowding occurs in those various populations. In his classic rat studies Calhoun pointed out the disastrous consequences of overcrowding (Calhoun, 1962). He found that high density resulted in the disruption of the rats' nesting patterns because of changes in their normal social behavior. Crowding intensified social interaction and the competition for resources. Infant mortality increased with the disturbance of nesting patterns, and some of the young were even consumed by other rats. Aggressive attacks became more frequent.

The relevance of these animal studies to the human situation has been questioned. For example, Freedman (1975) pointed out the following:

> . . . it is both difficult and risky to generalize directly from the behavior of one animal to that of another. It would be a mistake to conclude that dogs act a particular way just because cats do or that monkeys act the same way as lions; and it is of course much more difficult to conclude anything about humans from other animals. Humans are more intelligent, have language, and have an extremely complex social structure, are much more flexible and innovative than other animals . . . there is enough difference between humans and the rest of the animal world to make it difficult to conclude anything about humans from what other animals do.

Of course, as Freedman also noted, ". . . work on animals is not only extremely interesting but can also be a source of ideas and suggestions about how humans behave" (Freedman, 1975, 41). A knowledge of animal behavior may suggest hypotheses about human behavior, but we need to test them. Though some of the animal studies may not be transferable to people, we cannot help but wonder about the quality of life that can be maintained as ever more people are crowded into cities.

Crowding is to fill by pressing or thronging together.

Attempts have been made to correlate high crime rates with population density, partly because researchers have suggested that aggression increases with **crowding** and partly because it has been shown that urban crime rates are higher than rural crime rates. For example, in the United States the crime rate per 100,000 residents is over five times greater in the largest cities than it is in rural areas, whereas intermediate crime rates prevail in suburbs and small towns. Still, large populations do not necessarily mean high population densities. There have been few well-designed studies of the effect of population density on crime. After reviewing the results of several of those studies, Freedman (1975, 69) concluded that ". . . there are a great many reasons why people commit crimes, [and there are] many factors in modern, complex society that cause crime, but there is no

Attempts have been made to correlate high crime rates with population density, partly because researchers have suggested that aggression increases with **crowding** and partly because it has been shown that urban crime rates are higher than rural crime rates.

evidence that crowding is one of them." Part of the problem with most of the studies is the difficulty of isolating the effect of crowding from other variables such as poverty, ethnic composition, and educational levels.

In an attempt to analyze the effects of cramming more and more people into cities, the organization, Zero Population Growth, devised an "urban stress test" and applied it to all 192 United States cities with populations over 100,000 people. The test looked at eleven criteria associated with urban blight and found that the twenty-two cities with the best scores averaged 116,000 people, with about 3,700 people per square mile and the twenty worst cities averaged 1,154,000 people with 8,200 people per square mile. According to Ehrlich and Ehrlich (1990, 156), "The message seems clear: measured either by social and environmental indicators together or by environmental indicators alone, more people mean more problems in American cities."

Regardless of the exact nature of the effect of crowding on human behavior, there is growing evidence that the quality of people's lives in many places is being diminished by population growth and the increased densities that often accompany it. For example, consider the following statement by Brown, McGrath, and Stokes (1976, 42):

> Aerial photographs of Java reveal that people are actually moving into the craters of occasionally active volcanoes in their search for land and living space. Periodic evacuations and loss of life result. In Bangladesh, people are driven by population pressure into floodprone lowlands and onto low coastal islands previously uninhabited because of the danger of tidal waves and typhoons. *The New York Times* of November 15, 1970, reported more than 168, 000 people killed by a tidal wave that swept the coastal area. Described as one of the worst natural disasters of the century, this loss of life is more accurately attributed to overcrowding than to any "natural" phenomenon.

Beyond the likely, if difficult to "prove," relationship between crowding and quality of life, population growth may be increasing levels of violence in other ways as well.

Beyond the likely, if difficult to "prove," relationship between crowding and quality of life, population growth may be increasing levels of violence in other ways as well. For example, rapidly increasing populations in many less developed countries are likely to intensify violent conflicts because of increasing resource scarcities. As Homer-Dixon, Boutwell, and Rathjens (1993, 38) noted:

> . . . scarcities of renewable resources are already contributing to violent conflicts in many parts of the developing world. These conflicts may foreshadow a surge of similar violence in coming decades, particularly in poor countries where shortages of water, forests and, especially, fertile land, coupled with rapidly expanding populations, already cause great hardship.

Aside from the obvious problems associated with population pressure on nonrenewable resources, the authors argue that growing problems over renewable resources are likely as well. They argue that human actions can generate scarcities of renewable resources in at least three different ways:

1. Use or degradation can occur because resources are used faster than they can be renewed.
2. Common resources such as water may have to be divided among too many users.
3. Changes in the way resources are distributed within a society, mainly concentration in a few hands.

Strong political and economic systems in less developed countries could certainly mitigate many potential conflicts, but such countries are often lacking in those very strengths. Help from the more developed countries is going to be needed, yet most of them are increasingly fighting their own battles to survive in a globalizing economic system that is creating new sets of winners along with vast numbers of losers. So far, the euphoria of having unfettered free markets diffused around the world has not led many political or corporate leaders to consider global capitalism's downside, those too poor and voiceless to fend for themselves. As Weatherford (1994, 277) put it, "We are caught up in an artificial negative tension between a theory of global economics and a reality in which people live. And we do live in real places."

An even broader view of the Third World (and parts of the rest as well) suggests that the very social and political fabric of societies is being shredded by overpopulation, resource scarcity, increasing crime, and the rapid spread of contagious diseases such as AIDS. Unsafe streets are common in many American cities at night, but they are nothing compared to the streets of West African cities after the sun goes down. In a thought-provoking article, journalist Robert Kaplan (1994) begins with a discussion of the breakdown of order in West Africa, then moves through other parts of the world, arguing that anarchy and chaos are becoming increasingly common as authorities lose control of growing populations. Kaplan sees a world increasingly confronted with more refugees, increasing interethnic conflicts (severely straining multiethnic nation-states), powerful international drug cartels, private armies, and security firms. Among his conclusions, the following (with blatant geographic appeal) is worth considerable reflection:

> Imagine cartography in three dimensions, as if in a hologram. In this hologram would be the overlapping sediments of group and other identities atop the merely two-dimensional color markings of city-states and the remaining nations, themselves confused in places by shadowy tentacles, hovering overhead, indicating the power of drug cartels, mafias, and private security agencies. Instead of borders, there would be moving "centers" of power, as in the Middle Ages. Many of these layers would be in motion. Replacing fixed and abrupt lines on a flat space would be a shifting pattern of buffer entities, like the Kurdish and Azeri buffer entities between Turkey and Iran, the Turkic Uighur buffer entity between Central Asia and Inner China . . . and the Latino buffer entity replacing a precise U.S.-Mexican border. To this protean cartographic hologram one must add other factors, such as migrations of populations, explosions of birth rates, vectors of disease. . . . This future map—in a sense, the "Last Map"—will be an ever-mutating representation of chaos (Kaplan, 1994, 75).

Pessimistic? Yes, but sobering as well. Before this article was published the debacle in Bosnia was well underway; however, the horror in Rwanda came later. In 2004 and subsequent years there was genocide in Darfur, as well as wars in Afghanistan and Iraq. A chaos-free future seems highly unlikely.

Global Warming

Global warming, euphemistically referred to by many Washingtonians today simply as climate change, has risen to a high level of concern in recent years, though we've known about it for at least two decades. In the press it is talked about much like a religion, with discussions about whether people "believe" in it or not—some do, some don't. To illustrate the two ends of this religious spectrum, consider the positions of Vice President Al Gore and Oklahoma's Republican Senator James Inhofe on global warming. On the Senate floor on July 28, 2003, Senator Inhofe described global warming as "The greatest hoax ever perpetrated on the American people." On March 21, 2007, Al Gore described global warming as "A true planetary emergency." We suggest that between those extremes lies reality—an Earth that is now warming up, and generally has been since the end of the last Ice Age about 11,000 years ago, though with a few notable fluctuations (Figure 9–1).

Few serious scientists question either whether we are in a warming period or whether we are in part contributing to that warming, mainly by burning fossil fuels, thus increasing the content of carbon dioxide in the atmosphere (Figure 9–2). Estimates show that the level

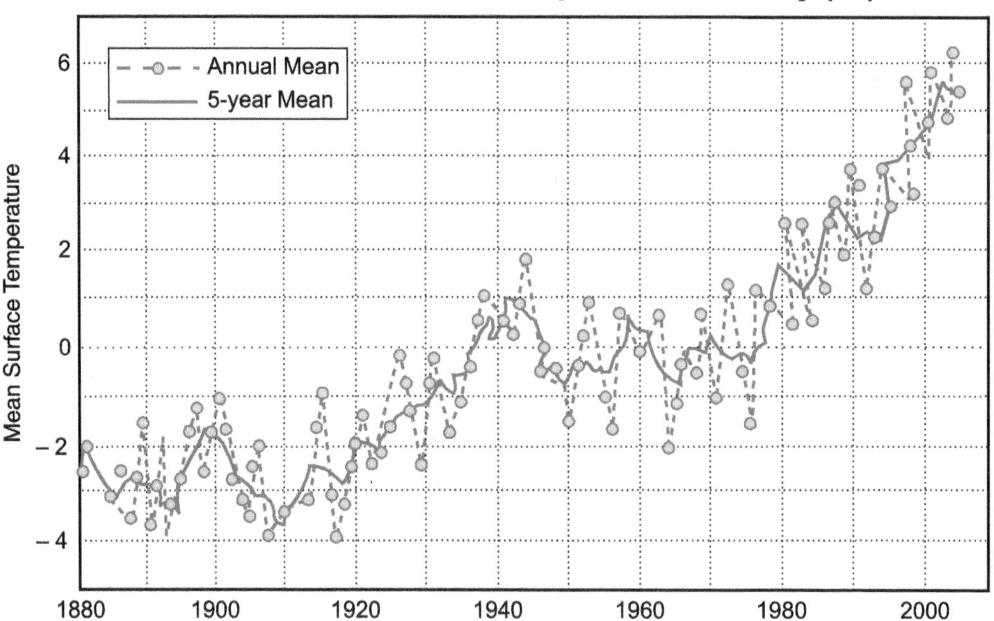

Figure 9–1 Global-Mean Surface Temperature Anomaly (°C)

Source: NASA, Goddard Institute for Space Studies, Research News, February 8, 2007.

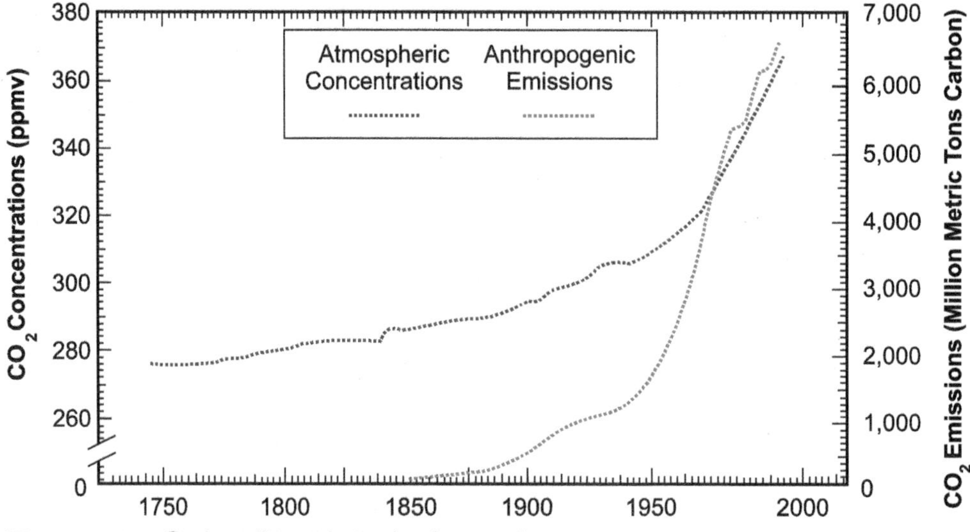

Figure 9–2 Carbon Dioxide in the Atmosphere

Source: Oak Ridge National Laboratory, Carbon Dioxide Information Analysis Center, http://odac.asd.omi.gov/.

of atmospheric carbon dioxide before the Industrial Revolution was around 280 parts per million. Today it is around 398 parts per million, and, if current rates of growth in carbon emissions continue, then it may reach 800 parts per million by the end of this century. Other "greenhouse gases" are accumulating in the atmosphere as well, including methane. Over the past 100 years or so Earth's average temperature has risen about 1.1 degrees Fahrenheit; future warming seems inevitable and is likely to accelerate. This is a fact, not a hoax.

Before going on, let's consider some recent changes that at least strongly suggest that our global atmosphere has been warming up:

- Spring has been arriving about one week earlier in the Northern Hemisphere.
- The length of the growing season in some places has increased by a week since 1980.

- The amount of vegetation in North America, Europe, and Siberia may have increased by as much as 20 percent since 1960.
- The carbon dioxide content of the atmosphere has increased steadily since monitoring began in 1958.
- Human activities contribute more than 6 billion metric tons of carbon dioxide to the atmosphere annually.
- After Hurricane Andrew in 1992 cost them some $16 billion, many insurance companies began to examine possible implications of global warming for their business. A real gut-check came with Hurricane Katrina's destruction of New Orleans in 2005 and was further confirmed by Hurricane Sandy's devastation of the U.S. East Coast in October 2012 (Barrett, 2012).
- In early 2002 the Larsen B ice shelf on the Antarctic Peninsula collapsed and disintegrated. The Greenland ice sheet is melting faster than expected.
- A study in Queensland, Australia, found that the number of days with frost had been declining since 1900.
- The Aletsch Glacier in the Swiss Alps has been retreating since 1865.
- Worldwide, glaciers are retreating almost everywhere, and the rate of retreat seems to be accelerating.
- Sea ice is thinning in the Arctic.
- The 1990s were the warmest decade since records started being kept around the world, and the 2000s have been as warm or warmer.
- January 2010 was the warmest ever recorded; before that January 2002 was the warmest.
- As of the end of 2012, the eight warmest years on record were, in order, 2010, 2005, 1998, 2007, 2002, 2003, 2006 and 2009 (Harvey, 2012).

All of these observations are compatible with, if not substantial confirmation of, the idea that the earth's atmosphere has been warming for more than a century now. Less well documented, but still likely, is that humans are helping the atmosphere to warm, mainly by burning fossil fuels and adding other "greenhouse" gases to the atmosphere, including methane and chlorofluorocarbons. As Fagan (2000, 209) described it:

> We live on a benign planet, protected by the heat absorbing abilities of the atmosphere, the so-called "greenhouse effect." Energy from the sun heats the surface of the earth and so drives world climate. The earth, in turn, radiates energy back into space. Like the glass windows of a greenhouse, atmospheric gases such as water vapor and carbon dioxide trap some of this heat and reradiate it downward. . . . But the effect is no longer purely natural. Atmospheric concentrations of carbon dioxide have now increased nearly 30 percent since the beginning of the Industrial Revolution; methane levels have more than doubled; and nitrous oxide concentrations have risen by about 15 percent. These increases have enhanced the heat-trapping capabilities of the atmosphere.

According to more recent estimates, atmospheric carbon dioxide grew by nearly 27 percent (from 310 to 392.92 parts per million) between 1958 and November 2012. Moreover, "The concentrations of CO_2 in the atmosphere are increasing at an accelerating rate from decade to decade" and while the "upper safety limit for atmospheric CO_2 is 350 parts per million (ppm) . . . [these] levels have stayed higher than 350 ppm since early 1988" (CO2Now.org, 2012, no pp).

At the local level, temperature and rainfall patterns are affected by human occupance. Agricultural and urban-induced dust pollution and increased carbon dioxide in the atmosphere due to the burning of fossil fuels have led to changes in local rainfall and temperature regimes; urban heat islands and downwind rainfall plumes have been identified in a variety of places. Dust particles in the atmosphere act as condensation nuclei and thereby increase rainfall. It has also been hypothesized that increasing quantities of airborne dust act as insulation by reflecting the sun's rays away from the earth and thus lowering temperatures.

On the one hand, atmospheric dust has a cooling effect; but on the other hand carbon dioxide has a warming influence because it traps the earth's heat. Chlorofluorocarbons (CFCs, also known as halocarbons) and methane gas are also effective atmospheric heaters (via the greenhouse effect). At the local level the results of increased population, coupled with the increased burning of fossil fuels, will maintain urban areas as heat islands. Worldwide, the concentration of carbon dioxide in the atmosphere will continue its upward climb, whether we act to control it or not. Earth is going to get warmer.

There is some debate among scientists about the precise nature and extent of this warming. The United States National Academy of Sciences suggested that the warming over the next century will be from 1.5 degrees Celsius to 4.5 degrees Celsius. According to Stephen H. Schneider (1990, 30), head of Interdisciplinary Climate Systems at the National Center for Atmospheric Research, "The earth has not been more than 1 to 2 degrees Celsius warmer during the 10,000-year era of human civilization. The previous Ice Age, in which mile-high ice sheets stretched from New York to Chicago, was only 5 degrees Celsius colder than now."

International concern for global warming led to formation of the Intergovernmental Panel for Climate Change (IPCC) in 1988. Its purpose was to review the state of knowledge with respect to human-induced climate change and to assess possible responses. Composition of the IPCC is international, with hundreds of scientists making contributions to its work. IPCC's first report appeared in 1990, and it served as an important basis for discussions of climate change at the United Nations Conference on Environment and Development (UNCED) in Rio de Janeiro.

IPCC's second report was published in 1995. It suggested that the world was likely to experience a warming of 2 degrees centigrade by 2100, along with a rise in sea level during that time of between 0.13 and 0.94 meters. It also for the first time went on record as implicating humans in the global warming process.

IPCC's third report was published early in 2001. Compared to the 1995 report, the 2001 report suggested a somewhat larger average temperature increase was likely by 2100 (between 1.4 and 5.8 degrees centigrade), but with a somewhat smaller rise in sea level (between 0.09 to 0.88 meters) than the panel had predicted in 1995. The panel also suggested that long-term patterns of changing temperatures and sea levels were likely to continue for thousands of years.

IPCC's fourth report appeared in 2007. Its projected warming by 2100 ranged from a low of 1.8 degrees Celsius to a high of 6.4 degrees Celsius, depending on which of several scenarios was considered in the models. Different assumptions about future greenhouse gas emissions lead to different degrees of warming. As Meehl, et al. (2007, 749) noted, "Possible future variations in natural forcings (e.g., a large volcanic eruption) could change those values somewhat, but about half of the early 21st-century warming is committed in the sense that it would occur even if atmospheric concentrations were held fixed at year 2000 values." As the century progresses, and we are already slightly over a decade into it, more of the projected warming will be committed if we do not alter our use of fossil fuels and find other ways of diminishing greenhouse gas emissions. Meehl, et al. (2007, 749) pointed out that "Geographical patterns of projected SAT [Surface Air Temperature] warming show greatest temperature increases over land (roughly twice the global average temperature increase) and at high northern latitudes, and less warming over the southern oceans and North Atlantic, consistent with observations during the latter part of the 20th century. . . ."

According to the IPCC's 2007 report, along with other research, it seems likely that not only global temperatures but also local and regional weather patterns will be altered. More heat waves, such as the one that killed an estimated 35,000 people in Europe in

2003, are likely. In 2003 temperatures in France reached 104 degrees Fahrenheit and in Germany 105 degrees Fahrenheit; nearly 15,000 people died in France and about 7,000 people died in Germany. On August 10, 2003, London recorded its first ever triple-digit Fahrenheit temperature.

More and more species of animals and plants are on the move, at least partly in response to rising temperatures and changing precipitation patterns. From Edith's checkerspot butterfly to red ants, there is a poleward march going on that may not be obvious to lay people but are attracting the attention of scientists around the world. Warmer ocean waters are destroying coral reefs, in turn affecting both fisheries and the tourism industry in many locales. Extinction will become a distinct possibility for species of plants and animals that cannot shift their locations swiftly enough.

Average precipitation may change as well as the Earth warms up. Expectations are for heavier precipitation in tropical areas that already get lots of rainfall, mainly because warmer air will evaporate and carry more moisture, as well as decreases in the subtropics, which already suffer from water deficits in many places, and increases in the high latitudes. In mountainous areas more precipitation may come as rain rather than snow, leaving smaller snow packs to provide spring and summer runoff. Rainfall events may intensify when they do occur, and droughts are likely to become more prolonged. Droughts can lead to famines, famines to starvation and conflict. There is little doubt that underlying the problems in Darfur, Sudan, for example, are the lack of rainfall and the scarcity of food.

Feedback loops may exacerbate global warming. For example, as the planet warms, the land and oceans will not be able to absorb as much anthropogenic carbon dioxide, thus increasing its level in the atmosphere at an accelerating rate. As Meehl, et al., (2007, 750) point out, "The higher the stabilization scenario [for anthropogenic carbon dioxide], the larger the climate change, the larger the impact on the carbon cycle, and hence the larger the required emission reduction. Worse yet, a recent study has shown that the southern ocean is becoming saturated with carbon dioxide, which means that it may not be able to absorb much more, leaving more to accumulate in the atmosphere. Another example is the changing albedo in the high latitudes, especially in the northern hemisphere. Snow cover has a very high albedo, so it reflects 90 percent or more of incoming solar energy. However, as snow melts over wider areas, the albedo decreases, allowing Earth to absorb more incoming solar energy, warming its surface and sending more infrared energy outward to be captured by the greenhouse effect.

Tipping points are also mentioned more frequently by climatologists. These are sizable feedback loops that may be triggered by relatively small changes in warming. Disintegration of the West Antarctic ice sheet and Arctic sea ice are often-cited examples. On July 16, 2007, a British swimmer jumped into open water at the North Pole and swam a kilometer in the freezing water just to draw public attention to the fact that the ice had melted. Unfortunately, it was the same day that American actress and singer Lindsay Lohan was released from a rehabilitation center, so that got all the media attention in the United States.

Though the carbon dioxide content of the atmosphere has steadily increased since the Industrial Revolution really got underway in the 19th century, one paleoclimatologist has suggested that humans have probably been affecting the carbon dioxide content of the atmosphere for much longer than that. According to Ruddiman (2005, 5), "Carbon dioxide concentrations began their slow rise 8,000 years ago when humans began to cut and burn forests in China, India, and Europe to make clearings for croplands and pastures." He also argued (2005, 5) that "methane concentrations began a similar rise 5,000 years ago when humans began to irrigate for rice farming and tend livestock in unprecedented numbers." If he is correct, then humans have been contributing to global warming since the early centuries of the agricultural revolution.

Future scenarios are just that—scenarios. They include the possibility of increased storm intensity, more coastal floods such as those caused by Hurricanes Katrina and Sandy, shifting agricultural and forest belts, more tropical diseases such as malaria in higher latitudes, intensified monsoon rains, ecosystem damage, and crop failures in many locales. Another scenario that has many scientists concerned today is the possibility of a substantial change in the circulation of oceanic waters, especially in the northern Atlantic, and climatic repercussions that such a change might have. The northward flowing Gulf Stream carries warm water and heat poleward, resulting in a Western Europe that is much warmer than it would be otherwise. In the higher latitudes of the north Atlantic, water carried there by the so-called oceanic conveyor belt loses its heat, becomes saltier, hence denser, and finally sinks, to return southward as a submerged countercurrent. This sinking of denser water in turn makes room for more warm water to come poleward, hence keeping in motion the transfer of heat from tropical oceanic waters to the higher latitudes. Today more fresh water in the north Atlantic, some of it from glacial melting in the Arctic and in Greenland, is decreasing the density of water in the region. As its density decreases, it is less likely to sink. If it finally fails to sink, then it can't be replaced by warmer water from the south, and the entire conveyor system comes to a gradual halt. If that were to happen, then northeastern North America and northwestern Europe would quickly cool off. Such a cooling in western Europe would in turn disrupt grain production and the entire food supply system over much of the region.

The result of global warming at such an unprecedented scale could be catastrophic. Two of the most significant changes would be a rise in sea level, which could flood low-lying river deltas throughout the world (especially in less developed countries, which could not afford to erect dikes and other barricades), and changes in agricultural productivity that might be associated with shifting climate patterns. Christopher Flavin (1991, 82) believes that if global warming is permitted to continue, it may ". . . soon affect economies and societies worldwide. Indeed, it can be compared to nuclear war for its potential to disrupt a wide range of human and natural systems." More recently journalist Ross Gelbspan (2004, 176) was even more emphatic, stating that ". . . climate change is not just another issue in this complicated world of proliferating issues. It is *the* issue that, unchecked, will swamp all other issues." Bill McKibben (2004, 33) seems to agree and tells us that dealing with global warming ". . . will demand nothing less than the overhaul of the entire global economy, which is currently based on the very fossil fuels whose combustion we can no longer afford, but whose replacement remains technologically, economically, and politically more challenging than perhaps any transition in modern human history." It is no wonder that politicians shy away from the issue whenever they can.

On July 23, 2001, in Bonn, Germany, the 180-nation U.N. Climate Change Conference reached a milestone compromise agreement to reduce greenhouse-gas emissions that affect global warming to levels below the 1990 baseline between 2005 and 2012. After his rejection of the Kyoto Protocol on domestic economic grounds and its failure to hold China and India to the same mandatory greenhouse gas emissions caps as the industrial nations, former President George W. Bush chose not to commit the United States to subsequent climate accords even though his administration continued to argue that it took the issue of global warming seriously—a position hardly taken seriously by the 191 other nations that supported and eventually ratified the Kyoto Protocol. Initially Japan, Canada, Russia, and Australia opposed the protocol to some degree but they eventually ratified it—Japan and Canada in 2002, Russia in 2004, and Australia in 2007. The United States signed but never ratified it and Canada eventually withdrew from it in 2011. As the Kyoto Protocol's first commitment period ended in December 2012, U.N. member states met in Doha and renewed the Kyoto framework until 2020 when the next agreement is expected to come into force. Overall, Kyoto had, at best, given its many compromises and exclusions a modest influence on global warming. It:

> . . . covers about 15 percent of emissions worldwide. Canada, Japan, New Zealand and Russia renounced obligations under a second commitment period beginning in 2013. The U.S. never ratified the pact, and developing nations from China to India . . . never set binding goals. Developing countries see it as an important step by [the] rich nations [that are] most responsible for global warming [to] move first toward a solution (Morales and Krukowska, 2012, no pp).

Nevertheless, Kyoto did provide a framework for future climate agreements, with the hope that the United States would sooner or later provide needed leadership on the issue. Such U.S. leadership is yet to materialize even from the Obama White House because of significant domestic opposition to energy policies that might drive up electricity costs and harm the still relatively weak U.S. economy. Moreover, many Americans have yet to "connect the dots between extreme weather . . . [and] climate change" (Chipman and Morales, 2012, no pp).

Despite this, there is evidence of growing awareness of global warming in the U.S. and of the need to take action. For instance, in 2006, Al Gore's movie, *An Inconvenient Truth*, was released, and its impact on Americans has been measurable. After seeing it, some of them may even have swapped their SUVs for Toyota Priuses and got on the bandwagon to oppose global warming, or at least feel that they were making a contribution. But their numbers were only a trickle, when a flood is needed. Next on the global warming entertainment calendar was Live Earth, a 24-hour, 7-continent concert series that took place on 7-7-07. More than 100 music artists performed, hoping to solve the climate crisis. Solve the climate crisis? Pardon our cynicism, but we're just not sure that one more concert series is going to get everyone out of their SUVs, back into small houses, and living close to their jobs. Most of the performers at the concert series wore their carbon footprints on their sleeves— personal jets, multiple houses, gas-guzzling SUVs, and much more. More than a little of the reaction to global warming in the United States smacks a bit of elitism—Al Gore tearing around the country on planes, but paying for his carbon in "offsets," university professors giving up their Volvos for Priuses (but still flying off to conferences around the world), and writers living "locally" by going to farmers' markets (usually for a limited time period just to show that it can be done, though they did need to bring in a supply of not-so-local wine from France and maybe a few truffles from Italy). Caught up in their daily commutes, raising their children, and running on a wage/cost treadmill, most Americans don't yet seem ready for any more sacrifices.

Control of rapid population growth has the potential to make a major contribution to raising living standards and to easing environmental problems like greenhouse warming. For this reason, the United States has long been central to global population control efforts through its Global Health Initiative which, among other things, provides support for global family planning and reproductive health programs more so in poor countries. The U.S. is also a major contributor to the United Nations Population Fund (UNFPA) though funding for this program and other international population control programs varies by administration. It is often lower under conservative (Republican) administrations such as that of former President George W. Bush and higher under liberal (Democratic) administrations such as that of current President Barack Obama.

It is not that there are no possible remedies to slow or stop global climate change. Rather, it is reluctance on the part of corporations, especially big oil, big coal, and autos, to change their ways, and, of course, reluctance on the part of politicians to force such changes. President Bush, for example, used jobs and the economy as his excuse for not approving the Kyoto Protocol. The weak U.S. economy has also significantly muted the Obama administration's environmental initiatives (Chipman and Morales, 2012).

Perhaps the most serious look at what to do about global warming and whether it would be worthwhile or not was what became known as the Stern Report (Stern, 2007), named after its author, Sir Nicholas Stern. It is a lengthy and thorough look at both the likelihood of global warming and the many reasons why it would be better for us to act now rather than later or not at all. Nobel Prize-winning economist, Robert M. Solow, said of the Stern Report that "If the world is waiting for a calm, reasonable, carefully argued approach to climate change, Nick Stern and his team have produced one. They outline a feasible adjustment policy at tolerable cost beginning now. Sooner is much better" (Testimony on Cambridge University Press web site). Among the Stern Report's conclusions are the following:

- There is still time to avoid the worst impacts of climate change, if we take strong action now.
- Climate change could have very serious impacts on growth and development, so consideration must be given to adaptation strategies as well as to trying to slow or reverse it.
- The costs of stabilizing the climate are significant but manageable, though delays will be dangerous and costly.
- All countries need to be involved (which would be a real first).
- A range of options exists to cut emissions of greenhouse gases, if agreement can be reached on international goals and frameworks for action.

While the Stern Report argued that the benefits of doing something about continued global warming outweighed the costs, not everyone agreed. In fact many economists have been critical of the study, mainly for its choice of discount rates and other details.

Some more general criticisms about global warming strategies have arisen as well. For example, many have pointed out that if Earth is warming up then there are going to be both winners and losers. Not everyone in Alaska, Canada, and Siberia is really unhappy with the thought of shorter winters and warmer summers. On the other hand, losers may include farmers in the world's major grain belts, where less rainfall may drive up the cost of producing food for a still-growing world. Even wine drinkers may experience some serious shifts in the quality of products from some regions and the opening up of new ones for the cultivation of premium wine grapes (Brown, 2007). Furthermore, as Brown pointed out (2007, F6), "Wine is the canary in the climate-change coal mine, according to climatologists. Even slight changes in climate can wreak havoc on high-quality wine, making it particularly vulnerable to global warming."

Finally, despite the Stern Report's analysis, there remain real questions about whether and how much should be done about controlling greenhouse gas emissions, who should do the controlling and how do we encourage them to do it, and whether money could be invested in better ways. Considerable uncertainties exist, and cannot be discounted. For example, we really don't know which end of the scale of temperature increases we're going to end up on by 2100. Another degree or two might be quite acceptable, whereas another five degrees might not. Mitigating carbon dioxide emissions, especially in a world that continues to add nearly 80 million residents each year, is not going to be either easy or free. Climatologists want big changes now, whereas economists are in favor of a more gradual approach. We also don't know how much the demand for energy can be decreased by simple measures such as building more energy-efficient buildings and using more energy-efficient appliances. Nor do we know how fast the cost of renewable energy technologies will fall, which would in turn affect their rate of adoption. Most politicians favor some kind of cap-and-trade system to limit carbon dioxide emissions, but most economists favor carbon taxes. The latter are more efficient, can be easily adjusted, and could be traded off against other taxes in a way that would make them revenue-neutral.

If we are to shift away from fossil fuels, then much can be done realistically, though not immediately. Socolow and Pacala (2006) provided an excellent overview of some of the possibilities and introduced "the wedge concept." They focused on carbon reduction over a 50 year period, 2006-2056, viewed it graphically, looked at a triangle created by two lines, a begin action line now and a delay action until 2056 line, and the straight line between the two in 2056. That triangle, then, represented the difference between doing a lot now and doing nothing for another 50 years. They then divided the triangle into seven wedges, each representing 25 billion tons of carbon over the 50 year period. Each wedge, then, could represent one strategy for controlling carbon emissions over the 50 year period. They then suggested 15 ways to make a wedge. Among their suggestions were the following: expand conservation tillage to 100 percent of cropland, stop all deforestation, drive two billion cars on ethanol, increase wind power 80-fold to make hydrogen for cars, replace 1,400 large coal-fired power plants with gas-fired ones, and cut electricity use in buildings by 25 percent. They even suggested that controlling fertility could form a wedge, if we could keep the 2056 population down to eight billion as opposed to the projected nine billion. Most scientists agree that we can reduce our carbon emissions over time by using everything from solar energy to carbon sequestration, from windmills to nuclear power plants, from hybrid autos to those powered by fuel cells. The future of nuclear power plants had been looking brighter, but that may have changed on July 16, 2007, when a magnitude 6.8 rattled Japan and created numerous problems at the world's largest nuclear reactor, Kashiwazaki-Karina, which automatically shut down. Socolow and Pacala (2006) created a schematic way of looking at possible solutions, but only for a 50 year period, after which more things would need to be done, especially if world population continued to grow. They ended up saying (2006, 57) that "Critically, a planetary consciousness will have grown. Humanity will have learned to address its collective destiny—and to share the planet." Though we hope they are correct, nothing so far in the history of the world would support such an outcome, especially in so short a period of time.

Then there are new ideas on the technological fringe (Fleming, 2007). These include using reflective nanoparticles or orbiting giant mirrors to reflect solar energy back into space. One scientist, Alfred Wong, has suggested that Earth's magnetic field could be tapped to act as a conveyor belt that would carry carbon dioxide into outer space. However, as Fleming (2007, 46) mentions, "Today's aspiring climate engineers wildly exaggerate what is possible, and they scarcely consider political, military, and ethical implications of attempting to manage the world's climate—with potential consequences far greater than any [of] their predecessors were likely to face."

Global leadership remains behind the curve when it comes to global climate change. Gasoline consumption could be curbed considerably and quickly, for example, by combining a carbon tax on fuel with incentives for consumers to move to gas-electric hybrid autos. Several of these are on the market already, including the Toyota Prius, Honda Insight, and Ford's hybrid version of the Escape SUV. Coupled with some disincentives for the purchase of giant SUVs and increased auto fuel consumption mandates, even more could be accomplished. Encouragingly, unlike the Bush administration that tarried and tiptoed around the global warming issue for fear of angering oil, coal, and auto interests, in August 2012, the Obama administration announced "strict new vehicle fuel-efficiency standards . . . requiring that the U.S. auto fleet average 54.5 miles per gallon by 2025, an uncontroversial move that, unlike other administration energy policies, was endorsed by industry and environmentalists alike" (Eilperin, 2012a, no pp). This is because the new mandate promises to reduce harmful auto emissions even as it harmonizes the nation's auto emissions standards thereby making it easier and cost-effective for auto makers to design and manufacture compliant vehicles. Moreover, before the promulgation of the new standards, auto makers had been contending with piecemeal greenhouse gas emissions standards in California and other states; standards that led to costly court battles between these states and the auto industry. While the new auto emission standards represent a major step in reducing the United States' greenhouse gas emissions, the reality is that in order to slow down global warming, Americans and the rest of humanity may have to make real lifestyle changes, and not just nibble around the edges of the problem.

The United States is a key part of the solution to global warming because it is the second largest producer of greenhouse gases; it produces about one-fifth of the world's greenhouse gases, even though it has slightly less than five percent of the world's population. SUVs, trucks, long-distance commuting, and energy-consuming McMansions may please people, but they may need to be sacrificed if we are really to burn fewer fossil fuels. Are Americans ready for that? In China, which now leads the world in greenhouse gas emissions (United States Environmental Protection Agency, 2012), equally painful lifestyle changes will need to be made just as cars are becoming popular, as are larger houses, energy-consuming appliances, and even malls full of goods. It is, perhaps, easier to understand the enormity of the changes that the United States and China will have to make if we bear in mind that both countries have large coal reserves, and that coal-fired power plants will continue to be important power sources for both countries in the foreseeable future. Worse yet, it is unlikely that either country will be able to build substantial numbers of clean coal power plants (those that have built-in sequestering units that would remove carbon dioxide to underground or other safe storage areas) in the short term. If we are to slow the atmosphere's warming, then the United States and China must lead the way, and so far this is not happening.

Beyond that, wind power could be encouraged as a replacement for other forms of electrical energy generation. New wind turbines, some of them 300 feet tall, are more efficient than earlier models. Wind power is widely distributed geographically, completely renewable, environmentally clean, climate-neutral, and becoming cheaper all the time. Unlike such alternative energy sources as fuel cells, wind power requires no new additional infrastructure—it can be delivered throughout current electrical grids. Windmills would make much more sense, environmentally and economically, across vast areas of the Great Plains in the United States and Canada than corn-based ethanol production. Ethanol is promoted as an economic fuel additive but only because corn is a major subsidized crop in the United States. Farm state politicians are much more responsive to cries for more subsidies for existing crops than to suggestions for entirely new ways to "grow" energy in this vast region.

Beyond these two fixes, already available, a mix of other energy sources could be developed as well, including more solar, geothermal, and biomass units for electrical generation and, ultimately, fuel cells. In the meantime, improvements could be made to systems that use fossil fuels, which are not going to disappear any time soon but from which we should be gradually weaned in favor of more abundant and less polluting sources of energy. People must also recognize that subsidies of various sorts have kept fossil fuel energy much cheaper than it would have been otherwise. Higher health care and military expenses, the loss of forests and fisheries, and even climate change have been ways of keeping fossil fuel prices low (Greenstone and Looney, 2012).

Deforestation

The primary causes of **deforestation**, clearing land for agriculture and the gathering of wood for fires, are directly related to population growth.

The increase in the world's population over the centuries has brought about a concomitant decrease in forested areas, as we suggested earlier. Trees have been cut down for a variety of purposes such as home building and firewood. The primary causes of **deforestation**, clearing land for agriculture and the gathering of wood for fires, are directly related to rapid population growth. In some areas of the world, such as Algeria, Tunisia, and Morocco, forests at one time covered over 30 percent of the total land area, whereas today they cover only about 10 percent. In Haiti an estimated 98 percent of forests are gone, leaving little or no topsoil to hold back runoff when it rains. This was made starkly apparent in 2004, when Tropical Storm Jeanne killed hundreds of Haitians and left the poverty-stricken island nation an utter mess. Overpopulation exacerbates all of Haiti's other problems.

Forests are being cut down at a faster rate than they are being replanted. It has been known for centuries that deforestation results in floods, local changes in climate, and heavy soil erosion, but little has been done about it. Certain areas that were once densely settled, such as the Middle East, were long ago deforested. Many poor countries in other areas of the world are passing through the same stages of forest destruction, but at a more rapid pace. Hines, in discussing China (1973, 2), stated that:

> The Yellow River, which flows through an area of great population, rides high on silt from the surrounding lands and periodically floods the region through which it travels. Through the ages, overcutting of forests and exploitative farming practices have reduced China's cultivable land so severely that today the despoiled areas, together with those climatically and physically unsuited to agriculture, constitute 80 percent of its total land—all now useless for farming.

The continuous growth of China's population, and the clearing and burning of forest tracts in order to enlarge land for cultivation, has placed considerable stress on China's forest resources. According to Jing-Neng Li (1991, 256), a vicious circle (Figure 9–3) appears to characterize the relationship between deforestation and population growth, and ". . . in the final analysis the most important thing the Chinese government can do to break the vicious circle of overpopulation and deforestation is to promote the practice of family planning and to strictly control population growth." Geping and Jinchang (1994) agree that rapid population growth is the major restraint that China faces in solving its growing environmental problems.

Throughout many tropical regions shifting cultivation takes its toll on forest lands. The impact of agriculture, along with those of logging and ranching, seriously threatens to destroy the tropical forest ecosystem (Richards, 1973, 67). There has been a clear downward trend in per capita forest production. Almost a billion hectares of forests and

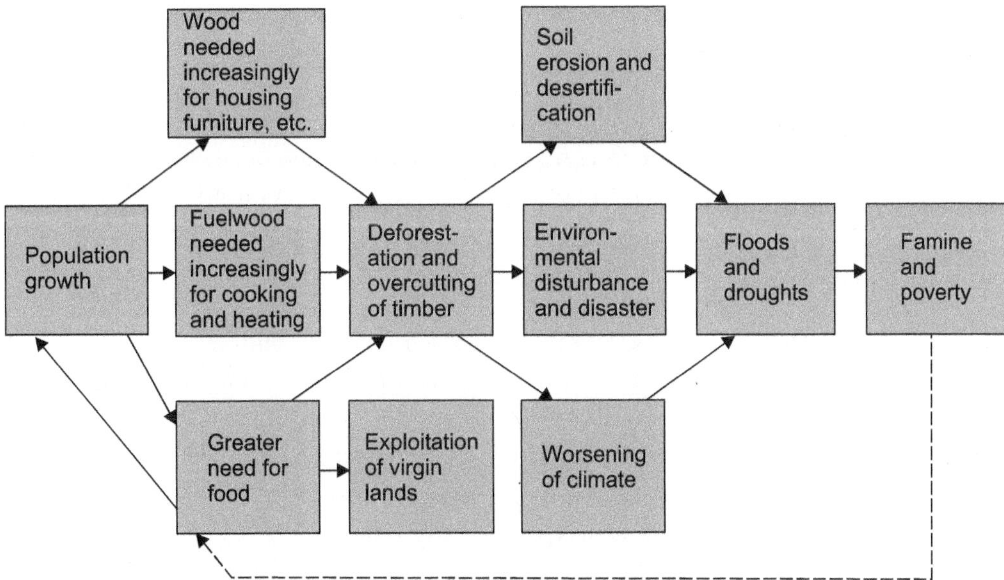

Figure 9–3 Population Growth and Deforestation

Source: Excerpted from *Resources, Environment and Population: Present Knowledge, Future Options* edited by Davis and Bernstam. Copyright © 1991. Reprinted by permission of Population Council.

woodlands have been cleared since 1850. Although much of this deforestation took place in temperate forests, there has been a shift since World War II, with the locus of deforestation currently in tropical areas. According to FAO estimates, more than 11 million hectares, an area the size of the State of Pennsylvania, was deforested each year in tropical regions during the 1980s. Population growth and energy price increases have intensified the pressures on fuelwood and timber supplies in less developed countries. The open savannas of the Sahel region of northern Africa are also being rapidly stripped of their remaining trees. Annual fuelwood consumption alone exceeded annual tree growth by an estimated 30 percent (Repetto, 1987, 17). The worldwide estimate by the FAO is that 1.5 billion of the 2 billion people who rely mostly on fuelwood are cutting wood faster than it is growing back (World Resources Institute (WRI) and International Institute for Environment and Development (IIED), 1987, 272).

Tropical rainforests are being cut down at an alarming rate throughout the world. Although these forests only cover around 7 percent of the earth's land area, they play a disproportionate role in the biosphere. Rainforests are home to one-half of all the species found on earth and are a gigantic storehouse for carbon. Although exact figures are hard to find, tropical deforestation and burning probably contribute between 7 and 31 percent of the carbon dioxide released into the atmosphere each year (Durning, 1991, 169). Thus, increased global deforestation is a growing contributor to global warming (Houghton, 2005, 13).

Estimates are that somewhere between one-third and one-half of the world's original tropical rainforests are now gone. The remaining area is slightly larger than the United States. In many instances population pressure has forced people to move to marginal lands where they clear and burn the forests in order to provide agricultural land. Unfortunately, this provides only a temporary solution as land stripped of forests erodes rapidly and soils are quickly depleted of their nutrients. Saving the rainforests will involve efforts at both the national and international levels (Table 9–1). Worldwide, the Earth loses more than 35 million acres of forest each year (Williams, 2003), a figure that has major implications for global warming (Mbatu and Otiso, 2012, 147).

Table 9–1 Saving Rainforests: Action Priorities
At the International Level
• Encourage sustainable development programs that delve into the basic causes of poverty, including the disproportionate possession of farmland. • Poor countries should have family-planning programs. • In exchange for forest protection laws, international debt would be reduced.
At the National Level
• Acknowledge and uphold native people's rights. Give them the right to maintain control over traditional lands and the natural resources the lands contain. • Implement strong land reform and family planning programs. Increase participatory programs to alleviate poverty. • Put an end to antiforest bias by changing land tenure, credit, subsidy, and tax policies. • Use information on ecological capacities of soils and biological diversity to create thorough landuse planning. Set apart and protect biological reserves, extractive reserves, sustainable forestry land, and agricultural land based on the plan. • Create and distribute tree-garden methods based on native models.

Source: Adapted from *THE WORLD WATCH READER ON GLOBAL ENVIRONMENTAL ISSUES*, Editor: Lester R. Brown.

Though periods of deforestation have occurred throughout much of human history, it is clear that, as McNeill (2000, 236) tells us:

> The late twentieth century was a great age of deforestation, like that of the Roman Mediterranean, Song China, or North America in the railroad age. The scale in the late twentieth century was larger, the ecological effects quite different, and the technologies employed radically different, but the motives much the same: arable or grazing land and marketable timber. Alarm about the obliteration of forest peoples, the lost ecosystem services, and the contribution of deforestation to greenhouse gas accumulation, while considerable after about 1980, weighed little in the balance against these motives. The political power of the beneficiaries of deforestation was far too great.

The good news with deforestation, as with many other environmental problems, is that we can, if given the will, probably reverse at least some of the damage. As writer Bill McKibben (1995) has pointed out, reforestation has been a success story in the eastern United States. Evidence of success includes more than just trees, as McKibben (1995, 64) made clear in the following comment:

> The proof that what is happening is significant lies in the recovery not only of the forests themselves but of much of the life they always supported. Perhaps 40,000 black bears roam the East. In 1972, thirty-seven wild turkeys were introduced into western Massachusetts. By now the population exceeds 10,000.

Indeed, by the end of 2007, biologists put the number of wild turkeys in Massachusetts at 20,000. However, the growing presence of these birds in many urban and suburban areas of the state is beginning to worry some because they can be aggressive toward people (O'Brien, 2007).

Nonetheless, the eastern United States is not the wet tropics, and worldwide forest losses are likely to continue. More generally, wealthy countries can more easily afford environmental improvements at home by partly outsourcing their negative environmental impacts (such as deforestation) to poor nations. An example of this is China's "export" of its deforestation and other environmental problems to Cameroon (Mbatu and Otiso, 2012, 154).

Decreasing Ecological Diversity

Consider the following list of biological creatures: the ivory-billed woodpecker, the Israel painted frog, the West Indian monk seal, the Xerces blue butterfly, the Tasmanian tiger, the Bali tiger, the Falkland Island wolf, the Aurochs, and the Moa. All have one thing in common—they are now extinct. Each extinction in turn affects the ecosystem of which it was a part.

As population growth forces people to seek out new areas for farming, mostly on increasingly marginal lands, wildlife habitats are being destroyed. Where they are not obliterated, natural habitats are often separated into "islands" cut off from each other and are no longer parts of a whole ecosystem. In 2007, for example, the Audubon Society reported what it called a "sharp and startling" decline in some common bird populations in the United States over the last four decades. Twenty bird species were counted and the average decline was 68 percent. Bobwhites declined from 31 million to 5.5 million, for example, a look at what the road to extinction looks like. Human population growth has not been kind to other species, unless, of course they are of use to us. Even as the populations of bobwhites, field sparrows, common grackles, and others have declined, the populations of chickens and turkeys have grown considerably. Wild habitat loss is often a result of creating habitats for domestic animals, from chickens and turkeys to cattle and pigs.

Ecological diversity refers to the diversity of a place at the level of ecosystems.

In addition to habitat destruction and habitat formation, further threats to **ecological diversity** include overkill and the introduction of exotic species.

In recent years the United States has also experienced a decline in honey bee populations, and those bees play a critical role in pollination, including pollination of many of our food crops. Especially in the wet tropics, ecosystems are rich in species, many of which have yet to be identified and cataloged. In addition to habitat destruction and habitat formation, further threats to **ecological diversity** include overkill and the introduction of exotic species. Even the extinction of one species in an ecosystem may in turn lead to other extinctions as the ecosystem adjusts.

Biologists argue that habitat destruction is resulting in an escalation of extinctions; we could be losing as many as 150 species a day according to some, though no actual number is really known. In fact, we don't even know how many species of life there are on the planet, and we probably never will. Since 1980 over 20 species have gone extinct; over one hundred others have not been seen for years and are probably gone as well. The major inference, however, is that if we destroy ecosystems, we are going to destroy species as well (Wilson, 1993). Even polar bears are threatened with gradual extinction as sea ice thins and food supplies dwindle.

In addition, threats to species may come from such environmental changes as global warming and increased ultraviolet radiation (because of depletion of the ozone layer). Concerns among biologists are aroused when similar biological changes are noted in disparate regions of the world. Two such changes deserve mention here: declining amphibian populations and declining human sperm counts.

Amphibians, especially frogs and toads, are in trouble on every continent (Blaustein and Wake, 1995). In California alone, foothill yellow-legged frogs, Yosemite toads, Cascade frogs, and leopard frogs are disappearing. According to Phillips (1994), the suspected cause of the decline in amphibian populations is the increased stress caused by exposure to ultraviolet-B radiation. Are frogs the proverbial "canaries in the mine" that warn us of our deteriorating and potentially harmful environmental conditions? More research is needed, of course, but the frog's problems should make thoughtful humans pause and reflect a bit.

Declining human sperm counts—we're putting you on, right? Again, evidence is accumulating from around the world, as are hypotheses to explain this rather disturbing (though it could be a benefit for population control, if not for *machismo*) observation (Swan et al., 2003; Jorgensen et al., 2006). After sorting through a ream of studies, journalist Lawrence Wright (1996) found that not only were sperm counts declining around the globe but that a high proportion of sperm were damaged or misshapen as well. Urban living, stress, and environmental chemicals that masquerade as estrogen are among possible suspects as causes. As Wright (1996, 55) concluded:

> The most important unanswered question among the many theories that purport to explain the falling sperm count is whether the decline is permanent and irreparable. The hope, of course, is that some modern condition or habit has somehow waylaid the production of sperm, and that the cause needs only to be discovered and removed for the count to rebound. Unfortunately, the truth about the sperm count is that it is under attack from many different sources. . . . It is as if manhood itself were waging a losing campaign against forces as yet unknown but frighteningly overwhelming.

Overgrazing is the exceedance of the carrying capacity of a pasture. It results from overstocking a pasture with livestock for too long a period of time to the point of hindering a pasture's ability to regenerate.

Overgrazing

The **overgrazing** impact of a growing population of humans and livestock is similar to that of deforestation. As the human population increases, particularly in the poorer countries, there is a growing demand for livestock that serves as food, security, repository of family wealth, and as power to pull agricultural implements. Increasing numbers of livestock, like cattle or goats, can quickly denude a landscape of its natural grass cover.

Denudation in turn will increase runoff, accelerate erosion, and increase siltation. There are many historic examples of overgrazing. North Africa is presently largely unproductive and barren, although at one time it was considered the granary of the Roman Empire. The Tigris-Euphrates Valley, known as the Fertile Crescent, probably supports fewer people today than it did during the pre-Christian era. According to Brown, McGrath, and Stokes (1976, 41), overgrazing is not new, but its scale and rate of acceleration are. Damage that formerly took place over centuries is now being compressed into years by the fateful arithmetic of rapid growth in human and animal populations. These populations are, in effect, outgrowing the biological systems that sustain them.

The expansion of arid areas, desertification, is at least partly due to overgrazing (Goudie, 1986, 46–50). As the size of the world's livestock herds increases, degradation of rangeland occurs. When the size of the livestock herds surpass the carrying capacity of perennial grasses on the range, the plant cover starts to diminish, leaving the land exposed to the ravages of both water and wind. According to Postel (1991, 27), "In the most severe stages, animal hooves trample nearly bare ground into a crusty layer no roots can penetrate, causing erosion to accelerate. The appearance of large gullies or sand dunes signals that desertification can claim another victory." Even if overgrazing were not a problem, growing herds of cattle pose another environmental concern—they generate significant quantities of methane gas, which is one culprit in global warming.

The Ehrlich-Commoner Debate

One of the most widely discussed topics among ecologists and other scientists concerned with population growth is the exact nature of the relationship between the size of a human population and its effect on the ecology of an area. Foremost among advocates of two different views on this question are Paul Ehrlich of Stanford University and Barry Commoner, who was an ecologist, and one of the founders of the environmental movement. Ehrlich proposes that environmental deterioration is a direct consequence of population growth. Commoner, on the other hand, believed that although population plays a role in environmental deterioration, it is not the major determinant of the environmental crisis. He believed that other variables, primarily technology, play a much more significant role in the ecological crisis. Thus, those concerned with the relationship between environmental problems and population growth are served well by the confrontation of viewpoints held by Ehrlich and Commoner.

Ehrlich proposes that environmental deterioration is a direct consequence of population growth. Commoner, on the other hand, believes that although population plays a role in environmental deterioration, it is not the major determinant of the environmental crisis. He believes that other variables, primarily technology, play a much more significant role in the ecological crisis.

The Ehrlich Viewpoint

Ehrlich's basic argument is centered on five theorems that he believes provide a realistic framework for analysis. These five theorems originally were published in *Science* (Ehrlich and Holdren, 1971, 1212), and they are that:

1. Population growth causes a *disproportionate* negative impact on the environment.
2. Problems of population size and growth, resource utilization and depletion, and environmental deterioration must be considered jointly and on a global basis.
3. Population density is a poor measure of population pressure, and redistributing population would be a dangerous pseudo-solution to the population problem.
4. Environment must be broadly construed to include such things as the physical environment of urban ghettos, the human behavioral environment, and the epidemiological environment.
5. Theoretical solutions to our problems are often not operational and sometimes are not solutions. Ehrlich believes that each person has a negative impact on his environment. In order to meet his needs, man simplifies ecological systems through the establishment of agriculture and is involved in the utilization of nonrenewable and renewable resources.

In a restatement of these ideas, Ehrlich and Ehrlich (1990, 39) state that the "population connection" is the key to understanding the root causes of our environmental problems. The number of people in an area relative to the carrying capacity of that area is used to determine if an area is over populated. If the long-term carrying capacity of an area is clearly being degraded by its current human occupants, then that area is overpopulated. By this standard, the Ehrlichs believe that virtually every nation today is overpopulated.

According to Ehrlich and Ehrlich (1990, 58):

> The impact of any human group on the environment can be usefully viewed as the product of three different factors. The first is the number of people. The second is some measure of the average person's consumption of resources (which is also an index of affluence). Finally, the product of those two factors—population and its per-capita consumption—is multiplied by an index of environmental disruptiveness of the technologies that provide the goods consumed . . . In short,
>
> Impact = Population × Affluence × Technology, or I = PAT

The Ehrlichs believe that this I = PAT equation is the key part of the population connection and useful in understanding our environmental crisis. Under this formulation, environmental problems are found throughout the world because rich nations with relatively small populations and poor nations with large populations have a significant environmental impact (Espenshade, 1991, 332).

The Commoner Viewpoint

Commoner, Corr, and Stamler (1971) suggested two ways to test the validity of Ehrlich's assumption that population growth is the crucial variable. The first way is to quantify the variables in the equation. The second is to analyze a specific environmental problem and determine the exact nature of the impact of population growth.

The time period between 1946 and 1968 was chosen for analysis because many of the present-day environmental problems—for example, pollution from detergents, photochemical smog, and pollution from synthetic pesticides—began after World War II. In addition, many new production techniques were introduced during this period. Pollution levels in the United States between 1946 and 1968 increased 200–1000 percent, while the increase in the U.S. population for the same time period was approximately 43 percent. Commoner, Corr, and Stamler argued that population growth alone could not account for such large increases in pollution levels and that the population component was not large enough to balance Ehrlich's equation.

Their next step was to see whether economic growth and increased per capita consumption could account for the increases in pollution levels. However, the income data for that period show that income alone did not account for the increased pollution levels.

After analyzing changes in per capita consumption of a variety of selected products, Commoner concluded that increases in the consumption of some products were counterbalanced by decreases in consumption of others. This led him to conclude that the most important factor was the nature of technologies used to produce various goods, and the impact of those technologies on the environment.

The largest increases in per capita consumption were related to products that turned out to be important causes of pollution. New technologies were the major culprits in rising pollution levels, not an increase in population. In a recent restatement of his position, Commoner (1990, 14) developed his own equation to measure total pollution as follows:

> The total amount of pollution generated can . . . be expressed by multiplying the pollution per unit good (technology factor) by the total amount of good produced. Finally, the latter figure can be broken down into the product of two factors: good produced per capita (the affluence factor) multiplied by the size of the population. In this way, the total amount of pollution can be expressed numerically in the form of an equation:
>
> Total Pollution = Pollution per good × Good per capita × Population.

With this equation the total amount of pollution can increase when any of the factors increase. Thus, total pollution can be increased because of population increases (Ehrlich) or because of high pollution technologies (Commoner).

Commoner went on to look at these three factors relative to post-1950 production technologies and concluded (1990, 151) that:

> . . . the data both from an industrial country like the United States and from developing countries show that the largest influence on pollution levels is the pollution-generating tendency of the system of industrial and agricultural production, and the transportation and power systems. In all countries, the environmental impact of the technology factor is significantly greater than the influence of population size or of affluence.

This debate between Ehrlich and Commoner outlines two of the major viewpoints regarding the role of population in environmental degradation. Probably the truth, as often is the case in such complex areas, embodies elements from both sides.

In another study of the importance of population growth and its role in environmental degradation, Ridker concluded that three generalizations can be made about this interrelationship (Ridker, 1980, 116):

1. Most of the environmental effects of changes in assumptions about population growth are relatively small.
2. The resource and environmental impacts of changes in per capita income are significantly larger than those of an equal-percentage change in population in early years, but the latter impacts grow over time, so that by the year 2025 they are roughly equal in magnitude.
3. Other determinants, such as the extent of recycling, technological changes, changes in availability of resources, and specific policies directly aimed at reducing the emission of a particular pollutant, generally have larger impacts than either population or economic growth rates.

Population and Resources

The interrelationship between population and resources has been the subject of study for many natural and social scientists. With population growth and the concomitant increase in demand for goods and services, the pressure on resources also increases.

In an evaluation of the need to balance population and resources, Hinrichsen (1991, 27) concluded that, “It is increasingly evident that many developing countries, struggling with rapidly growing populations and dwindling stocks of natural resources, must evolve strategic development plans that incorporate population and resource concerns.” There are different opinions, however, about the impact of population growth on resource depletion. However, we remain well aware of the danger of trying to predict future trends.

As Wallach (2005, 269) pointed out, "Fears of resource shortages have circulated for longer than anyone now alive can remember, and they have always proven alarmist." There is an old adage that even paranoids have enemies. We might also say that sooner or later that wolf at the door might have real teeth.

The Population-Resource Region

Geographers have always been concerned with regions and have spent a considerable amount of time and energy examining the problems of resource adequacy and population growth. If we try to define a population-resource region, then the concept of technology must be examined. Thus, the state of a country's technology is an important measure of the availability of its resources. In general, the greater a country's level of technology, the greater is its ability to exploit its resources.

A useful regional classification, based on population-resource relationships and technology, was developed by Ackerman (Ackerman, 1967). He divided the world into five categories:

1. European Type—technology-source areas of high population-potential resource ratios
2. United States Type—technology-source areas of low population-potential resource ratios
3. Brazil Type—technology-deficient areas of low population-resource ratios
4. India-China Type—technology-deficient areas of high population-resource ratios
5. Arctic-Desert Type—technology-deficient with few potential food-producing resources

Ackerman's discussion of population-resource regions pointed out some interesting observations. More than half of the world's population lives in areas that could be considered technologically deficient and have high ratios of population to potential resources. The remaining population is rather evenly divided among three of the other population-resource types. The technology-deficient countries of low population-potential resource ratios, including much of Latin America and Africa, have about one-sixth of the world's population. The Western European countries and Japan, which have another sixth of the world's population, are examples of regions where "industrial organization and technology permit them to extend their resource base through world trade, thus effectively meeting the deficiency of their low domestic per capita resource production" (Ackerman, 1967, 87). The remaining portion of the world's population lives in technically advanced societies with low ratios of population to potential resources. These countries include the United States, Canada, Australia, and Russia.

Since the 1970s, rapid economic development, globalization, and technological change in the Global South have significantly altered Ackerman's Brazil and India-China Type population-resource regions even as they have put Brazil, India and China on the path of joining the ranks of the world's more developed countries in the twenty-first century. Africa's economic and technological development prospects have also improved considerably over the last three decades. For instance, between 2000 and 2010, the continent had six of the ten fastest growing economies in the world and over the last thirty years, Africa's middle class has tripled in size to over 310 million people. Forty percent of the continent's over one billion people now live in urban areas, compared to just 28 percent in the 1980s (Cuñat, 2012). Moreover, Africa is in the midst of a major technological revolution that has also shown the continent to be capable of significant technological innovation in areas like mobile telephony (Mark, 2012). All of these developments have profound implications on Africa's (and much of the Global South's) ability to support its population. In short, population-resource regions are dynamic rather than static.

The Limits to Growth

One of the principal objectives of **The Limits to Growth** study was to examine the longterm outlook for the world.

One of the most ambitious attempts to bring together forecasts of resource depletion and its relationship to population growth was the publication of *The Limits to Growth* (Meadows, et al., 1972). Through the use of system dynamics, the authors developed a model with which they could examine the following five factors: population, agricultural production, industrial production, natural resources, and pollution.

One of the principal objectives of *The Limits to Growth* study was to examine the long-term outlook for the world. The various levels of human concern are depicted in Figure 9–4. All human concerns can be found somewhere on the graph, depending upon how far in time we wish to go forward and the amount of geographical space we are interested in. The concern of the vast majority of humankind would be concentrated in the lower left-hand corner of the graph. That is, most people are concerned with their immediate family on a day-by-day basis. However, the concern of the authors is depicted

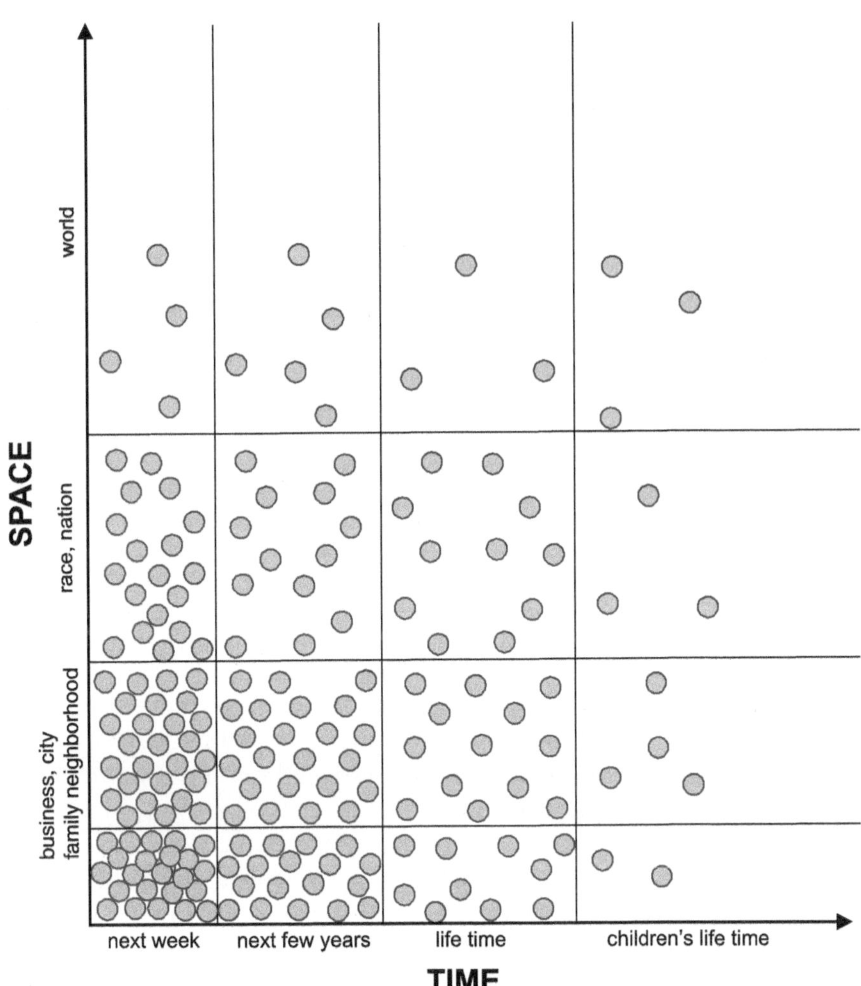

Although the perspectives of the world's people vary in space and in time, every human concern falls somewhere on the space-time graph. The majority of the world's people are concerned with matters that affect only family or friends over a short period of time. Others look farther ahead in time or over a larger area—a city or a nation. Only a very few people have a global perspective that extends far into the future.

Figure 9–4 Human Perspectives

Source: From *THE LIMITS TO GROWTH: A Report for THE CLUB OF ROME'S Project on the Predicament of Mankind,* Second Edition, By Donella H. Meadows, Dennis L. Meadows, Jorgen Randers, William H. Behrens III. A Potomac Associates Book published by Universe Books, New York, 1974. Graphics by Potomac Associates. Used by permission.

in the upper right-hand corner of the graph. They examined the long-range future, of the children's lifetime, of the entire world.

In order to examine this long-range future, a formal mathematical model was developed. The advantages of such a model are twofold:

1. All of the assumptions of the model are written out in a precise form and are thus open for inspection and criticism.
2. After all of the assumptions have been analyzed and revised to agree with the most current ideas, a computer can be used to determine the nature of the complex interactions.

The construction of the model followed four principal steps (Meadows, et al., 1972, 98):

1. Professionals in many fields (economics, demography, geology, and nutrition) were contacted and the literature was searched in order to identify the important causal relationships among the five levels of the model.
2. Using global data where it was available, each relationship was quantified as accurately as possible. If global data was not available local data was used.
3. A computer was used to calculate the simultaneous operation of all the relationships over time. The effects of the numerical changes were then tested in the basic assumptions in order to determine the principle and critical determinants of the system's behavior.
4. The fourth and final step was to test the effect of various policies that were currently being proposed on the global system. In order to apply their model to a single nation, or even a smaller areal unit, each relationship in the structure had to be quantified with numbers characteristic of that nation. In order for the model to represent the world, data on world characteristics had to be used.

After quantifying all of the above variables and their relationships, a computer was used to make graphs of the changes that would occur in each of the five factors from 1900 to 2100. The standard world model assumed that there were no major changes in the economic, physical, and social relationships that have historically determined the development of the world system. The mode of behavior for the world system in the standard run is that of overshoot and collapse. According to the authors of *The Limits to Growth* (Meadows, et al., 1972, 129):

> Food, industrial output, and population grow exponentially until the rapidly diminishing resource base forces a slowdown in industrial growth. Because of natural delays in the system, both population and pollution continue to increase for some time after the peak of industrialization. Population growth is finally halted by a rise in the death rate due to decreased food and medical services.

In addition to the standard model, several other assumptions about the world system were also developed, and models of the system under those assumptions were run through the computer. Included were the following assumptions:

- doubling of natural resources;
- unlimited resources;
- unlimited resources and pollution controls;
- unlimited resources, pollution controls, and increased agricultural productivity;
- unlimited resources, pollution controls, and perfect birth control; and
- unlimited resources, pollution controls, increased agricultural productivity, and perfect birth control

We're Flat Being Overrun

The Colorado Plateau of America's Southwest is a land of spectacular beauty with a rich cultural heritage. It includes a vast area of national forest and other public lands and has the highest concentration of National Park Service units in the nation, including such crown jewels of our park system as Grand Canyon, Bryce, Zion, Arches, and Mesa Verde National Parks. However, in recent years tourism in this region has literally exploded. Millions of people from around the country and from abroad have made the Colorado Plateau an increasingly more popular tourist destination.

The number of visitors to the 27 National Park Service units on the Colorado Plateau has increased dramatically. For example, between 1980 and 1995 visitor days increased from 9 million to more than 16 million, or about 80 percent. During that same time period visitor days to all National Park Service units increased by only 10 percent. Close to 4.3 million visitors gazed into the gaping jaws of the Grand Canyon in 2011; while slightly over 1 million visited it in 1960. Consequently, the park is working to reduce overcrowding and its negative impact on the park's ecosystem.

Rapid visitor growth has also outrun the regional infrastructure's ability to provide accommodations. For example, visitor days in Canyonlands increased by 160 percent between 1981 and 1993; employment in the park during the same period increased by only 43 percent. Similarly, a visitor day increase of 68 percent in Grand Canyon during that time period was accompanied by only a 38 percent increase in employment within the park and an increase of only 23 percent in constant dollars of support.

Other public lands in the Southwest are also being inundated by visitors, including millions of acres in the region that are managed by either the Forest Service or the Bureau of Land Management. Annual visitor user days in the Grand Gulch Primitive Area in Utah, for example, grew from 6,500 in 1977 to 70,000 in 1995—more than ten-fold. As a result, the Bureau of Land Management had to institute controls on visitation. As BLM archaeologist Dale Davidson, in charge of cultural resources at Grand Gulch, put it, "We're not just being impacted, we're flat being overrun."

The nature of the regional economy has altered as well, with recreation and tourism looming ever larger as a percentage of the economic base. As a result, protecting the region's natural amenities and resources while accommodating expanding opportunities for users has become a challenge for the region's residents. (Readers may want to look at two publications from The Grand Canyon Trust: *Beyond the Boundaries: The Human and Natural Communities of the Greater Grand Canyon* and *Charting the Colorado Plateau: An Economic and Demographic Exploration.*)

Based on their analysis of the world model under varying assumptions, the authors of *The Limits to Growth* arrived at the following conclusions (Meadows, et al., 1972, 29):

1. If the present growth trends in world population, industrialization, pollution, food production, and resource depletion continue unchanged, then the limits to growth on this planet will be reached sometime within the next one hundred years (that is, by 2072). The most probable result will be a rather sudden and uncontrollable decline in both population and industrial capacity.
2. It is possible to alter these growth trends and to establish a condition of ecological and economic stability that is sustainable far into the future. A state of global equilibrium could be designed so that the basic material needs of each person on earth would be satisfied and each person would have an equal opportunity to realize his individual human potential.
3. If the world's people decide to strive for this second outcome rather than for the first, the sooner they begin working to attain it, the greater will be their chances of success.

The Limits to Growth model has received considerable attention and criticism throughout the academic world. For example, a group of thirteen scientists set out to critically examine *The Limits to Growth* model and published the results of their analysis in a book entitled *Models of Doom: A Critique of the Limits to Growth.* Their criticism of *The Limits to Growth* model was centered on the following three essential differences (Cole, et al., 1973, 10):

1. *The Limits to Growth* emphasized purely physical limits, whereas the critics felt that a greater emphasis should be placed on the social and political limits to growth.
2. *The Limits to Growth* was criticized because it is said to have underestimated the possibilities of continuous technological progress, which is by its very nature difficult to predict. For example, the critics pointed out that a forecast made in 1870 would have omitted the principal source of energy in 1970, oil. It would have excluded not only all the synthetic materials, fibers, and rubbers, but probably aluminum and sundry other metals as well.
3. The third essential difference is skepticism that world models developed from System Dynamics are useful tools for forecasting and making policy decisions. Golub forcefully argued that the world model approach is inherently dangerous and encourages self-delusion in five ways (Cole, et al., 1973, 12):
    i. By giving the spurious appearance of precise knowledge of quantities and relationships which are unknown and in many cases unknowable.
    ii. By encouraging the neglect of factors which are difficult to quantify, such as policy changes or value changes.
    iii. By stimulating gross oversimplification, because of the problem of aggregation and the comparative simplicity of our computers and mathematical techniques.
    iv. By encouraging the tendency to treat some features of the model as rigid and immutable.
    v. By making it extremely difficult for the nonnumerate or those who do not have access to computers to rebut what are essentially tendentious and rather naive political assumptions.

In a response to *Models of Doom* the authors of *The Limits to Growth* expanded on the disagreements between the two views. A basic difference turns out to be different concepts of people, and they conclude (Cole, et al., 1973, 240) that they see no objective way of resolving these very different views of humankind and its role in the world. It seems to be possible for either side to look at the same world and find support for its view.

Technological optimists see only rising life expectancies, more comfortable lives, the advance of human knowledge, and improved wheat strains. Malthusians see only rising populations, destruction of the land, extinct species, urban deterioration, and increasing gaps between the rich and the poor.

Computer models provide a way to look at the dynamics of an ecological system. By altering variables and equations it is possible to observe the consequences of user's assumptions or proposed policies (Muir, 1991, 113). In an evaluation of *The Limits to Growth* model, demographer Kingsley Davis (1991, 12) concluded that:

> It is now two decades since *The Limits to Growth* was published. This is too brief a period to test the accuracy of even the short-term theoretical predictions, much less the long-term ones that were the main focus of interest in the project. Nevertheless, one cannot ignore developments in the last two decades that tend to support the study's findings. . . . Thus the grizzly truth may turn out to be that *Limits* was more prophetic than its detractors and even some of its defenders thought possible.

Limits to Growth was updated in 2004 (Meadows et al., 2004). Turner (2008) and others compared the thirty years between the original work and update of Meadows et al., and found that its predictions in terms of changes in industrial production, food production, and pollution are all in line with the book's predictions of economic and societal collapse in the current century. Its critics, however, still abound.

Again the truth probably lies in some intermediate zone between the Technologists and the Malthusians. The relationships between population growth, environmental degradation, and resource depletion are difficult to define and understand, as evidenced by the considerable disagreement between social and physical scientists.

Finally, we sometimes forget that those very same humans who cause so many of our environmental problems are also capable of fixing them, a point made much more frequently by economists than by ecologists. The United States provides an excellent example. In the late 1960s Americans read about rivers on fire—the Kaw in Kansas and the notorious Cuyahoga in Cleveland. Also in the 1960s, eye-searing smog in Los Angeles brought tears to the eyes of the most vigorous opponents of environmental laws, and in our National Forests timber was clear-cut almost at will. These and similar observations led to the first Earth Day (in 1970) and to creation of the Environmental Protection Agency and a long series of environmental laws, starting with the Clean Air Act (1970) and Clean Water Act (1972). People cared enough to make a difference, and Americans still want an environment that is cleaner and better even though they seldom agree on how to achieve it.

The Global 2000 Study

Another attempt to look at population, resource, and environmental problems was the Global 2000 Study. This study, prepared at the request of President Carter, used a variety of analytical techniques and computer models in order to focus on trends and changes that would take place ". . . in the world's population, natural resources, and environment through the end of the century" (Barney, 1980, 6).

The conclusions of the Global 2000 Study were similar to those of the *Limits to Growth* study. If present trends continue, according to the study, then

> . . . the world in 2000 will be more crowded, more polluted, less stable ecologically, and more vulnerable to disruption than the world we live in now. Serious stresses involving population, resources, and environment are clearly visible ahead. Despite greater material output, the world's people will be poorer in many ways than they are today.
>
> For hundreds of millions of the desperately poor, the outlook for food and other necessities of life will be no better. For many it will be worse. Barring revolutionary advances in technology, life for most people on earth will be more precarious in 2000 than it is now—unless the nations of the world act decisively to alter current trends (Barney, 1980, 1).

Unfortunately, this plea for decisive action was not acted upon and the Reagan and first Bush administrations virtually ignored the recommendations of the Global 2000 Study. Subsequent administrations did little as well although the current Obama administration has been lauded for implementing ". . . the most sweeping attack on air pollution in U.S. history . . . [by imposing] the first carbon-dioxide limits on new power plants, [tightening] fuel-efficiency rules . . . [and steering] billions of federal dollars to clean-energy projects" (Eilperin, 2012a, no pp).

Among both scholars and politicians there is growing interest in "sustainable development"—though many would argue that it is an oxymoron. Former President Clinton established a Presidential Council on Sustainable Development; the current Obama administration has pledged to promote it; the United Nations has a Commission on Sustainable Development; and there is even a Business Council for Sustainable Development. So, we might ask, what is it?

The idea is both simple and complex. Simply stated, it involves a world in which economic growth can be achieved without either environmental degradation or impoverishment of a sizable segment of the population. As geographer Thomas Wilbanks (1994, 543) said about sustainable development, "Clearly, the concept revolves around our capacity for meeting the basic needs of the world's population—especially if that population continues to grow—without running into environmental limits." Current trends, many of which we have already discussed in this chapter, are probably unsustainable—the rising price of environmental deterioration may result in curtailing economic progress if we are unable to change our ways.

The concept of sustainable development, important though it may be for the future of humans on the planet, is complicated not only by its ambiguity but also because it runs head long into a diversity of conflicts. Ultimately the idea is political; growth versus conservation and individual freedom versus control are among the potential areas of conflict. Wilbanks (1994, 553) has these final words for geographers:

> In addition to integrating knowledge in order to meet pressing social needs and helping to unify our various traditions as a discipline, sustainable development focuses our attention on a great problem of mutual concern that can help to integrate the various pieces of our individual professional lives—to integrate them in the interest of a problem that we care enough about to go that extra mile to do extraordinarily well, not only in our scholarship but in every aspect of the ways that we live as experts in something the world needs very badly.

One of the most troubling issues facing acceptance of sustainable development is likely to be environmental protection, already under attack in the United States. A number of environmental concerns were pointed out by Heinrich von Lersner (1995), president of Germany's *Umweltbundesamt* (the German equivalent of the United States' Environmental Protection Agency). He suggested that advances in technology, for example, will be absolutely essential to future sustainability; more self-sufficient housing and better transportation systems, for example, will help conserve energy. Agriculture, he suggests, may be the most serious threat; clean water is already getting scarcer. Waste is still another serious problem, though it could be mitigated by better packaging (and often less packaging). He ends up saying that, ". . . the greatest demand in the future will not be for coal, oil or natural gas; it will be for the time we need to adapt our laws, behaviors and technologies to the new requirements" (von Lersner, 1995, 188).

Some Final Comments

If we fail to find solutions to the combined threats of rapid human population growth, environmental degradation, and the potential for divisiveness and violent conflict, the quality of life for most of us can only get worse. As geographer Crispin Tickell (1993, 226) reminds us, "Life itself is so robust that the human experience could soon become no more than a tiny episode. Nature is not fragile. But we are." A somewhat different perspective was suggested by David Quammen, who argued that people will continue to survive, but believed that "What will increase most dramatically as time proceeds . . . won't be generalized misery or futuristic modes of consumption but the gulf between two global classes experiencing those extremes" (Quammen, 1998, 69).

Perhaps we have just been lucky so far. Despite a quadrupling of population during the twentieth century, we have dodged most environmental catastrophes. At the same time, however, economic and technological progress during the past century has greatly increased our control over the environment and our capacity to do great harm. We have noted numerous problems that can no longer be avoided, including global warming. Add to that list the world's growing appetite for oil, disappearing wetlands, the building of mega-dams, the bleaching of coral reefs, overfishing, and the chronic problem of nuclear waste (Klesius, 2002). No one can predict the future, but it seems to us that we should proceed more cautiously now than we have in the past and that we need to slow down our population growth as soon as possible. New forces on the development horizon, especially China, India, Brazil, and Africa, cannot be denied better futures for their populations, but supporting the growing wealth of those demographic giants will strain the planet's capacity more than ever.

CHAPTER 10

Population and Food Supply

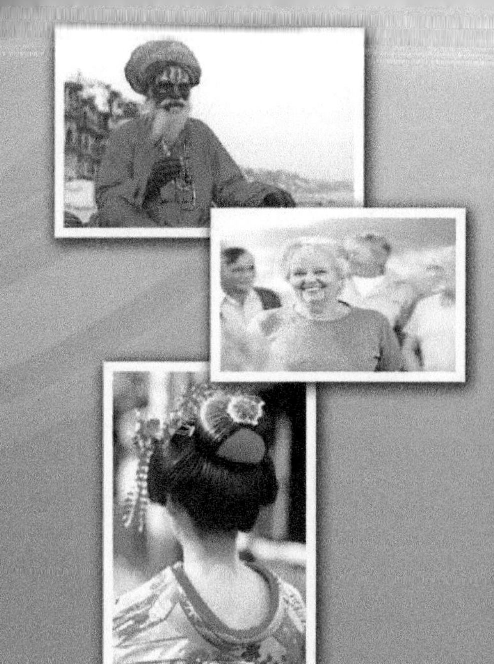

Key Terms

arable lands	triage
"Green Revolution"	lifeboats

Nowhere do we hear more frequent echoes of early Malthusian arguments than in discussions of the relationship between population growth and the supply of food. This idea was summarized in his famous Principle of Population. "This power of population is indefinitely greater than the power of the earth to produce subsistence for man." Since Malthus proposed this principle in 1798, many developments, unforeseen by Malthus, have occurred. For example, world cropland has more than doubled, new agricultural technologies have been developed to more than quadruple yields achieved by traditional farming methods, and international trade networks and communication links have led to a degree of global interdependence inconceivable in the eighteenth century. As in the days of Malthus, however, debate continues. Are there too many people? Can we eliminate hunger? Emotion often dominates such discussions, usually obscuring both the known facts and the complexity of people/food relationships. Yet we must still see that food availability affects human carrying capacity (Hopfenberg, 2003).

> *"This power of population is indefinitely greater than the power of the earth to produce subsistence for man."*

In an earlier discussion of population growth, it was argued that throughout most of human history the number of people was limited by the supply of food, among other things. With the beginning of the Agricultural Revolution, perhaps 10,000 years ago, humans began to domesticate plants and animals (Kurlansky, 2002). At the same time they also began a long process of diminishing the variety in their diets (though not necessarily to the extent of today's American teenager!). Of the hundreds of different animals and thousands of different plants that were once consumed, only a selected few were domesticated. With the emergence of cities, variety in the diet was even further diminished, until finally, as Harlan (1976, 89) commented, "The supermarket and quick-food services have drastically restricted the human diet in the U.S., and their influence is beginning to be felt abroad." Golden Arches, Burger Kings, and Pizza Huts are strewn throughout the cityscapes of Europe and parts of Asia, for example. Harlan (1976,

89) also seriously questioned the revolutionary nature of early agricultural changes and went so far as to point out that "... agriculture is not an invention or a discovery and is not as revolutionary as we had thought; furthermore, it was adopted slowly and with great reluctance." Gilland (2006) argued that the carrying capacity of a place must be seen as a function of per capita food consumption.

One way or another, in response to an increasing ability to wrest a living from their environments, populations began to grow somewhat more rapidly, though growth rates were far below those that we are experiencing in the world today. In turn, population growth created pressure for further increases in food production. Today, even with signs of a decrease in the rate of world population growth, the number of new mouths to feed is rapidly increasing. Each year an additional 83 million or more people are added to a planet on which most physical resources—land, clean water, and breathable air—are obviously finite. The growing number of people, combined with the increasing affluence of many of the world's residents, is generating an enormous increase in the demand for food.

The relationship between more people and increased food needs is intuitively obvious. More subtle, however, is the effect of income increases, which operate through changes in the nature of the diet. Mainly, increasing affluence leads to an increase in meat consumption, as is apparent in Table 10–1. Between them, China and India have more than 2.6 billion residents, and the economies of both nations are growing. Economic growth is rapidly expanding the demand for meat, especially in China. India, which consumes low amounts of meat for religious reasons, has one of the world's largest cattle herds. It accounts for twenty-five percent of the world's beef exports (Table 10–1; United States Department of Agriculture, 2012).

Increasing meat consumption naturally leads to the production of more meat. In turn, this requires more cereals unless animals are raised entirely on refuse or on the open range (where they can do a considerable amount of environmental damage). Animals, however, are not efficient converters of grain into calories that people consume. We find that in the United States per capita consumption of cereals is nearly five times what it is in the less developed countries, though in the United States most of that cereal is consumed only indirectly, in the form of meat. Furthermore, the linkages between meat consumption and diseases—including heart disease and cancer of the colon—are well established. All of us, as well as the planet, would be better off if humans, especially those in the wealthier countries, ate lower down on the food chain.

New challenges are at hand if world food production is to continue to meet the increasing demands for food. Lester Brown (2003, 7) may have been right when he warned us that, "The sector of the economy that seems likely to unravel first is food. Eroding soils, deteriorating rangelands, collapsing fisheries, falling water tables, and rising temperatures are converging to make it more difficult to expand food production fast enough to keep up with demand." In the remainder of this chapter some of the major prospects for increasing food supplies are discussed, along with some of their associated problems and consequences.

Current Trends in Food Production

According to Evans (1998, 226):

> The world population has long since passed the point where reliance on a self-sufficient agriculture is possible. Reaching three billion was the turning point. Since then the increase in food production has not relied on the further clearing of land for agriculture but, rather, on increases in yield from the conjunction of cheaper nitrogenous fertilizers, dwarf varieties, effective herbicides and irrigation.

In recent years crops have been grown on perhaps three billion acres of land, approximately 10 percent of the earth's entire land surface. About two-thirds of this

Table 10–1 Per Capita World Meat Supply in Selected Countries (Kilograms per Year)

Africa	1993	1994	1995	1996	1997	1998	1999	2000	2001	2002	2003
Botswana	30	24	32	31	27	22	20	22	21	27	22
Chad	14	12	13	13	14	15	14	13	13	13	12
Egypt	16	18	19	19	21	22	22	23	20	21	21
Ethiopia	8	7	7	8	7	7	7	7	7	8	8
Kenya	13	13	13	13	13	13	13	12	13	13	14
Nigeria	7	7	7	7	8	8	7	7	7	7	7
South Africa	39	40	37	36	37	35	36	41	38	39	41
Asia											
China	32	35	38	37	42	46	47	48	48	49	50
Hong Kong	103	125	121	115	113	120	123	126	118	113	112
India	4	4	4	4	4	4	4	4	4	4	4
Iran	22	22	22	23	24	24	23	23	25	25	28
Kuwait	61	64	66	67	68	65	62	67	70	59	63
Kyrgyzstan	46	43	37	37	39	41	39	38	38	37	35
Malaysia	49	51	52	52	52	49	45	45	46	47	48
Mongolia	90	85	87	96	104	108	111	101	86	72	56
Pakistan	13	14	14	11	12	11	11	11	11	12	12
Philippines	22	22	23	25	27	27	28	29	29	30	31
Americas											
Brazil	61	66	75	72	73	72	76	79	76	78	79
Canada	91	94	93	91	92	98	102	101	101	100	98
Chile	48	53	57	59	61	63	62	65	65	64	67
Colombia	32	34	37	35	33	34	33	33	34	34	34
Mexico	42	45	44	43	46	50	52	55	57	59	58
United States of America	115	117	117	117	116	119	123	121	120	124	122
Europe											
Austria	105	109	106	112	110	114	114	113	109	110	110
Bulgaria	62	57	59	58	57	66	57	57	56	54	46
Finland	56	57	61	63	63	66	67	65	65	67	67
Norway	53	56	57	61	58	59	57	60	61	61	62
Romania	57	54	54	51	50	54	50	48	48	55	60
France	93	94	97	98	99	101	99	100	102	99	98
United Kingdom	72	73	73	73	73	76	77	77	78	80	83
Oceania											
New Zealand	104	89	122	108	107	99	107	92	101	105	107
Australia	108	108	105	104	104	109	109	109	108	108	117

Source: Reprinted by permission of Food and Agriculture Organization of the United Stations, 2012, http://faostat.fao.org/

cultivated cropland is planted to cereals, which provide just over half of the entire human food-energy intake. Wheat and rice are at the top of the list; cultural preferences and environmental constraints help explain variations in the geographic patterns of these basic crops. Typically, for any country, as income rises cereals make up an increasingly smaller proportion of total calories in the diet, and the share provided by livestock increases. Rising meat consumption has been a hallmark of modernization but it also creates some undesirable consequences (Smil, 2002).

It is one thing to look at world agricultural output and talk about it in terms of world per capita supplies of cereals and other foods. It is another, and significantly different, thing to talk about geographic differences in the production and consumption of agricultural commodities. Both Hart (2003) and Fitzgerald (2003) have contributed, in different ways, to our understanding of how the scale and operation of agriculture in the United States has changed. It has become far more productive, despite the decline in both cropland and the number of farms. Larger farms and an emphasis on economic efficiency have allowed productivity to increase. As Hart said (2003, 2), "Entrepreneurs have driven this change in scale. Many people seem to assume that things just happen, but things do not just happen, they happen only because someone makes them happen. Things happen, places are changed, and new systems are created by the decisions and by the initiatives of individual entrepreneurs." Hart suggests that in the United States a "new tripartite macrogeography" describes the regional patterns of agriculture quite well. These patterns include cash-grain farming in a region in the Midwestern Heartland, from Ohio to Nebraska, livestock-producing regions around this cash-grain core, and specialization in the coastal states, which produce fruits, vegetables, horticultural products, and cotton. Beyond the United States, Naylor, et al. (2005, 1621) pointed out that "Virtually all of the growth in livestock production is occurring in industrial systems—a trend that has been evident in the United States for several decades. . . . The most striking feature of this geographic concentration is the delinking of livestock from the supporting natural resource base." As we would expect, and as is apparent in Figure 10–1, there are considerable variations in the average daily calorie supply from one place to the next. The relative abundance of food in the affluent countries is not necessarily a solution to the problems of food shortages or famines in the less developed world. Nor is the surfeit of food in those countries always healthy; obesity is a growing problem in the United States, for example, as are anorexia and bulimia (which affect females almost exclusively). Paradoxically, in the United States, Crawford, et al. (2004, 12) noted, "Although overweight is often considered a problem of overeating rather than hunger and scarcity, low-income adults and children have gained the most weight in recent decades." This is because low-income people consume more "energy-dense foods composed of refined grains, added sugars, or fats" because they are more affordable (Drewnowski and Specter, 2004, 6). Moreover, "poverty and food insecurity are associated with lower food expenditures, low fruit and vegetable consumption, and lower-quality diets" (*ibid.*).

That the geographic distribution of food supplies is in turn related to the distribution of wealth is hardly a new idea. The venerable Chinese poet Tu Fu (A.D. 712–770), writing twelve centuries ago (though still appropriate today in many lands), remarked that:

> Behind those vermilion gates meat and wine go to waste
> While out on the road lie the bones of men frozen to death. . . .

Global food production between 1950 and 2007, usually measured by cereal grain production, increased approximately 3.3 times, exceeding the 2.6 times increase in the population during the same period (from 2.5 billion to 6.6 billion). But while grain production per person in that period peaked in 1984 at 342 kilograms, it has since declined because of population growth (Figure 10-2). As the 1990s ended and the 2000s began, annual world grain harvests were in a fairly narrow range, and not increasing very much.

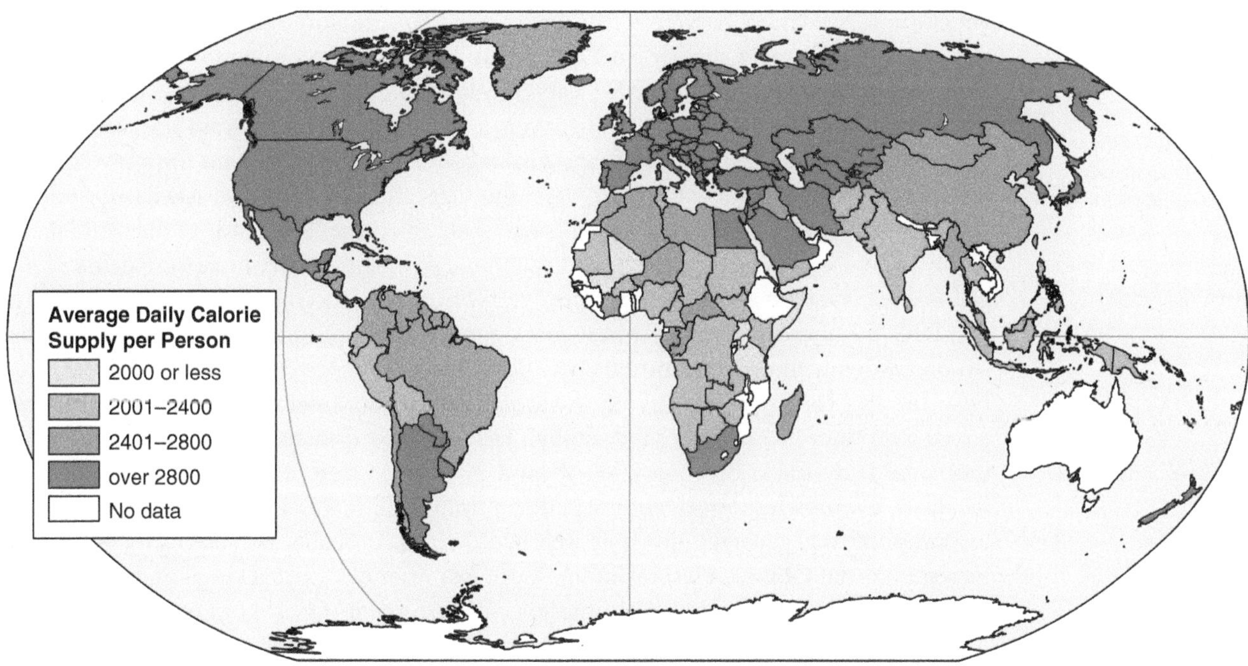

Figure 10–1 Average Daily Calorie Supply Per Person

Source: Peter Hendry, (1988) "Food and Population: Beyond Five Billion," *Population Bulletin* 43 (2):6. Courtesy of the Population Reference Bureau, Inc.

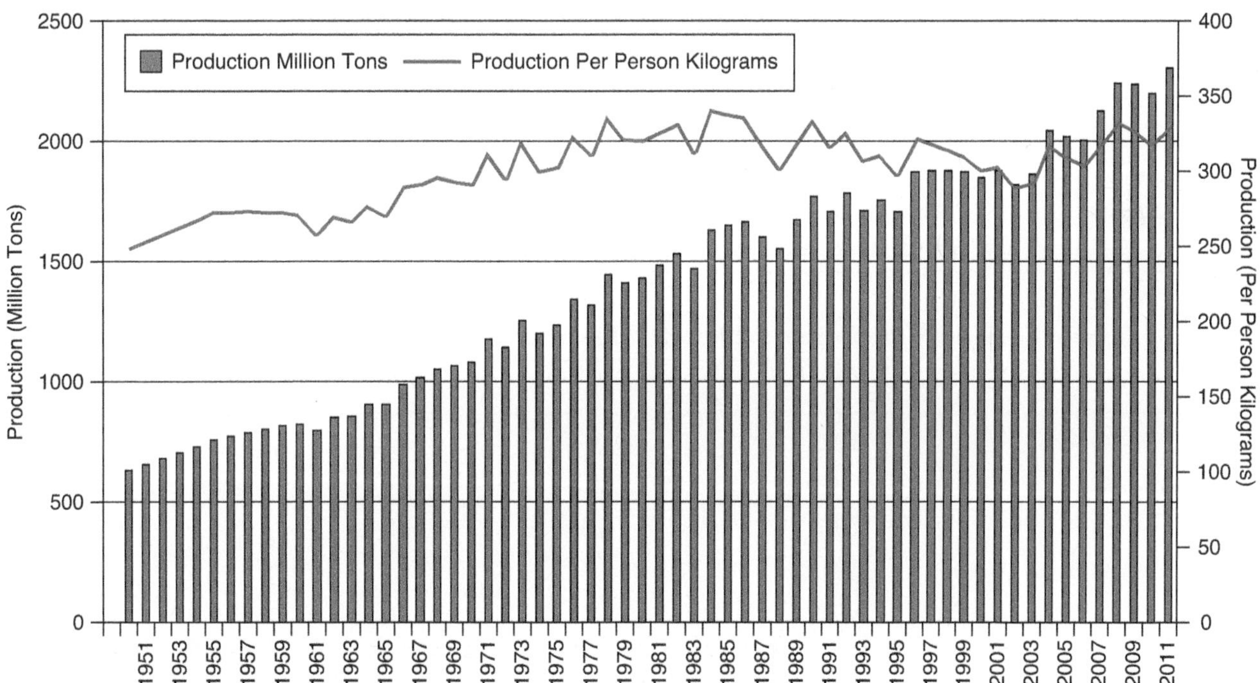

Figure 10–2 World Grain Production and Production Per Person, 1950–2011

Source: Authors. Based on data compiled by Earth Policy Institute with 1950–59 from Worldwatch Institute, Signposts 2000, CD-Rom (Washington, DC: 2000); and with 1960–2011 from U.S. Department of Agriculture, *Production, Supply, & Distribution*, electronic database, at www.fas.usda.gov/psdonline, updated 11 July 2012; population from U.N. Population Division, *World Population Prospects: The 2010 Revision*, electronic database, at esa.un.org/unpd/wpp/index .htm, updated 3 May 2011.

For example, the largest harvest in those years was 1,879 million tons, in 1997. In 2001 the total grain harvest was a slightly smaller 1,875 million tons. According to the United States Department of Agriculture, the 2001 harvest was about 40 million tons short of consumption for the year, a deficit that was filled from a shrinking supply of grain reserves. In part low grain prices have discouraged farmers from planting more grains and from investing in better technologies. At the same time, water shortages and droughts have also taken a toll on grain production. China, according to Brown (2002, 3), is the "wild card" in the world grain market. Even as population growth and economic development are increasing demand for grains in China, Chinese farmers in recent years have experienced lower total grain harvests.

Table 10–2 shows data for wheat production and consumption in China during the last decade, with an estimate for the 2007/08 crop year. There are some notable items here. First, the area harvested in 2007/08 was lower than it was a decade earlier, though it had grown a bit since 2004/05. Second, though yields increased since the middle of the 1990s, they were fairly stable between 2004/05 and 2007/08. Third, as a result of the previous two facts, overall wheat production was down from what it was a decade earlier. Domestic consumption was generally down as well by 2007/08, probably because other items had increased in the Chinese diet, including meat. For example, during the duration of Table 10–2, both animal and human consumption of coarse grains such as corn rose by quite a lot.

As noted earlier, since 1984 grain production per capita has actually fallen (Figure 10–2), and production increases have failed to keep up with population growth in many areas. While global grain production increased substantially between 1950 and 2007, per capita production has not returned to the 1984–85 level. Also, the apparent progress in increasing food supplies in less developed countries can be mainly attributed to significant production increases in one country, China, which accounts for 35 percent of the Third World's food output and 30 percent of its population. If China is removed from this calculation, food production increases in the other less developed countries is barely matched by population growth and in some areas it was significantly less than population growth. In the mid-1990s China, with its improving economy and growing population, began to scour the world for grains to import. Providing that Asian giant with additional

Table 10–2 China: Wheat Supply and Demand (Millions of Metric Tons/Hectares) 1996–2008

	Area Harvested	Yield	Production	Imports	Exports	Feed Dom. Consumption	Domestic Consumption	Ending Stocks
Wheat								
1996/97	29.6	3.7	110.6	2.7	1.0	3.4	107.6	81.2
1997/98	30.1	4.1	123.3	1.9	1.2	4.9	109.1	96.2
1998/99	29.8	3.7	109.7	0.8	0.5	5.0	108.3	97.9
1999/00	28.9	3.9	113.9	1.0	0.5	6.5	109.3	102.9
2000/01	26.7	3.7	99.6	0.2	0.6	10.0	110.3	91.9
2001/02	24.6	3.8	93.9	1.1	1.5	9.0	108.7	76.6
2002/03	23.9	3.8	90.3	0.4	1.7	6.5	105.2	60.4
2003/04	22.0	3.9	86.5	3.7	2.8	6.0	104.5	43.3
2004/05	21.6	4.3	92.0	6.7	1.2	4.0	102.0	38.8
2005/06	22.8	4.3	97.5	1.0	1.4	3.5	101.0	34.9
2006/07	23.2	4.5	104.0	0.5	2.5	4.0	101.0	35.8
2007/08	23.1	4.5	105.0	0.5	2.5	4.0	100.5	38.3

Source: United States Department of Agriculture, Foreign Agricultural Service.

cereals has affected world grain prices and disturbed traditional relationships between countries that have had grain surpluses—for example, Australia, Canada, and the United States—and the less developed countries that may be unable to compete with the Chinese in the world grain market (Brown, 1995 and Brown, 2003).

Thus, there is a deterioration of diets and an increase in hunger in sub-Saharan Africa and South Asia. Moreover, both the absolute number and the proportion of hungry people in these regions have increased. More than any other world region, Africa has seen a significant decline in per capita food production since the 1970s. Thus, most of the malnourished countries in the world at the beginning of the last century (that is, Peru, Chad, the Central African Republic, Angola, Ethiopia, Kenya, Afghanistan, and Bangladesh) were in Africa (Tempest, 1997). Rather than introduce large-scale irrigation projects and dams, many of these malnourished nations would be helped more, and more quickly, if the rich countries would do more small-scale things that would directly benefit food producers. These might include help on the farm itself, in the form of small amounts of fertilizer, better seeds, and maybe pumps for irrigation. Beyond that, health could be boosted at low cost by providing vitamins, medicines for common illnesses and infections, and malaria bed nets. Though these costs would be small, their benefits in local regions would be significant, making them good investments.

Instead, what we are seeing is mostly disinterest on the part of the rich countries and large-scale commercial agricultural enterprises such as the 100,000 acre soybean operation in Mato Grosso, Brazil, which employs only one person for each 400 acres (Wallace, 2007). Brazil may soon surpass the United States in soybean production, thanks to the introduction of heat-resistant varieties in 1997. According to Wallace (2007, 69), "Last summer [2006], Cargill and other big soy traders agreed to a two-year moratorium on buying soy grown on newly deforested land in the Amazon. The agreement is sending a signal to soy producers that the environmental impact of their operations is increasingly important in the world marketplace."

Although some writers have been led by ideology and "demographic quackery" to believe that there is little relationship between population growth and food availability, many demographic scholars believe the real battle to feed the expanding human population may be just beginning. As Hinrichsen and Marshall (1991, 26) observed:

> There are genuine fears that the world may be reaching a watershed in food production. There is not much more good land available for agricultural expansion; water suitable for irrigation is shrinking as demand grows; in many cases current cropland has been pushed to its productive limits; and additional inputs of fertilizers and pesticides are proving counterproductive.

A survey of 93 less developed countries, conducted by the Food and Agricultural Organization (FAO), determined that to meet food production requirements at the end of the 1900s, another 83 million hectares would have to be added to the existing 770-million-hectare area of farmable land (FAO, 1987). Although such expansion seemed modest, the FAO study cautioned that most of these new **arable lands** contained only marginal soils (Figure 10–3) and some were located in areas of unreliable rainfall. According to Tarrant (1990, 235):

Arable land is an agricultural term meaning land that can be used for growing crops.

> Pressure to conserve rainforests, difficulties and the expense of developing further agricultural areas, and the loss of agricultural land to other uses, erosion, desertification, and salination, are together likely to mean that an expansion of the area of cultivation will continue to make only a small contribution to extra food production over the next decade.

Although a few countries, such as Brazil, will be able to add more arable land, most will not; therefore, food for the nearly one billion people that will be added to the world's population during this decade will primarily come from raising land productivity. Most

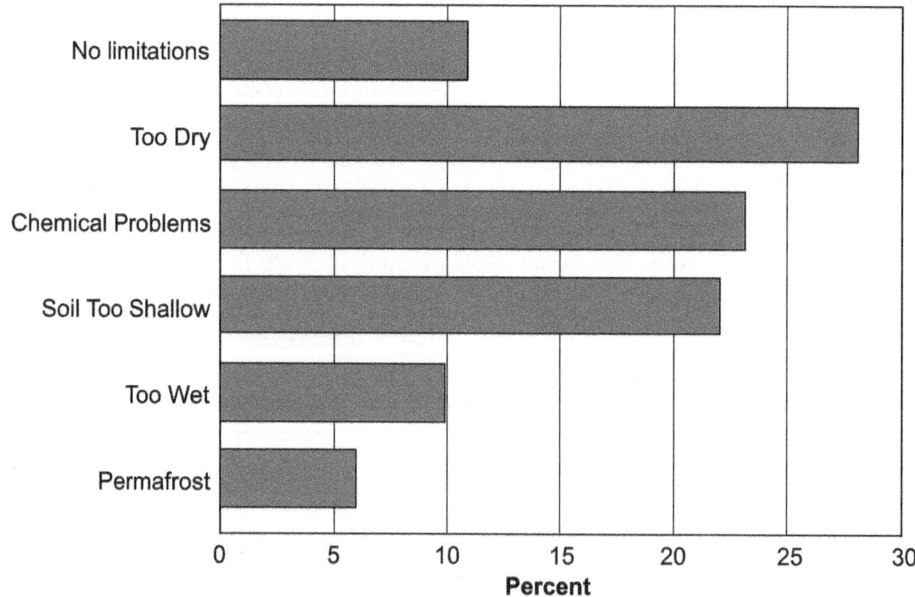

Figure 10–3 World Soil Suitable for Agriculture
Source: United Nations, Food and Agricultural Organization, New York.

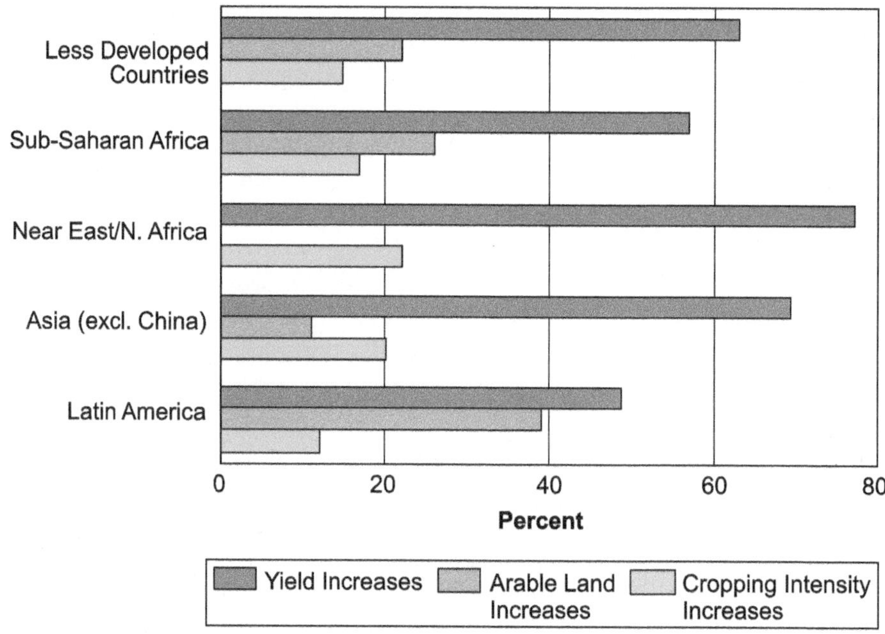

Figure 10–4 Possibilities of Increased Food Production: 1982/1984–2000
Source: Peter Hendry. (1988) "Food and Population: Beyond Five Billion," *Population Bulletin* 43(2):9. Courtesy of the Population Reference Bureau, Inc.

less developed countries will have to increase the yield from the same land in order to substantially increase food production (Figure 10–4). In those same countries, of course, pressure to move onto marginal lands will be most intense, natural ecosystems will be most threatened, and pleas for help from the more developed countries will be the loudest.

Perhaps the most thorough and balanced look at the world's food needs and ways to meet them in the years ahead was done by geographer Vaclav Smil (2000). He tells us (Smil, 2000, xxvii) that "There are no guarantees that we will succeed: irrational policies and misplaced priorities may lead us astray. . . . But we do have the tools needed to steer a more encouraging course . . . there appear to be no insurmountable biophysical reasons why we could not feed humanity in decades to come while at the same time easing the burden that modern agriculture puts on the biosphere."

Smil arrived at his conclusions only after careful and exhaustive study of the many components of food production, including photosynthesis and crop productivity, land, water, nutrients, agroecosystems and biodiversity, soils, climate change, fertilizer efficiency, farming systems, harvest and post-harvest food losses, and human dietary needs. Smil's review of a vast literature suggests that we can, if we wish, make decisions that will aid us in feeding a few billion more people, and feeding them adequate diets. This does not mean that we will.

By better using our current resources, improving the efficiency of agricultural production systems, reducing waste, and encouraging people to make wiser dietary choices, Smil believes that success is possible. It is not, however, guaranteed. Nor is there any guarantee that additional food supplies will reach the world's poorest mouths and stomachs—a reminder that political economy, not food supply limitations, result in a current world in which perhaps 800 million people have either nutritionally inadequate diets or not enough food to eat. Hunger amidst plenty still exists, and it will so long as grinding poverty leaves millions unable to purchase enough food for their needs, even if it is readily available.

An analysis of nutrition in less developed countries by the World Bank provided useful information for determining food security trends. This study was based on Food and Agriculture Organization/World Health Organization (FAO/WHO) calculations that are used to determine the amount of food a person would need in order to function at full capacity in all daily activities. The FAO/WHO looked at people whose food consumption was 80 percent and 90 percent of this established level. The World Bank estimated that 340 million people fell short of the 80 percent consumption level, which results in stunted growth and serious health risks. Women and children are most at risk; malnourished infants are less likely to live than others, and if they do live, are less likely to grow to full size and to develop normal brain capacities. Half of these 340 million people were living on the Indian subcontinent and one-fourth in sub-Saharan Africa.

Another important measure of the current world food situation is the food security associated with carry-over stocks (Table 10–3)—grain in storage when the new crop begins

Table 10–3 World Cereal Stocks[1] (Million Tonnes)

	2008	2009	2010	2011	2012 estimate	2013 forecast
TOTAL CEREALS	**419.7**	**503.1**	**527.5**	**499.9**	**511.8**	**547.6**
Wheat	**140.4**	**174.4**	**197.4**	**188.4**	**192.4**	**181.2**
held by: - main exporters[2]	41.5	65.7	76.0	69.0	73.1	62.9
- others	98.9	108.7	121.4	119.4	119.3	118.3
Coarse grains	**165.4**	**200.2**	**195.7**	**170.4**	**166.7**	**200.7**
held by: - main exporters[2]	77.4	105.2	102.7	71.7	65.5	95.3
- others	88.0	95.0	93.0	98.7	101.2	105.4
Rice (milled basis)	**113.9**	**128.5**	**134.4**	**141.0**	**152.7**	**165.8**
held by: - main exporters[2]	28.3	35.4	32.2	32.1	36.7	41.4
- others	85.6	93.1	102.2	108.9	116.0	124.4

(Continued)

Table 10–3 World Cereal Stocks[1] (Million Tonnes) (Continued)						
	2008	**2009**	**2010**	**2011**	**2012 estimate**	**2013 forecast**
BY REGION						
More developed countries	**126.2**	**175.8**	**188.1**	**150.3**	**145.2**	**162.6**
Australia	5.5	6.2	6.6	9.0	9.7	8.5
Canada	8.5	13.0	13.6	10.8	9.1	10.2
European Union	30.3	46.9	44.0	32.5	32.2	29.9
Japan	4.8	4.6	4.8	4.9	4.7	4.7
Russian Federation	5.2	17.7	20.0	15.7	12.9	10.7
South Africa	1.8	2.7	3.6	4.5	3.2	2.4
Ukraine	4.9	8.0	6.7	5.2	11.1	7.7
United States	54.3	65.9	75.9	57.3	46.2	72.0
Less developed countries	**293.5**	**327.3**	**339.4**	**349.6**	**366.6**	**385.0**
Asia	**247.4**	**272.6**	**284.5**	**291.3**	**308.2**	**325.2**
China	145.1	158.5	168.0	172.8	182.6	195.6
India	40.9	47.9	43.3	44.2	48.8	53.1
Indonesia	6.1	7.4	8.7	10.9	11.9	11.6
Iran (Islamic Republic of)	3.0	5.6	5.4	5.2	5.6	5.2
Korea, Republic of	3.0	2.9	4.1	4.1	4.2	3.9
Pakistan	3.2	3.5	4.1	2.2	3.0	3.2
Philippines	3.2	4.2	5.0	4.1	3.5	3.2
Syrian Arab Republic	3.6	2.5	3.0	1.6	1.0	1.0
Turkey	5.2	4.1	4.2	4.2	4.6	4.3
Africa	**24.1**	**26.3**	**30.6**	**33.6**	**33.3**	**31.1**
Algeria	3.4	2.7	3.6	3.9	3.7	3.9
Egypt	3.3	5.6	6.9	6.4	7.5	7.6
Ethiopia	0.7	0.8	1.5	1.6	1.9	2.0
Morocco	1.9	1.3	2.7	3.2	3.9	2.8
Nigeria	1.2	1.6	1.6	1.8	1.8	1.2
Tunisia	1.9	1.5	1.5	1.0	1.3	1.3
Central America	**5.3**	**5.9**	**4.4**	**5.5**	**4.0**	**4.5**
Mexico	3.2	4.1	2.7	3.6	2.2	2.5
South America	**16.3**	**22.2**	**19.6**	**18.8**	**20.8**	**23.8**
Argentina	7.3	3.7	2.2	5.3	7.1	5.8
Brazil	2.3	10.9	10.2	6.6	6.6	10.7

Note: Based on official and unofficial estimates. Totals computed from unrounded data.

[1] Stocks data are based on an aggregate of carryovers at the end of national crop years and do not represent world stock levels at any point in time.

[2] Major Wheat Exporters are Argentina, Australia, Canada, the EU, Kazakhstan, Russian Fed., Ukraine and the United States; Major Coarse Grain Exporters are Argentina, Australia, Brazil, Canada, the EU, Russian Fed., Ukraine and the United States; Major Rice Exporters are India, Pakistan, Thailand, the United States, and Vietnam.

Source: FAO (June 2012), *Crop Prospects and Food Situation*, No. 2, June 2012: 35, http://www.fao.org/docrep/015/al990e/al990e00.pdf (Accessed December 22, 2012)

to come in and is readily available if transportation facilities were provided to ship the grain to needed areas. Whereas, annual carry-over stocks averaged 16 percent between 1995 and 2000, in the 2002 to 2012 period they rose slightly to 21 percent. Everyone is in favor of having large carry-over stocks of food but, unfortunately, few are willing to pay for them. Such stocks are very expensive to acquire and maintain. Although recent data indicate adequate supplies of carry-over stocks, great fluctuations can occur from year to year and reserves can be rapidly depleted in times of poor harvest. A discussion of strategies for increasing food security is included in Table 10–4.

Carry-over stocks are grain in storage when the new crop begins to come in and readily available if transportation facilities were provided to ship the grain to needed areas.

Oceanic fisheries and aquaculture also play a significant role in feeding humankind. According to Querna (2004, 62), "The science of raising fish or aquatic plants for consumption, aquaculture is the fastest-growing segment of U.S. agriculture and one of the fastest-growing industries in the world." Between 1950 and 2012 the world fish catch quintupled or grew five times, though such gains are unlikely to continue because the oceanic fisheries, which account for around 90 percent of the total catch, may be nearing their maximum sustainable yield (which has already been exceeded for many individual species).

Between 1950 and 2012, the world fish catch averaged nearly 80 million tons (Earth Policy Institute, 2012). While most of that catch was from wild sources, the gap between the wild and farmed fish catch has narrowed rapidly since the late 1980s as oceanic and other wild fisheries have been fished beyond their sustainable yields (Larsen, 2002). While

Table 10–4 Strategies for Increasing Food Security

To stem falling per capita food production and resource degradation and to improve agricultural production, hard-pressed less developed countries should undertake the following:
• Establish comprehensive national population programs; • Provide integrated planning for future food needs that takes account of population growth, distribution, and rural-urban migration patterns; • Implement sustainable development strategies that combat soil erosion and impoverishment, deforestation, falling agricultural output, and water mismanagement; • Establish rural agricultural extension schemes that provide credit, seeds, fertilizers and advice to poorer farmers, whether men or women; • Provide special agricultural and environmental extension services for and available to women, who do most of the land and water management in poorer areas of the less developed world; • Ensure that women can inherit, buy, and have full legal title to land and that they have access to credit and marketing facilities; • Emphasize education for women and girls in rural areas. Better educated women are more effective as farmers and environmental managers, and have smaller families; • Encourage community development strategies, including the setting up of agricultural cooperatives, where essential services such as purchasing, marketing, soil and water conservation, water supply, health care and family planning, sanitation, housing and education can be integrated; • Establish comprehensive and accessible programs of maternal and child healthcare and family planning to reduce the size of families and to improve the health and well-being of the entire community; • Support research on the integration of traditional and emerging technologies for food production.

Source: Don Hinrichsen and Alex Marshall. (1991) "Population and the Food Crisis," *Populi* 16(2):32.

the wild fish catch increased more than five-fold from 17.2 to 90.1 million tons, the farmed fish catch grew 135-fold from 0.5 to 67.3 million tons between 1950 and 2012 (Figure 10–5). Since the difference between the wild and farmed fish catch in 2012 is only 23 million tons, it is likely that the world will get most of its fish from farmed sources in the near future. As close to a billion people around the world depend on fish as their primary protein source, many governments have sought to subsidize fishing, especially in the North Atlantic. However, as Larsen (2002, 2) points out, "Subsidies hide the fact that current fishing practices are unsustainable, both economically and ecologically."

Oceanic fisheries are also far from evenly distributed around the globe and in 2010, the top ten wild fish producers in the world were (in order of decreasing order of importance) China, Indonesia, India, United States, Peru, Russia, Japan, Burma (Myanmar), Chile, and Norway (Earth Policy Institute, 2012). Because of the inequitable distribution of fisheries around the world, controversies about fishing rights have become common as sophisticated trawlers range ever farther in search of their catch. Many fisheries are being depleted, including the salmon runs in California, Oregon, and Washington. Canadian cod fisheries have closed in recent years as well, and a "dead zone" has been identified in the Gulf of Mexico (Beardsley, 1997).

Recent research suggests that the ocean and its inhabitants have long been affected by humans. Commercial fishing and assorted pollutants hurt coastal ecosystems today, but even early humans who settled near the sea had significant impacts on local oceanic populations. With the help of such technology as on-board sonar, today's fishermen have become too efficient, and political issues make the management of oceanic fisheries extremely difficult. Recent research also suggests that ciguatera poisoning occurs from eating exotic fish that are overloaded with toxic poisons of various sorts. Scientists estimate that 50,000 or more cases a year occur, though the vast majority go unreported. Most sufferers recover, but not after experiencing some grim symptoms in their digestive systems, along with numbness in their limbs. Fish species most affected are those that live in warmer waters in close proximity to coral reefs. The source of the problem is a toxic algae, which seems to be increasing in abundance around coral reefs as ocean waters warm up. Smaller fish consume the toxins, then are eaten by larger fish,

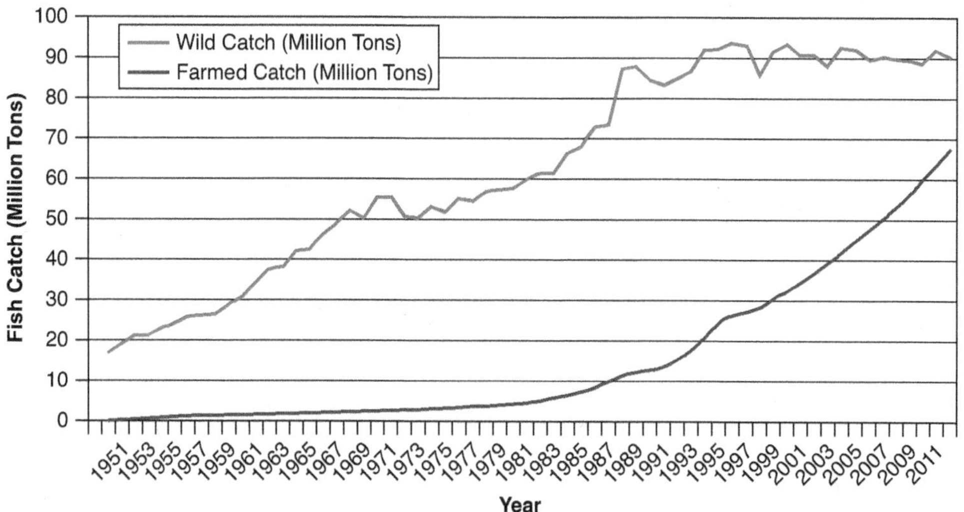

Figure 10–5 World Wild Fish Catch and Farmed Fish Production, 1950–2012

Source: Authors. Based on data compiled by Earth Policy Institute with 1950–2010 from U.N. Food and Agriculture Organization (FAO), *Global Capture Production and Global Aquaculture Production*, electronic databases, at www.fao.org/fishery/topic/16140/en, updated March 2012; estimates for 2011 and projections for 2012 from FAO, *Food Outlook* (Rome: November 2012), p. 10.

such as barracuda and grouper, which then concentrate the toxins further, producing what might be a meal to remember for unsuspecting diners. The worst cases may end in paralysis and death.

Aquaculture at first glance might seem to be an excellent supplement to our fish supply. However, it is often limited by pollution in populated coastal areas. In addition, if fish are "farmed" they must be fed and most of that food comes from land sources, including cereals and legumes that could also feed animals or even people. Aside from fish, shrimp aquaculture has grown considerably in recent decades, though it has run into a number of environmental problems (Boyd and Clay, 1998). McGinn (1998) argued that aquaculture's success rests in large part on resisting the use of chemicals and hormones and minimizing energy inputs. Larsen (2002) noted that raising carnivorous fish such as salmon still puts pressure on oceanic fisheries to provide fish meal and fish oil. Moreover, aquaculture is not without significant environmental costs. For instance, according to Markham (2006), a farm with 200,000 salmon produces as much fecal matter as a city of 65,000 people.

Increasing Yields on Land Already Under Cultivation

There are four major means of raising crop yields:

1. chemical fertilizers
2. pesticides and herbicides
3. irrigation
4. genetic improvement of plants

These are not new developments, but their use in many places, especially in the less developed countries, has come about only in recent decades.

Fertilizer and Crop Yields

The use of fertilizers for improving yields is hardly new. For centuries farmers have added manure to fields to increase productivity; even human wastes have been used by some farmers. In parts of East Asia it is referred to as "night soil." In 1847, Justus von Liebig, a German agricultural chemist, established the technological foundation for the use of chemical fertilizers when he demonstrated that all the nutrients needed by plants for growth could be added in chemical form. However, the escalating use of chemical fertilizers has mainly occurred since the mid-1950s and early 1960s (Figure 10–6), and mainly in the more developed countries.

Chemical fertilizers will improve yields rather dramatically under favorable circumstances as shown by growing grain production in the 1961–2011 period (Figure 10–6). However, there are often decreasing returns as the amount of fertilizer is increased, as is apparent in Figure 10–7 and from the declining trend in *grain production per ton of fertilizer* in Figure 10–6. Between 1950 and 1989, world fertilizer use rose from 14 million tons to an estimated 146 million tons (Brown, 1989, 35). Since 1980, there has been a reduction in the rate of growth of fertilizer use. Also, in the 1970s many Third World governments encouraged fertilizer use through subsidies, but these subsidies have recently been eliminated entirely or significantly reduced. Fertilizer prices have risen considerably during the last two decades, making their use in Third World countries ever more difficult.

Obviously, given the nature of the fertilizer response curve shown in Figure 10–7, farmers in the less developed countries could benefit considerably from the increased use of chemical fertilizers. Yet, these are the very farmers least able to afford such fertilizers in many cases. A major redistribution of world fertilizer supplies would probably increase

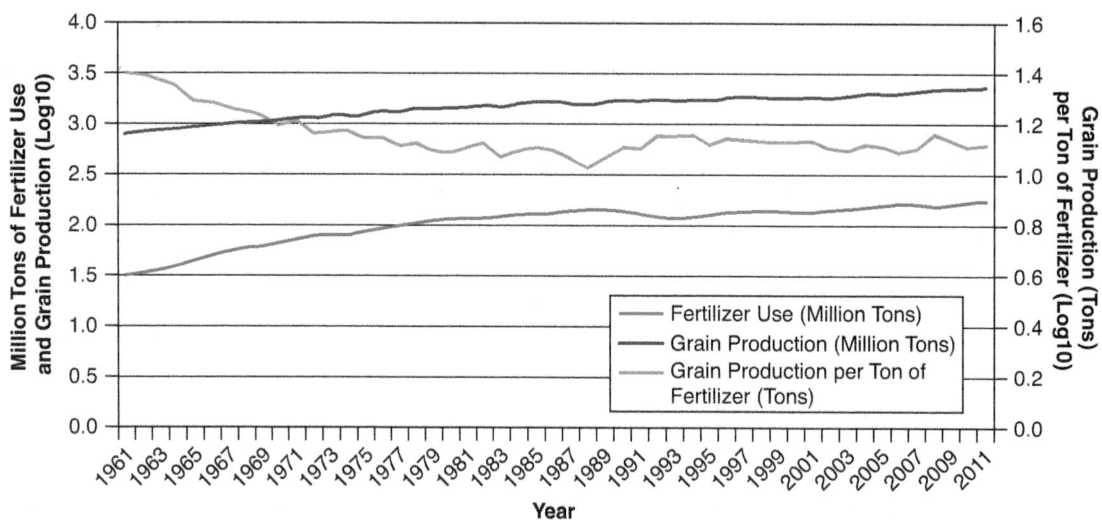

Figure 10–6 World Fertilizer Use, Grain Production, and Grain Production per Ton of Fertilizer, 1961–2011

Source: Authors. Based on data compiled by Earth Policy Institute (http://www.earth-policy.org/data_center/C24) with 1961–2006 fertilizer data from International Fertilizer Industry Association (IFA), IFADATA, electronic database, at www.fertilizer.org/ifa/ifadata/search, downloaded 28 August 2012; with 2007–2011 fertilizer data from Patrick Heffer, Medium-Term Outlook for World Agriculture and Fertilizer Demand 2011/12 – 2016/17 (Paris: IFA, June 2012), p. 26; and with grain data from U.S. Department of Agriculture, Production, Supply and Distribution, electronic database, at fas.usda.gov/psdonline, updated 10 August 2012.

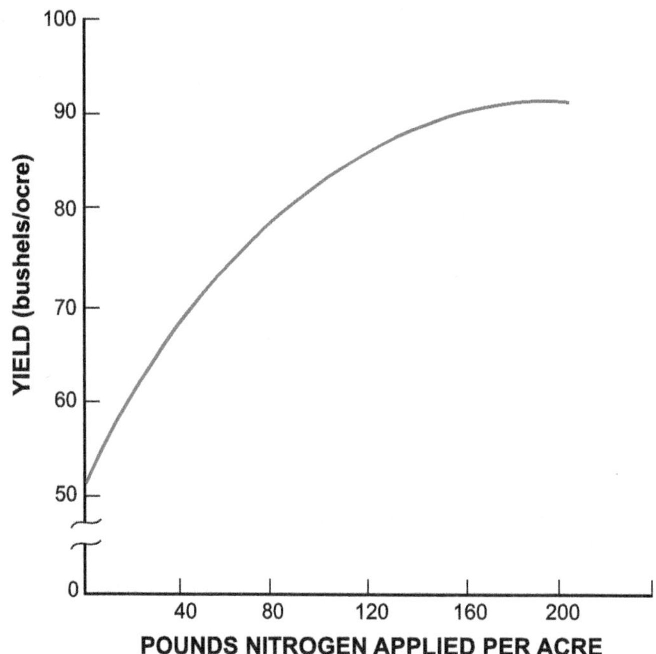

Figure 10–7 Typical Fertilizer Response Curve (Corn Yield in Iowa, 1964)

Source: U.S. Department of Agriculture.

world agricultural output, but it is unlikely to occur for both political and economic reasons. Golf courses and lawns in the rich countries will remain lush and green while Indian farmers harvest their meager yields of wheat and rice! In 1950, when 14 million tons of fertilizer were used, grain production was 624 million tons, with a response ratio (kilograms of additional crop produced per kilogram of additional plant nutrient applied) of 45. With the increasing use of fertilizer between 1950 and 1980, the response ratio fell from 45 to 12. Between 1980 and 2011, the annual response ratio has ranged between 11 and 13. According to Brown (1987, 130), "Growth in world fertilizer use is likely to remain slow in the absence of either a substantial improvement in the grain/fertilizer price ratio or technological advances that boost the fertilizer responsiveness of grain." Less developed countries are also starting to experience diminishing returns in fertilizer use in some cases. For instance, India's food grain production peaked in the 1980s; it dropped by two percent per annum in the 1990s and early 2000s even as population has continued to grow by nearly 20 million per annum. Thus, feeding the country's growing population is a growing challenge (National Academy of Agricultural Sciences, 2006, 1).

This reliance on chemical fertilizers will also have an environmental impact in both the near and longer term. Hendry (1988, 25) noted that, "Beginning with Rachel Carson's seminal 'Silent Spring' in 1962, a growing body of literature documented the environmental and health hazards likely to arise from agricultural systems that depended too heavily on chemical inputs." Nonetheless, it is difficult to argue with geographer Vaclav Smil (1997, 81), who recently commented that "Barring some surprising advances in bioengineering, virtually all the protein needed for the growth of another two billion people to be born during the next two generations will come from the same source—the Haber-Bosch synthesis of ammonia (a process making nitrogen nutritionally available to plants)."

Herbicides and Pesticides

Some estimates suggest that as much as half of the world's food production is lost, mainly to hungry animals and insects. Chemical control of pests would certainly help save some of this tragic loss, as would more adequate storage facilities. Yet, pesticides raise an environmental dilemma. Many such chemicals remain in the environment; and some are concentrated as they move up through the food chain. DDT (*dichlorodiphenyltrichloroeth ane*), a white, crystalline solid, tasteless, and almost odorless organochlorine insecticide, is only one example.

Weeds are another menace to better crop production. Weeds compete with food plants for sunlight and nutrients; hence, their control is another way to improve yields. Again, the major means of control is the use of chemicals, and there is again the possibility of environmental problems. In the less developed countries, tradeoffs are necessary. An increase in food supplies is essential and some risks are acceptable.

Irrigation

Although many large areas of the world receive sufficient rainfall to meet their agricultural requirements, many of the countries that are the most densely populated are found in the drier climatic regions. Many of these countries are poor and don't have the resources to purchase food from other areas. Therefore, in order for many of these countries to produce enough food to feed their populations, it will be absolutely essential that irrigation facilities be developed. In the several thousand years since irrigation was first developed in the Middle East, it has diffused gradually throughout the world. Much of the increase in irrigated lands, however, has taken place in the past seven or so decades. Between 1800 and 2010 the world's irrigated land grew 40 times from 8 million hectares in 1800, to 48 million hectares in 1900, 94 million hectares in 1950, 198 million hectares in 1970, 216 million hectares in 1980, 252 million hectares in 1990, 285 million hectares in 2000, and 312 million hectares in 2010 (FAO, 1992; Earth Policy Institute, 2012). This has been a major factor in the impressive growth of world food production in that period. About 75 percent of this irrigated land is in less developed countries.

Irrigated agricultural land plays an important role in feeding the world's growing population and in providing an adequate living for rural populations as well. For example, in the 1990s, 65 percent of Pakistan's farmland was under irrigation and was producing

80 percent of that country's food supply. In the same period, 30 percent of India's cultivated land was under irrigation and was producing 55 percent of its food output. Similarly, 50 percent of China's croplands were irrigated and were producing 70 percent of all the food produced in that nation (Hinrichsen and Marshall, 1991, 30).

The irrigated area grew most rapidly during the fifties and sixties, with an annual rate of almost 4 percent. But between the 1970 and 2009, the rate of expansion slowed to an annual average of 1 percent (Figure 10–8). After the initial rapid expansion of irrigated land in the 1950s and 1960s, the amount of land being taken out of production was rising while there was a decline in the amount of new land being put under irrigation. The reason for taking irrigated land out of production is primarily associated with what the FAO calls the three "silent enemies,"—alkalization, salinization, and waterlogging. It is equally worrying that the amount of irrigated acreage per thousand people has been declining since the 1960s (Figure 10–8).

Only about 10 percent of the earth's cultivated land is currently being irrigated (that is 312 million hectares out of 3.11 billion hectares), although approximately one-half of the world's food comes from these areas. Thus, it appears that there is a significant potential for expanding irrigation more so in Africa (Frenken, 2005, 34). However, the expansion of irrigation is limited both by the uneven distribution of water sources and the high capital costs of major irrigation projects such as dams and canals.

Water will, however, be a key factor in raising agricultural output. Many scholars have attributed at least part of the failure of the Green Revolution to inadequate water supplies. According to Falkenmark and Suprapto (1992, 35):

> It will be fundamental to match rapidly enough the reduced per capita water availability and the increasing per capita demands. In order to keep even the present level of water demands, more water-resource structures are needed to make more water accessible for use. The financial aspects of water-resource development projects that are needed to supply the rapidly growing population in low-income countries, require serious attention.

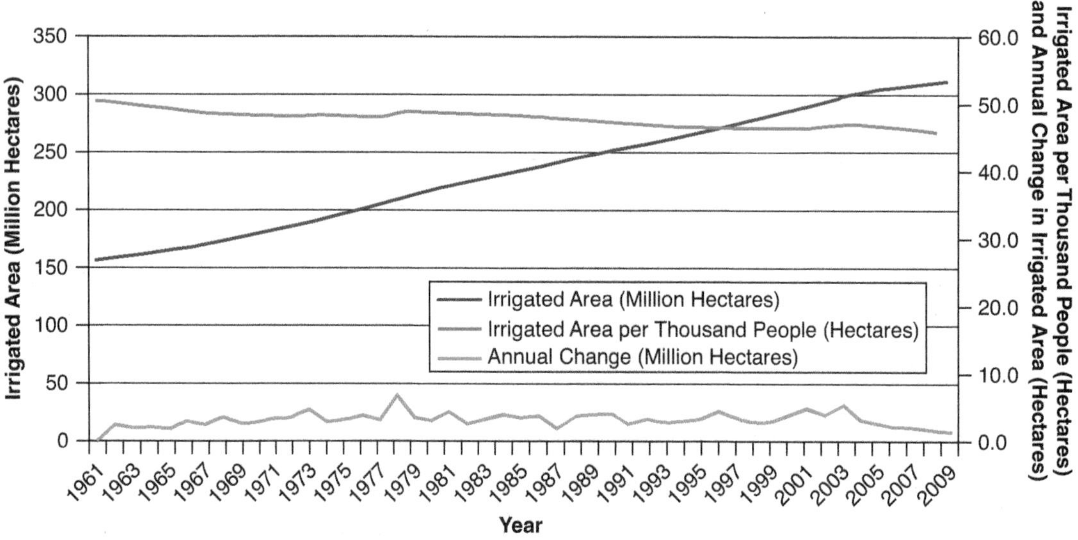

Figure 10–8 World Irrigated Cropland, Annual Change in Irrigated Area, and Irrigated Area per Thousand People, 1961–2009

Source: Authors. Based on data compiled by Earth Policy Institute from U.N. Food and Agriculture Organization, FAOSTAT, electronic database, at faostat.fao.org, updated 21 July 2011; population from U.N. Population Division, World Population Prospects: The 2010 Revision, electronic database, at esa.un.org/unpd/wpp/index.htm, updated 3 May 2011 (Obtained from http://www.earth-policy.org/data_center/C24, Accessed December 24, 2012).

Water from the oceans could conceivably be desalinated to meet growing demands for drinking water. It could also be used to irrigate such salt-tolerant plants as glasswort, which can be consumed by livestock and has seeds that can be pressed for oil (Glenn, Brown, and O'Leary, 1998).

Clean, fresh water is likely to become scarcer in the decades ahead, leading to rising prices and potential conflicts between different users—farmers and urban dwellers for example. Nonetheless, economic reality will go a long ways toward forcing people to adapt, especially if we add another 3 billion people over the next 50 years. As Gleick (2001, 45) tells us:

> Addressing the world's basic water problems requires fundamental changes in how we think about water, and such changes are coming about slowly. Rather than trying endlessly to find enough water to meet hazy projections of future desires, it is time to find a way to meet our present and future needs with the water that is already available, while preserving the ecological cycles that are so integral to human well-being.

According to Brown (2003, 23), "The world is incurring a vast water deficit—one that is largely invisible, historically recent, and growing fast." More recently Pimentel and Wilson (2005, 2) reiterated the problems that water is creating in some places:

> The availability of adequate supplies of fresh water for human direct use and agriculture is already critical in many regions, especially the Middle East and parts of North Africa where low rainfall is endemic. Surface waters, for instance, are often poorly managed, resulting in water shortages and pollution, both of which threaten humans and aquatic biota. Groundwater—rainfall lying in underground aquifers—is another vital source of water for agriculture; it too is often used profligately . . . groundwater resources must be carefully managed to prevent overuse and depletion, but this wisdom is often ignored.

New Strains of Plants and the Green Revolution

Since the beginning of the Agricultural Revolution, improvements in crops and livestock have made notable contributions to increasing the quantity and improving the quality of the human diet. Revolutions of varying dimensions have occurred many times over the past 10,000 years. Farmers have gradually kept and improved a few major plant species, while myriads of others have long since disappeared from the diets of most of the world's people.

Among such "revolutions" would be found the exchange of crops between the Old and New World, an exchange that began 500 years ago and continues even today, and one that certainly increased the world's food-producing capacity. (Consider Ireland before the potato, for instance.)

More recently, a number of "revolutions" have resulted from discoveries in plant genetics, discoveries which have led to the development of higher-yielding cereals. Hybrid plants may also be developed to better tolerate environmental conditions such as cold or drought, or to be more resistant to problematic diseases, or to be more responsive to fertilizers. Improved strains of corn have been grown in the United States for several decades with good results. The breeding of short-season corn has allowed the northern limit on corn production to be extended as much as 500 miles.

The "**Green Revolution**" that has attracted the world's attention has been that of the development of so-called "miracle strains" of wheat and rice that were developed in the mid-1960s.

But the "**Green Revolution**" that has attracted the world's attention has been that of the development of so-called "miracle strains" of wheat and rice that were developed in the mid-1960s. These new strains are shorter and stiffer-strawed than traditional varieties. They are also highly responsive to chemical fertilizers, mature earlier than traditional varieties, and have a reduced sensitivity to variations in the length of days. This latter property creates increased opportunities for multiple cropping—the raising of more than one crop per calendar year. In Pakistan, for example, some farmers get three crops over a 14-month period.

The spread of these high-yielding varieties of wheat and rice occurred quite rapidly. Among the major countries to benefit from these new strains were India, Pakistan, the Philippines, Indonesia, and Mexico. Countries sometimes went rapidly from being net grain importers to being net grain exporters.

Throughout the 1970s food production at least kept pace with population growth, thanks mainly to the crops from the new seeds. Some countries, such as India, Pakistan, the Philippines, and Thailand, proved that yields could be substantially increased by peasant farmers when conditions were favorable. By the end of the 1970s new varieties of corn, sorghum, and soybeans were available along with the new strains of rice and wheat, but other crops still await significant improvement especially those that are native to many of the world regions with food deficits.

The Green Revolution was a major advance in international agricultural development, yet the agricultural progress made possible by the Green Revolution was not evenly distributed. Aggregate statistics hide a significant group of farmers who have not benefited from the new technologies. Most of these farmers bypassed by the Green Revolution were subsistence farmers who were raising food for their families on rain-fed, marginal land. According to Dalrymple (1986), only about one-third of the land planted to cereal grains in the Third World uses the high-yielding varieties. The rates of adoption also vary widely by region, with Africa having the lowest adoption rate, at only 1 percent of the grain area, Latin America with 22 percent, and Asia and the Middle East with the highest percentage of adoption, 36 percent of the grain area. Thanks to the work of one man, Monty Jones, there is more hope now than before that food production might improve in Africa. He has been working on a group of high-productivity rices that would be well-adapted for growth in West Africa. With some luck his work will inspire other researchers in Africa along the path toward a more African Green Revolution. At the same time, rising temperatures and diminishing rainfall are likely to compound the problem of providing ample diets for Africans.

As Brown (1981, 34) noted, "Traditional farmers must also be trained and encouraged to use new techniques: how and when to plant the new seeds; how to control weeds, insects, and disease; and how to conserve and manage water." However, training programs are too often inefficient and poorly financed. Also, many countries still fail to provide sufficient incentives to encourage small farmers to adopt new seeds and more efficient farming techniques. Farmers in Third World countries are often penalized by deliberate government policies that attempt to keep food prices low in their burgeoning cities.

Though it undoubtedly helped in the battle against hunger, the Green Revolution was not a panacea for food problems in the less developed countries. Complacency about population growth sometimes resulted from the rapid advances in agricultural production. Populations have continued to grow, often eating up the benefits of increased production in true Malthusian fashion, reducing per capita gains to little or nothing.

Young (1992) believes that although the biotechnology industry could be an important weapon in the future fight to eradicate hunger, the control of the industry is by corporations who have little interest in plant research related to Third World food crops. He noted (1992, 97) that, "Firmer tomatoes and herbicide-tolerant corn, not hardier rice or cassava, are their objectives. These companies are not sinking money into less-profitable plant research that might help close the widening gap between the growth in world food supplies and that in human numbers."

World food production varies from year to year in response to weather, among other factors. Each swing upward or downward in production tends to produce a new wave of optimism or pessimism. The world currently produces enough food to provide everyone with adequate, or at least nearly adequate diets, yet hundreds of millions, mainly in the less developed countries, are undernourished. Even when diets are adequate in terms of calories, they are often inadequate in terms of nutrients, especially protein. Such inadequacies lead to nutrient deficiency diseases such as kwashiorkor, which results from severe protein deficiency.

The specter of Malthus is regularly called upon in discussions of population growth versus food supplies in the less developed countries. Although for reasons that are not always in agreement, experts generally agree that slowing the rate of world population growth is, if not an absolute necessity, at least a way of easing other problems. Of the world's 7.1 billion people at the end of 2012, around 1 billion (1 in 7 people) were undernourished or lacked enough food to be healthy and lead an active life. According to World Food Programme, the result of factors such as natural disasters, war, poverty, stripping the environment of natural resources, insufficient agricultural methods, financial difficulties and economic crises is hunger. More than AIDS, malaria and tuberculosis combined, hunger and malnutrition are the single greatest risk to human health in the modern world. Most of the undernourished are poor and tired.

Genetically Modified Foods

Though the Green Revolution provided the world with more food as a result of plant hybridization and different agricultural practices, the newer methods of genetically modifying plants represent a new phase in crop alteration. As a result, genetic engineering—the production of transgenic plants—has become a source of considerable controversy, as has the successful cloning of farm animals.

People have modified plants and animals to improve their usefulness to humans for thousands of years. What is new about today's genetically modified organisms (GMOs) is the way in which new genetic traits are introduced into plants or animals. New genes from a variety of sources (including, but not limited to, other plants) can be spliced into existing plants, for example, to make crops that are resistant to droughts, diseases, and pests, or even to improve the nutritional quality of the plant.

Suppose you would like to have a new strain of corn that would be resistant to certain common insect pests? Such a corn has been genetically engineered by isolating a gene from *Bacillus thuringiensis* (a bacterium) that directs cells to produce a protein that is toxic to certain insects. The Bt genes are inserted into cells, along with a "marker" gene, so that Bt's presence in cells can subsequently be identified. Cells that have the Bt gene will then be allowed to grow into corn plants, which will produce the Bt toxin in their cells, in turn killing insects that consume them. It is much neater than applying insecticides to fields to control pests.

However, GMOs have created a firestorm of controversy. Though there are perhaps 110 million acres of genetically altered crops growing today, led by corn, cotton, and soybeans, opponents have grown both more vocal and more numerous in recent years. Commercial planting of genetically altered plants began with tobacco in China in 1992, and two years later the FlavrSavr tomato became the first commercially produced GMO in the United States. By 2000 the United States accounted for about two-thirds of total world production of GMOs.

Critics denounce GMOs as "Frankenfoods" and argue that they are injurious to everything from people and insects to the environment. Among their many concerns are whether harm will come to consumers of genetically altered crops, whether innocent creatures will be harmed by them, whether "superweeds" will arise, and whether such crops might suddenly fail because of new resistant strains of insects and weeds. So far evidence has been scanty, but some environmental links have been identified, including one study that found harm to monarch butterflies that consumed Bt corn. More needs to be done before scientists are convinced, however.

Advocates of GMOs see them as the next essential step in allowing us to feed a burgeoning human population—providing safe (and sometimes more nutritious) foods with less damage to the environment than traditional crops cause. American farmers alone use nearly a billion pounds of pesticides annually, a figure that could be reduced by using Bt crops, especially cotton, corn, and potatoes. A new strain of "golden" rice, rich in beta carotene, has the potential for decreasing vitamin A deficiency in children who consume large quantities of rice as their staple diet. Is it a new "miracle" rice or a "Frankenfood?" Should we keep children in the Third World from having it, especially if an early death may be the alternative for one or two million of them a year?

Unfortunately, sound scientific answers to questions raised by critics of genetically engineered foods are still hard to find because few large-scale tests have been conducted over sufficiently long periods of time. In the meantime indications of the widespread GMO products on the market have become more common. In 2000, for example, Star-link corn (a genetically modified variety not approved for human consumption) was found in the human food supply in the United States. So far, however, no harm has come to humans from consuming it, though it did become clear to everyone that GMOs were more widely distributed than many people had thought. Among concerns for humans are the possible introduction of allergens into crops.

So far, despite numerous protests and discussions, there has not even been a law passed in the U.S. to require labeling products that contain GMOs, though a few manufacturers are now doing the opposite. They are carefully noting that their products do not contain genetically modified foods of any sort. At the same time, genetically modified soybeans, corn, and canola have become so widely distributed that it is unlikely that they would disappear in the short run no matter what we decide. You can be certain, however, that more studies will be done, more protests will be held, and more people will be eating the products of genetic engineering, often without even knowing it. Finally, as geographer Michael Winter (2004, 669) pointed out, "Both the public debate and academic discourses surrounding the GM issue have tended to focus on crop genetics. It is worth just noting that in the medium to long term the animal genomics debate is likely to be every bit as sensitive, and interesting to researchers, as the debate on crops."

Food and People

Despite the impressive gains in agriculture brought about by better technology, irrigation, genetic engineering, and chemical fertilizers, there still appears, often among knowledgeable and competent scientists, the growing fear of creeping Malthusianism. On the other hand, there are scientists who feel that continued advances will keep the growth in food supplies somewhat ahead of population growth for a while longer. Most would agree that a decline in the rate of population growth would ease the strain on resources and possibly allow adequate diets for an increasing proportion of the world's people. Some of these differing views are discussed in the following section.

The Paddocks and Triage

Three decades ago William and Paul Paddock, in a revised edition of an earlier book, made the following statement (1976, 8–9):

> A locomotive is roaring full-throttle down the track. Just around the bend an impenetrable mudslide has oozed across the track. There it lies, inert, static, deadly. Nothing can stop the locomotive in time. Catastrophe is foredoomed. Miles back up the track the locomotive could have been warned and stopped. Years ago the mud-soaked hill could have been shored up to forestall the landslide. Now it is too late.

The locomotive roaring straight at us, of course, is the population explosion. The unmovable landslide across the tracks is the stagnant production of food in the less developed nations, the very nations where the population increases are greatest.

The collision is inevitable. The famines are inevitable. After accepting the veracity of the above statement, the Paddocks suggested that the most reasonable solution to the situation is to employ the concept of triage as a means of allocating America's food surplus to the "hungry nations."

The term **triage** has been used mainly in military medicine (remember MASH?). It is a method of assigning treatment priorities to the wounded by classifying them into three categories:

1. those so seriously wounded that they will die regardless of the treatment they receive
2. those who will survive without treatment
3. those who will live provided they receive prompt medical attention

Scarce medical resources, including doctors, may then be best allocated by treating those in the third category, thus allowing doctors to save the maximum number of lives. The Paddocks suggest that such a system should be used to classify the hungry nations of the world into the following three categories:

1. those who cannot be saved because of population growth and the lack of agricultural potential and leadership ability
2. those who have sufficient resources, in agriculture and/or foreign exchange, to adequately deal with their population growth
3. those in which there is an imbalance between population growth and food supply which would be manageable if aid were received

Triage is the idea that the world's nations can be put into three categories relative to their food resources: (1) those nations that cannot be saved regardless of food aid from other countries, (2) those nations that will survive without food aid from other countries, and (3) those nations that could feed their people with some aid from the more developed countries.

Thus, the Paddocks suggested that America's scarce foreign aid resources can be best allocated if they are directed toward the countries in the third category. As an approach to the solution of the world food crisis, triage seems attractive to a significant number of people. Too often, however, those same people have an inadequate understanding of the relationship between population growth and food supplies. Furthermore, the argument may be morally questionable. *The New York Times* referred to this use of triage as "one of the most pessimistic and threadbare intellectual positions to be advanced since the demise of the Third Reich" (cite in Simon, 1975, 36). Despite the Paddocks's dire warning, the world's population has grown by close to 2.5 billion since 1976, and a vast majority of humanity is better off now than it was then. However, this vast expansion of earth's population was aided by the heavy use of fossil fuels to produce more food.

Hardin, the Lifeboat Ethic, and the Tragedy of the Commons

Garrett Hardin (1915–2003), an eminent biological scientist, popularized a view of the rich and the poor countries in an analogy using **lifeboats**. The rich countries are viewed as lifeboats filled almost to capacity. The addition of more people threatens to sink them, hence drowning everyone. The poor, of course, are struggling to enter the lifeboats and, hence, must be pushed away if the rich countries are to survive. In other words, in Hardin's view, feeding those in the hungry nations threatens those in the lifeboats, so it is not the thing to do.

The rich countries are viewed as **lifeboats** filled almost to capacity. The addition of more people threatens to sink them, hence drowning everyone. The poor, of course, are struggling to enter the **lifeboats** and, hence, must be pushed away if the rich countries are to survive.

Hardin's most cited (and by many most-ignored) work was an essay entitled "The Tragedy of the Commons." He explained the essence of that strategy as follows:

> The Tragedy of the Commons develops in this way. Picture a pasture open to all. It is to be expected that each herdsman will try to keep as many cattle as possible on the commons. Such an arrangement may work reasonably satisfactorily for centuries because tribal wars, poaching, and disease keep the numbers of both

In Hardin's view, feeding those in the hungry nations threatens those in the lifeboats, so it is not the thing to do.

> man and beast well below the carrying capacity of the land. Finally, however, comes the day of reckoning, that is, the day when the long-desired goal of social stability becomes a reality. At this point, the inherent logic of the commons remorselessly generates tragedy (Hardin, 1995, 18).

Beyond that point, Hardin noted, a rational herdsman will still want to add more animals to his herd. The dilemma, however, is that though he will gain from his additional animal, the commons will be harmed by that same animal's overgrazing. Each herdsman, acting in his own self-interest, will increase his herd. The result, pointed out by Hardin (1995, 18), is clear: "Therein is the tragedy. Each man is locked into a system that compels him to increase his herd without limit—in a world that is limited." His conclusion (1995, 18) is that "Ruin is the destination toward which all men rush, each pursuing his own best interest in a society that believes in the freedom of the commons. Freedom in a commons brings ruin to all."

When applied to population growth, the tragedy of the commons is clear enough, even though many are unhappy with its moral implications. Since most couples in the world today are not faced with the entire costs of their fertility decisions, many of them have more children than they otherwise would. The costs of those children, then, must be born by others, so the parents, like the herdsman in the example, are exploiting the commons for their own gain, even if it makes a nation worse off. As Hardin (1995, 30) noted, "The most important aspect of necessity that we must now recognize is the necessity of abandoning the commons in breeding. . . . Freedom to breed will bring ruin to all." Though it may sound harsh, it squares with the obvious reality that the human population cannot expand its numbers forever—there are limits.

Behind the lifeboat and triage theories lie some frightening, unspoken assumptions about the value of life. Is a hungry, impoverished Asian child less human, with less reason to live and less right to live, than children born in the U.S. or Europe? Playing God in this regard is especially dangerous because we are easily seduced by answers that protect our own advantages, even if they cost human lives. On the other hand, excessive population growth on a finite planet is clearly not sustainable.

Is There Hope?

There is no denying that hundreds of millions in today's world are hungry and malnourished, nor is there any doubt that everywhere it is the poor (and especially women and children) who suffer the most. But is it necessary to take such a dim view of this large share of humanity that we simply turn our heads and disregard its existence? Many think not.

No serious view of the future suggests that solving food and population problems would be easy or automatic. However, in his essay on raising agricultural productivity, Wolf (1987, 156) concluded that:

> The world is far from having solved the problems of agricultural productivity. The conventional approach to raising productivity—combining new crop varieties with fertilizers, pesticides, and heavy use of energy—succeeded dramatically in increasing food production in industrial countries and in parts of the Third World. But new approaches are needed to reach farmers who could not afford to follow this path, as well as to correct inequities in the distribution of environmental problems. Complementing the use of conventional resources with innovative biological technologies that maximize agriculture's internal resources can begin to achieve affordable and sustainable gains in agricultural productivity.

Though mass starvation appears unlikely in the near future, hunger and malnutrition remain widespread, especially in the less developed countries of Africa, Latin America, and Asia. The specter of Malthusianism is unlikely to disappear in the decades ahead. More people implies an obvious need for more food. Food supplies can be increased, though probably not without threatening the environment. Raising yields on land already under cultivation remains the best alternative for expanding food supplies in the years ahead.

Care should therefore be taken to minimize misuse and waste of available human food supplies. For instance, the significant conversion of corn to ethanol in the United States is not encouraging for hungry people. Consequently, the UN has urged the United States to cut ethanol production from corn (Blas and Meyer, 2012). Even as many Americans may see ethanol from corn as a way to have their SUVs and drive them too, as more corn is diverted to ethanol production, prices for not only corn but many other foodstuffs, including meat that is produced from animals raised on corn, are rising also. Worse yet, as Runge and Senauer (2007, 48) pointed out, "Biofuels may have even more devastating effects in the rest of the world, especially on the prices of basic foods. If oil prices remain high—which is likely (in the foreseeable future)—the people most vulnerable to the price hikes brought on by the biofuel boom will be those in countries that both suffer food deficits and import petroleum." According to Wald (2007) there are a number of problems with the ethanol from corn program in the United States. The production of ethanol from corn requires large amounts of fossil fuels and ethanol burned in vehicles does little to reduce pollution. Only by making ethanol from cellulose, including cornstalks and various grasses such as switchgrass, would ethanol become more attractive economically or environmentally. However, that process is currently not up and running on a large scale because of problems in the use of natural enzymes. In conclusion Wald (2007, 49) wrote that "Relying on ethanol from corn is an unsustainable strategy: agriculture will never be able to supply nearly enough crop, converting it does not combat global warming, and socially it can be seen as taking food off people's plates. Backers defend corn ethanol as a bridge technology to cellulose ethanol, but for the moment it is a bridge to nowhere." But remember that the United States Congress has funded "bridges to nowhere" before for political reasons.

At the same time, as was suggested in the previous chapter, the concept of sustainable development needs to be more carefully spelled out and encouraged. Plucknett and Winkelmann (1995) have discussed various aspects of the technology necessary for sustainable development. Also, geographers Richard Le Heron and Michael Roche (1995) found that in New Zealand the idea of sustainability is being worked through in a way that fits well with current business practices. Finally, in the United States sustainable ideas in agriculture are also receiving considerable attention. In many California vineyards, for example, owls and hawks are replacing pesticides and ground covers are being grown to attract "good" bugs, those that feed on undesirable insect pests.

Though it is scientifically possible to feed a growing population, given today's technology, the principal determinants of success are going to be political, social, and economic. Moral, rather than scientific, questions must be answered as well. For example, should the more developed countries help feed the less fortunate ones? How can an equitable distribution of food be assured? Does each person in the world have a *right* to food? Such questions cannot be answered by science alone, but they must be answered. In Garrett Hardin's view, our faith in perpetual growth must give way to acceptance of life in a finite world. As he concluded (Hardin, 1999), "If the dream of perpetual growth is now near its end, then it is time to explore the possibilities of living in a non-growing but sustainable world, a world in which temperance in global ambitions is a virtue." In a similar vein Lester Brown (2003, 176) noted that "As a species, our failure to control our numbers is taking a frightening toll. Slowing population growth is the key to eradicating poverty and its distressing symptoms. . . . As journalist Gregg Easterbrook (2003, 317) concluded, "It is never too late to change the world. . . . It is ours to decide what the future will hold. And if we decide well, the future may hold an ever-better life. . . . We hope that he is right, and we hope that we can ultimately decide well.

References

Introduction

Bailey, Adrian (2005) *Making Population Geography*. London: Hodder Education.

Barrett, Paul M. (2012) "It's Global Warming, Stupid," *Bloomberg Businessweek*, November 01, 2012, http://www.businessweek.com/articles/2012-11-01/its-global-warming-stupid (Accessed December 20, 2012).

Beaujeu-Garnier, Jacqueline (1966) *Geography of Population*. New York: St. Martin's Press.

Bogue, Donald J. (1969) *Principles of Demography*. New York: John Wiley and Sons, Inc.

Boyle, Paul (2003) "Population Geography: Does Geography Matter in Fertility Research?" *Progress in Human Geography* 27(5):615–626.

Brunn, Stanley (1992) "Are We Missing Our 'Forests' and our 'Trees'? It's Time for a Census," *Annals of the Association of American Geographers* 82:1–2.

Buzar, Stefan, Ogden, Philip E., and Hall, Ray (2005) "Households Matter: The Quiet Demography of Urban Transformation," *Progress in Human Geography* 29(4):413–436.

Clarke, John I. (1972) *Population Geography*. Second Edition. Oxford: Pergamon Press.

Clarke, John I. ed. (1984) *Geography and Population: Approaches and Applications*. Oxford: Pergamon Press.

Congdon, P. and Batey, P. eds. (1989) *Advances in Regional Demography*. London: Belhaven Press.

Courgeau, D. (1976) "Quantitative, Demographic, and Geographic Approaches to Internal Migration," *Environment and Planning* A, 8:261–269.

Demko, George J., Rose, Harold M., and Schnell, George A., eds. (1970) *Population Geography: A Reader*. New York: McGraw-Hill Book Company.

Findlay, Allan M. (1991) "Population Geography," *Progress in Human Geography* 15:64–72.

Findlay, Allan M. (1993) "Population Geography: Disorder, Death and Future Directions," *Progress in Human Geography* 17: 73–83.

Findlay, Allan M. and Graham, Elspeth (1991) "The Challenge Facing Population Geography," *Progress in Human Geography* 15:149–162.

Gober, Patricia (1992) "Urban Housing Demography," *Progress in Human Geography* 16:171–189.

Gober, Patricia and Tyner, James A. (2003), "Population Geography," in Gary L. Gaile and Cort J. Willmott, eds. *Geography in America at the Dawn of the 21st Century*. New York: Oxford University Press, pp. 185–199.

Gore, Al (2006) *An Inconvenient Truth: The Planetary Emergency of Global Warming and What We Can Do About It*. New York: Rodale.

Gore, Senator Al (1992) *Earth in the Balance: Ecology and the Human Spirit*. Boston: Houghton Mifflin Company.

Graham, Elspeth (2000) "What Kind of Theory for What Kind of Population Geography?" *International Journal of Population Geography* 6: 257–272.

Graham, Elspeth (2004) "The Past, Present and Future of Population Geography: Reflections on Glenn Trewartha's Address Fifty Years On," *Population, Space and Place* 10(4):289–294.

Heyman, Rich (2006) "Withered Geography," *The Professional Geographer* 58(1):104–105.

Hugo, Graeme (2006) "Population Geography," *Progress in Human Geography* 30(4): 513–523.

Hugo, Graeme (2007) "Population Geography," *Progress in Human Geography* 31(1):77–88.

James, Preston E. (1954) "The Geographic Study of Population," in Preston E. James and Clarence F. Jones, eds., *American Geography: Inventory and Prospect.* Syracuse, N.Y.: Association of American Geographers, pp. 106–122.

Longman, Phillip (2004) *The Empty Cradle: How Falling Birthrates Threaten World Prosperity (And What to Do About It).* New York: Basic Books.

McFalls, Joseph A., Jr. (2007) "Population: A Lively Introduction," *Population Bulletin* 62(1):1–31.

McNicoll, Geoffrey (1999) "Population Weights in the International Order," *Population and Development Review* 25(3): 411–442.

Myers, Dowell (1992) *Analysis with Local Census Data: Portraits of Change.* Boston: Academic Press.

Nash, Alan (1994) "Population Geography," *Progress in Human Geography* 18:385–395.

Ogden, Philip E. (2000) "Weaving Demography into Society, Economy and Culture: Progress and Prospect in Population Geography," *Progress in Human Geography* 24(4): 627–640.

Pandit, Kavita (2004) "Introduction: The Trewartha Challenge," *Population, Space and Place* 10(4):277–278.

Petersen, William (1975) *Population.* Third Edition. New York: Macmillan Publishing Company, Inc.

Stoddart, D. R. (1987) "To Claim the High Ground: Geography for the End of the Century," *Transactions of the Institute of British Geographers* 12:327–336.

Thomlinson, Ralph (1976) *Population Dynamics: Causes and Consequences of World Demographic Change.* Second Edition. New York: Random House, Inc.

Trewartha, Glenn T. (1953) "A Case for Population Geography," *Annals of the Association of American Geographers* 43:71–97.

Trewartha, Glenn T. (1969) *A Geography of Population: World Patterns.* New York: John Wiley and Sons, Inc.

White, Stephen E., et al. (1989) "Population Geography," in Gaile, Gary L. and Willmott, Cort J. eds. *Geography in America.* Columbus, Ohio: Merrill Publishing Co., 258–289.

Woods, Robert and Rees, P. eds. (1986) *Population Structure and Models: Developments in Spatial Demography.* London: Allen and Unwin.

Zelinsky, Wilbur (1966) *A Prologue to Population Geography.* Englewood Cliffs, N.J.: Prentice-Hall, Inc.

Chapter 1

Barlett, Donald L. and Steele, James B. (2004) " America's Border: Who Left the Door Open?" *Time,* 164(12) http://www.barlettandsteele.com/journalism/time_border_1.php

Barinaga, Marcia (1992) " 'African Eve' Backers Beat a Retreat," *Science* 255:686–687.

Berelson, Bernard and Freedman, Ronald (1974) "The Human Population," in *Scientific American*, eds. *The Human Population*. San Francisco: W. H. Freeman and Co., pp. 3–11.

Bianchi, S. M. (1990) "America's children: Mixed prospects." *Population Bulletin*, 45, 1–39.

Bouvier, Leon F. (1976) "On Population Growth," *Intercom* 4: 8–9.

Bouvier, Leon F. (1980) "America's Baby Boom Generation: The Fateful Bulge." *Population Bulletin*. 35(1).

Bouvier, Leon F. and DeVita, Carol J. (1991) "The Baby Boom—Entering Midlife" *Population Bulletin* 46(3):1–33.

Callahan, Daniel (1971) *Ethics and Population Limitation*. New York: Population Council.

Cann, R. L., Stoneking, M., and Wilson, A.C. (1987) "Mitochondrial DNA and Human Evolution," *Nature* 325:31–36.

Cipolla, C. M. (1974) *The Economic History of World Population*. Baltimore: Penguin Books.

Cohen, Joel E. (1995a) "Population Growth and the Earth's Carrying Capacity," *Science* 269:341–346.

Cohen, Joel E. (1995b) *How Many People Can the Earth Support?* New York: W. W. Norton and Company.

Darden, Joe T. (1975) "Population Control or a Redistribution of Wealth: A Dilemma of Class and Race," *Antipode* 7:50–52.

Deevey, Edward S. (1960) "The Human Population," *Scientific American* 203:3–9.

Denevan, William M. (1996) "Carl Sauer and Native American Population Size," *Geographical Review* 86(3):385–397.

Diamond, Jared (1997) *Guns, Germs, and Steel: The Fates of Human Societies*. New York: W. W. Norton.

Diamond, Jared (2005) *Collapse: How Societies Choose to Fail or Succeed*. New York: Viking.

England, Robert (1987) "The Senior Citizen Secret," *Insight*, (March 2):8–11.

Flannery, Tim (2005) *The Weather Makers: How Man is Changing the Climate and What it Means for Life on Earth*. New York: Atlantic Monthly Press.

Francese, Paula A. and Renaghan, Leo M. (1991) "Finding the Customer," *American Demographics* 13(1):48–51.

Gibson, Campbell (1977) "Population Projections for the United States," *Intercom* 5:7–9.

Goldstein, Joshua R. and Schlag, Wilhelm (1999) "Longer Life and Population Growth," *Population and Development Review* 25(4):741–747.

Gugliotta, Guy (2002) "Study Challenges Date of First Primates," *The Sacramento Bee* (April 19): A9.

Hardin, Garrett (1999) *The Ostrich Factor: Our Population Myopia*. New York: Oxford University Press.

Haub, Carl (2002) "How Many People Have Ever Lived on Earth? 2002 Update," *Population Today* 30(8):3–4.

Hoefer, Michael, Nancy Rytina and Bryan C. Baker, 2011. "Estimates of the Unauthorized Immigrant Population Residing in the United States: January 2010," Office of Immigration Statistics, Policy Directorate, U.S. Department of Homeland Security, http://www.dhs.gov/xlibrary/assets/statistics/publications/ois_ill_pe_2010.pdf

Hollmann, F. W., Mulder, T. J., and Kallan, J. E. (2000) *Methodology and Assumptions for the Population Projections of the United States: 1999–2100*. Population Division Working Paper No. 38. Washington, D.C.: U.S. Census Bureau.

Hooyman, Nancy R. and Kiyak, H. Asuman (1988) *Social Gerontology: A Multidisciplinary Perspective*, Boston, Ma.: Allyn and Bacon.

Hopfenberg, Russell (2003) "Human Carrying Capacity is Determined by Food Availability," *Population and Environment* 25(2):109–117.

Humphreys, J. M. (2010) *The Multicultural Economy 2010.* Atlanta: Selig Center for Economic Growth, University of Georgia.

Hyatt, James (1979) *Changing Demographics and How They Affect the Future of Business*, FOB Series No. 9, Washington, D.C.: Center for Strategic and International Studies.

Jacobs, Jane (2004) *Dark Age Ahead.* New York: Random House.

Johanson, D. C. and Edey, M. A. (1981) Lucy: *The Beginnings of Humankind.* New York: Simon and Schuster.

Laws, Glenda (1991) "Aging Population: A Planning Dilemma," *Earth and Mineral Sciences* 60(2):32–36.

Lazer, William (1985) "Inside the Mature Market," *American Demographics* 7(3):3–25.

Leakey, Meave (1995) "The Dawn of Humans: The Farthest Horizon," *National Geographic* 188(3):38–51.

Leakey, Meave and Walker, Alan (1997) "Early Hominid Fossils from Africa," *Scientific American* 276(6)74–79.

Lee, Ronald (2000) "Long–term Population Projections and the US Social Security System," *Population and Development Review* 26(1):137–143.

Livi-Bacci, Massimo (2006) "The Depopulation of Hispanic America after the Conquest," *Population and Development Review* 32(2):199–232.

Longman, Phillip (2004) *The Empty Cradle: How Falling Birthrates Threaten World Prosperity (And What to do About It).* New York: Basic Books.

Lord, Lewis (1997) "How Many People Were Here Before Columbus?" *U.S. News and World Report* 123(7):68–70.

McEvedy, C. and Jones, R. (1978) *Atlas of World Population History.* Middlesex, England: Penguin Books, Ltd.

McNeill, J. R. and McNeill, William H. (2003) *The Human Web: A Bird's-Eye View of World History.* New York: W. W. Norton and Company.

New American Dimensions. (2009) *First African Consumer Segment Study.* Los Angeles: New American Dimensions.

Nortman, Dorothy, assisted by Hofstater, Ellen (1975) "Population and Family Planning Programs: A Factbook," *Reports on Population/Family Planning.* No. 2. Seventh Edition. New York: The Population Council.

Olson, Steve (2002) *Mapping Human History: Genes, Race, and Our Common Origins.* Boston: Mariner Books.

Ornstein, Robert and Ehrlich, Paul (1989) *New World, New Mind.* New York: Doubleday.

Reinhardt, Hazel H. (1979) "The Ups and Downs of Education," *American Demographics* 1(6):9–11.

Shreeve, James (2006) "The Greatest Journey," *National Geographic* 209(3):60–73.

Sloan, Christopher P. (2006) "The Origin of Childhood," *National Geographic* 210(5):148–159.

Soldo, Beth J. and Agree, Emily M. (1988) "America's Elderly," *Population Bulletin* 43(3):1–53.

Stengel, Richard. (2006) "Tracking America's Journey." *Time*, (October 30):8.

Stringer, Christopher B. (1990) "The Emergence of Modern Humans," *Scientific American* 263(6):98–104.

Tarver, James D. (1995) *The Demography of Africa*. Westport, CT: Praeger Publishers.

Tattersall, Ian (1997) "Out of Africa Again . . . and Again?" *Scientific American* 276(4):60–67.

Tattersall, Ian (2000) "Once We Were Not Alone," *Scientific American* 282(1):56–62.

Thorne, Alan G. and Wolpoff, Milford H. (1992) "The Multi-regional Evolution of Humans," *Scientific American* 266(4): 76–83.

Tickell, Crispin (1993) "The Human Species: A Suicidal Success?" *The Geographical Review* 159(2):219–226.

United Nations Population Division. (2004) *World Population to 2300*. New York: Population Division, Department of Economic and Social Affairs, United Nations.

United Nations Population Division (2011) *World Population Projections: The 2010 Revision*. York: Population Division, Department of Economic and Social Affairs, United Nations.

Wade, Nicholas (2004) "New Species Revealed: Tiny Cousins of Humans," http://www.nytimes.com/2004/10/28/science/ 28tiny.html?adxnnl51&adxnnlx=10989757 (Accessed October 28, 2004).

Waldrop, Judith (1991) "The Baby Boom Turns 45," *American Demographics* 13(1):22–27.

Wallerstein, Immanuel (1999) *The End of the World as We Know It: Social Science for the Twenty-First Century*. Minneapolis: University of Minnesota Press.

Weatherford, Jack (1994) *Savages and Civilization: Who Will Survive*? New York: Crown Publishers, Inc.

Wenke, Robert J. (1990) *Patterns in Prehistory*. Third Edition. Oxford: Oxford University Press.

Wilson, Allan C. and Cann, Rebecca L. (1992) "The Recent African Genesis of Humans," *Scientific American* 266(4):68–73.

Witeck-Combs Communications. (2010) *The Gay and Lesbian Market in the U.S.: Trends and Opportunities in the LGBT Community*, 6th Ed. Washington, DC: Packaged Facts.

Wong, Kate (2000) "Who Were the Neandertals?" *Scientific American* 282(4):98–107.

Young, Louise B., ed. (1968) *Population in Perspective*. New York: Oxford University Press.

Zinsser, Hans (1967) *Rats, Lice, and History*. New York: Bantam Books, Inc.

Chapter 2

Anderson, Margo J. (1988) *The American Census: A Social History*. New Haven, CT: Yale University Press.

Anderson, Margo J. and Fienberg, Stephen E. (1999) *Who Counts? The Politics Of Census-Taking in Contemporary America*. New York: Russell Sage.

Blacker, J. G. C. (1969) "Some Unsolved Problems of Census and Demographic Works in Africa," in *International Population Conference*. Vol. 1. London: United Nations.

Bogue, Donald J. (1969) *Principles of Demography*. New York: John Wiley and Sons, Inc.

Caldwell, J. and Igun, A. A. (1971) "An Experiment with Census-Type Age Enumeration in Nigeria," *Population Studies* 25: 287–302.

Carr-Saunders, A. M. (1936) *World Population*. Oxford: The Clarendon Press.

Fitzpatrick, John C. ed. (1939) *The Writings of George Washington*. Washington, D.C.: United States Government Printing Office.

Graham, David and Waterman, Stanley (2005) "Underenumeration of the Jewish Population in the UK 2001 Census," *Population, Space and Place* 11:89–102.

Hock, Saw Swee (1967) "Errors in Chinese Age Statistics," *Demography* 4:859–875.

Howenstine, Erick (1993) "Measuring Demographic Change: The Split Tract Problem," *The Professional Geographer* 45(4): 425–430.

Kahn, E. J., Jr. (1974) *The American People*. Baltimore: Penguin Books, Inc.

Martin, David (2006) "Last of the Censuses? The Future of Small Area Population Data," *Transactions of the Institute of British Geographers* NS 31(6):6–18.

Office of Management and Budget (2010) "2010 Standards for Delineating Metropolitan and Micropolitan Statistical Areas; Notice" *Federal Register*, June 28, 75(123).

Petersen, William (1975) *Population*. Third Edition. New York: MacMillan Publishing Co., Inc.

Prewitt, Kenneth (2000) "The US Decennial Census: Political Questions, Scientific Answers," *Population and Development Review* 26(1):1–16.

Ryder, N. (1964) "Notes on the Concept of a Population," *American Journal of Sociology* 69:447–463.

Seltzer, William (1973) *Demographic Data Collection: A Summary of Experience*. New York: The Population Council.

Shryock, H. S., Jr. and Siegel, J. (1971) *The Methods and Materials of Demography*. Washington, D.C.: United States Government Printing Office.

Skerry, Peter (2000) *Counting on the Census? Race, Group Identity, and the Evasion of Politics*. Washington, D.C.: Brookings Institution Press.

Thomlinson, Ralph (1976) *Population Dynamics: Causes and Consequences of World Demographic Change*. Second Edition. New York: Random House.

Tukey, John W. (1960) "Conclusions vs. Decisions." *Technometrics*, 2(4): 423-433.

United Nations (2008) *Principles and Recommendations for Population and Housing Censuses Revision 2*. Department of Economic and Social Affairs Statistical Paper Series M no.67/Rev2. New York: United Nations.

United States Bureau of the Census (1971) *U.S. Census of Population: 1970*. "Number of Inhabitants: United States Summary." PC(1)-A1. Washington, D.C.: United States Government Printing Office.

Whetten, Nathan L. (1961) *Guatemala: The Land and the People*. Caribbean Series 4. New Haven: Yale University Press.

Zelinsky, Wilbur (1966) *A Prologue to Population Geography*. Englewood Cliffs, NJ: Prentice-Hall, Inc.

Chapter 3

Adepoju, Aderanti and Oppong, Christine. eds. (1994) *Gender, Work, and Population in Sub-Saharan Africa*. London: James Currey.

Alba, Richard D., et al (1995) "Neighborhood Change under Conditions of Mass Immigration: The New York City Region, 1970–1990," *International Migration Review* 29(3): 625–656.

Allen, James Paul and Turner, Eugene James (1988) *We the People: An Atlas of America's Ethnic Diversity*. New York: Macmillan Publishing Company.

Allen, James Paul and Turner, Eugene James (1997) *The Ethnic Quilt: Population Diversity in Southern California*. Northridge: The Center for Geographical Studies, California State University, Northridge.

Andrews, Gavin J. and Phillips, David R. eds. (2005) *Ageing and Place: Perspectives, Policy, Practice*. London and New York: Routledge.

Attané, Isabelle (2006) "The Demographic Impact of a Female Deficit in China, 2000–2050," *Population and Development Review* 32(4):755–770.

Beaujeu-Garnier, J. (1966) *Geography of Population*. New York: St. Martin's Press.

Bondi, Liz (1992) "Gender and Dichotomy," *Progress in Human Geography* 16:98–104.

Bondi, Liz (1993) "Gender and Geography," *Progress in Human Geography* 17:241–246.

Bourne, Larry S. and Ley, David F. (1993) *The Changing Social Geography of Canadian Cities*. Montreal: McGill-Queen's University Press.

Bouvier, Leon and de Vita, Carol J. (1991) "The Baby Boom—Entering Midlife," *Population Bulletin* 46(3):1–34.

Bouvier, Leon, Atlee, Elinore, and McVeigh, Frank (1975) "The Elderly in America," *Population Bulletin* 30(3):1–36.

Brimelow, Peter (1995) *Alien Nation: Common Sense About America's Immigration Disaster*. New York: Random House.

Brown, Lawrence A. and Chung, Su-Yeul (2006) "Spatial Segregation, Segregation Indices and the Geographical Perspective," *Population, Space and Place* 12:125–143.

Chant, Sylvia (2007) *Gender, Generation and Poverty: Exploring the "Feminisation of Poverty" in Africa, Asia and Latin America*. Cheltenham/Northampton, MA: Edward Elgar.

Clarke, John J. (1972) *Population Geography*. Second Edition. Oxford: Pergamon Press.

Coale, Ansley J. (1964) "How a Population Ages or Grows Younger," in Ronald Freedman, ed., *Population: The Vital Revolution*. Garden City, New York: Anchor Books, Doubleday and Co., Inc., pp. 47–58.

Cowgill, Donald O. (1978) "Residential Segregation by Age in American Metropolitan Areas," *Journal of Gerontology* 33: 446–453.

Crimmins, Eileen M. (2001) "Americans Living Longer, Not Necessarily Happier, Lives," *Population Today* 29(2):5 and 8.

Danson, Mike and Hardill, Irene (2006) "Introduction: Demography, Ageing and Activity," *Population, Space and Place* 12: 317–321.

DeFrancis, Marc (2002) "U.S. Elder Care Is in Fragile State," *Population Today* 30(1):1–3.

Doyle, Rodger (1997) "By the Numbers: Female Illiteracy World-wide," *Scientific American* 276(5):20.

Doyle, Rodger (1998) "By the Numbers: Women in Politics Throughout the World," *Scientific American* 278(1): 35.

Fortuijn, Joos Droogleever, et al. (2006) "The Activity Patterns of Older Adults: A Cross-Sectional Study in Six European Countries," *Population, Space and Place* 12:353–369.

Frazier, John W., Tetty-Fio, Eugene L., and Henry, Norah F. (2011) *Race, Ethnicity and Place in a Changing America*. Eds. 2nd Edition. Albany: State University of New York Press.

Frazier, John W., Margai, Florence, M. and Tetty-Fio, Eugene (2003) *Race and Place: Equity Issues in Urban America*. Boulder, CO: Westview Press.

Friedland, Robert B. and Summer, Laura (2010) *Demography is Not Destiny Revisited*. Commonwealth Fund Publication # 789, Washington DC: Center on an Aging Society Georgetown University.

Frey, William H. (2010) "Baby Boomers and the New Demographics of America's Seniors," *Journal of the American Society on Aging* 34(3): 28–37.

Fry, Richard. (2011) *Hispanic College Enrollment Spikes, Narrows Gaps with Other Groups. Washington*, D.C.: Pew Hispanic Center. http://www.pewhispanic.org/files/2011/08/146.pdf

Fuchs, Lawrence H. (1990) *The American Kaleidoscope: Race, Ethnicity, and the Civic Culture.* Hanover, NH: The University Press of New England.

Gonzales, Juan L., Jr. (1990) *Racial and Ethnic Groups in America.* Dubuque, IA: Kendall/Hunt Publishing Company.

Gonzales, Juan L., Jr. (1991) *The Lives of Ethnic Americans.* Dubuque, IA: Kendall/Hunt Publishing Company.

Gonzalez, Juan (2000) *Harvest of Empire: A History of Latinos in the Americas.* New York: Viking.

Goodman, Allen C. (1987) "Using Lorenz Curves to Characterize Urban Elderly Populations," *Urban Studies* 24:77–80.

Himes, Christine L. (2001) "Elderly Americans," *Population Bulletin* 56(4):1–40.

Hirschman, Charles (2004) "The Origins and Demise of the Concept of Race," *Population and Development Review* 30(3): 385–415.

Hudson, Valerie M. and Den Boer, Andrea M. (2004) *Bare Branches: The Security Implications of Asia's Surplus Male Population.* Cambridge, MA: The MIT Press.

Hussain, Shahnaz, Khan, Amanat, and Momsen, Janet (2005) *Gender Atlas of Bangladesh.* Dhaka: USAID.

Ingoldsby, Bron B. and Smith, Suzanna. eds. (1995) *Families in Multicultural Perspective.* New York: The Guilford Press.

Isbister, John (1996) *The Immigration Debate: Remaking America.* West Hartford, CN: Kumarian Press.

Johnston, Ron, Poulsen, Michael, and Forrest, James (2006) "Blacks and Hispanics in Urban America: Similar Patterns of Residential Segregation?" *Population, Space and Place* 12: 389–406.

Jones, John Paul III, Nast, Heidi J., and Roberts, Susan M. (1997) *Thresholds in Feminist Geography: Difference, Methodology, Representation.* Lanham, MD: Rowman and Littlefield.

Kent, Mary M. (2007) "Immigration and America's Black Population," *Population Bulletin* 6(24):1–16.

Lamphere, Louise. ed. (1992) *Structuring Diversity: Ethnographic Perspectives on the New Immigration.* Chicago: The University of Chicago Press.

Lee, Ronald and Haaga, John (2002) "Government Spending in an Older America," *PRB Reports on America* 3(1):1–16.

Little, Jani S. and Rogers, Andrei (2007) "What Can the Age Composition of a Population Tell Us About the Age Composition of Its Out-Migrants?" *Population, Space and Place* 13:23–39.

Logan, J. R., et. al. (2004) "Segregation of Minorities in the Metropolis: Two Decades of Change," *Demography* 41(1):1–22.

Longhurst, Robyn (2007) *Maternities: Gender, Bodies and Space.* London and New York: Routledge.

Martí-Henneberg, Jordi (2005) "Empirical Evidence of Regional Population Concentration in Europe, 1870–2000," *Population, Space and Time* 11:269–281.

McKee, Jesse O. (1985) *Ethnicity in Contemporary America: A Geographical Appraisal.* Dubuque, IA: Kendall/Hunt Publishing Company.

Monk, Janice (1994) "International Perspectives on Feminist Geography," *The Professional Geographer* 46:277–288.

Monk, Janice and Hanson, Susan (1982) "On Not Excluding Half of the Human in Human Geography," *The Professional Geographer* 34:11–23.

Murdock, Steve H. (1995) *An America Challenged: Population Change and the Future of the United States.* Boulder, CO: Westview Press.

Olson, Steve (2002) *Mapping Human History: Genes, Race, and Our Common Origins.* Boston: Mariner Books.

Otterstrom, Samuel M. (2001) "Trends in National and Regional Population Concentration in the United States from 1790 to 1990: From the Frontier to the Urban Transformation," *The Social Science Journal* 38(3):393–407.

Peters, Julie and Wolper, Andrea. eds. (1995) *Women's Rights, Human Rights: International Feminist Perspectives.* New York: Routledge.

Peterson, Peter G. (1999) *Gray Dawn: How the Coming Age Wave Will Transform America—and the World.* New York: Times Books.

Radcliffe, Sarah (2006) "Development and Geography III: Gendered Subjects in Development Processes and Interventions," *Progress in Human Geography* 30(4):1633–1651.

Riley, Nancy E. (2004) "China's Population: New Trends and Challenges," *Population Bulletin* 59(2):1–36.

Roberts, Sam (1993) *Who We Are: A Portrait of America Based on the Latest U.S. Census.* New York: Times Books.

Rodriguez, Gregory (2001) "The Future Americans," *Los Angeles Times* (March 18):M1 and M6.

Roy, Avok, S. A. (2011) "Medicare Reform: Who Decides?" *The American Spectator* (July/August): 48–51.

Samarasinghe, Vidya (2007) *Female Sex Trafficking.* London and New York: Routledge.

Small, Christopher and Cohen, Joel E. (2004) "Continental Physiography, Climate, and the Global Distribution of Human Population," *Current Anthropology* 45(1): 269–277.

Smith, D. I., Spraggins, R. E., & U.S. Census Bureau. (2001) *Gender, 2000.* Washington, D.C.: U.S. Dept. of Commerce, Economics and Statistics Administration, U.S. Census Bureau.

Smith, T. Lynn and Zopf, Paul E., Jr. (1976) *Demography: Principles and Methods.* Port Washington, NY: Alfred Publishing Co., Inc.

Soldo, Beth J. (1980) "America's Elderly in the 1980s," *Population Bulletin* 35(4):1–47.

Squires, Gregory D. and Kubrin, Charis E. (2005) "Privileged Places: Race, Uneven Development and the Geography of Opportunity in Urban America," *Urban Studies*, 42(1): 47-68.

Taylor, J. Edward and Martin, Philip L. (2000) "Central Valley Evolving into Patchwork of Poverty and Prosperity," *California Agriculture* 54(1):26–32.

Trewartha, Glenn J. (1969) *A Geography of Population: World Patterns.* New York: John Wiley and Sons, Inc.

Vafeidis, A., Neumann, B., Zimmermann, J. and Nicholls, R.J. (2011) "Migration and Global Environmental Change: MR9: Analysis of land area and population in the low-elevation coastal zone (LECZ)," London: UK Government's Foresight Project, Migration and Global Environmental Change http://www.bis.gov.uk/assets/bispartners/foresight/

docs/migration/modelling/11-1169-mr9-land-and-population-in-the-low-elevation-coastal-zone.pdf. Retrieved July 6, 2012.

Vernooy, Ronnie (2006) *Social and Gender Analysis in Natural Resource Development: Learning Studies and Lessons from Asia*. Thousand Oaks, CA: Sage.

Waldman, Carl (2009) *The Atlas of the North American Indian.* 3rd Edition. New York: Checkmark Books.

Weeks, John R. (1992) *Population: An Introduction to Concepts and Issues*. Fifth Edition.

Belmont, CA Wadsworth Publishing Company.

White, Michael (1986) "Segregation and Diversity Measures in Population Distribution," *Population Index* 52(2):198–221.

Wilmoth, Janet M. (2006) "Demographic Trends that Will Shape US Policy in the Twenty-first Century" *Research on Aging*. 28(3): 269-288.

Wright, Melissa (2006) *Disposable Women and Other Myths of Global Capitalism*. London and New York: Routledge.

Chapter 4

Bach, Robert L. (1980) "The New Cuban Immigrants: Their Background and Prospects," *Monthly Labor Review* 103(10):39–46.

de Beauvoir, Simone (1972) *The Second Sex*, transl. by H. M. Parshley, New York: Vintage.

Boserup, Ester (1965) *The Conditions of Agricultural Growth: The Economics of Agrarian Change under Population Pressure*. Chicago: Aldine Publishing Company.

Boserup, Ester (1981) *Population and Technological Change: A Study of Long-Term Trends*. Chicago: The University of Chicago Press.

Burch, Thomas K. (2003) "Demography in a New Key: A Theory of Population Theory," *Demographic Research* 19(11): 264–282.

Caldwell, John C. (1976) "Toward a Restatement of Demographic Transition Theory," *Population and Development Review* 2: 321–366.

Caldwell, John C. (2004) "Demographic Theory: A Long View. (Notes and Commentary)," *Population and Development Review* 30(2):297–316.

Carlson, Elwood and Omori, Megumi (1998) "Fertility Regulation in a Declining State Socialist Economy: Bulgaria, 1976—1995" *International Family Planning Perspectives* 24(4): 184–187.

Clark, Kenneth (1969) *Civilisation: A Personal View*. New York: Harper and Row.

Coale, Ansley J. and Watkins, Susan Cotts. Editors. (1986) *The Decline of Fertility in Europe*. Princeton, NJ: Princeton University Press.

Collver, O. Andrew (1965) *Birth Rates in Latin America: New Estimates of Historical Trends and Fluctuations*. Berkeley, CA: Institute of International Studies, University of California.

Dahal, Dilli Ram (1983) *Poverty or Plenty: Innovative Responses to Population Pressure in an Eastern Nepalese Hill Community.* Unpublished Ph.D. dissertation. Honolulu: University of Hawaii.

Danielson, Ross (1979) *Cuban Medicine*. New Brunswick, NJ: Transaction.

Davis, Kingsley. (1963) "The Theory of Change and Response in Modern Demographic History," *Population Index* 29:345–366.

Diaz-Briquets, Sergio (1977) "Mortality in Cuba: Trends and Determinants, 1880–1971," Unpublished Ph.D. dissertation. Philadelphia: University of Pennsylvania.

Diaz-Briquets, Sergio and Perez, Lisandro (1981) "Cuba: The Demography of Revolution," *Population Bulletin* 36(1):1–43.

Diaz-Briquets, Sergio (2006) "Cuba and Miami: Migration and the Future" *Cuban Affairs Quarterly Electronic Journal* 1(2): 1–16. http://www.cubanaffairsjournal.org

Ehrlich, Paul (1968) *The Population Bomb*. New York: Ballantine Books.

Ehrlich, Paul R. and Ehrlich, Anne H. (1990) *The Population Explosion*. New York: Simon and Schuster.

Eversley, David R. (1959) *Social Theories of Fertility and the Malthusian Debate*. London: Oxford University Press.

Feeney, G. F., Wang, F., Zhou, M. K., and Xiao, B. Y. (1989) "Recent Fertility Dynamics in China: Results from the 1987 One Percent Survey," *Population and Development Review* 15: 297–322.

Feng, Wang (1989) "China's One-Child Policy: Who Complies and Why?" (Paper delivered at the Annual Meeting of the Association for Asian Studies, Washington, D.C., March 17–19.)

Findlay, A.M. and Graham, E. (1991) "The Challenge Facing Population Geography," *Progress in Human Geography* 15: 149–162.

Freedman, Ronald (1979) "Theories of Fertility Decline: A Reappraisal," in Philip M. Hauser, ed., *World Population and Development: Challenges and Prospects*. Syracuse: Syracuse University Press, pp. 63–79.

Goldstein, S. and Goldstein, A. (1991) *Permanent and Temporary Migration Differentials in China*. Papers, East-West Population Institute 117, Honolulu: East-West Center.

Goodkind, Daniel (2011) "Child Underreporting, Fertility and Sex Ratio Imbalance in China," *Demography* 48(1): 291–316.

Gonzalez, Gerardo, Correa, German, Errazuriz, Margarita M., and Tapia, Raul (1978) *Development Strategy and Demographic Transition: The Case of Cuba*. Santiago, Chile: Centro Latinoamericano de Demografia.

Graham, Elspeth (2000) "What Kind of Theory for What Kind of Population Geography?" *International Journal of Population Geography* 6: 257–272.

Greenhalgh, Susan (1990) "Socialism and Fertility in China," *Annals of the American Academy of Political and Social Science* 510:73–86.

Harrison, Paul (1980) "Lessons for the Third World," *People* 7: 2–20.

Hernandez, Jose (1974) *People, Power, and Policy: A New View on Population*. Palo Alto, CA: National Press Books.

Hollerbach, Paula E. (1980) "Recent Trends in Fertility, Abortion, and Contraception in Cuba," *International Family Planning Perspectives* 6:97–106.

Jacobson, Jodi L. (1991) "China's Baby Budget," in Lester R. Brown ed. *The World Watch Reader on Global Environmental Issues*. New York: W. W. Norton and Company.

Johansson, Sten and Nygren, Ola (1991) "The Missing Girls of China: A New Demographic Account," *Population and Development Review* 17:35–53.

Kennedy, Bingham, Jr. (2001) "Dissecting China's 2000 Census," http://prb.org/*Regions/asia_near__east/Dissecting Chinas2000 Census.html.*

Kindleberger, Charles P. and Herrick, Bruce (1977) *Economic Development*. Third Edition. New York: McGraw-Hill Book Company.

Longman, Phillip (2004) *The Empty Cradle: How Falling Birthrates Threaten World Prosperity (And What to Do About It)*. New York: Basic Books.

Malthus, T. R. (1798) *An Essay on the Principle of Population, as it Affects the Future Improvement of Society.* London: J. Johnson.

Malthus, T. R. (1803) *An Essay on the Principle of Population; or, A View of Its Past and Present Effect on Human Happiness.* London: J. Johnson.

National Academy of Sciences (1999) *Science and Creationism: A View from the National Academy of Sciences.* 2nd ed. Washington, DC: National Academy of Sciences.

National Bureau of Statistics China (1990) Annual data. http://www.stats.gov.cn/english/

National Bureau of Statistics China (2010) Annual data. http://www.stats.gov.cn/english/

Pannell, Clifton W. and Torguson, Jeffrey S. (1991) "Interpreting Spatial Patterns from the 1990 China Census," *The Geographical Review* 81(3):304–317.

Perez, Lisandro (1977) "The Demographic Dimensions of the Educational Problem in Socialist Cuba," *Cuban Studies* 7:33–57.

Petersen, William (1999) *Malthus: Founder of Modern Demography.* New Brunswick, NJ: Transaction Publishers.

Riley, Nancy E. (2004) "China's Population: New Trends and Challenges," *Population Bulletin* 59(2):1–36.

Smil, Vaclav (2000) *Feeding the World: A Challenge for the Twenty-First Century.* Cambridge, MA: MIT Press.

Smil, Vaclav (2005a) "The Next 50 Years: Fatal Discontinuities," *Population and Development Review* 31(2):201–236.

Smil, Vaclav (2005b) "The Next 50 Years: Unfolding Trends," *Population and Development Review* 31(4):605–643.

Song, J., H. Tuan, and J. Yu (1985) *Population Control in China.* New York: Praeger.

State Statistical Bureau (1988) *Tabulations of China One Percent Population Survey, National Volume.* Beijing: China Statistical Press.

State Statistical Bureau (1990) "Geographical Distribution, Density and Natural Growth Rate of China's Population," *Beijing Review* 33(51):25–27.

Strangeland, Charles E. (1904) *Pre-Malthusian Doctrines of Population.* New York: Columbia University Press.

Taleb, Nassim N. (2007) *The Black Swan: The Impact of the Highly Improbable.* New York: Random House.

Teitelbaum, Michael S. and Winter, Jay M. Eds. (1989) *Population and Resources in Western Intellectual Traditions.* Cambridge: Cambridge University Press.

Tien, H. Yuan (1983) "China: Demographic Billionaire," *Population Bulletin* 38(2):1–43.

Tien, H. Yuan (1988) "A Talk with China's Wang Wei," *Population Today* 16(1):6–8.

Tien, H. Yuan (1990) "China's Population Planning After Tiananmen," *Population Today* 18(9):6–8.

Tomaselli, Sylvana (1989) "Moral Philosophy and Population Questions in Eighteenth Century Europe," in Teitelbaum, Michael S. and Winter, Jay M. Editors. *Population and Resources in Western Intellectual Traditions.* Cambridge: Cambridge University Press, pp. 7–29.

U.S. Congressional Research Service *Cuban Migrants to the United States: Policy and Trends* (R40566; June 2, 2009) by Ruth Ellen Wasem.

U.S. Department of Homeland Security. (2012) *Yearbook of Immigration Statistics: 2011.* Washington, D.C.: U.S. Department of Homeland Security, Office of Immigration Statistics.

Vallin, Jacques (2002) "The End of the Demographic Transition: Relief or Concern?" *Population and Development Review* 28(1):105–120.

Vance, Rupert (1952) "Is Theory for Demographers?" *Social Forces* 31(1): 9–13.

van de Walle, Etienne and Knodel, John (1980) "Europe's Fertility Transition: New Evidence and Lessons for Today's Developing World," *Population Bulletin* 34(6):1–43.

van de Walle, Etienne and Muhsam, Helmut V. (1995) "Fatal Secrets and the French Fertility Transition," *Population and Development Review* 21(2):261–279.

Watkins, Susan Cotts (1990) "From Local to National Communities: The Transformation of Demographic Regimes in Western Europe, 1870–1960," *Population and Development Review* 16: 241–272.

White, P. and Jackson, P. (1995) "(Re)theorizing Population Geography," *International Journal of Population Geography* 1:111–123.

Wrigley, E. A. (1989) "The Limits to Growth: Malthus and the Classical Economists," in Teitelbaum, Michael S. and Winter, Jay M. Editors. *Population and Resources in Western Intellectual Traditions*. Cambridge: Cambridge University Press, pp. 30–48.

Zeng, Y. (1989) "Population Policy in China: New Challenge and Strategies," in J. M. Eekelaar and D. Pearl (eds.) *An Aging World*. Oxford: Oxford University Press, pp. 61–73.

Zeng, Yi, et al. (1991) "A Demographic Decomposition of the Recent Increase in Crude Birth Rates in China," *Population and Development Review* 17(3):435–459.

Chapter 5

Antecol, Heather and Bedard, Kelly (2006) "Unhealthy Assimilation: Why Do Immigrants Converge to American Health Status Levels?" *Demography* 43(2):337–360.

Birdsall, Stephen (1991) "Medical Geography," in H. J. deBlij and Peter O. Muller, *Geography: Regions and Concepts*. 6th Edition. New York: John Wiley and Sons, Inc., pp. 392–393.

Bongaarts, John (2006) "How Long Will We Live?" *Population and Development Review* 32(4):605–628.

Boyle, P., Muir, C. S., and Grundman, P. Eds. (1989) *Mapping and Cancer: Recent Results in Cancer Research 114*. Berlin: Springer Verlag.

Brink, Susan (2004) "AIDS: Darkening in America," *U.S. News and World Report* 137(1):132 and 134.

Brown, J. Larry and Pollitt, Ernesto (1996) "Malnutrition, Poverty and Intellectual Development," *Scientific American* 274(2): 38–43.

Buettner, Dan (2005) "New Wrinkles on Aging," *National Geographic* 208(5):2–27.

Centers for Disease Control and Prevention (2011) *Diagnoses of HIV Infection in the United States and Dependent Areas*, 2011. www.cdc.gov/hiv/library/reports/surveillance/2011/surveillance_Report_vol_23.html

Cliff, A. D. and Haggett, P. (1988) *Atlas of Disease Distributions*. Oxford: Basil Blackwell.

Committee for Planning and Investment (CPI) (2007) *Lao Reproductive Health Survey 2005*. Vientiane Capital, Lao PDR: National Statistics Center (NSC).

Condran, Gretchen A. and Cheney, Rose A. (1982) "Mortality Trends in Philadelphia: Age- and Cause-Specific Death Rates 1870–1930," *Demography* 18:94–124.

Cotter, John V. and Patrick, Larry L. (1981) "Disease and Ethnicity in an Urban Environment," *Annals of the Association of American Geographers* 71:40–49.

Crawford, Patricia B. (2007) "Key Partners Working Together to Stem Obesity Epidemic," *California Agriculture* 61(3):98.

Crimmins, Eileen M. (1981) "The Changing Pattern of American Mortality Decline, 1940–77, and Its Implications for the Future," *Population and Development Review* 7:229–254.

Donovan, Bill (2008) "Year of Fear: Medicine Men Knew Cause of Hantavirus," *Flutracker* February http://www.gallupindependent.com/2008/May/051708hantavirus.html.

Dunavan, Claire Panosian (2005) "Tackling Malaria," *Scientific American* 293(6):76–83.

Dutt, Ashok K. et al. (1987) "Geographical Patterns of AIDS in the United States," *The Geographical Review* Vol. 77, No. 4: 456–471.

Elo, Erma T. (2009) "Social Class Differentials in Health and Mortality: Patterns and Explanations in Comparative Perspective," *Annual Review of Sociology* 35: 553–572.

Evans, Bethan (2006) "'Gluttony or Sloth': Critical Geographies of Bodies and Morality in (Anti) Obesity Policy," *Area* 38(3): 259–267.

Ezzel, Carol (2000) "Care for a Dying Continent," *Scientific American* 282(5):96–105.

Franz, Jennifer S. and FitzRoy, Felix (2006) "Child Mortality and Environment in Developing Countries," *Population and Environment* 27(3):263–284.

Gakidou, Emmanuela, Cowling, Krycia, Lozano, Raphael, Murray, Christopher L. (2010) "Increased Educational Attainment and Its Effect on Child Mortality in 175 Countries Between 1970 and 2009: A Systematic Analysis," *The Lancet* 376: 959–974.

Garrett, Laurie (1994) *The Coming Plague: Newly Emerging Diseases in a World Out of Balance*. NY: Farrar, Strauss and Giroux.

Gehrmann, Rolf (2002) "Infant Mortality in Town and Countryside: Northern Germany, ca. 1750–1800," *The History of the Family* 7(4):545–556.

Gibbs, W. Wayt and Soares, Christine (2005) "Preparing for a Pandemic," *Scientific American* 293(5):44–55.

Ginter, Emil and Simko, Vladimir (2010) "Health Differences Between the United States of America and the European Union," *Central European Journal of Public Health*. 18(4): 215–218.

Gould, P. (1991) "Editorial Report," *Science* 234:1022.

Gould, P., Kabel, J., Gorr, W., and Golub, A. (1991) "AIDS: Predicting the Next Map," *Interfaces* 21:80–92.

Gould, W. T. S. (2005) "Vulnerability and HIV/AIDS in Africa: From Demography to Development," *Population, Space and Place* 11:473–484.

Grant, James P. (1990) *The State of the World's Children 1990*. New York and Oxford: Oxford University Press for UNICEF.

Hayward, M. D., et. al. (2004) "The Long Arm of Childhood: The Influence of Early-Life Social Conditions on Men's Mortality," *Demography* 41(1):87–107.

Heisler, Elayne J. (2012) *The U.S. Infant Mortality Rate: International Comparisons, Underlying Factors, and Federal Programs* Congressional Research Service 7-5700 www.crs.gov R41378.

Hopkin, Karen (2004) "Making Methuselah," *Scientific American* 14(3):12–17.

Horton, Richard (2004) "AIDS: The Elusive Vaccine," *New York Review of Books* LI(14):53–57.

House, J. S., Landis, K. R., and Umberson, D. (1988) "Social Relationships and Health," *Science* 241:540–545.

Jia, Zhongwie et al. (2011) "Tracking the Evolution of HIV/AIDS in China from 1989-2009 to Inform Future Prevention and Control Efforts," *PLoS ONE* 6(10): e25671. doi:10.1371/journal.pone.0025671.

Jones, Kelvyn and Moon, Graham (1991) "Medical Geography," *Progress in Human Geography* 15:437–443.

Kahn, Jennifer (2007) "Healing the Heart," *National Geographic* 211(2):40–65.

Kalben, Barbara B. (2000) "Why Men Die Younger: Causes of Mortality Differences by Sex," *North American Actuarial Journal* 4(4): 83–114.

Kaufman, Marc (2004) "Cigarettes Cut About 10 Years Off Life, 50–Year Study Shows," http://www.washingtonpost.com/ac2/ wp-dyn/A61981–2004Jun22 (Accessed June 24, 2004).

Kearns, Robin A. (1993) "Place and Health: Toward a Reformed Medical Geography," *The Professional Geographer* 45(2): 139–147.

Keil, Roger and Ali, S. Harris (2006) "The Avian Flu: Some Lessons Learned from the 2003 SARS Outbreak in Toronto," *Area* 38(1):107–109.

Koch, Tom (2005) Cartographies of Disease: *Maps, Mapping, and Medicine.* Redlands, Calif. : ESRI Press.

Koch, Tom (2011) Disease Maps: *Epidemics on the Ground.* Chicago: University of Chicago Press.

Kohler, Hans-Peter, Behrman, Jere R., and Watkins, Susan C. (2007) "Social Networks and HIV/AIDS Risk Perceptions," *Demography* 44(1):1–33.

Kolivras, Korine N. (2006) "Mosquito Habitat and Dengue Risk Potential in Hawaii: A Conceptual Framework and GIS Application," *The Professional Geographer* 58(2):139–154.

Ladusingh, Laishram and Singh, Chungkham Holendro (2006) "Place, Community Education, Gender and Child Mortality in North-East India," *Population, Space and Place* 12:65–76.

Lamptey, Peter, et al. (2002) "Facing the HIV/AIDS Pandemic," *Population Bulletin* 57(3):1–39.

Lance P, Angeles G, Islam S. (2008) Baseline Urban Bangladesh Smiling Sun Franchise Program (BSSFP) Evaluation Survey. Dhaka, Bangladesh and Chapel Hill, NC, USA: Mitra and Associates and MEASURE Evaluation; 2009.

Langridge, William H. R. (2000) "Edible Vaccines," *Scientific American* 283(3):66–71.

Le Guenno, Bernard (1995) "Emerging Viruses," *Scientific American* 273(4):56–62.

Madigan, Francis C. (1957) "Are Sex Mortality Differentials Biologically Caused?" *Milbank Memorial Fund Quarterly* 35: 202–223.

Manton, Kenneth G., Gu, XiLiang, and Lamb, Vicki L. (2006) "Long-Term Trends in Life Expectancy and Active Life Expectancy in the United States," *Population and Development Review* 32(1):81–105.

McKeown, Thomas and Brown, R. G. (1969) "Medical Evidence Related to English Population Changes in the Eighteenth Century," in Michael Drake, Ed., *Population in Industrialization.* London: Methuen and Co., Ltd.

Ministry of Health Republic of Indonesia (2004) *Indonesia Health Profile 2004.* Jakarta: Ministry of Health RI 2006.

Mohan, J. (1988) "Restructuring, Privitisation and the Geography of Health Care Provision in England 1983–1987," *Transactions of the Institute of British Geographers* 13:449–465.

Mokdad, Ali H., et al. (2004) "Actual Causes of Death in the United States, 2000," http://jama.ama-assn.org/cgi/content/ abstract/291/10/1238 (Accessed October 4, 2004).

Mosley, W. Henry and Cowley, Peter (1991) "The Challenge of World Health," *Population Bulletin* 46:1–39.

National Bureau of Statistics (NBS) 2007, Nigeria Multiple Indicator Cluster Survey 2007 Final Report. Abuja: Nigeria.

Naím, Moisés (2007) "The Hidden Pandemic," *Foreign Policy* Jul-Aug(161): 95,96.

Nestle, Marion (2006) *What to Eat*. New York: North Front Press.

Niranjan, S. and Madhusudana, Battala (2006) "Spatial Variations in Elevated Blood Lead Levels Among Young Children in Mumbai, India," *Population, Space and Place* 12:243–255.

Olshansky, S. Jay and Alt, Brian (1986) "The Fourth Stage of the Epidemiologic Transition: The Stage of Delayed Degenerative Diseases," *The Milbank Quarterly* 64(3):355–391.

Olshansky, S. Jay, Hayflick, Leonard, and Carnes, Bruce A. (2004) "No Truth to the Fountain of Youth," *Scientific American* 14(3): 98–102.

Omran, Abdel (1971) "The Epidemiological Transition: A Theory of the Epidemiology of Population Change," *The Milbank Memorial Fund Quarterly* 49:509–538.

Omran, Abdel (1977) "Epidemiologic Transition in the U.S.: The Health Factor in Population Change," *Population Bulletin* 32(2):1–42.

Osterholm, Michael T. (2007) "Unprepared for a Pandemic," *Foreign Affairs* 86(2):47–57.

Pandey, Arvind, Battacharya, B.N., Sahu, D. and Sultana, Rehena (2005) Components of under-five mortality trends, current stagnation and future forecasting levels. *NCMH Background Papers·Burden of Disease in India.* http://s3.amazonaws.com/zanran_storage/whoindia.org/ContentPages/14744110.pdf#page=10.

Phillips, David R. (1990) *Health and Health Care in the Third World.* New York: John Wiley and Sons, Inc.

Pisani, Elizabeth (2000) "AIDS into the 21st Century: Some Critical Considerations," *Reproductive Health Matters* 8(15): 36–76.

Pollan, Michael (2006) The *Omnivore's Dilemma: The Natural History of Four Meals*. New York: Penguin Press.

Preston, Richard (1994) *The Hot Zone*. New York: Random House.

Preston, Samuel and Wang, Haidong (2006) "Sex Mortality Differences in the United States: The Role of Cohort Smoking Patterns," *Demography* 43(4):631–646.

Preston, Samuel H. and Stokes, Andrew (2011) "Contribution of Obesity to International Differences in Life Expectancy," *American Journal of Public Health* 101(11): 2137–2143.

Pyle, Gerald F. and Rees, Philip H. (1971) "Modeling Patterns of Death and Disease in Chicago," *Economic Geography* 47(4): 475-488.

Reher, Davis S. (2001) "In Search of the Urban Penalty: Exploring Urban and Rural Mortality Patterns in Spain During the Demographic Transition," *International Journal of Population Geography* 7(2):105–127).

Ritchie, Lorrene D., et al. (2007) "Preventing Obesity: What Should We Eat," *California Agriculture* 61(3):112–118.

Rockett, Ian R. H. (1999) "Population and Health: An Introduction to Epidemiology," *Population Bulletin* 54(4):1–44.

Rogers, Richard G. and Hackenberg, Robert (1987) "Extending Epidemiologic Transition Theory: A New Stage," *Social Biology* 34(3–4):234–243.

Rutstein, Shea O. (1991) "Levels, Trends, and Differentials in Infant and Child Mortality in the Less Developed Countries." Paper presented at the seminar on Child Survival

Interventions: Effectiveness and Efficiency at The Johns Hopkins University School of Hygiene and Public Health, Baltimore, MD.

Sachs, Jeffrey D. (2007) "The Neglected Tropical Diseases," *Scientific American* 296(1):33A.

Salomon, Joshua A. and Murray, Christopher J. L. (2002) "The Epidemiologic Transition Revisited: Compositional Models for Causes of Death by Age and Sex," *Population and Development Review* 28(2):205–228.

Sapolsky, Robert (2005) "Sick of Poverty," *Scientific American* 293(6):92–99.

Senior, M. and Williamson, S. (1990) "An Investigation into the Influence of Geographical Factors on Attendance for Cervical Cytology Screening," *Transactions of the Institute of British Geographers* 15:421–434.

Smil, Vaclav (2005) "The Next 50 Years: Fatal Discontinuities," *Population and Development Review* 31(2):201–236.

Smith, David W. and Bradshaw, Benjamin S. (2006) "Variation in Life Expectancy During the Twentieth Century in the United States," *Demography* 43(4):647–657.

Smyth, Fiona (2005) "Medical Geography: Therapeutic Places, Spaces and Networks," *Progress in Human Geography* 29(4): 488–495.

Stamp, L. Dudley (1964) *The Geography of Life and Death*. Ithaca, NY: Cornell University Press.

Thomlinson, Ralph (1976) Population Dynamics: *Causes and Consequences of World Demographic Change*. Second Edition. New York: Random House.

Thomas, Kevin J. A. (2007) "Child Mortality and Socioeconomic Status: An Examination of Differentials by Migration Status in South Africa," *International Migration Review* 41(1):40–74.

Thomlinson, Ralph (1976) *Population Dynamics: Causes and Consequences of World Demographic Change*. Second Edition. New York: Random House.

Torrey, Barbara Boyle (2004) "A Comparison of US and Canadian Mortality in 1998," *Population and Development Review* 30(3): 519–530.

UNAIDS (2010) *UNAIDS Report on the Global AIDS Epidemic 2010*. New York: Joint United Nations Programme on HIV/AIDS (UNAIDS). http://www.unaids.org/globalreport/Global_report.htm

UNAIDS (2012) *UNAIDS Report on the Global AIDS Epidemic 2012*. New York: Joint United Nations Programme on HIV/AIDS (UNAIDS). http://www.unaids.org/globalreport/Global_report.htm

UNICEF (2011) Levels and Trends

UNICEF (2011) *Levels and Trends in Child Mortality: Report 2011*. New York: United Nations Inter-agency Group for Child Mortality Estimation.

UNICEF (2012) *Pneumonia and Diarrhea: Tackling the Deadliest Diseases for the World's Poorest Children*. www.childinfo.org/publications

Vatanoglu, Ermine E. and Ataman, Ahmet D. (2011) "A Sexually Transmitted Disease: History of Disease Through Philately," *Journal of the Turkish German Gynecological Association* 12:192–196.

Wart, Paula J. (2004) "Adult Obesity Ups Death Risk," http://vanderbiltowc.wellsource.com/dh/content (Accessed October 21, 2004).

Wilson, James L. (1993) "Mapping the Geographical Diffusion of a Finnish Smallpox Epidemic from Historical Population Records," *The Professional Geographer* 45(3):276–285.

World Health Organization (2012) "Child Health Epidemiology" http://www.who.int/maternal_child_adolescent/epidemiology/child/en/index.html.

Chapter 6

Aassve, Arnstein, Billari, Francesco C. and Spéder, Zsolt (2006) "Societal Transition, Policy Changes and Family Formation: Evidence from Hungary," *European Journal of Population* 22(2):127–152.

Agyei-Mensah, Samuel (2006) "Fertility Transition in Ghana: Looking Back and Looking Forward," *Population, Space and Place* 12:461–477.

Andersson, Gunnar, et al. (2006) "Gendering Family Composition: Sex Preferences for Children and Childbearing Behavior in the Nordic Countries," *Demography* 43(2):255–267.

Anonymous (2004) "Where Have all the Bambini Gone?" http: //www.telegraph.co.uk/news/main.jhtml?xml5/news/2004/04/18/wbamb18.xml&sSheet5/news/2004/04/18/ixworld.html (Accessed November 2, 2004).

Axmon, A. et al., (2006) "Time to Pregnancy as a Function of Male and Female Serum Concentrations of 2,2' 4,4' 5,5' -hexachlorobiphenyl (CB-153) and 1,1-dichloro-2,2-bis (p-chlorophenyl)-ethylene (p, p'-DDE)," *Human Reproduction* 21(3): 657–665.

Bledsoe, Caroline (1990) "Transformations in Sub-Saharan African Marriage and Fertility," *Annals of the American Academy of Political and Social Science* 510:115–125.

Boldrin, Michele and Jones, Larry E. (2002) "Mortality, Fertility, and Saving in a Malthusian Economy," *Review of Economic Dynamics* 5(4):775–814.

Bongaarts, John and Feeney, Griffith (1998) "On the Quantum and Tempo of Fertility," *Population and Development Review* 24(2):271–291.

Bongaarts, John (1986) "The Transition in Reproductive Behavior in the Third World," in Jane Menken, ed. (1986) *World Population and U.S. Policy.* New York: W. W. Norton, pp.105–132.

Boyle, Paul (2003) "Population Geography: Does Geography Matter in Fertility Research?" *Progress in Human Geography* 27(5):615–626.

Brea, Jorge A. (2003) "Population Dynamics in Latin America," *Population Bulletin* 58, no. 1 Washington, DC: Population Reference Bureau.

Brownridge, Douglas A. (2004) "Understanding Women's Heightened Risk of Violence in Common Law Unions," *Violence Against Women* 10(6):626–651.

Bryant, John (2007) "Theories of Fertility Decline and the Evidence from Development Indicators," *Population and Development Review* 33(1):101–127.

Carr, David L., Pan, William K. Y., and Bilsborrow, Richard E. (2006) "Declining Fertility on the Frontier: The Ecuadorian Amazon," *Population and Environment* 28(1):17–39.

Coleman, David (2006) "Immigration and Ethnic Change in Low-Fertility Countries: A Third Demographic Transition," *Population and Development Review* 32(3):401–446.

Davis, Kingsley and Blake, Judith (1956) "Social Structure and Fertility: An Analytic Framework," *Economic Development and Cultural Change*, 4:211–235.

Davis, Kingsley, Bernstam, Mikhail S., and Ricardo-Campbell, Rita. eds. (1987) *Below-Replacement Fertility in Industrial Societies: Causes, Consequences, Policies.* Cambridge: Cambridge University Press.

De Broe, Sofie and Hinde, Andrew (2006) "Diversity in Fertility Patterns in Guatemala," *Population, Space and Place* 12: 435–459.

Doring, Gerhard K. (1969) "The Incidence of Anovular Cycles in Women," *Journal of Reproduction and Fertility*, Supplement 6:77–81.

Doskoch, P. (2010) "Why Is Kenya's Fertility Rate Still High? HIV Epidemic May Be a Factor," *International Perspectives on Sexual & Reproductive Health* 36(4): 211–212.

Dye, Jane L. (2008) "Fertility of Women: 2006" *Current Population Reports no. P-20-558.* Washington, DC: US Census Bureau.

Eloundou-Enyegue, Parfait M. and Williams, Lindy B. (2006) "Family Size and Schooling in Sub-Saharan African Settings: A Reexamination," *Demography* 43(1):25–52.

Espenshade, Thomas J. (1977) "The Value and Cost of Children," *Population Bulletin* 32(1): 1–47.

Ford, Nicholas and Bowie, Cameron (1989) "Urban-Rural Variations in the Level of Heterosexual Activity of Young People," *Area* 21:237–248.

Foster, Caroline (2000) "The Limits to Low Fertility: A Biosocial Approach," *Population and Development Review* 26(2): 209 234.

Frejka, Tomas and Sardon, Jean-Paul (2004) *Childbearing Trends and Prospects in Low-Fertility Countries: A Cohort Analysis.* Dordrecht: Kluwer Academic Publishers.

Gavin, Lorrie et al. (2009) "Sexual and Reproductive Health of Persons Aged 10-24—United States, 2002-2007," *MMWR Surveillance Summaries* 58(SS-6): 1–60.

Gober, Patricia (1994) "Why Abortion Rates Vary: A Geographical Examination of the Supply of and Demand for Abortion Services in the United States in 1988," *Annals of the Association of American Geographers* 84(2):230–250.

Gray, Jo Anna, Stockard, Jean, and Stone, Joe (2006) "The Rising Share of Nonmarital Births: Fertility Choice or Marriage Behavior?" *Demography* 43(2):241–253.

Guilmoto, Christophe Z. and Rajan, S. Irudaya (2001) "Spatial Patterns of Fertility Transition in Indian Districts," *Population and Development Review* 27(4): 713–738.

Hajnal, J. (1965) "European Marriage Patterns in Perspective," in David V. Glass and D. E. C. Eversley, eds. *Population in History.* Chicago: Aldine Publishing Company, pp. 101–143.

Hawthorn, Geoffrey (1970) *The Sociology of Fertility.* London: Collier-Macmillan Limited.

Hirsch, Jennifer S. et al. (2007) "The Inevitability of Infidelity: Sexual Reputation, Social Geographies, and Marital HIV Risk in Rural Mexico," *American Journal of Public Health* 97(6): 968–996.

Hoffman, Lois W. and Hoffman, Martin L. (1973) "The Value of Children to Parents," in James W. Fawcett, ed. *Psychological Perspectives on Population.* New York: Basic Books, pp. 19–76.

Hussain, Athar (2002) "Demographic Transition in China and its Implications," *World Development* 30(10):1823–1834.

James, W. H. (1966) "The Effect of Altitude on Fertility in Andean Countries," *Population Studies,* 20:97–101.

Jonsson, S. H., et. al. (2004) "The Fertility Contribution of Mexican Immigration to the United States," *Demography* 41(1): 129–150.

Jorgensen, Neils, Asklund, Camillea, Carlsen, Elisabeth, Skakkebaek, Neils (2006) "Coordinated European Studies of Semen Quality: Results from Studies of Scandinavian Young Men is a Matter of Concern" *International Journal of Andrology* 29: 54–61.

Kahn, Joan R. and Anderson, Kay E. (1992) "Intergenerational Patterns of Teenage Fertility," *Demography* 29:39–57.

Kashiwase, Haruna (2002) "Shotgun Weddings a Sign of the Times in Japan, *Population Today* 30(5):1 and 4.

Kenney, Catherine T. and McLanahan, Sara S. (2006) *Demography* 43(1):127–140.

Krishnan, Gopal (1989) "Fertility and Mortality Trends in Indian States," *Geography* 74:53–56.

Kulczycki, Andrzej (1995) "Abortion Policy in Postcommunist Europe: The Conflict in Poland," *Population and Development Review* 21(3):471–505.

Kulu, Hill (2006) "Fertility of Internal Migrants: Comparison Between Austria and Poland," *Population, Space and Place* 12: 147–170.

Leete, Richard ed. (1999) *Dynamics of Values in Fertility Change*. Oxford: Oxford University Press.

Leibenstein, Harvey (1963) *Economic Backwardness and Economic Growth*. New York: John Wiley and Sons, Inc.

Liu, Lee (2005) "Fertility Trends in China's More Developed Urban Districts: The Case of Four Cities," *Population, Space and Place* 11:411–423.

Livingston, Gretchen and Cohn, D'Vera (2012, November 29) "U.S. Birth Rate Falls to a Record Low; Decline is Greatest Among Immigrants," *Social and Demographic Trends* Washington, DC: Pew Hispanic Center.

Longman, Phillip (2004) *The Empty Cradle: How Falling Birthrates Threaten World Prosperity (And What to Do About It)*. New York: Basic Books.

Lutz, Wolfgang, Testa, Maria Rita, and Penn Dustin J. (2006) "Population Density is a Key Factor in Declining Human Fertility," *Population and Environment* 28(2):69–81.

MacInnes, John and Díaz, Julio Pérez (2007) "'Low' Fertility and Population Replacement in Scotland," *Population, Space and Place* 13:3–21.

Mamdani, Mahmood (1972) *The Myth of Population Control: Family, Caste, and Class in an Indian Village*. New York: Monthly Review Press.

Martin, Molly (2006) "Family Structure and Income Inequality in Families with Children, 1976 to 2000," *Demography* 43(3): 421–445.

McKenzie, Richard B. and Tullock, Gordon (1989) *The Best of the New World of Economics . . . and Then Some*. Fifth Edition. Homewood, IL: Richard D. Irwin.

Messina, Jane P. et al. (2010) "Spatial and Socio-behavioral Patterns of HIV Prevalence in the Democratic Republic of the Congo," *Social Science and Medicine* 71(8): 1428-1435.

Michielin, Francesca (2004) "Lowest Low Fertility in an Urban Context: The Role of Migration in Turin, Italy," *Population, Space and Place* 10(4):331–347.

Morgan, S. Philip (2003) "Is Low Fertility a Twenty-First Century Demographic Crisis?" *Demography* 40(4):589–603.

Mott, Frank L. and Mott, Susan H. (1980) "Kenya's Record Population Growth: A Dilemma of Development," *Population Bulletin* 35(3):1–43.

Mulder, Clara H., Clark, William A. V. and Wagner, Michael (2006) "Resources, Living Arrangements and First Union Formation in the United States, the Netherlands and West Germany," *European Journal of Population* 22(1):3–35.

Murthi, Mamta (2002) "Fertility Change in Asia and Africa," *World Development* 30(10): 1769–1778.

Nahmias, Petra and Stecklov, Guy (2007) "The Dynamics of Fertility Amongst Palestinians in Israel from 1980 to 2000," *European Journal of Population* 23(1)71–99.

Nortman, Dorothy (1977) "Changing Contraceptive Patterns: A Global Perspective," *Population Bulletin* 32(3):1–37.

Ogden, Philip E. and Schnoebelen, François (2005) "The Rise of the Small Household: Demographic Change and Household Structure in Paris," *Population, Space and Place* 11:251–268.

Orenstein, Peggy (2001) "Japan's 'Parasite Singles,'" *The Sacramento Bee* (July 15):L1 and L6.

Population Reference Bureau Staff (2004) "Transitions in World Population," *Population Bulletin* 59(1):1–40.

Rendall, Michael S. and Bachieva, Raisa A. (1998) "An Old-Age Security Motive for Fertility in the United States?" *Population and Development Review* 24(2):293–307.

Robertson, A. F. (1991) *Beyond the Family: The Social Organization of Reproduction.* Berkeley and Los Angeles: University of California Press.

Robey, Bryant, Rutstein, Shea O., and Morris, Leo (1993) "The Fertility Decline in Developing Countries," *Scientific American* 269(6):60–67.

Saxton, G. A., Jr. and Serwada, D. M. (1969) "Human Birth Interval in East Africa," *Journal of Reproduction and Fertility*, Supplement 6:83–88.

Schreffler, Karina M., Dodoo, F. Nii-Amoo (2009) "The Role of Intergenerational Transfers, Land, and Education in Fertility Transition in Rural Kenya: The Case of Nyeri District," *Population and Environment* 30(3): 75–92.

Swan, Shanna H. (2006) "Does Our Environment Effect Our Fertility? Some Examples to Help Reframe the Question," *Seminars in Reproductive Medicine* 24: 142–146.

Swan S.H. Kruse R.L. Liu F. et al. (2003) "Semen Quality in Relation to Biomarkers of Pesticides Exposure" *Environmental Health Perspectives* 111: 1478–1484.

Teitelbaum, Michael S. and Winter, Jay M. (1985) *The Fear of Population Decline.* Orlando, FL: Academic Press.

Thornton, Arland and Lin, Hui-Sheng (1994) *Social Change and the Family in Taiwan.* Chicago: University of Chicago Press.

Tietze, Christopher (1977) "Legal Abortions in the United States: Rates and Ratios by Race and Age, 1972–1974," *Family Planning Perspectives* 9:12–15.

Uthman, Olalekan A. (2008) "Geographical Variations and Contextual Effects on Age of Initiation of Sexual Intercourse Among Women in Nigeria: A Multilevel and Spatial Analysis," *International Journal of Health Geographics* 7(27): no pp. www.biomedcentral.com/content/pdf/1476-072x-7-27.pdf

Ulmann, Andre, Teutsch, Georges, and Philibert, Daniel (1990) "RU 486," *Scientific American* 262(6):42–48

United Nations, Department of Economic and Social Affairs, Population Division (2011). World Contraceptive Use 2010 (POP/DB/CP/Rev2010).

Weeks, John, et al. (2004) "The Fertility Transition in Egypt: Intraurban Patterns in Cairo," *Annals of the Association of American Geographers* 94(1):74–93.

Westlake, Adam (2012, May 3) "Over 3 Million 'Parasite Singles' in Japan: Over 35, Still Living with Parents," *The Japan Daily Press.* Accessed on December 4, 2012, http://japandailypress.com/over-3-million-parasite-singles-in-japan-over-35-still-living-with-parents-031659

Yamada, Masahiro (1999) *The Age of Parasite Singles. Tokyo: Chikuma Shob̄o.*

Zhang, Guangyu and Zhao, Zhongwei (2006) "Reexamining China's Fertility Puzzle: Data Collection and Quality over the Last Two Decades," *Population and Development Review* 32(2):293–321.

Zhao, Zongwei and Xiaomu, Zhang (2010) "China's Recent Fertility Decline: Evidence from Reconstructed Fertility Statistics," *Population* 65(3): 451–478.

Chapter 7

Aassve, Arnstein and Altankhuyag, Gereltuya 2002 "Changing Patterns of Mongolian Fertility at a Time of Social and Economic Transition," *Studies in Family Planning* 33(2): 165–172.

Bertrand, Jane (2011) *USAID Graduation from Family Planning Assistance: Implications for Latin America.* Washington, DC: Population Institute and Tulane University School of Public Health and Tropical Medicine.

Butler, Steven (1998) "Japan's Baby Bust," *U.S. News and World Report* 125(13):42–44.

Billingsley, Doc M. (2005) *The Role of Family Planning Programs in the Thai Fertility Decline: An Anthropological Perspective on Successful Population Control* Accessed on December 20, 2012 from http://www.olemiss.edu/pubs/amsa/pdfs/AMSA%201_1_%20 Billingsley%20-%20Thai%20Fertility%20Decline.pdf

Blanc, Ann K., Curtis, Sian L., and Croft, Trevor N. (2002) "Monitoring Contraceptive Continuation: Links to Fertility Outcomes and Quality of Care," *Studies in Family Planning* 33(2):127–140.

Bongaarts, John (2008) *Fertility Trends in Developing Countries: Progress or Stagnation?* Working Paper No. 7, New York: Population Council www.popcouncil.org/pdfs/wp/pgy/007.pdf

Bongaarts, John and Sindig, Steven W. (2009) "A Response to Critics of Family Planning Programs," *International Perspectives on Sexual and Reproductive Health* 35(1): 39–44.

Caldwell, John C. (1994) "Fertility in Sub-Saharan Africa: Status and Prospects," *Population and Development Review* 20(1): 179–187.

Claeys, Vicky (2010) "Brave and Angry—The Creation and Development of the International Planned Parenthood Federation," *European Journal of Contraceptive and Reproductive Health Care* 15(S2): S67–S76.

David, Henry P. (1982) "Eastern Europe: Pronatalist Policies and Private Behavior," *Population Bulletin* 36(6):1–48.

Donaldson, Peter J. and Tsui, Amy Ong (1990) "The International Family Planning Movement," *Population Bulletin* 45:.

Dow, Thomas E., Jr., et al. (1994) "Wealth Flow and Fertility Decline in Kenya, 1981–92," *Population and Development Review* 20(2):343–364.

Freedman, Ronald (1990) "Family Planning Programs in the Third World," *The Annals of the American Academy of Political and Social Science* 510:33–43.

Gillespie, Duff G. and Seltzer, Judith R. (1990) "The Population Assistance Program of the U.S. Agency for International Development," in Helen M. Wallace and Kanti Giri (eds.) *World Population and U.S. Policy*. New York: W. W. Norton and Company, pp. 75–206.

Gupta, Neeru, Katende, Charles, and Bessinger, Ruth (2003) "Associations of Mass Media Exposure with Family Planning Attitudes and Practices in Uganda," *Studies in Family Planning* 34(1):19–31.

Haub, Carl (2010) *Did South Korea's Policy Work Too Well?* Washington, DC: Population Reference Bureau http://www.prb.org/Articles/2010/koreafertility.aspx

Khanna J., Van Look P.F.A., Griffin P.D. (1992) *Reproductive Health: A Key to a Brighter Future*. Geneva: WHO *Countries: Barriers, Benefits, and Policies*. Baltimore: Johns Hopkins University Press.

Le, Linh Cu, et al. (2004) "Reassessing the Level of Unintended Pregnancy and Its Correlates in Vietnam," *Studies in Family Planning* 35(1):15–26

London Summit on Family Planning (4 December 2012) *Summaries of Commitments*. Accessed December 20, 2012 at http://www.londonfamilyplanningsummit.co.uk/

Magadi, Monica and Curtis, Sian (2003) "Trends and Determinants of Contraceptive Method Choice in Kenya," *Studies in Family Planning* 34(3):149–159.

Maynard-Tucker, Gisele (2009) "HIV/AIDS and Family Planning Services Integration: Review of Prospects for a Comprehensive Approach in Sub-Saharan Africa (caps for major words)Saharan Africa," *African Journal of AIDS Research* 8(4): 465–472.

Newman, Karen and Helzner, Judith F. (1999) "IPPF Charter on Sexual and Reproductive Rights," *Journal of Women's Health and Gender-Based Medicine* 8(4): 459-463.

Nortman, Dorothy (1973) "Population and Family Planning: A Factbook," *Reports on Population/Family Planning*. New York: The Population Council.

Oshio, Takahashi (2008) "The Declining Birthrate in Japan," *Japan Economic Currents* No. 69 Washington, DC: Kezai Koho Center.

Population Reference Bureau (2012) *2012 World Population Data Sheet*. Washington, DC: Population Reference Bureau www.prb.org

Potts, Malcolm (2000) "The Unmet Need for Family Planning," *Scientific American* 282(1):88–93.

RamaRao, Saumya and Mohanam, Raji (2003) "The Quality of Family Planning Programs: Concepts, Measurements, Interventions, and Effects," *Studies in Family Planning* 34(4):227–248.

Roberts, Godfrey Ed. (1990) *Population Policy: Contemporary Issues*. New York: Praeger.

Robinson, Warren C. and Ross, John A. eds. (2007) *The Global Family Planning Revolution: Three Decades of Population Policies and Programs*. New York: World Bank Publications.

Ross, John A., et al. (1993) *Family Planning and Population: A Compendium of International Statistics*. New York: The Population Council.

Schroeder, Richard C. (1976) "Policies on Population Around the World," *Population Bulletin* 29(6):1–36.

Seltzer, Judith R. (2002) *The Origins and Evolution of Family Planning Programs in Developing Countries*. California: Rand.

Sen, Gita (2010) "Integrating Family Planning with Sexual and Reproductive Health and Rights: The Past as Prologue?" *Studies in Family Planning*. 41(2): 143–146.

Sibanda, Amson, et al. (2003) "The Proximate Determinants of the Decline to Below-Replacement Fertility in Addis Ababa, Ethiopia," *Studies in Family Planning* 34(1):1–7.

Singh, Harbans (1990) "India's High Fertility Despite Family Planning: An Appraisal," in Godfrey Roberts (ed), *Population Policy: Contemporary Issues*. New York: Praeger.

Singh, Susheela and Darroch, Jacqueline E. (2012) *Adding It Up: Costs and Benefits of Contraceptive Services—Estimates for 2012*, New York: Guttmacher Institute and United Nations Population Fund (UNFPA), 2012, http://www.guttmacher.org/pubs/AIU-2012-estimates.pdf.

Thomlinson, Ralph (1976) *Population Dynamics: Causes and Consequences of World Demographic Change*. Second Edition. New York: Random House.

United Nations (1969) "World Population Situation," Note by the Secretary General. Geneva, Switzerland: United Nations Population Commission.

United Nations (2011) *World Contraceptive Use 2001*. New York: United Nations Department of Economic and Social Affairs Population Division www.unpopulation.org

United Nations (2012) *By Choice Not By Chance: Family Planning, Human Rights and Development. State of the World Population 2012, November 14*. New York: United Nations Population Fund.

Visaria, L. and Visaria, P. (1981) "India's Population: Second and Growing," *Population Bulletin* 36(4): 1-57.

Witte, James C. and Wagner, Gert G. (1995) "Declining Fertility in East Germany After Unification: A Demographic Response to Socioeconomic Change," *Population and Development Review* 21(2):387–397.

World Bank (1991) *World Development Report 1991.* New York: Oxford University Press.

World Health Organization (2009) *WHO, Strategic Considerations for Strengthening the Linkages Between Family Planning and HIV/AIDS Policies, Programs, and Services* Geneva: World Health Organization.

Chapter 8

African Economic Outlook (2013) "Remittances," *African Economic Outlook*, http://www.africaneconomicoutlook.org/en/outlook/financial_flows/remittances/ (Accessed January 4, 2013).

Alba, R., Schmidt, P., and Wasmer, M. eds. (2003) *Germans or Foreigners? Attitudes Toward Ethnic Minorities in Post-Reunification Germany.* New York: Palgrave-Macmillan.

Alba, Richard D., et al. (1995) "Neighborhood Change under Conditions of Mass Immigration: The New York City Region, 1970–1990," *International Migration Review* 24(3): 625–656.

Asiedu, Alex (2005) "Some Benefits of Migrants' Return Visits to Ghana," *Population, Space and Place* 11:1–11.

Bala, Raj (1986) *Trends in Urbanisation in India, 1901–1981.* Jaipur, India: Rawat Publication.

Bartlett, Donald L. and Steele, James B. (2004) "Who Left the Door Open," *Time* 164(12):51–66.

Bean, Frank D., Edmonston, Barry, and Passel, Jeffrey S. eds. (1990) *Undocumented Migration to the United States: IRCA and the Experience of the 1980s.* Santa Monica, CA: Rand Corporation.

Bean, Frank, et al. (2004) "Immigration and Fading Color Lines in America," http://www.prb.org/Template.cfm?Section5 PRB&template5/Content/ContentGroups/04_Ar . . . (Accessed August 4, 2004).

Beaujeu-Garnier, J. (1966) *Geography of Population.* New York: St. Martin's Press.

Beier, George J. (1976) "Can Third World Cities Cope?" *Population Bulletin* 31(4):1–36.

Berry, Brian J. L. (1973) *The Human Consequences of Urbanization.* New York: St. Martin's Press.

Berry, Brian J. L. (1980) "Urbanization and Counterurbanization in the United States," *Annals of the American Academy of Political and Social Sciences* 451:13–20.

Berry, Brian J. L. (1990) "Urban Systems by the Third Millennium: A Second Look," *Journal of Geography* 81(3):98–101.

Berry, Brian J. L. (1993) "Transnational Urbanward Migration, 1830–1980," *Annals of the Association of American Geographers* 83(3):389–405.

Billari, Francesco C. and Liefbroer, Aart C. (2007) "Should I stay or Should I Go? The Impact of Age Norms on Leaving Home," *Demography* 44(1):181–198.

Black, Richard (1991) "Refugees and Displaced Persons: Geographical Perspectives and Research Directions," *Progress in Human Geography* 15:281–298.

Black, Richard, et al. (2006) "Routes to Illegal Residence: A Case Study of Immigration Detainees in the United Kingdom," *Geoforum* 37:552–564.

Bloom, Stephen G. (2006) "The New Pioneers," *Wilson Quarterly* XXX(3):60–68.

Bogue, Donald J. (1969) *Principles of Demography.* New York: John Wiley and Sons, Inc.

Bolan, Marc (1997) "The Mobility Experience and Neighborhood Attachment," *Demography* 34(2):225–237.

Bongaarts, John (2004) "Population Aging and the Rising Cost of Public Pensions," *Population and Development Review* 30(1): 1–23.

Boswell, Christina (2007) "Theorizing Migration Policy: Is There a Third Way?" *International Migration Review* 41(1):75–100.

Bouvier, Leon F., Shryock, Henry S., and Henderson, Harry W. (1977) "International Migration: Yesterday, Today, and Tomorrow," *Population Bulletin* 32(4):1–42.

Brockerhoff, Martin P. (2000) "An Urbanizing World," *Population Bulletin* 55(3):1–44.

Bruegmann, Robert (2005) *Sprawl: A Compact History*. Chicago: University of Chicago Press.

Calavita, Kitty (2006) "Gender, Migration, and Law: Crossing Borders and Bridging Disciplines," *International Migration Review* 40(1):104–132.

Carletto, Calogero, et al. (2006) "A Country on the Move: International Migration in Post-Communist Albania," *International Migration Review* 40(4):767–785.

Carter, Sean (2005) "The Geopolitics of Diaspora," *Area* 37(1): 54–63.

Castles, Stephen (2006) "Guestworkers in Europe: A Resurrection?" *International Migration Review* 40(4):741–766.

Castles, Stephen and Miller, Mark J. (1993) *The Age of Migration: International Population Movements in the Modern World*. New York: The Guilford Press.

Cavounidis, Jennifer (2006) "Labor Market Impact of Migration: Employment Structures and the Case of Greece," *International Migration Review* 40(3):635–660.

Clark, W. A. V. (1986) *Human Migration*. Beverly Hills: Sage Publications.

Clark, William A. V. and Huang, Youqin (2006) "Balancing Move and Work: Women's Labour Market Exits and Entries after Family Migration," *Population, Space and Place* 12:31–44.

Cohen, Jeffrey H. and Rodriguez, Leila (2005) "Remittance Outcomes in Rural Oaxaca, Mexico: Challenges, Options and Opportunities for Migrant Households," *Population, Space and Place* 11:49–63.

Collyer, Michael (2006) "The Search for Solutions: Achievements and Challenges," *International Migration Review* 40(2): 451–459.

Cooke, Thomas J. (2005) "Migration of Same-Sex Couples," *Population, Space and Place* 11:401–409.

Costello, Lauren (2007) "Going Bush: The Implications of Urban-Rural Migration," *Geographical Research* 45(1):85–94.

De Haas, Hein (2006) "Migration, Remittances and Regional Development in Southern Morocco," *Geoforum* 37:565–580.

DeBardeleben, Joan, ed. (2005) *Soft or Hard Borders? Managing the Divide in an Enlarged Europe*. Burlington, VT: Ashgate Publishing.

Desbarats, Jacqueline (1985) "Indochinese Resettlement in the United States," *Annals of the Association of American Geographers* 75(4):522–538.

DeVoretz, Don J. (2006) "Immigration Policy: Methods of Economic Assessment," *International Migration Review* 40(2): 390–418.

Donato, Katharine M., et al. (2006) "A Glass Half Full? Gender in Migration Studies," *International Migration Review* 40(1): 3–26.

Eberstadt, Nicholas (2004) "Power and Population in Asia," http: //www.policyreview.org/feb04/eberstadt.html (Accessed November 9, 2004).

Ellis, Mark and Goodwin-White, Jamie (2006) "1.5 Generation Internal Migration in the U.S.: Dispersion from States of Immigration?" *International Migration Review* 40(4):899–926.

Epstein, Gil S. and Nitzan, Shmuel (2006) "The Struggle Over Migration Policy," *Journal of Population Economics* 19(4): 703–723.

Everitt, John (1991) "Refugees on Mainland Middle America," *The Canadian Geographer* 35(2):194–195.

Exline, Christopher H., Peters, Gary L., and Larkin, Robert P. (1982) *The City: Patterns and Processes in the Urban Ecosystem*. Boulder, CO: Westview Press.

Fairchild, Henry P. (1925) *Immigration*. New York: The Macmillan Company.

Fitzgerald, David (2006) "Inside the Sending State: The Politics of Mexican Emigration Control," *International Migration Review* 40(2):259–293.

Folger, John K. and Nam, Charles B. (1967) *Education of the American Population*. A 1960 Census Monograph. Washington, D.C.: United States Government Printing Office.

Frey, William H. (1990) "Metropolitan America: Beyond the Transition," *Population Bulletin* 45(2):1–51.

Fuguitt, Glenn V. and Zuiches, James J. (1975) "Residential Preferences and Population Distribution," *Demography* 12:491–504.

Gallent, Nick (2007) "Second Homes, Community and a Hierarchy of Dwelling," *Area* 39(1):97–106.

Gao, Jia (2006) "Organized International Asylum-Seeker Networks: Formation and Utilization by Chinese Students," *International Migration Review* 40(2):294–317.

Garcia, Angel Solano (2006) "Does Illegal Immigration Empower Rightist Parties?" *Journal of Population Economics* 19(4): 649–670.

Garreau, Joel (1991) *Edge City: Life on the New Frontier*. New York: Doubleday. Gat, Azar (2007) "The Return of Authoritarian Great Powers," *Foreign Affairs* 86(4):59–70.

Gat, Azar (2007) "The Return of Authoritarian Great Powers," *Foreign Affairs*, Jul–Aug 2007, 86(4): 59-69.

Geddes, Andrew (2003) *The Politics of Migration and Immigration in Europe*. London: Sage Publications.

Gerber, Theodore P. (2006) "Regional Economic Performance and Net Migration Rates in Russia, 1993–2002," *International Migration Review* 40(3):661–697.

Germain, Kate Saint and Stevens, Carly J. (1996) "1996 Illegal Immigration Reform & Immigrant Responsibility Act," US immigration legislation online, http://library.uwb.edu/guides/usimmigration/1996_illegal_immigration_reform_and_immigrant_responsibility_act.html (Accessed January 30, 2013).

Ghandnoosh, Nazgol and Waldinger, Roger (2006) "Strangeness at the Gates: The Peculiar Politics of American Immigration," *International Migration Review* 40(3):719–734.

Gibson, Campbell (1975) "The Contribution of Immigration to United States Population Growth: 1790–1970," *International Migration Review* 9:157–177.

Giddens, Anthony (1982) *Profiles and Critiques in Social Theory*. Berkeley: University of California Press.

Giddens, Anthony (1984) *The Constitution of Society: Outline of the Theory of Structuration*. Cambridge: Polity Press.

Glasser, Jeff (2001) "A Broken Heartland," *U.S. News and World Report* (May 7):16–22.

Glytsos, Nicholas P. (2005) "Stepping from Illegality to Legality and Advancing towards Integration: The Case of Immigrants in Greece," *International Migration Review* 39(4): 819–840.

Goldscheider, Calvin (1971) *Population, Modernization, and Social Structure*. Boston: Little, Brown and Company.

Goldstein, Joshua, Lutz, Wolfgang, and Scherbov, Sergei (2003) "Long-term Population Decline in Europe: The Relative Importance of Tempo Effects and Generational Change," *Population and Development Review* 29(4):699–707.

Gordon, Linda W. (2005) "Trends in the Gender Ratio of Immigrants to the United States," *International Migration Review* 39(4):796–818.

Goss, Jon and Lindquist, Bruce (1995) "Conceptualizing International Labor Migration: A Structuration Perspective," *International Migration Review* 22(2):317–351.

Gray, Paul (1994) "Looking for Work? Try the World," *Time* (Sept. 19):44–46.

Greenwood, Michael J. (1975) "Research on Internal Migration in the United States: A Survey," *Journal of Economic Literature* 12:397–433

Grieco, Elizabeth M., Acosta, Yesenia D., de la Cruz, G. Patricia, Gambino, Christine, Gryn, Thomas, Larsen, Luke J., Trevelyan, Edward N. and Walters, Nathan P. (2012) "The Foreign-Born Population in the United States: 2010, ACS-19," Washington, DC: U.S. Department of Commerce, Economics and Statistics Administration, U.S. Census Bureau.

Greenhill, Jim (2010) "National Guard troops to deploy to U.S.-Mexico Border," U.S. Army, July 19, 2010, http://www.army.mil/article/42515/national-guard-troops-to-deploy-to-us-mexico-border/ (Accessed January 2, 2013).

Gugler, Josef Ed (1988) *The Urbanization of the Third World*. Oxford: Oxford University Press. Haggett, Peter (1979) *Geography: A Modern Synthesis*. Third Edition. New York: Harper and Row, Publishers.

Halfacree, Keith (2004) "A Utopian Imagination in Migration's *Terra Incognita*? Acknowledging the Non-Economic Worlds of Migration Decision-Making," *Population, Space and Place* 10(3):239–253.

Halfacree, Keith H. and Boyle, Paul J. (1993) "The Challenge Facing Migration Research: The Case for a Biographical Approach," *Progress in Human Geography* 17(3):333–348.

Hall, Michael and Müller, Dieter K. eds. (2004) *Tourism, Mobility and Second Homes: Between Elite Landscape and Common Ground*. Clevedon: Channel View Publications.

Hart, John Fraser (1991) "The Perimetropolitan Bow Wave," *The Geographical Review* 81(1):35–51.

Haubert, Jeannie and Fussell, Elizabeth (2006) "Explaining Pro-Immigrant Sentiment in the U.S.: Social Class, Cosmopolitanism, and Perceptions of Immigrants," *International Migration Review* 40(3):489–507.

History Channel (2013) "U.S. Immigration Since 1965," http://www.history.com/topics/us-immigration-since-1965 (Accessed January 29, 2013).

Hix, Simon and Noury, Abdul (2007) "Politics, not Economic Interests: Determinants of Migration Policies in the European Union," *International Migration Review* 41(1):182–205.

Hochschild, Arlie Russell (2000) "The Nanny Chain," *The American Prospect* 11(4):32–36.

Howell, Clark F. (1965) *Early Man*. New York: Time-Life Books.

Hugo, Graeme J. (2006) "Immigration Responses to Global Change in Asia: A Review," *Geographical Research* 44(2):155–172.

Huntington, Samuel P. (2004) *Who Are We? The Challenges to America's National Identity*. New York: Simon and Schuster.

International Fund for Agricultural Development - IFAD (2013) "Africa," http://www.ifad.org/remittances/maps/africa.htm (Accessed January 4, 2013).

Jacoby, Tamar (2006) "Immigration Nation," *Foreign Affairs* 85(6):50–66.

Johnson, Hans P. (2006) *Illegal Immigration*. San Francisco: Public Policy Institute of California.

Johnson, Richard (2004) "Economic Policy Implications of World Demographic Change," *Economic Review (Federal Reserve Bank of Kansas City)* Q(I):39–64.

Keddie, Philip D. and Joseph, Alun E. (1991) "The Turnaround of the Turnaround? Rural Population Change in Canada, 1976 to 1986," *The Canadian Geographer* 35:367–379.

Kennedy, Edward M. (1981) "Refugee Act of 1980," *International Migration Review* 15 (1/2), Refugees Today (Spring–Summer, 1981): 141–156.

Kenzer, Martin S. (1991) "The African Refugee Crisis in Context," *The Canadian Geographer* 35(2):197–200.

Kliot, N. (1987) "The Era of Homeless Man," *Geography* 72: 109–121.

Kotkin, Joel (2005) *The City: A Global History*. New York: The Modern Library.

Lee, Everett S. (1966) "A Theory of Migration," *Demography* 3: 47–57.

Legal Information Institute (2010) "Illegal Immigration Reform and Immigration Responsibility Act," http://www.law.cornell.edu/wex/illegal_immigration_reform_and_immigration_responsibility_act (Accessed January 31, 2013).

Leiden, Warren R. and Neal, David L. (1990) "Highlights of the U.S. Immigration Act of 1990," *Fordham International Law Journal* 14 (1) 1990, Article 14: 328–339.

Leinbach, Thomas R., et al. (1992) "Employment Behavior and the Family in Indonesian Transmigration," *Annals of the Association of American Geographers* 82(1):23–47.

Lewis, Peirce (1985) "Beyond Description," *Annals of the Association of American Geographers* 75:465–477.

Lindgren, Urban (2003) "Who is the Counter-Urban Mover? Evidence from the Swedish Urban System," *International Journal of Population Geography* 9(5):399–418.

Los Angeles Homeless Services Authority (2011) *2011 Greater Los Angeles Homeless Count Report: Including Detailed Geography Reports*, Los Angeles: Los Angeles Homeless Services Authority, http://www.lahsa.org/docs/2011-Homeless-Count/HC11-Detailed-Geography-Report-FINAL.PDF (Accessed January 25, 2013).

Lucas, David, Amoateng, Acheampong Yaw, and Kalule-Sabiti, Ishmael (2006) "International Migration and the Rainbow Nation," *Population, Space and Place* 12:45–63.

Luciuk, Lubomyr Y. (1991) "A Landscape of Despair: Comments on the Geography of the Contemporary Afghan Refugee Situation," *The Canadian Geographer* 35(2): 195–197.

Ludden, Jennifer (2006) "1965 Immigration Law Changed Face of America," NPR, May 09, 2006, http://www.npr.org/templates/story/story.php?storyId=5391395 (Accessed January 29, 2013).

Mahler, Sarah J. and Pessar, Patricia R. (2006) "Gender Matters: Ethnographers Bring Gener from the Periphery Toward the Core of Migration Studies," *International Migration Review* 40(1):27–63.

Martin, George (2007) "Global Motorization, Social Ecology and China," *Area* 39(1):66–73.

Martin, Philip and Midgley, Elizabeth (2006) "Immigration: Shaping and Reshaping America, Revised and Updated 2nd Edition," *Population Bulletin* 61(4):1–28.

Martin, Philip and Widgren, Jonas (2002) "International Migration: Facing the Challenge," *Population Bulletin* 57(1):1–40.

Massey, Douglas S. (2006) "America's Never-Ending Debate: A Review Essay," *Population and Development Review* 32(3): 573–584.

McHugh, Kevin E. (1989) "Hispanic Migration and Population Redistribution in the United States," *Professional Geographer* 41(4):429–439.

McNeill, J. R. and Mcneill, William H. (2003) *The Human Web: A Bird's-Eye View of World History*. New York: W. W. Norton and Company.

Meenan, J. F. (1958) "Eire," in Brinley Thomas, ed. *Economics of International Migration*. London: Macmillan.

Millar, Jane and Salt, John (2007) "In Whose Interests? IT Migration in an Interconnected World Economy," *Population, Space and Place* 13:41–58.

Mirdal, Gretty M. (2006) "Stress and Distress in Migration: Twenty Years After," *International Migration Review* 40(2):375–389.

Moore, Eric G. (1972) *Residential Mobility in the City*. Resource Paper No. 13. Washington, D.C.: Association of American Geographers.

Mora, Marie T. (2006) "Self-Employed Mexican Immigrants Residing along the U.S.-Mexico Border: The Earnings Effect of Working in the U.S. versus Mexico," *International Migration Review* 40(4):885–898.

National Underground Railroad Freedom Center (2004-2012) "Invisible: Slavery Today," The National Underground Railroad Freedom Center, http://www.freedomcenter.org/slavery-today/ (Accessed January 2, 2013).

Nelson, P. (1959) "Migration, Real Income, and Information," *Journal of Regional Science* 1:43–74.

Niedomysl, Thomas (2005) "Tourism and Interregional Migration in Sweden: An Explorative Approach," *Population, Space and Place* 11:187–204.

Norris, Kathleen (1993) *Dakota: A Spiritual Geography*. New York: Houghton Mifflin Company.

Northam, Ray M. (1975) *Urban Geography*. New York: John Wiley and Sons, Inc.

Oberhauser, Ann M. (1991) "The International Mobility of Labor: North African Migrant Workers in France," *Professional Geographer* 43(4):431–445.

Oberlander, Hans (1992) "Mord an der Seele," *Stern* 14(26.Marz): 20–32.

OECD (2006). International Migrant Remittances and their Role in Development. International *Migration Outlook*, SOPEMI 2006 Edition, http://www.oecd.org/els/internationalmigrationpoliciesanddata/38840502.pdf (Accessed January 3, 2013).

OECD (2011), "Detailed aid statistics: Official and private flows", OECD International Development Statistics (database), doi: 10.1787/data-00072-en (Accessed on 05 January 2013).

Otiso, Kefa M. (2003) "State, voluntary and private sector partnerships for slum upgrading and basic service delivery in Nairobi City, Kenya" *Cities* 20 (4): 221–229.

Packer, George (2006) "The Megacity: Decoding the Chasos of Lagos," *The New Yorker* (Nov. 13):62–75.

Parfit, Michael (1998) "Human Migration," *National Geographic* (4):6–35. Parnwell, Mike. (1993) *Population Movements and the Third World*. London: Routledge.

Petersen, William (1958) "A General Typology of Migration," *American Sociological Review* 23:256–265.

Petersen, William (1975) *Population*. Third Edition. London: The Macmillan Company.

Phizacklea, A. ed. (1983) *One Way Ticket? Migration and Female Labor*. London: Routledge and Kegan Paul.

Piper, Nicola (2006) "Gendering the Politics of Migration," *International Migration Review* 40(1):133–164.

Plaza, Sonia and Ratha, Dilip (2011) "Harnessing Diaspora Resources for Africa," in Sonia Plaza and Dilip Ratha (eds), *Diaspora for Development in Africa*, Washington DC: The International Bank for Reconstruction and Development / The World Bank.

Population Reference Bureau (2013) "Urban Population to Become the New Majority Worldwide," Washington, DC: Population Reference Bureau, http://www.prb.org/Articles/2007/UrbanPopToBecomeMajority.aspx (Accessed January 24, 2013).

Portes, Alejandro, Escobar, Cristina, and Radford, Alexandria Walton (2007) "Immigrant Transnational Organizations and Development: A Comparative Study," *International Migration Review* 41(1):242–281.

Potts, Deborah (2006) "Rural Mobility as a Response to Land Shortages: The Case of Malawi," *Population, Space and Place* 12:291–311.

Ravenstein, Edward G. (1889) "The Laws of Migration," *Journal of the Royal Statistical Society* 52:241–305.

Raymer, James, Bonaguidi, Alberto, and Valentini, Alessandro (2006) "Describing and Projecting the Age and Spatial Structures of Interregional Migration in Italy," *Population, Space and Place* 12:371–388.

Refugee Council USA (2013) "Asylum and Detention," http://www.rcusa.org/index.php?page=asylum-and-detention (Accessed January 30, 2013).

Reimers, David M. (2005) *Other Immigrants: The Global Origins of the American People.* New York: New York University Press.

Richburg, Keith B. (2004) "No Longer Just Nordic: Growing Role of Immigrant Communities is Changing What it Means to be a Swede," http://www.washingtonpost.com/ac2/wp-dyn/ A52672–2004Oct21 (Accessed October 21, 2004).

Ritchey, P. Neal (1976) "Explanations of Migration," *Annual Review of Sociology* 2:363–404.

Robinson, Isaac (1986) "Blacks Move Back to the South," *American Demographics* 8(6): 40–43.

Rogge, John R. (1991) "Refugees in Southeast Asia: An Uncertain Future?" *The Canadian Geographer* 35(2):190–193.

Roseman, Curtis C. (1971) "Migration as a Spatial and Temporal Process," *Annals of the Association of American Geographers* 61:589–598.

Roseman, Curtis C. (1977) *Changing Migration Patterns Within the United States.* Resource Papers for College Geography No. 77–2. Washington, D.C.: Association of American Geographers.

Rosholm, Michael, Scott, Kirk, and Husted, Leif (2006) "The Times They Are A-Changin': Declining Immigrant Employment Opportunities in Scandinavia," *International Migration Review* 40(2):318–347.

Rossi, Peter (1955) *Why Families Move.* Glencoe, IL: Free Press.

Rudzitis, Gundars (1991) "Migration, Sense of Place, and Nonmetropolitan Vitality," *Urban Geography* 12(1):80–88.

Sachs, Aaron (1994) "The Last Commodity: Child Prostitution in the Developing World," *World Watch* (July/August):24–30. Salt, John (1985) "Europe's Foreign Labour Migrants in Transition," *Geography* 70:151–158.

Scheer, Robert (1998) "The Dark Side of the New World Order," *Los Angeles Times* (Jan. 13):B7.

Schou, Poul (2006) "Immigration, Integration and Fiscal Sustain-ability," *Journal of Population Economics* 19(4):671–689.

Schwind, Paul J. (1971) "Spatial Preferences of Migrants for Regions: The Example of Maine," *Proceedings of the Association of American Geographers* 3:150–156.

Shaw, R. Paul (1975) *Migration Theory and Fact*. Philadelphia: Regional Science Research Institute.

Sijuwade, Philip O. (2010) "The Economic Implications of Rapid Urban Growth in the Third World Countries," *Anthropologist* 12 (2): 79–85.

Silvey, Rachel (2004) "On the Boundaries of a Subfield: Social Theory's Incorporation in Population Geography," *Population, Space and Place* 10(4):303–308.

Silvey, Rachel (2006) "Geographies of Gender and Migration: Spatializing Social Difference," *International Migration Review* 40(1):64–81.

Simendinger, Alexis (2013) "Obama Presses Congress on Immigration Reform," Real Clear Politics, January 30, 2013, http://www.realclearpolitics.com/articles/2013/01/30/obama_presses_congress_on_immigration_reform_116850.html (Accessed January 30, 2013).

Simkins, Paul D. (1970) "Migration as a Response to Population Pressure: The Case of the Philippines," in Wilbur Zelinsky, Leszek A. Kosinski, and Mansell R. Prothero, eds. *Geography and a Crowding World*. New York: Oxford University Press, pp. 259–267.

Simkins, Paul D. (1978) "Characteristics of Population in the United States and Canada," in Glenn T. Trewartha, ed. *The More Developed Realm: A Geography of Its Population*. Oxford: Pergamon Press, pp. 189–220.

Sinke, Suzanne M. (2006) "Gender and Migration: Historical Perspectives," *International Migration Review* 40(1):82–103.

Sjaastad, L. A. (1962) "The Costs and Returns of Human Migration," *The Journal of Political Economy*. 70:80–93.

Skeldon, Ronald (2006) "Interlinkages Between Internal and International Migration and Development in the Asian Region," *Population, Space and Place* 12:15–30.

Smith, Marian L. (2009) "Overview of INS History to 1998," http://www.uscis.gov/portal/site/uscis/menuitem.5af9bb95919f35e66f614176543f6d1a/?vgnextoid=b7294b0738f70110VgnVCM1000000ecd190aRCRD&vgnextchannel=bc9cc9b1b49ea110VgnVCM1000004718190aRCRD (Accessed January 29, 2013).

Stockdale, Aileen (2006) "The Role of a 'Retirement Transition' in the Repopulation of Rural Areas," *Population, Space and Place* 12:1–13.

Streisand, Betsy (2006) "The City of Angels Struggles to Deal with a Devil of a Place," *U. S. News and World Report* 141(23): 50–54.

Taylor, Matthew J., Moran-Taylor, Michelle J., and Ruiz, Debra Rodman (2006) "Land, Ethnic, and Gender Change: Transnational Migration and its Effects on Guatemalan Lives and Landscapes," *Geoforum* 37:41–61.

Thomas, G. Scott (2012) "Minorities form racial majority in 106 U.S. cities," *The Business Journal*, Feb 27, 2012, http://www.bizjournals.com/bizjournals/on-numbers/scott-thomas/2012/02/minorities-form-racial-majority-in-106.html (Accessed January 28, 2013). Republished with permission of American City Business Journals; permission conveyed through Copyright Clearance Center.

Thomlinson, Ralph (1976) *Population Dynamics: Causes and Consequences of World Demographic Change*. Second Edition. New York: Random House.

Trewartha, Glenn T. (1969) *A Geography of Population*. New York: John Wiley and Sons, Inc.

Tyler, Charles (1991) "The World's Manacled Millions," *Geographical Magazine* 58(1):30–35.

Tyner, James A. (1996) "Constructions of Filipina Migrant Entertainers," *Gender, Place and Culture* 3(1):77–93.

Tyner, James A. (1998) "Asian Labor Recruitment and the World Wide Web," *The Professional Geographer* 50(3):331–344.

United Nations (1991) *World Urbanization Prospects.* New York: United Nations.

United Nations (2012) *World Urbanization Prospects: The 2011 Revision Highlights,* ESA/P/WP/224, March 2012. New York: Department of Economic and Social Affairs, Population Division, United Nations.

United Nations High Commission for Refugees – UNCHR (2011) *UNHCR Statistical Yearbook 2010: Ten Years of Statistics,* 10th edition, Geneva: UNCHR.

U.S. Census Bureau (2012) "Growth in Urban Population Outpaces Rest of Nation, Census Bureau Reports," http://www.census.gov/newsroom/releases/archives/2010_census/cb12-50.html (Accessed January 28, 2013).

U.S. Census Bureau (2013, January 10) "Huron (city), California," http://quickfacts.census.gov/qfd/states/06/0636084.html (Accessed January 31, 2013).

U.S. Citizenship and Immigration Services – USCIS (2012) "E-Verify," http://www.uscis.gov/portal/site/uscis/menuitem.eb1d4c2a3e5b9ac89243c6a7543f6d1a/?vgnextoid=75bce2e261405110VgnVCM1000004718190aRCRD&vgnextchannel=75bce2e261405110VgnVCM1000004718190aRCRD (Accessed January 31, 2013).

U.S. Citizenship and Immigration Services – USCIS (2013a) "Immigration and Nationality Act," http://www.uscis.gov/portal/site/uscis/menuitem.eb1d4c2a3e5b9ac89243c6a7543f6d1a/?vgnextoid=f3829c7755cb9010VgnVCM10000045f3d6a1RCRD&vgnextchannel=f3829c7755cb9010VgnVCM10000045f3d6a1RCRD (Accessed January 29, 2013).

U.S. Citizenship and Immigration Services – USCIS (2013b) "Public Laws Amending the INA," http://www.uscis.gov/portal/site/uscis/menuitem.eb1d4c2a3e5b9ac89243c6a7543f6d1a/?vgnextoid=0c829c7755cb9010VgnVCM10000045f3d6a1RCRD&vgnextchannel=0c829c7755cb9010VgnVCM10000045f3d6a1RCRD (Accessed January 29, 2013).

U.S. Citizenship and Immigration Services – USCIS (2013c) "Immigration Reform and Control Act of 1986 (IRCA)," http://www.uscis.gov/portal/site/uscis/menuitem.5af9bb95919f35e66f614176543f6d1a/?vgnextchannel=b328194d3e88d010VgnVCM10000048f3d6a1RCRD&vgnextoid=04a295c4f635f010VgnVCM1000000ecd190aRCRD (Accessed January 29, 2013).

U.S. Citizenship and Immigration Services – USCIS (2013d) "Immigration Act of 1990," http://www.uscis.gov/portal/site/uscis/menuitem.5af9bb95919f35e66f614176543f6d1a/?vgnextoid=84ff95c4f635f010VgnVCM1000000ecd190aRCRD&vgnextchannel=b328194d3e88d010VgnVCM10000048f3d6a1RCRD (Accessed January 29, 2013).

United States Department of Homeland Security (2012a) "History," http://www.dhs.gov/history (Accessed January 3, 2013).

United States Department of Homeland Security (2012b) *Yearbook of Immigration Statistics: 2011.* Washington, D.C.: U.S. Department of Homeland Security, Office of Immigration Statistics

United States Department of Labor (1977) Employment and Training Administration. *Why Families Move: A Model of the Geographic Mobility of Married Couples.* R and D Monograph 48. Washington, D.C.: United States Government Printing Office.

Vernez, Georges and Ronfeldt, David (1991) "The Current Situation in Mexican Immigration," *Science* 251 (4998): 1189–1193.

Werner, Carrie A. (2011) "The Older Population: 2010," 2010 Census Briefs - 2010BR-09, US Census Bureau, http://www.census.gov/prod/cen2010/briefs/c2010br-09.pdf (Accessed January 4, 2013).

West, Darrell (2010) *Brain Gain: Rethinking U.S. Immigration Policy*. Washington, DC: Brookings Institution Press.

Williams, Allan M. (2006) "Lost in Translation? International Migration, Learning and Knowledge," *Progress in Human Geography* 30(5):588–607.

Williamson, Celia, Perdue, Tasha, Belton, Lisa and Burns, Olivia (2012) "Domestic Sex Trafficking in Ohio," Ohio Human Trafficking Commission, August, 8 2012, Columbus, Ohio.

Withers, S. D. (1997) "Methodological Considerations in the Analysis of Residential Mobility: A Test of Duration, State Dependence, and Associated Events," *Geographical Analysis* 29(4):354–374.

Withers, Suzanne Davies and Clark, William A. V. (2006) "Housing Costs and the Geography of Family Migration Outcomes," *Population, Space and Place* 12:273–289.

Wolpert, Julian (1965) "Behavioral Aspects of the Decision to Migrate," *Papers and Proceedings of the Regional Science Association* 15:159–169.

Wolpert, Julian (1966) "Migration as an Adjustment to Environmental Stress," *Journal of Social Issues* 22:92–102.

Zelinsky, Wilbur (1971) "The Hypothesis of the Mobility Transition," *The Geographical Review* 61:219–249.

Zimmerman, Klaus F. ed. (2005) *European Migration: What do We Know*? New York: Oxford University Press.

Zolberg, Aristide R. (2006) *A Nation by Design: Immigration Policy in the Fashioning of America*. New York: Russell Sage Foundation.

Zolberg, Aristide R., Suhrke, Astri, and Aguayo, Sergio (1989) *Escape from Violence: Conflict and the Refugee Crisis in the World*. Oxford: Oxford University Press.

Chapter 9

Abernethy, Virginia (1992) "Editorial: Lessons from the Past," *Population and Environment* 13(4):235–236.

Ackerman, Edward A. (1967) "Population, Natural Resources, and Technology," *Annals of the American Academy of Political and Social Science* 369:84–97.

American Lung Association (2013) *State of the Air 2013* Washington,D.C.: American Lung Association. www.lung.org/about-us/our-impact/top-stories/state-of-the-air-much-progress-cut-challenges.html.

Barney, Gerald O. Ed. (1980) *The Global 2000 Report to the President of the U.S.* Vol. 1, New York: Pergamon Press.

Barrett, Paul M. (2012) "It's Global Warming, Stupid," *Bloomberg Businessweek*, November 01, 2012,http://www.businessweek.com/articles/2012-11-01/its-global-warming-stupid (Accessed December 20, 2012).

Blaustein, Andrew R. and Wake, David B. (1995) "The Puzzle of Declining Amphibian Populations," *Scientific American* 272(10):52–57.

Boulding, Kenneth E. (1966) *Human Values on the Spaceship Earth*. New York: National Council of Churches.

Brown, Corie (2007) "A Scorching Future," *Los Angeles Times* (Jan. 24):F1 and F6.

Brown, Lester R., McGrath, Patricia L., and Stokes, Bruce (1976) *Twenty-Two Dimensions of the Population Problem*. World-watch Paper No. 5, New York: Worldwatch Institute.

Calhoun, John B. (1962) "Population Density and Social Pathology," *Scientific American* 206:139–146.

Chipman, Kim and Morales, Alex. (2012) "Climate Treaty Hinges on Obama Making Case, Ex-Aides Say," *Bloomberg*, December 10, 2012.

CO2Now (2012), "Atmospheric CO2 for November 2012," www. http://co2now.org/ (Accessed December 20, 2012).

Cole, H. S. D., Freeman, Christopher, Jahoda, Marie, and Pavitt, K.L. R. Eds. (1973) *Models of Doom: A Critique of the Limits to Growth*. New York: Universe Books.

Collins, Charles O. and Scott, Steven L. (1993) "Air Pollution in the Valley of Mexico," *The Geographical Review* 83(2): 119–133.

Commoner, Barry (1990) *Making Peace With The Planet*. New York: Pantheon Books.

Commoner, Barry, Corr, Michael, and Stamler, Paul J. (1971) "The Causes of Pollution," *Environment* 13:2–19.

Cuñat, Javier (2012) "Africa Means Business: Opportunities in Frontier Markets," *INSEAD Knowledge*, October 4, 2012, http://knowledge.insead.edu/world/emerging-markets/africa-means-business-opportunities-in-frontier-markets-2288 (Accessed December 21, 2012).

Davis, Kingsley (1991) "Population and Resources: Fact and Interpretation," in Kingsley Davis and Mikhail S. Bernstam (ed), *Resources, Environment, and Population*. New York: Oxford University Press, pp. 1–25. From *Resources, Environment, and Population: Present Knowledge, Future Options* by Davis and Bernstam (Eds.). Copyright © 1991 Population Council. Reprinted by permission.

de Sherbinin, Alex and Kalish, Susan (1994) "Population-Environment Links: Crucial, but Unwieldy," *Population Today* 22(1): 1–2.

Diamond, Jared (2005) *Collapse: How Societies Choose to Fail or Succeed*. New York: Viking.

Durning, Alan (1991) "Cradles of Life," in Lester R. Brown (ed), *The Worldwatch Reader on Global Environmental Issues*. New York: W. W. Norton, pp. 167–188.

Eckholm, Erik P. (1976) *Losing Ground: Environmental Stress and World Food Prospects*. New York: W. W. Norton and Co.

Ehrlich, Paul R. and Ehrlich, Anne H. (1990) *The Population Explosion*. New York: Simon and Schuster.

Ehrlich, Paul R. and Holdren, John P. (1971) "Impact of Population Growth," *Science* 171:1212–1217.

Eilperin, Juliet (2012a) "EPA issues new fuel-efficiency standard; Autos must average 54.5 mpg by 2025," *The Washington Post*, Tuesday, August 28, 2012, http://articles.washingtonpost.com/2012-08-28/national/35490347_1_fuel-efficiency-fuel-standards-vehicle-fuel-efficiency-standards (Accessed December 20, 2012). © 2012 *The Washington Post*. All rights reserved. Used by permission and protected by the Copyright Laws of the United States. The printing, copying, redistribution, or retransmission of the material without express written permission is prohibited.

Eilperin, Juliet (2012b) "A crystal clear agenda for a clean environment," *The Washington Post*, Monday, October 22, 2012, A-Section, Pg. A07.

Espenshade, Thomas J. (1991) "Book Review of The Population Explosion," in *Population and Development Review* 17(2): 331–339.

Fagan, Brian (2000) *The Little Ice Age: How Climate Made History 1300–1850*. New York: Basic Books.

Flavin, Christopher (1991) "The Heat is On," in Lester R. Brown (ed) *The Worldwatch Reader on Global Environmental Issues*. New York: W. W. Norton, pp. 75–96.

Fleming, James R. (2007) "The Climate Engineers," *The Wilson Quarterly*, Spring 2007: 45-60.

Freedman, Jonathan L. (1975) *Crowding and Behavior*. San Francisco: W. H. Freeman and Co.

Gelbspan, Ross (2004) *Boiling Point: How Politicians, Big Oil and Coal, Journalists, and Activists Have Fueled the Climate Crisis—and What We Can Do to Avert Disaster*. New York: Basic Books.

Geping, Qu and Jinchang, Li (1994) *Population and the Environment in China*. Boulder: Lynne Rienner Publishers.

Goudie, Andrew (1986) *The Human Impact on the National Environment*, Second Edition. Cambridge, MA.: The MIT Press.

Greenstone, Michael and Looney, Adam. (2012) "Paying Too Much for Energy? True Costs of Our Energy Choices," *Daedelus* 141(2): 10-30.

Hall, Edward T. (1969) *The Hidden Dimension*. Garden City, NJ: Anchor Books.

Hanink, Dean M. (1995) "The Economic Geography in Environmental Issues: a Spatial-Analytic Approach," *Progress in Human Geography* 19(3):372–387.

Harvey, Fiona (2012) "2012 Expected to be Ninth Warmest Year on Record," *The Guardian*, Wednesday, 28 November, 2012, http://www.guardian.co.uk/environment/2012/nov/28/2012-ninth-warmest-year-record (Accessed December 20, 2012).

Hern, Warren M. (1990) "Why Are There So Many of Us? Description and Diagnosis of a Planetary Ecopathological Process," *Population and Development* 12(1):9–40.

Hines, Lawrence G. (1973) *Environmental Issues; Population, Pollution and Economics*. New York: W. W. Norton and Co.

Hinrichsen, Don (1991) "The Need to Balance Population with Resources," *Populi* 18(3): 27–38.

Homer-Dixon, Thomas F., Boutwell, Jeffrey H., and Rathjens, George W. (1993) "Environmental Change and Violent Conflict," *Scientific American* 268(2):38–45.

Houghton, Richard A. (2005) "Tropical deforestation as a source of greenhouse gas emissions," in Paulo Moutinho and Stephan Schwartzman, eds. *Tropical Deforestation and Climate Change*, Belém - Pará - Brazil: IPAM - Instituto de Pesquisa Ambiental da Amazônia; Washington DC - USA : Environmental Defense, pp. 13-21.

Jorgensen, Neils, Asklund, Camillea, Carlsen, Elisabeth, Skakkebaek, Neils. (2006) "Coordinated European Studies of Semen Quality: Results from Studies of Scandinavian Young Men is a Matter of Concern" *International Journal of Andrology* 29: 54–61.

Kaplan, Robert D. (1994) "The Coming Anarchy," *The Atlantic Monthly* 273(2):44–76.

Keyfitz, Nathan (1972) "Population Theory and Doctrine: A Historical Survey," in William Petersen (ed.), *Readings in Population*, New York: Macmillan.

Keyfitz, Nathan (1994) "Demographic Discord," *The Sciences* 34(5):21–27.

Klesius, Michael (2002) "The State of the Planet," *National Geographic* 202(3):102–115.

Lents, James M. and Kelly, William J. (1993) "Clearing the Air in Los Angeles," *Scientific American* 269(4):32–39.

Li, Jing-Neng (1991) "Population Effects on Deforestation and Soil Erosion in China," in Kingsley Davis and Mikhail S. Bernstam (eds), *Resources, Environment, and Population*. New York: Oxford University Press, pp. 254–258.

Mark, Monica (2012) "West Africa's technological revolution driven by mobile phones," *The Guardian*, Monday, 24 September, 2012, http://www.guardian.co.uk/global-development/2012/sep/24/nigeria-mobile-phones-success-technology (Accessed December 21, 2012).

Mbatu, Richard S. and Otiso, Kefa M. Otiso (2012) "Chinese economic expansionism in Africa: a theoretical analysis of the environmental Kuznets Curve Hypothesis in the Forest Sector in Cameroon," *African Geographical Review*, 31 (2): 142–162.

McKibben, Bill (1995) "An Explosion of Green," *The Atlantic Monthly* 275(4):61–83.

McKibben, Bill (2004) "Crossing the Red Line," *New York Review of Books* LI(10):32–36.

McNeill, J. R. (2000) *Something New under the Sun: An Environmental History of the Twentieth-Century World.* New York: W. W. Norton.

Meadows, Donella H., Meadows, Dennis L., Randers, Jorgen, and Behrens, William W., III (1972) *The Limits to Growth.* New York: Universe Books.

Meadows, Donella H., Meadows, Dennis L., and Randers, Jorgen. (2004) *Limits to Growth The 30-Year Update.* White River Junction, VT: Chelsea Green Publishing.

Meehl, G. A., et al. (2007) "Global Climate Projections," in S. Solomon, et al., eds. *Climate Change 2007: The Physical Science Basis.* Cambridge: Cambridge University Press.

Morales, Alex and Krukowska, Ewa (2012) "Pollution Limits Renewed with UN Push for Climate Aid," *Bloomberg,* December 8, 2012, http://www.bloomberg.com/news/2012-12-08/pollution-limits-extended-in-un-global-warming-pact.html (Accessed December 20, 2012).

Muir, Donald E. (1991) "Using Computers to Explore Ecological Issues: A Simple Limits-to Growth Educational Program," *Population and Environment* 13(2):113–117.

O'Brien, Keith (2007) "Turkeys take to cities, towns," The Boston Globe, Tuesday, October 23, 2007, http://www.boston.com/news/local/articles/2007/10/23/turkeys_take_to_cities_ towns/?page=full, (Accessed December 20, 2012).

Park, Chris (1993) "Environmental Issues," *Progress in Physical Geography* 17(4):473–483.

Phillips, Kathryn (1994) *Tracing the Vanishing Tree Frogs: An Ecological Mystery.* New York: St. Martin's Press.

Postel, Sandra. (1991) "Restoring Degraded Land," In Lester R. Brown (ed) *The World Watch Reader on Global Environmental Issues.* New York and London: W. W. Norton & Company.

Quammen, David (1998) "Planet of Weeds: Tallying the Losses of Earth's Animals and Plants," *Harper's Magazine* 297(1781): 57–69.

Repetto, Robert (1987) "Population, Resources, Environment: An Uncertain Future," *Population Bulletin* 42(2):1–44

Richards, Paul W. (1973) "The Tropical Rain Forest," *Scientific American* 232(6):58–67.

Ridker, Ronald G. (1980) *Resource and Environmental Consequences of Population and Economic Growth.* Reprint 172, Washington, D.C.: Resources for the Future, Inc.

Ruddiman, William F. (2005) *Plows, Plagues and Petroleum: How Humans Took Control of Climate.* Princeton, NJ: Princeton University Press.

Sauri-Punjol, David (1993) "Putting the Environment Back into Human Geography: A Teaching Experience," *Journal of Geography in Higher Education* 17(1):3–9.

Schneider, Stephen H. (1990) "Cooling It," *World Monitor,* July 1990, pp. 30–38.

Socolow, Robert H. and Pacala, Stephen W. (2006) "A Plan to Keep Carbon in Check," *Scientific American* 295(3):50–59.

Stern, Nicholas (2007) *The Economics of Climate Change: The Stern Review.* Cambridge: Cambridge University Press.

Swan S.H., Kruse R.L., Liu F., et al. (2003) "Semen Quality in Relation to Biomarkers of Pesticides Exposure" *Environmental Health Perspectives* 111: 1478–1484.

Tickell, Crispin (1993) "The Human Species: A Suicidal Success," *The Geographical Journal* 159(2):219–226.

Turner, Graham. (2008) "A Comparison of 'The Limits to Growth' with Thirty Years of Reality," *Global Environmental Change* 18(3): 397-411.

United States Environmental Protection Agency - EPA (2012) "Global Emissions," EPA, Thursday, June 14, 2012, http://www.epa.gov/climatechange/ghgemissions/global.html#three (Accessed December 20, 2012).

United States Office of Technology Assessment (1984) *Acid Rain and Transported Air Pollutants: Implications for Public Policy*, Washington, D.C.: U.S. Government Printing Office.

von Lersner, Heinrich (1995) "Commentary: Outline for an Ecological Economy," *Scientific American* 273(3):188.

Wallach, Bret (2005) *Understanding the Cultural Landscape*. New York: The Guilford Press.

Weatherford, Jack (1994) *Savages and Civilization: Who Will Survive*? New York: Crown Publishers, Inc.

Wilbanks, Thomas J. (1994) "'Sustainable Development' in Geographic Perspective," *Annals of the Association of American Geographers* 84(4):541–556.

Williams, Michael (2003) *Deforesting the Earth: From Prehistory to Global Crisis*. Chicago: University of Chicago Press.

Wilson, Edward O. (1993) *The Diversity of Life*. New York: W.W. Norton.

World Health Organization (2010) "Access to safe drinking water improving; sanitation needs greater efforts," World Health Organization Media Centre, http://www.who.int/mediacentre/news/releases/2010/water_20100315/en/index.html (Accessed December 20, 2012).

World Resources Institute and International Institute for Environment and Development (1987) *World Resources 1987*. New York: Basic Books.

Wright, Lawrence (1996) "Silent Sperm," *The New Yorker* (January 15):41–54.

Chapter 10

Beardsley, Tim (1997) "In Focus: Death in the Deep," *Scientific American* 277(5):17–18.

Blas, Javier and Meyer, Gregory (2012) "UN urges US to cut ethanol production," The Financial Times, August 9, 2012, http://www.ft.com/cms/s/0/8808b144-e240-11e1-8e9d-00144feab49a.html#axzz2G5gcVlo8 (Accessed December 24, 2012).

Boyd, Claude E. and Clay, Jason W. (1998) "Shrimp Aquaculture and the Environment," *Scientific American* 278(6): 58–65.

Brown, Lester R. (1981) "World Food Resources and Population: The Narrowing Margin," *Population Bulletin* 36(3): 1–43.

Brown, Lester (1987) *State of the World*—1987, New York: W. W. Norton and Company.

Brown, Lester R. (1989) "Feeding Six Billion," *Worldwatch*, September/October: 32–40.

Brown, Lester R. (1995) *Who Will Feed China*? New York: W. W. Norton and Company.

Brown, Lester (2002) "Grain Harvest Growth Growing," http: //www.earth-policy.org/Indicators/indicator6_print. htm (Accessed December 1, 2004).

Brown, Lester (2003) *Plan B: Rescuing a Planet under Stress and a Civilization in Trouble*. New York: W. W. Norton and Company.

Crawford, Patricia B. et al. (2004) "How Can Californians be Overweight and Hungry?" California Agriculture 58(1):12–17

Dalrymple, Dana G. (1986) *Development and Spread of High-yielding Rice Varieties in Developing Countries*, Washington, D.C.: U.S. Agency for International Development.

Drewnowski, Adam and Specter, S. E. (January 2004) "Poverty and obesity: the role of energy density and energy costs," *American Journal of Clinical Nutrition*, 79(1): 6–16

Earth Policy Institute (2012) "Food and Agriculture" (http://www.earth-policy.org/data_center/C24 (accessed December 24, 2012).

Easterbrook, Gregg (2003) *The Progress Paradox: How Life Gets Better While People Feel Worse*. New York: Random House.

Evans, L. T. (1998) *Feeding the Ten Billion: Plants and Population Growth*. Cambridge: Cambridge University Press.

Falkenmark, Malin and Suprapto, Riga Adiwoso (1992) "Population-Landscape Interactions in Development: A Water Perspective to Environmental Sustainability," *Ambio* 21(1):31–36.

FAO (1987) *Agriculture: Toward 2000*. Rome, Italy: FAO.

FAO (1992) *The use of saline waters for crop production*, http://www.fao.org/docrep/T0667E/T0667E00.htm (Accessed December 24, 2012).

Fitzgerald, Deborah (2003) *Every Farm a Factory: The Industrial Ideal in American Agriculture*. New Haven, CN: Yale University Press.

Frenken, Karen, ed. (2005) *Irrigation in Africa in Figures – AQUASTAT Survey – 2005*. Rome: Food and Agriculture Organization of the United Nations. ftp://ftp.fao.org/agl/aglw/docs/wr29_eng.pdf (Accessed December 24, 2012).

Gilland, Bernard (2006) "Population, Nutrition and Agriculture," *Population and Environment* 28(1):1–16.

Gleick, Peter H. (2001) "Making Every Drop Count," *Scientific American* 284 (2): 40-45.

Glenn, Edward P., Brown, J. Jed, and O'Leary, James W. (1998) "Irrigating Crops with Seawater," *Scientific American* 279(2): 76–81.

Hardin, Garrett (1995) *The Immigration Dilemma: Avoiding the Tragedy of the Commons*. Washington, D.C.: Federation for American Immigration Reform.

Hardin, Garrett (1999) *The Ostrich Factor: Our Population Myopia*. New York: Oxford University Press.

Harlan, Jack R. (1976) "The Plants and Animals that Nourish Man," *Scientific American* 235(3):88–97.

Hart, John Fraser (2003) *The Changing Scale of American Agriculture*. Charlottesville: University of Virginia Press.

Hendry, Peter (1988) "Food and Population: Beyond Five Billion," *Population Bulletin* 43(2):1–40.

Hinrichsen, Don and Marshall, Alex (1991) "Population and the Food Crisis," *Populi* 16(2):24–34.

Hopfenberg, Russell (2003) "Human Carrying Capacity is Determined by Food Availability," *Population and Environment* 25(2):109–117.

Kurlansky, Mark (2002) *Salt: A World History*. New York: Penguin Books.

Larsen, Janet (2002) "Fish Catch Leveling Off," http://www.earthpolicy.org/Indicators/indicator3_print.htm (Accessed December 1, 2004).

Le Heron, Richard and Roche, Michael (1995) "A 'Fresh' Place in Food's Space," *Area* 27(1):23–33.

Markham, Victoria D. with Steinzor, Nadia (2006) *U. S. National Report on Population and the Environment*. New Canaan, CT: Center for Environment and Population.

McGinn, A. P. (1998) "Blue Revolution: The Promises and Pitfalls of Fish Farming," *World Watch* 11(2):10–19.

National Academy of Agricultural Sciences (2006) "Low and Declining Crop Response to Fertilizers," Policy Paper No. 35. New Delhi: National Academy of Agricultural Sciences, pp. 8, http://naasindia.org/Policy%20Papers/policy%2035.pdf (Accessed December 24, 2012).

Naylor, Rosamond, et al. (2005) "Losing the Links Between Livestock and Land," *Science* 310:1621–22.

Paddock, William and Paddock, Paul (1976) *Time of Famines: America and the World Food Crisis.* Boston: Little, Brown and Company.

Pimentel, David and Wilson, Anne (2005) "World Population, Agriculture, and Malnutrition," http://www.energybulletin.net/ 3834.html (Accessed July 17, 2007).

Plucknett, Donald L. and Winkelmann, Donald L. (1995) "Technology for Sustainable Agriculture," *Scientific American* 273(3):182–186.

Querna, Elizabeth (2004) "Fixing Fish Farms," *U.S. News and World Report* 137(5):62.

Runge, C. Ford and Senauer, Benjamin (2007) "How Biofuels Could Starve the Poor," *Foreign Affairs* 86(3):41–53.

Simon, Arthur (1975) *Bread for the World.* New York: Paulist Press.

Smil, Vaclav (1997) "Global Population and the Nitrogen Cycle," *Scientific American* 277(1):76–81.

Smil, Vaclav (2000) *Feeding the World: A Challenge for the Twenty-First Century.* Cambridge, MA: MIT Press.

Smil, Vaclav (2002) "Eating Meat: Evolution, Patterns, and Consequences," *Population and Development Review* 28(4): 599–639.

Tarrant, John R. (1990) "World Food Prospects for the 1990's," *Journal of Geography* 89(6):234–238.

Tempest, Rone (1997) "The Coexistence of Feast, Famine," *Los Angeles Times* (Jan. 22) A1 and A16–A17.

United States Department of Agriculture (2012) "Indian Beef Exports Surge to Continue in 2013," *Livestock and Poultry: World Markets and Trade,* http://www.fas.usda.gov/psdonline/circulars/livestock_poultry.pdf (Accessed December 25, 2012).

Wald, Matthew L. (2007) "Is ethanol the long haul?" *Scientific American* Jan:42-49.

Wallace, Scott (2007) "Last of the Amazon," *National Geographic* 211(1):40–71.

Winter, Michael (2004) "Geographies of Food: Agro-Food Geographies—Farming, Food and Politics," *Progress in Human Geography* 28(5):664–670.

Wolf, Edward C. (1987) "Raising Agricultural Productivity," in Lester Brown, ed., *State of the World*—1987, New York: W. Norton.

World Food Programme (2012) "World Hunger," http://www.wfp.org/hunger (Accessed December 25, 2012).

Young, John E. (1992) "Bred for the Hungry," in Robert M. Jackson, (ed) *Global Issues 1992/1993,* Guilford, CT: Dushkin Publishing Company, pp. 97–104.

Web Site Listings

General Population Studies

The Alan Guttmacher Institute
www.guttmacher.org/

Association of Population Centers
www.popcenters.org

The Brookings Institution
www.brook.edu/

Center for Comparative Migration Studies
www.ccis.ucsd.edu/

Center for Immigration Studies
www.cis.org/

Center for Migration Studies
http://cmsny.org/

Migration Studies Project
www.migrationstudiesproject.psu.edu/

Population Association of America
www.popassoc.org

The Population Council
www.popcouncil.org

Population Reference Bureau
www.prb.org/

United Nations

Food and Agriculture Organization
www.fao.org/

Population Division (POPIN)
www.un.org/popin/wdtrends.htm

Population Fund (UNFPA)
www.unfpa.org/

United States

Census Bureau
www.census.gov/

National Bureau of Economic Research, Center for Aging Research
www.nber.org/aging

National Institute on Aging, Behavioral and Social Research
www.nia.nih.gov/research/extramural/behavior

National Institute on Child Health and Human Development, Demographic and Behavioral Sciences Branch
www.nichd.nih.gov/about/cpr/dbs/dbs.htm

International

U.S. Bureau of Labor Statistics—National Statistical Agencies of Other Countries and International Organizations
www.bls.gov/bls/other.htm

Statistics Canada
http://www.statcan.ca/

U.S. Census' International Statistical Agencies' Database (i.e., Other Countries Census Bureaus)
www.census.gov/population/international/links/stat_int.html

Bowling Green State University, Center for Family and Demographic Research
www.bgsu.edu/organizations/cfdr/

Brown University, Population Studies and Training Center
www.pstc.brown.edu

Duke University, DuPRI: Duke Population Research Institute
www.dupri.duke.edu/

FHI360
www.fhi360.org/en/index.htm

Florida State University, Center for Demography and Population Health
www.fsu.edu/?popctr

Georgetown University, Center for Population and Health
www.cph.georgetown.edu

Harvard Center for Population and Development Studies
www.hsph.harvard.edu/hcpds

Johns Hopkins University, Hopkins Population Center
www.web.jhu.edu/popcenter

Measure DHS, Demographic and Health Surveys
www.measuredhs.com

Minnesota Population Center
www.pop.umn.edu

University of Chicago, Center on Demographics and Economics of Aging
www.spc.uchicago.edu/coa

The Ohio State University, Institute for Population Research
www.ipr.osu.edu

Pennsylvania State University, Population Research Institute
www.pop.psu.edu

Princeton University, Office of Population Research
www.opr.princeton.edu

RAND, Center for the Study of Aging
www.rand.org/labor/aging

RAND, Labor and Population
www.rand.org/labor

Stanford University, Center for Population Research
www.iriss.stanford.edu/scpr

University at Albany, Center for Social and Demographic Analysis
www.albany.edu/csda

UCLA, California Center for Population Research
www.ccpr.ucla.edu

University of Chicago and National Opinion Research Center, Population Research Center
www.spc.uchicago.edu/orgs/prc

University of Colorado, Institute of Behavioral Studies, Population Program
www.colorado.edu/IBS/PAC

University of Maryland, Population Research Center
www.popcenter.umd.edu

University of Michigan, Michigan Center on the Demography of Aging
micda.psc.isr.umich.edu

University of Michigan, Population Studies Center
www.psc.isr.umich.edu

University of North Carolina–Chapel Hill, Carolina Population Center
www.cpc.unc.edu

University of Pennsylvania, Population Aging Research Center
www.parc.pop.upenn.edu/

University of Pennsylvania, Population Studies Center
www.pop.upenn.edu/

University of Southern California and UCLA, Center of Biodemography and Population Health
http://gero.usc.edu/CBPH/

University of Texas at Austin, Population Research Center
www.prc.utexas.edu

University of Washington, Center for Studies in Demography and Ecology
csde.washington.edu

University of Wisconsin, Center for Demography and Ecology
www.ssc.wisc.edu/cde

The Urban Institute, Center on Labor, Human Services & Population
www.urban.org/content/PolicyCenters/Population/Overview.htm

University of Wisconsin, Center for Demography of Health and Aging
www.ssc.wisc.edu/cdha

Appendix A

Census Regions and Divisions of the United States

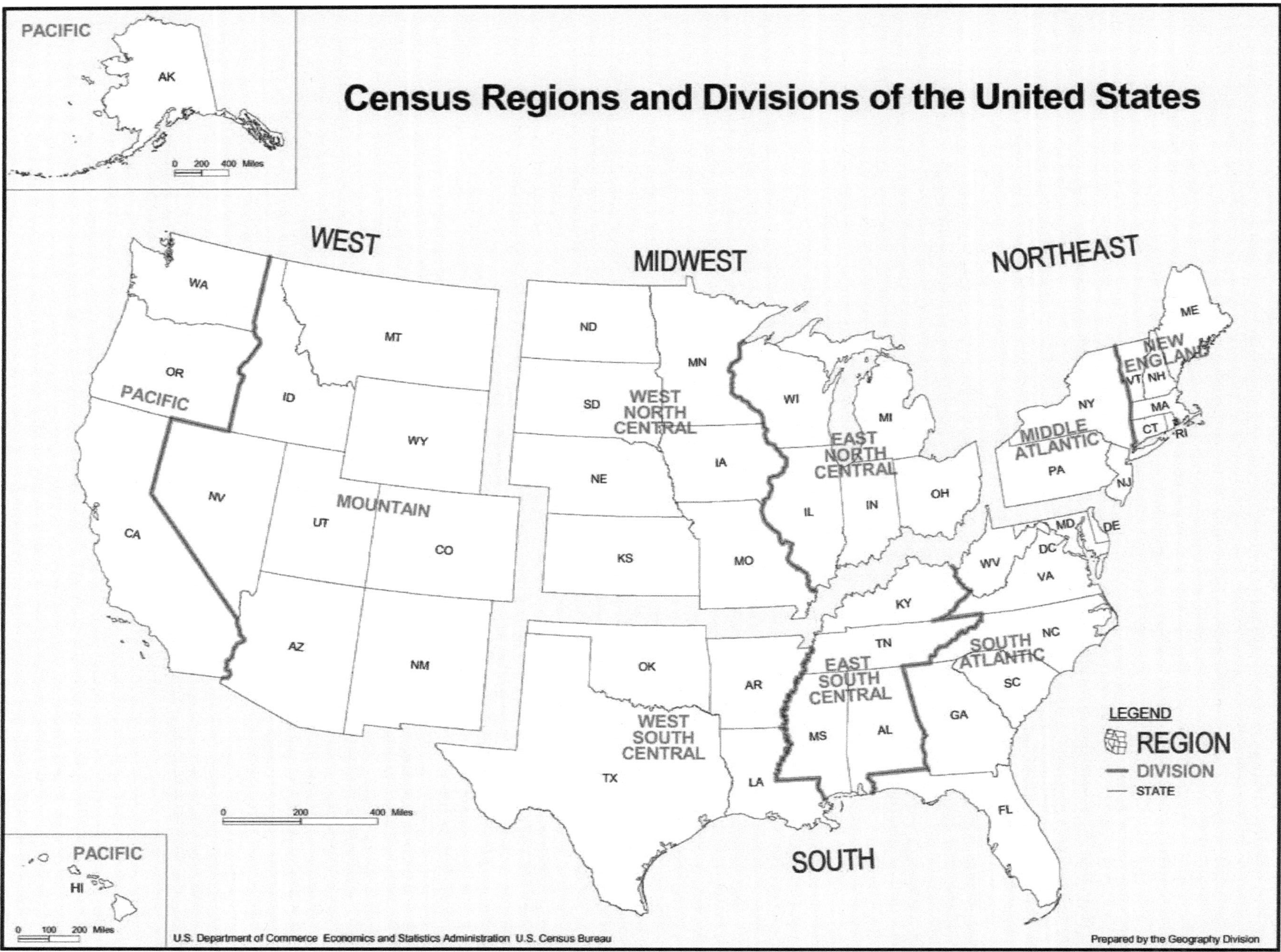
PACIFIC
AK
0 200 400 Miles
Census Regions and Divisions of the United States
WEST
MIDWEST
NORTHEAST
WA
OR
PACIFIC
ID
MT
WY
NV
UT
MOUNTAIN
CO
CA
AZ
NM
ND
SD
WEST NORTH CENTRAL
MN
NE
IA
KS
MO
WI
MI
EAST NORTH CENTRAL
IL
IN
OH
ME
NEW ENGLAND
VT
NH
MA
CT
RI
NY
MIDDLE ATLANTIC
PA
NJ
MD
DE
DC
WV
VA
KY
TN
NC
SOUTH ATLANTIC
SC
GA
FL
OK
AR
EAST SOUTH CENTRAL
MS
AL
WEST SOUTH CENTRAL
TX
LA
LEGEND
REGION
DIVISION
STATE
0 200 400 Miles
PACIFIC
HI
0 100 200 Miles
SOUTH
U.S. Department of Commerce Economics and Statistics Administration U.S. Census Bureau
Prepared by the Geography Division

U.S. Census Bureau

Census Bureau Regions and Divisions with State FIPS Codes

Region I: Northeast

Division I: New England

Connecticut (09)
Maine (23)
Massachusetts (25)
New Hampshire (33)
Rhode Island (44)
Vermont (50)

Division 2: Middle Atlantic

New Jersey (34)
New York (36)
Pennsylvania (42)

Region 2: Midwest*

Division 3: East North Central

Indiana (18)
Illinois (17)
Michigan (26)
Ohio (39)
Wisconsin (55)

Division 4: West North Central

Iowa (19)
Kansas (20)
Minnesota (27)
Missouri (29)
Nebraska (31)
North Dakota (38)
South Dakota (46)

Region 3: South

Division 5: South Atlantic

Delaware (10)
District of Columbia (11)
Florida (12)
Georgia (13)
Maryland (24)
North Carolina (37)
South Carolina (45)
Virginia (51)
West Virginia (54)

Division 6: East South Central

Alabama (01)
Kentucky (21)
Mississippi (28)
Tennessee (47)

Division 7: West South Central

Arkansas (05)
Louisiana (22)
Oklahoma (40)
Texas (48)

Region 4: West

Division 8: Mountain

Arizona (04)
Colorado (08)
Idaho (16)
New Mexico (35)
Montana (30)
Utah (49)
Nevada (32)
Wyoming (56)

Division 9: Pacific

Alaska (02)
California (06)
Hawaii (15)
Oregon (41)
Washington (53)

**Prior to June 1984, the Midwest Region was designated as the North Central Region.*

Appendix B

Census Form 2000

United States Census 2000

U.S. Department of Commerce
Bureau of the Census

DC

This is the official form for all the people at this address. It is quick and easy, and your answers are protected by law. Complete the Census and help your community get what it needs – today and in the future!

Start Here

Please use a black or blue pen.

How many people were living or staying in this house, apartment, or mobile home on April 1, 2000?

Number of people

INCLUDE in this number:

- foster children, roomers, or housemates
- people staying here on April 1, 2000 who have no other permanent place to stay
- people living here most of the time while working, even if they have another place to live

DO NOT INCLUDE in this number:

- college students living away while attending college
- people in a correctional facility, nursing home, or mental hospital on April 1, 2000
- Armed Forces personnel living somewhere else
- people who live or stay at another place most of the time

Please turn the page and print the names of all the people living or staying here on April 1, 2000.

If you need help completing this form, *call 1±800±471±9424 between 8:00 a.m. and 9:00 p.m., 7 days a week. The telephone call is free.*

TDD ± *Telephone display device for the hearing impaired. Call 1±800±582±8330 between 8:00 a.m. and 9:00 p.m., 7 days a week. The telephone call is free.*

øNECESITA AYUDA? *Si usted necesita ayuda para completar este cuestionario llame al 1±800±471±8642 entre las 8:00 a.m. y las 9:00 p.m., 7 dóas a la semana. La llamada telef∞nica es gratis.*

The Census Bureau estimates that, for the average household, this form will take about 38 minutes to complete, including the time for reviewing the instructions and answers. Comments about the estimate should be directed to the Associate Director for Finance and Administration, Attn: Paperwork Reduction Project 0607-0856, Room 3104, Federal Building 3, Bureau of the Census, Washington, DC 20233.

Respondents are not required to respond to any information collection unless it displays a valid approval number from the Office of Management and Budget.

Form **D-2**

OMB No. 0607-0856: Approval Expires 12/31/2000

Please be sure you answered question 1 on the front page before continuing.

Please print the names of all the people who you indicated in question 1 were living or staying here on April 1, 2000.

Example – Last Name

JOHNSON

First Name MI

ROBIN J

Start with the person, or one of the people living here who owns, is buying, or rents this house, apartment, or mobile home. If there is no such person, start with any adult living or staying here.

Person 1 – Last Name

First Name MI

Person 2 – Last Name

First Name MI

Person 3 – Last Name

First Name MI

Person 4 – Last Name

First Name MI

Person 5 – Last Name

First Name MI

Person 6 – Last Name

First Name MI

Person 7 – Last Name

First Name MI

Person 8 – Last Name

First Name MI

Person 9 – Last Name

First Name MI

Person 10 – Last Name

First Name MI

Person 11 – Last Name

First Name MI

Person 12 – Last Name

First Name MI

Next, answer questions about Person 1.

FOR OFFICE USE ONLY

A. JIC1	B. JIC2	C. JIC3	D. JIC4

Form D-2

Your answers are important! Every person in the Census counts.

What is this person's name? *Print the name of Person 1 from page 2.*

Last Name

First Name MI

What is this person's telephone number? *We may contact this person if we don't understand an answer.*

Area Code + Number

What is this person's sex? *Mark* ☒ *ONE box.*

- ☐ Male
- ☐ Female

What is this person's age and what is this person's date of birth?

Age on April 1, 2000

Print numbers in boxes.

Month Day Year of birth

NOTE: Please answer BOTH Questions 5 and 6.

Is this person Spanish/Hispanic/Latino? *Mark* ☒ *the* ***"No"*** *box if* ***not*** *Spanish/Hispanic/Latino.*

- ☐ **No**, not Spanish/Hispanic/Latino
- ☐ Yes, Mexican, Mexican Am., Chicano
- ☐ Yes, Puerto Rican
- ☐ Yes, Cuban
- ☐ Yes, other Spanish/Hispanic/Latino – *Print group.*

What is this person's race? ***Mark*** ☒ ***one or more races*** *to indicate what this person considers himself/herself to be.*

- ☐ White
- ☐ Black, African Am., or Negro
- ☐ American Indian or Alaska Native – *Print name of enrolled or principal tribe.*
- ☐ Asian Indian
- ☐ Chinese
- ☐ Filipino
- ☐ Japanese
- ☐ Korean
- ☐ Vietnamese
- ☐ Other Asian – *Print race.*
- ☐ Native Hawaiian
- ☐ Guamanian or Chamorro
- ☐ Samoan
- ☐ Other Pacific Islander – *Print race.*
- ☐ Some other race – *Print race.*

What is this person's marital status?

- ☐ Now married
- ☐ Widowed
- ☐ Divorced
- ☐ Separated
- ☐ Never married

a. At any time since February 1, 2000, has this person attended regular school or college? *Include only nursery school or preschool, kindergarten, elementary school, and schooling which leads to a high school diploma or a college degree.*

- ☐ No, has not attended since February 1 → *Skip to 9*
- ☐ Yes, public school, public college
- ☐ Yes, private school, private college

Form D-2

2043

b. What grade or level was this person attending? *Mark ☒ ONE box.*

- ☐ Nursery school, preschool
- ☐ Kindergarten
- ☐ Grade 1 to grade 4
- ☐ Grade 5 to grade 8
- ☐ Grade 9 to grade 12
- ☐ College undergraduate years (freshman to senior)
- ☐ Graduate or professional school *(for example: medical, dental, or law school)*

What is the highest degree or level of school this person has COMPLETED? *Mark ☒ ONE box. If currently enrolled, mark the previous grade or highest degree received.*

- ☐ No schooling completed
- ☐ Nursery school to 4th grade
- ☐ 5th grade or 6th grade
- ☐ 7th grade or 8th grade
- ☐ 9th grade
- ☐ 10th grade
- ☐ 11th grade
- ☐ 12th grade, **NO DIPLOMA**
- ☐ **HIGH SCHOOL GRADUATE –** high school DIPLOMA or the equivalent *(for example: GED)*
- ☐ Some college credit, but less than 1 year
- ☐ 1 or more years of college, no degree
- ☐ Associate degree *(for example: AA, AS)*
- ☐ Bachelor's degree *(for example: BA, AB, BS)*
- ☐ Master's degree *(for example: MA, MS, MEng, MEd, MSW, MBA)*
- ☐ Professional degree *(for example: MD, DDS, DVM, LLB, JD)*
- ☐ Doctorate degree *(for example: PhD, EdD)*

What is this person's ancestry or ethnic origin?

(For example: Italian, Jamaican, African Am., Cambodian, Cape Verdean, Norwegian, Dominican, French Canadian, Haitian, Korean, Lebanese, Polish, Nigerian, Mexican, Taiwanese, Ukrainian, and so on.)

a. Does this person speak a language other than English at home?

- ☐ Yes
- ☐ No → *Skip to 12*

b. What is this language?

(For example: Korean, Italian, Spanish, Vietnamese)

c. How well does this person speak English?

- ☐ Very well
- ☐ Well
- ☐ Not well
- ☐ Not at all

Where was this person born?

- ☐ In the United States – *Print name of state.*
- ☐ Outside the United States – *Print name of foreign country, or Puerto Rico, Guam, etc.*

Is this person a CITIZEN of the United States?

- ☐ Yes, born in the United States → *Skip to 15a*
- ☐ Yes, born in Puerto Rico, Guam, the U.S. Virgin Islands, or Northern Marianas
- ☐ Yes, born abroad of American parent or parents
- ☐ Yes, a U.S. citizen by naturalization
- ☐ No, not a citizen of the United States

When did this person come to live in the United States? *Print numbers in boxes.*

Year

a. Did this person live in this house or apartment 5 years ago (on April 1, 1995)?

- ☐ Person is under 5 years old → *Skip to 33*
- ☐ Yes, this house → *Skip to 16*
- ☐ No, outside the United States – *Print name of foreign country, or Puerto Rico, Guam, etc., below; then skip to 16.*
- ☐ No, different house in the United States

Form D-2

b. Where did this person live 5 years ago?

Name of city, town, or post office

Did this person live inside the limits of the city or town?

☐ Yes
☐ No, outside the city/town limits

Name of county

Name of state

ZIP Code

Does this person have any of the following long-lasting conditions:

	Yes	No
a. Blindness, deafness, or a severe vision or hearing impairment?	☐	☐
b. A condition that substantially limits one or more basic physical activities such as walking, climbing stairs, reaching, lifting, or carrying?	☐	☐

Because of a physical, mental, or emotional condition lasting 6 months or more, does this person have any difficulty in doing any of the following activities:

	Yes	No
a. Learning, remembering, or concentrating?	☐	☐
b. Dressing, bathing, or getting around inside the home?	☐	☐
c. (Answer if this person is 16 YEARS OLD OR OVER.) Going outside the home alone to shop or visit a doctor's office?	☐	☐
d. (Answer if this person is 16 YEARS OLD OR OVER.) Working at a job or business?	☐	☐

Was this person under 15 years of age on April 1, 2000?

☐ Yes → *Skip to 33*
☐ No

a. Does this person have any of his/her own grandchildren under the age of 18 living in this house or apartment?

☐ Yes
☐ No → *Skip to 20a*

b. Is this grandparent currently responsible for most of the basic needs of any grandchild(ren) under the age of 18 who live(s) in this house or apartment?

☐ Yes
☐ No → *Skip to 20a*

c. How long has this grandparent been responsible for the(se) grandchild(ren)? *If the grandparent is financially responsible for more than one grandchild, answer the question for the grandchild for whom the grandparent has been responsible for the longest period of time.*

☐ Less than 6 months
☐ 6 to 11 months
☐ 1 or 2 years
☐ 3 or 4 years
☐ 5 years or more

a. Has this person ever served on active duty in the U.S. Armed Forces, military Reserves, or National Guard? *Active duty does not include training for the Reserves or National Guard, but DOES include activation, for example, for the Persian Gulf War.*

☐ Yes, now on active duty
☐ Yes, on active duty in past, but not now
☐ No, training for Reserves or National Guard only → *Skip to 21*
☐ No, never served in the military → *Skip to 21*

b. When did this person serve on active duty in the U.S. Armed Forces? *Mark ☒ a box for EACH period in which this person served.*

☐ April 1995 or later
☐ August 1990 to March 1995 (including Persian Gulf War)
☐ September 1980 to July 1990
☐ May 1975 to August 1980
☐ Vietnam era (August 1964–April 1975)
☐ February 1955 to July 1964
☐ Korean conflict (June 1950–January 1955)
☐ World War II (September 1940–July 1947)
☐ Some other time

c. In total, how many years of active-duty military service has this person had?

☐ Less than 2 years
☐ 2 years or more

2045

Form D-2

LAST WEEK, did this person do ANY work for either pay or profit? *Mark ☒ the "Yes" box even if the person worked only 1 hour, or helped without pay in a family business or farm for 15 hours or more, or was on active duty in the Armed Forces.*

- ☐ Yes
- ☐ No → *Skip to 25a*

At what location did this person work LAST WEEK? *If this person worked at more than one location, print where he or she worked most last week.*

a. Address (Number and street name)

(If the exact address is not known, give a description of the location such as the building name or the nearest street or intersection.)

b. Name of city, town, or post office

c. Is the work location inside the limits of that city or town?

- ☐ Yes
- ☐ No, outside the city/town limits

d. Name of county

e. Name of U.S. state or foreign country

f. ZIP Code

a. How did this person usually get to work LAST WEEK? *If this person usually used more than one method of transportation during the trip, mark ☒ the box of the one used for most of the distance.*

- ☐ Car, truck, or van
- ☐ Bus or trolley bus
- ☐ Streetcar or trolley car
- ☐ Subway or elevated
- ☐ Railroad
- ☐ Ferryboat
- ☐ Taxicab
- ☐ Motorcycle
- ☐ Bicycle
- ☐ Walked
- ☐ Worked at home → *Skip to 27*
- ☐ Other method

If "Car, truck, or van" is marked in 23a, go to 23b. Otherwise, skip to 24a.

b. How many people, including this person, usually rode to work in the car, truck, or van LAST WEEK?

- ☐ Drove alone
- ☐ 2 people
- ☐ 3 people
- ☐ 4 people
- ☐ 5 or 6 people
- ☐ 7 or more people

a. What time did this person usually leave home to go to work LAST WEEK?

: ☐ a.m. ☐ p.m.

b. How many minutes did it usually take this person to get from home to work LAST WEEK?

Minutes

Answer questions 25±26 for persons who did not work for pay or profit last week. Others skip to 27.

a. LAST WEEK, was this person on layoff from a job?

- ☐ Yes → *Skip to 25c*
- ☐ No

b. LAST WEEK, was this person TEMPORARILY absent from a job or business?

- ☐ Yes, on vacation, temporary illness, labor dispute, etc. → *Skip to 26*
- ☐ No → *Skip to 25d*

c. Has this person been informed that he or she will be recalled to work within the next 6 months OR been given a date to return to work?

- ☐ Yes → *Skip to 25e*
- ☐ No

d. Has this person been looking for work during the last 4 weeks?

- ☐ Yes
- ☐ No → *Skip to 26*

e. LAST WEEK, could this person have started a job if offered one, or returned to work if recalled?

- ☐ Yes, could have gone to work
- ☐ No, because of own temporary illness
- ☐ No, because of all other reasons *(in school, etc.)*

When did this person last work, even for a few days?

- ☐ 1995 to 2000
- ☐ 1994 or earlier, or never worked → *Skip to 31*

Form D-2

Industry or Employer – *Describe clearly this person's chief job activity or business last week. If this person had more than one job, describe the one at which this person worked the most hours. If this person had no job or business last week, give the information for his/her last job or business since 1995.*

a. For whom did this person work? *If now on active duty in the Armed Forces, mark ☒ this box →* ☐ *and print the branch of the Armed Forces.*

Name of company, business, or other employer

b. What kind of business or industry was this? *Describe the activity at location where employed. (For example: hospital, newspaper publishing, mail order house, auto repair shop, bank)*

c. Is this mainly – *Mark ☒ ONE box.*

- ☐ Manufacturing?
- ☐ Wholesale trade?
- ☐ Retail trade?
- ☐ Other *(agriculture, construction, service, government, etc.)*?

Occupation

a. What kind of work was this person doing? *(For example: registered nurse, personnel manager, supervisor of order department, auto mechanic, accountant)*

b. What were this person's most important activities or duties? *(For example: patient care, directing hiring policies, supervising order clerks, repairing automobiles, reconciling financial records)*

Was this person – *Mark ☒ ONE box.*

- ☐ Employee of a PRIVATE-FOR-PROFIT company or business or of an individual, for wages, salary, or commissions
- ☐ Employee of a PRIVATE NOT-FOR-PROFIT, tax-exempt, or charitable organization
- ☐ Local GOVERNMENT employee *(city, county, etc.)*
- ☐ State GOVERNMENT employee
- ☐ Federal GOVERNMENT employee
- ☐ SELF-EMPLOYED in own NOT INCORPORATED business, professional practice, or farm
- ☐ SELF-EMPLOYED in own INCORPORATED business, professional practice, or farm
- ☐ Working WITHOUT PAY in family business or farm

a. LAST YEAR, 1999, did this person work at a job or business at any time?

- ☐ Yes
- ☐ No → *Skip to 31*

b. How many weeks did this person work in 1999? *Count paid vacation, paid sick leave, and military service.*

Weeks

c. During the weeks WORKED in 1999, how many hours did this person usually work each WEEK?

Usual hours worked each WEEK

INCOME IN 1999 – *Mark ☒ the "Yes" box for each income source received during 1999 and enter the total amount received during 1999 to a maximum of $999,999. Otherwise, mark ☒ the "No" box.*

If net income was a loss, enter the amount and mark ☒ the "Loss" box next to the dollar amount.

For income received jointly, report, if possible, the appropriate share for each person; otherwise, report the whole amount for only one person and mark ☒ the "No" box for the other person. If exact amount is not known, please give best estimate.

a. Wages, salary, commissions, bonuses, or tips from all jobs – *Report amount before deductions for taxes, bonds, dues, or other items.*

☐ Yes Annual amount – *Dollars*

$, .00

☐ No

b. Self-employment income from own nonfarm businesses or farm businesses, including proprietorships and partnerships – *Report NET income after business expenses.*

☐ Yes Annual amount – *Dollars*

$, .00 ☐ Loss

☐ No

Form D-2

2047

c. **Interest, dividends, net rental income, royalty income, or income from estates and trusts –** *Report even small amounts credited to an account.*

☐ Yes Annual amount – *Dollars*

$ | | | , | | | .00 ☐ Loss

☐ No

d. **Social Security or Railroad Retirement**

☐ Yes Annual amount – *Dollars*

$ | | , | | | .00

☐ No

e. **Supplemental Security Income (SSI)**

☐ Yes Annual amount – *Dollars*

$ | | , | | | .00

☐ No

f. **Any public assistance or welfare payments from the state or local welfare office**

☐ Yes Annual amount – *Dollars*

$ | | , | | | .00

☐ No

g. **Retirement, survivor, or disability pensions –** *Do NOT include Social Security.*

☐ Yes Annual amount – *Dollars*

$ | | | , | | | .00

☐ No

h. **Any other sources of income received regularly such as Veterans' (VA) payments, unemployment compensation, child support, or alimony –** *Do NOT include lump-sum payments such as money from an inheritance or sale of a home.*

☐ Yes Annual amount – *Dollars*

$ | | | , | | | .00

☐ No

What was this person's total income in 1999? *Add entries in questions 31a–31h; subtract any losses. If net income was a loss, enter the amount and mark* ☒ *the "Loss" box next to the dollar amount.*

Annual amount – *Dollars*

☐ None OR $ | | | , | | | .00 ☐ Loss

Now, please answer questions 33–53 about your household.

Is this house, apartment, or mobile home –

☐ Owned by you or someone in this household with a mortgage or loan?
☐ Owned by you or someone in this household free and clear (without a mortgage or loan)?
☐ Rented for cash rent?
☐ Occupied without payment of cash rent?

Which best describes this building? *Include all apartments, flats, etc., even if vacant.*

☐ A mobile home
☐ A one-family house detached from any other house
☐ A one-family house attached to one or more houses
☐ A building with 2 apartments
☐ A building with 3 or 4 apartments
☐ A building with 5 to 9 apartments
☐ A building with 10 to 19 apartments
☐ A building with 20 to 49 apartments
☐ A building with 50 or more apartments
☐ Boat, RV, van, etc.

About when was this building first built?

☐ 1999 or 2000
☐ 1995 to 1998
☐ 1990 to 1994
☐ 1980 to 1989
☐ 1970 to 1979
☐ 1960 to 1969
☐ 1950 to 1959
☐ 1940 to 1949
☐ 1939 or earlier

When did this person move into this house, apartment, or mobile home?

☐ 1999 or 2000
☐ 1995 to 1998
☐ 1990 to 1994
☐ 1980 to 1989
☐ 1970 to 1979
☐ 1969 or earlier

How many rooms do you have in this house, apartment, or mobile home? *Do NOT count bathrooms, porches, balconies, foyers, halls, or half-rooms.*

☐ 1 room
☐ 2 rooms
☐ 3 rooms
☐ 4 rooms
☐ 5 rooms
☐ 6 rooms
☐ 7 rooms
☐ 8 rooms
☐ 9 or more rooms

How many bedrooms do you have; that is, how many bedrooms would you list if this house, apartment, or mobile home were on the market for sale or rent?

- ☐ No bedroom
- ☐ 1 bedroom
- ☐ 2 bedrooms
- ☐ 3 bedrooms
- ☐ 4 bedrooms
- ☐ 5 or more bedrooms

Do you have COMPLETE plumbing facilities in this house, apartment, or mobile home; that is, 1) hot and cold piped water, 2) a flush toilet, and 3) a bathtub or shower?

- ☐ Yes, have all three facilities
- ☐ No

Do you have COMPLETE kitchen facilities in this house, apartment, or mobile home; that is, 1) a sink with piped water, 2) a range or stove, and 3) a refrigerator?

- ☐ Yes, have all three facilities
- ☐ No

Is there telephone service available in this house, apartment, or mobile home from which you can both make and receive calls?

- ☐ Yes
- ☐ No

Which FUEL is used MOST for heating this house, apartment, or mobile home?

- ☐ Gas: from underground pipes serving the neighborhood
- ☐ Gas: bottled, tank, or LP
- ☐ Electricity
- ☐ Fuel oil, kerosene, etc.
- ☐ Coal or coke
- ☐ Wood
- ☐ Solar energy
- ☐ Other fuel
- ☐ No fuel used

How many automobiles, vans, and trucks of one-ton capacity or less are kept at home for use by members of your household?

- ☐ None
- ☐ 1
- ☐ 2
- ☐ 3
- ☐ 4
- ☐ 5
- ☐ 6 or more

Answer ONLY if this is a ONE-FAMILY HOUSE OR MOBILE HOME – All others skip to 45.

a. Is there a business (such as a store or barber shop) or a medical office on this property?

- ☐ Yes
- ☐ No

b. How many acres is this house or mobile home on?

- ☐ Less than 1 acre → *Skip to 45*
- ☐ 1 to 9.9 acres
- ☐ 10 or more acres

c. In 1999, what were the actual sales of all agricultural products from this property?

- ☐ None
- ☐ $1 to $999
- ☐ $1,000 to $2,499
- ☐ $2,500 to $4,999
- ☐ $5,000 to $9,999
- ☐ $10,000 or more

What are the annual costs of utilities and fuels for this house, apartment, or mobile home? *If you have lived here less than 1 year, estimate the annual cost.*

a. Electricity

Annual cost – *Dollars*

$ | , | | | .00

OR

- ☐ Included in rent or in condominium fee
- ☐ No charge or electricity not used

b. Gas

Annual cost – *Dollars*

$ | , | | | .00

OR

- ☐ Included in rent or in condominium fee
- ☐ No charge or gas not used

c. Water and sewer

Annual cost – *Dollars*

$ | , | | | .00

OR

- ☐ Included in rent or in condominium fee
- ☐ No charge

d. Oil, coal, kerosene, wood, etc.

Annual cost – *Dollars*

$ | , | | | .00

OR

- ☐ Included in rent or in condominium fee
- ☐ No charge or these fuels not used

2049

Form D-2

Answer ONLY if you PAY RENT for this house, apartment, or mobile home – All others skip to 47.

a. What is the monthly rent?

Monthly amount – *Dollars*

$ | | , | | | .00

b. Does the monthly rent include any meals?

☐ Yes
☐ No

Answer questions 47a–53 if you or someone in this household owns or is buying this house, apartment, or mobile home; otherwise, skip to questions for Person 2.

a. Do you have a mortgage, deed of trust, contract to purchase, or similar debt on THIS property?

☐ Yes, mortgage, deed of trust, or similar debt
☐ Yes, contract to purchase
☐ No → *Skip to 48a*

b. How much is your regular monthly mortgage payment on THIS property? *Include payment only on first mortgage or contract to purchase.*

Monthly amount – *Dollars*

$ | | , | | | .00

OR

☐ No regular payment required → *Skip to 48a*

c. Does your regular monthly mortgage payment include payments for real estate taxes on THIS property?

☐ Yes, taxes included in mortgage payment
☐ No, taxes paid separately or taxes not required

d. Does your regular monthly mortgage payment include payments for fire, hazard, or flood insurance on THIS property?

☐ Yes, insurance included in mortgage payment
☐ No, insurance paid separately or no insurance

a. Do you have a second mortgage or a home equity loan on THIS property? *Mark* ☒ *all boxes that apply.*

☐ Yes, a second mortgage
☐ Yes, a home equity loan
☐ No → *Skip to 49*

b. How much is your regular monthly payment on all second or junior mortgages and all home equity loans on THIS property?

Monthly amount – *Dollars*

$ | | , | | | .00

OR

☐ No regular payment required

What were the real estate taxes on THIS PROPERTY last year?

Yearly amount – *Dollars*

$ | | , | | | .00

OR

☐ None

What was the annual payment for fire, hazard, and flood insurance on THIS property?

Annual amount – *Dollars*

$ | | , | | | .00

OR

☐ None

What is the value of this property; that is, how much do you think this house and lot, apartment, or mobile home and lot would sell for if it were for sale?

☐ Less than $10,000
☐ $10,000 to $14,999
☐ $15,000 to $19,999
☐ $20,000 to $24,999
☐ $25,000 to $29,999
☐ $30,000 to $34,999
☐ $35,000 to $39,999
☐ $40,000 to $49,999
☐ $50,000 to $59,999
☐ $60,000 to $69,999
☐ $70,000 to $79,999
☐ $80,000 to $89,999
☐ $90,000 to $99,999
☐ $100,000 to $124,999
☐ $125,000 to $149,999
☐ $150,000 to $174,999
☐ $175,000 to $199,999
☐ $200,000 to $249,999
☐ $250,000 to $299,999
☐ $300,000 to $399,999
☐ $400,000 to $499,999
☐ $500,000 to $749,999
☐ $750,000 to $999,999
☐ $1,000,000 or more

Answer ONLY if this is a CONDOMINIUM –

What is the monthly condominium fee?

Monthly amount – *Dollars*

$ | | , | | | .00

Answer ONLY if this is a MOBILE HOME –

a. Do you have an installment loan or contract on THIS mobile home?

☐ Yes
☐ No

b. What was the total cost for installment loan payments, personal property taxes, site rent, registration fees, and license fees on THIS mobile home and its site last year? *Exclude real estate taxes.*

Yearly amount – *Dollars*

$ | | , | | | .00

Are there more people living here? If yes, continue with Person 2.

Form D-2

Census information helps your community get financial assistance for roads, hospitals, schools and more.

What is this person's name? *Print the name of Person 2 from page 2.*

Last Name

First Name MI

How is this person related to Person 1? *Mark [X] ONE box.*

- ☐ Husband/wife
- ☐ Natural-born son/daughter
- ☐ Adopted son/daughter
- ☐ Stepson/stepdaughter
- ☐ Brother/sister
- ☐ Father/mother
- ☐ Grandchild
- ☐ Parent-in-law
- ☐ Son-in-law/daughter-in-law
- ☐ Other relative – *Print exact relationship.*

If NOT RELATED to Person 1:

- ☐ Roomer, boarder
- ☐ Housemate, roommate
- ☐ Unmarried partner
- ☐ Foster child
- ☐ Other nonrelative

What is this person's sex?

- ☐ Male
- ☐ Female

What is this person's age and what is this person's date of birth?

Age on April 1, 2000

Print numbers in boxes.

Month Day Year of birth

NOTE: Please answer BOTH Questions 5 and 6.

Is this person Spanish/Hispanic/Latino? *Mark [X] the* ***"No"*** *box if* ***not*** Spanish/Hispanic/Latino.

- ☐ **No**, not Spanish/Hispanic/Latino
- ☐ Yes, Mexican, Mexican Am., Chicano
- ☐ Yes, Puerto Rican
- ☐ Yes, Cuban
- ☐ Yes, other Spanish/Hispanic/Latino – *Print group.*

What is this person's race? ***Mark [X] one or more races*** *to indicate what this person considers himself/herself to be.*

- ☐ White
- ☐ Black, African Am., or Negro
- ☐ American Indian or Alaska Native – *Print name of enrolled or principal tribe.*
- ☐ Asian Indian
- ☐ Chinese
- ☐ Filipino
- ☐ Japanese
- ☐ Korean
- ☐ Vietnamese
- ☐ Other Asian – *Print race.*
- ☐ Native Hawaiian
- ☐ Guamanian or Chamorro
- ☐ Samoan
- ☐ Other Pacific Islander – *Print race.*
- ☐ Some other race – *Print race.*

What is this person's marital status?

- ☐ Now married
- ☐ Widowed
- ☐ Divorced
- ☐ Separated
- ☐ Never married

2051

Form D-2

Appendix C

Census Form 2010

United States Census 2010

This is the official form for all the people at this address.
It is quick and easy, and your answers are protected by law.

U.S. DEPARTMENT OF COMMERCE
Economics and Statistics Administration
U.S. CENSUS BUREAU

Use a blue or black pen.

Start here

The Census must count every person living in the United States on April 1, 2010.

Before you answer Question 1, count the people living in this house, apartment, or mobile home using our guidelines.

- Count all people, including babies, who live and sleep here most of the time.

The Census Bureau also conducts counts in institutions and other places, so:

- Do not count anyone living away either at college or in the Armed Forces.
- Do not count anyone in a nursing home, jail, prison, detention facility, etc., on April 1, 2010.
- Leave these people off your form, even if they will return to live here after they leave college, the nursing home, the military, jail, etc. Otherwise, they may be counted twice.

The Census must also include people without a permanent place to stay, so:

- If someone who has no permanent place to stay is staying here on April 1, 2010, count that person. Otherwise, he or she may be missed in the census.

1. How many people were living or staying in this house, apartment, or mobile home on April 1, 2010?

Number of people = ☐

2. Were there any additional people staying here April 1, 2010 that you did not include in Question 1? *Mark ✗ all that apply.*

☐ Children, such as newborn babies or foster children
☐ Relatives, such as adult children, cousins, or in-laws
☐ Nonrelatives, such as roommates or live-in baby sitters
☐ People staying here temporarily
☐ No additional people

3. Is this house, apartment, or mobile home — *Mark ✗ ONE box.*

☐ Owned by you or someone in this household with a mortgage or loan? *Include home equity loans.*
☐ Owned by you or someone in this household free and clear (without a mortgage or loan)?
☐ Rented?
☐ Occupied without payment of rent?

4. What is your telephone number? *We may call if we don't understand an answer.*

Area Code + Number

☐ - ☐ - ☐

OMB No. 0607-0919-C: Approval Expires 12/31/2011.

Form **D-61** (1-15-2009)

5. Please provide information for each person living here. Start with a person living here who owns or rents this house, apartment, or mobile home. If the owner or renter lives somewhere else, start with any adult living here. This will be Person 1.

What is Person 1's name? *Print name below.*

Last Name ☐

First Name ☐ MI ☐

6. What is Person 1's sex? *Mark ✗ ONE box.*

☐ Male ☐ Female

7. What is Person 1's age and what is Person 1's date of birth?
Please report babies as age 0 when the child is less than 1 year old.

Print numbers in boxes.

Age on April 1, 2010 ☐ Month ☐ Day ☐ Year of birth ☐

→ **NOTE: Please answer BOTH Question 8 about Hispanic origin and Question 9 about race. For this census, Hispanic origins are not races.**

8. Is Person 1 of Hispanic, Latino, or Spanish origin?

☐ **No,** not of Hispanic, Latino, or Spanish origin
☐ Yes, Mexican, Mexican Am., Chicano
☐ Yes, Puerto Rican
☐ Yes, Cuban
☐ Yes, another Hispanic, Latino, or Spanish origin — *Print origin, for example, Argentinean, Colombian, Dominican, Nicaraguan, Salvadoran, Spaniard, and so on.*

9. What is Person 1's race? *Mark ✗ one or more boxes.*

☐ White
☐ Black, African Am., or Negro
☐ American Indian or Alaska Native — *Print name of enrolled or principal tribe.*

☐ Asian Indian
☐ Chinese
☐ Filipino
☐ Other Asian — *Print race, for example, Hmong, Laotian, Thai, Pakistani, Cambodian, and so on.*
☐ Japanese
☐ Korean
☐ Vietnamese
☐ Native Hawaiian
☐ Guamanian or Chamorro
☐ Samoan
☐ Other Pacific Islander — *Print race, for example, Fijian, Tongan, and so on.*

☐ Some other race — *Print race.*

10. Does Person 1 sometimes live or stay somewhere else?

☐ No ☐ Yes — *Mark ✗ all that apply.*

☐ In college housing
☐ In the military
☐ At a seasonal or second residence
☐ For child custody
☐ In jail or prison
☐ In a nursing home
☐ For another reason

→ **If more people were counted in Question 1, continue with Person 2.**

USCENSUSBUREAU

1. **Print name of Person 2**

Last Name

First Name MI

2. **How is this person related to Person 1?** *Mark X ONE box.*

- ☐ Husband or wife
- ☐ Biological son or daughter
- ☐ Adopted son or daughter
- ☐ Stepson or stepdaughter
- ☐ Brother or sister
- ☐ Father or mother
- ☐ Grandchild
- ☐ Parent-in-law
- ☐ Son-in-law or daughter-in-law
- ☐ Other relative
- ☐ Roomer or boarder
- ☐ Housemate or roommate
- ☐ Unmarried partner
- ☐ Other nonrelative

3. **What is this person's sex?** *Mark X ONE box.*

☐ Male ☐ Female

4. **What is this person's age and what is this person's date of birth?** *Please report babies as age 0 when the child is less than 1 year old. Print numbers in boxes.*

Age on April 1, 2010 Month Day Year of birth

→ **NOTE: Please answer BOTH Question 5 about Hispanic origin and Question 6 about race. For this census, Hispanic origins are not races.**

5. **Is this person of Hispanic, Latino, or Spanish origin?**

- ☐ **No,** not of Hispanic, Latino, or Spanish origin
- ☐ Yes, Mexican, Mexican Am., Chicano
- ☐ Yes, Puerto Rican
- ☐ Yes, Cuban
- ☐ Yes, another Hispanic, Latino, or Spanish origin — *Print origin, for example, Argentinean, Colombian, Dominican, Nicaraguan, Salvadoran, Spaniard, and so on.*

6. **What is this person's race?** *Mark X one or more boxes.*

- ☐ White
- ☐ Black, African Am., or Negro
- ☐ American Indian or Alaska Native — *Print name of enrolled or principal tribe.*
- ☐ Asian Indian
- ☐ Chinese
- ☐ Filipino
- ☐ Other Asian — *Print race, for example, Hmong, Laotian, Thai, Pakistani, Cambodian, and so on.*
- ☐ Japanese
- ☐ Korean
- ☐ Vietnamese
- ☐ Native Hawaiian
- ☐ Guamanian or Chamorro
- ☐ Samoan
- ☐ Other Pacific Islander — *Print race, for example, Fijian, Tongan, and so on.*
- ☐ Some other race — *Print race.*

7. **Does this person sometimes live or stay somewhere else?**

☐ No ☐ Yes — *Mark X all that apply.*

- ☐ In college housing
- ☐ In the military
- ☐ At a seasonal or second residence
- ☐ For child custody
- ☐ In jail or prison
- ☐ In a nursing home
- ☐ For another reason

→ **If more people were counted in Question 1 on the front page, continue with Person 3.**

1. **Print name of Person 3**

Last Name

First Name MI

2. **How is this person related to Person 1?** *Mark X ONE box.*

- ☐ Husband or wife
- ☐ Biological son or daughter
- ☐ Adopted son or daughter
- ☐ Stepson or stepdaughter
- ☐ Brother or sister
- ☐ Father or mother
- ☐ Grandchild
- ☐ Parent-in-law
- ☐ Son-in-law or daughter-in-law
- ☐ Other relative
- ☐ Roomer or boarder
- ☐ Housemate or roommate
- ☐ Unmarried partner
- ☐ Other nonrelative

3. **What is this person's sex?** *Mark X ONE box.*

☐ Male ☐ Female

4. **What is this person's age and what is this person's date of birth?** *Please report babies as age 0 when the child is less than 1 year old. Print numbers in boxes.*

Age on April 1, 2010 Month Day Year of birth

→ **NOTE: Please answer BOTH Question 5 about Hispanic origin and Question 6 about race. For this census, Hispanic origins are not races.**

5. **Is this person of Hispanic, Latino, or Spanish origin?**

- ☐ **No,** not of Hispanic, Latino, or Spanish origin
- ☐ Yes, Mexican, Mexican Am., Chicano
- ☐ Yes, Puerto Rican
- ☐ Yes, Cuban
- ☐ Yes, another Hispanic, Latino, or Spanish origin — *Print origin, for example, Argentinean, Colombian, Dominican, Nicaraguan, Salvadoran, Spaniard, and so on.*

6. **What is this person's race?** *Mark X one or more boxes.*

- ☐ White
- ☐ Black, African Am., or Negro
- ☐ American Indian or Alaska Native — *Print name of enrolled or principal tribe.*
- ☐ Asian Indian
- ☐ Chinese
- ☐ Filipino
- ☐ Other Asian — *Print race, for example, Hmong, Laotian, Thai, Pakistani, Cambodian, and so on.*
- ☐ Japanese
- ☐ Korean
- ☐ Vietnamese
- ☐ Native Hawaiian
- ☐ Guamanian or Chamorro
- ☐ Samoan
- ☐ Other Pacific Islander — *Print race, for example, Fijian, Tongan, and so on.*
- ☐ Some other race — *Print race.*

7. **Does this person sometimes live or stay somewhere else?**

☐ No ☐ Yes — *Mark X all that apply.*

- ☐ In college housing
- ☐ In the military
- ☐ At a seasonal or second residence
- ☐ For child custody
- ☐ In jail or prison
- ☐ In a nursing home
- ☐ For another reason

→ **If more people were counted in Question 1 on the front page, continue with Person 4.**

1. **Print name of Person 4**

Last Name

First Name MI

2. **How is this person related to Person 1?** *Mark ☒ ONE box.*

- ☐ Husband or wife
- ☐ Biological son or daughter
- ☐ Adopted son or daughter
- ☐ Stepson or stepdaughter
- ☐ Brother or sister
- ☐ Father or mother
- ☐ Grandchild
- ☐ Parent-in-law
- ☐ Son-in-law or daughter-in-law
- ☐ Other relative
- ☐ Roomer or boarder
- ☐ Housemate or roommate
- ☐ Unmarried partner
- ☐ Other nonrelative

3. **What is this person's sex?** *Mark ☒ ONE box.*

☐ Male ☐ Female

4. **What is this person's age and what is this person's date of birth?** *Please report babies as age 0 when the child is less than 1 year old.*

Print numbers in boxes.

Age on April 1, 2010 Month Day Year of birth

➜ **NOTE: Please answer BOTH Question 5 about Hispanic origin and Question 6 about race. For this census, Hispanic origins are not races.**

5. **Is this person of Hispanic, Latino, or Spanish origin?**

- ☐ **No,** not of Hispanic, Latino, or Spanish origin
- ☐ Yes, Mexican, Mexican Am., Chicano
- ☐ Yes, Puerto Rican
- ☐ Yes, Cuban
- ☐ Yes, another Hispanic, Latino, or Spanish origin — *Print origin, for example, Argentinean, Colombian, Dominican, Nicaraguan, Salvadoran, Spaniard, and so on.*

6. **What is this person's race?** *Mark ☒ one or more boxes.*

- ☐ White
- ☐ Black, African Am., or Negro
- ☐ American Indian or Alaska Native — *Print name of enrolled or principal tribe.*
- ☐ Asian Indian
- ☐ Chinese
- ☐ Filipino
- ☐ Other Asian — *Print race, for example, Hmong, Laotian, Thai, Pakistani, Cambodian, and so on.*
- ☐ Japanese
- ☐ Korean
- ☐ Vietnamese
- ☐ Native Hawaiian
- ☐ Guamanian or Chamorro
- ☐ Samoan
- ☐ Other Pacific Islander — *Print race, for example, Fijian, Tongan, and so on.*
- ☐ Some other race — *Print race.*

7. **Does this person sometimes live or stay somewhere else?**

☐ No ☐ Yes — *Mark ☒ all that apply.*

- ☐ In college housing
- ☐ In the military
- ☐ At a seasonal or second residence
- ☐ For child custody
- ☐ In jail or prison
- ☐ In a nursing home
- ☐ For another reason

➜ **If more people were counted in Question 1 on the front page, continue with Person 5.**

1. **Print name of Person 5**

Last Name

First Name MI

2. **How is this person related to Person 1?** *Mark ☒ ONE box.*

- ☐ Husband or wife
- ☐ Biological son or daughter
- ☐ Adopted son or daughter
- ☐ Stepson or stepdaughter
- ☐ Brother or sister
- ☐ Father or mother
- ☐ Grandchild
- ☐ Parent-in-law
- ☐ Son-in-law or daughter-in-law
- ☐ Other relative
- ☐ Roomer or boarder
- ☐ Housemate or roommate
- ☐ Unmarried partner
- ☐ Other nonrelative

3. **What is this person's sex?** *Mark ☒ ONE box.*

☐ Male ☐ Female

4. **What is this person's age and what is this person's date of birth?** *Please report babies as age 0 when the child is less than 1 year old.*

Print numbers in boxes.

Age on April 1, 2010 Month Day Year of birth

➜ **NOTE: Please answer BOTH Question 5 about Hispanic origin and Question 6 about race. For this census, Hispanic origins are not races.**

5. **Is this person of Hispanic, Latino, or Spanish origin?**

- ☐ **No,** not of Hispanic, Latino, or Spanish origin
- ☐ Yes, Mexican, Mexican Am., Chicano
- ☐ Yes, Puerto Rican
- ☐ Yes, Cuban
- ☐ Yes, another Hispanic, Latino, or Spanish origin — *Print origin, for example, Argentinean, Colombian, Dominican, Nicaraguan, Salvadoran, Spaniard, and so on.*

6. **What is this person's race?** *Mark ☒ one or more boxes.*

- ☐ White
- ☐ Black, African Am., or Negro
- ☐ American Indian or Alaska Native — *Print name of enrolled or principal tribe.*
- ☐ Asian Indian
- ☐ Chinese
- ☐ Filipino
- ☐ Other Asian — *Print race, for example, Hmong, Laotian, Thai, Pakistani, Cambodian, and so on.*
- ☐ Japanese
- ☐ Korean
- ☐ Vietnamese
- ☐ Native Hawaiian
- ☐ Guamanian or Chamorro
- ☐ Samoan
- ☐ Other Pacific Islander — *Print race, for example, Fijian, Tongan, and so on.*
- ☐ Some other race — *Print race.*

7. **Does this person sometimes live or stay somewhere else?**

☐ No ☐ Yes — *Mark ☒ all that apply.*

- ☐ In college housing
- ☐ In the military
- ☐ At a seasonal or second residence
- ☐ For child custody
- ☐ In jail or prison
- ☐ In a nursing home
- ☐ For another reason

➜ **If more people were counted in Question 1 on the front page, continue with Person 6.**

1. **Print name of Person 6**

Last Name

First Name MI

2. **How is this person related to Person 1?** *Mark ✗ ONE box.*

- ☐ Husband or wife
- ☐ Biological son or daughter
- ☐ Adopted son or daughter
- ☐ Stepson or stepdaughter
- ☐ Brother or sister
- ☐ Father or mother
- ☐ Grandchild
- ☐ Parent-in-law
- ☐ Son-in-law or daughter-in-law
- ☐ Other relative
- ☐ Roomer or boarder
- ☐ Housemate or roommate
- ☐ Unmarried partner
- ☐ Other nonrelative

3. **What is this person's sex?** *Mark ✗ ONE box.*

☐ Male ☐ Female

4. **What is this person's age and what is this person's date of birth?**
Please report babies as age 0 when the child is less than 1 year old.
Print numbers in boxes.

Age on April 1, 2010 Month Day Year of birth

➜ **NOTE: Please answer BOTH Question 5 about Hispanic origin and Question 6 about race. For this census, Hispanic origins are not races.**

5. **Is this person of Hispanic, Latino, or Spanish origin?**

- ☐ **No,** not of Hispanic, Latino, or Spanish origin
- ☐ Yes, Mexican, Mexican Am., Chicano
- ☐ Yes, Puerto Rican
- ☐ Yes, Cuban
- ☐ Yes, another Hispanic, Latino, or Spanish origin — *Print origin, for example, Argentinean, Colombian, Dominican, Nicaraguan, Salvadoran, Spaniard, and so on.*

6. **What is this person's race?** *Mark ✗ one or more boxes.*

- ☐ White
- ☐ Black, African Am., or Negro
- ☐ American Indian or Alaska Native — *Print name of enrolled or principal tribe.*
- ☐ Asian Indian
- ☐ Chinese
- ☐ Filipino
- ☐ Other Asian — *Print race, for example, Hmong, Laotian, Thai, Pakistani, Cambodian, and so on.*
- ☐ Japanese
- ☐ Korean
- ☐ Vietnamese
- ☐ Native Hawaiian
- ☐ Guamanian or Chamorro
- ☐ Samoan
- ☐ Other Pacific Islander — *Print race, for example, Fijian, Tongan, and so on.*
- ☐ Some other race — *Print race.*

7. **Does this person sometimes live or stay somewhere else?**

☐ No ☐ Yes — *Mark ✗ all that apply.*

- ☐ In college housing
- ☐ In the military
- ☐ At a seasonal or second residence
- ☐ For child custody
- ☐ In jail or prison
- ☐ In a nursing home
- ☐ For another reason

➜ **If more than six people were counted in Question 1 on the front page, turn the page and continue.**

➜ ***If more people live here, turn the page and continue.***

Form D-61 (1-15-2009)

Use this section to complete information for the rest of the people you counted in Question 1 on the front page. *We may call for additional information about them.*

Person 7 **Last Name** **First Name** **MI**

Sex ☐ Male ☐ Female
Age on April 1, 2010
Date of Birth Month Day Year
Related to Person 1? ☐ Yes ☐ No

Person 8 **Last Name** **First Name** **MI**

Sex ☐ Male ☐ Female
Age on April 1, 2010
Date of Birth Month Day Year
Related to Person 1? ☐ Yes ☐ No

Person 9 **Last Name** **First Name** **MI**

Sex ☐ Male ☐ Female
Age on April 1, 2010
Date of Birth Month Day Year
Related to Person 1? ☐ Yes ☐ No

Person 10 **Last Name** **First Name** **MI**

Sex ☐ Male ☐ Female
Age on April 1, 2010
Date of Birth Month Day Year
Related to Person 1? ☐ Yes ☐ No

Person 11 **Last Name** **First Name** **MI**

Sex ☐ Male ☐ Female
Age on April 1, 2010
Date of Birth Month Day Year
Related to Person 1? ☐ Yes ☐ No

Person 12 **Last Name** **First Name** **MI**

Sex ☐ Male ☐ Female
Age on April 1, 2010
Date of Birth Month Day Year
Related to Person 1? ☐ Yes ☐ No

Thank you for completing your official 2010 Census form.

FOR OFFICIAL USE ONLY
JIC1 **JIC2**

If your enclosed postage-paid envelope is missing, *please mail your completed form to:*

U.S. Census Bureau
National Processing Center
1201 East 10th Street
Jeffersonville, IN 47132

If you need help completing this form, *call 1-866-872-6868 between 8:00 a.m. and 9:00 p.m., 7 days a week. The telephone call is free.*

TDD — *Telephone display device for the hearing impaired. Call 1-866-783-2010 between 8:00 a.m. and 9:00 p.m., 7 days a week. The telephone call is free.*

¿NECESITA AYUDA? *Si usted necesita ayuda para completar este cuestionario, llame al 1-866-928-2010 entre las 8:00 a.m. y 9:00 p.m., 7 días a la semana. La llamada telefónica es gratis.*

The U.S. Census Bureau estimates that, for the average household, this form will take about 10 minutes to complete, including the time for reviewing the instructions and answers. Send comments regarding this burden estimate or any other aspect of this burden to: Paperwork Reduction Project 0607-0919-C, U.S. Census Bureau, AMSD-3K138, 4600 Silver Hill Road, Washington, DC 20233. You may e-mail comments to <Paperwork@census.gov>; use "Paperwork Project 0607-0919-C" as the subject.

Respondents are not required to respond to any information collection unless it displays a valid approval number from the Office of Management and Budget.

Glossary

A

Abortion rate The estimated number of abortions in a given year per 1000 women aged 15–44.

Age structure The percentage of people in various age groups in a population. It is best described by a population pyramid.

Age-sex structure The composition of a population as determined by the number or portion of males and females in each age category. This information is an essential prerequisite for the description and analysis of many other types of demographic data. See also *population pyramid.*

Age-specific rate Rate of population change obtained for a specific age group (for example, age-specific fertility rate, death rate, marriage rate, illiteracy rate, etc.).

AIDS Acquired Immune Deficiency Syndrome—a collection of symptoms and infections resulting from the specific damage to the immune system caused by the human immunodeficiency virus (HIV) in humans, and similar viruses in other species.

Antinatalist policy Population policy that aims to slow population growth by attempting to limit the number of births.

Arable land An agricultural term, meaning land that can be used for growing crops.

Arithmetic density Comparison measure between the total population of a place to its total area.

B

Baby boom The period following World War II from 1946–1964 characterized by a rapid increase in fertility rates and in the absolute number of births in the U.S., Europe, Canada, Australia, and New Zealand.

Baby bust The period immediately after the "baby boom" characterized by a rapid decline in U.S. fertility rates to record low levels.

Basic demographic equation An equation that ties together population change through the major demographic variables: deaths, births, and migration. The equation views the population of a place at some future date as a function of its present population plus births over the time interval, minus deaths over the time interval, plus in-migrants and minus out-migrants over the time interval.

Behavior The response of an individual, group, or species to its environment.

Birth control Practices adopted by couples that permit sexual intercourse with reduced likelihood of conception. The term is frequently used synonymously with such terms as contraception, family planning, and fertility control.

Brain drain The emigration of a significant proportion of a country's highly educated, highly skilled professional population, usually to other countries offering better social and economic opportunities.

C

Carrying capacity The number of people who can be sustained by a particular land area at a given technological level.

Case fatality rate The proportion of deaths within a selected population of cases over the course of a disease.

Census A total count of the population of a specified area, generally a nation.

Child-woman ratio The number of children under 5 years old per 1000 women aged 15–44 years in a population.

Circulation Activities such as shopping, commuting, and touring because each of these activities begins and ends at a person's place of residence.

Cohort A group of people sharing a common temporal demographic experience who are observed through time. For example, the birth cohort of 1950 would be the people born in that year. There are also marriage cohorts, school class cohorts, etc.

Completed fertility rate The number of children born per woman to a cohort of women by the end of their childbearing years.

Component methods Separately project births, deaths, and migration, them combine the components into an overall population projection.

Comstock Laws United States federal laws (1873) that made it illegal to send any "obscene, lewd, and/or lascivious" materials through the mail, including contraceptive devices and information.

Critical theory Emphasizes the reflective assessment and critique of society and culture through the application of knowledge from the social sciences and the humanities.

Crowding To fill by pressing or thronging together.

Crude birth rate The annual number of births per one thousand people in a given population.

Crude death rate The annual number of deaths per one thousand people in a given population.

D

De facto Expression that means "in fact" or "in practice" but not spelled out by law.

De jure An expression that means "based on law".

Deforestation Clearing land for agriculture and gathering of wood for fires.

Demographic inertia The tendency for current population parameters, such as growth rate, to continue for a period of time; there is often a delayed population response to gradual changes in birth and mortality rates.

Demographic transition The transition from a demographic regime characterized by high birth and death rates to one characterized by low birth and death rates.

Demography The scientific study of the human population, focusing on population composition, change, and distribution.

Demonstration effects Effects on the behavior of individuals caused by observation of the actions of others and their consequences.

Dependency ratio The ratio of persons in the dependent ages (under 20 and over 64 years) to those in the economically productive years (20 to 64 years) in a population.

Developed countries Countries that have industrialized and have relatively high levels of per capita income and per capita productivity.

Developing countries Countries that have not industrialized and have relatively low rates of per capita income and per capita productivity.

Differential migration The study of the selectivity of migration and the differing rates between various social, demographic, and economic groups.

Direct maintenance costs Actual monetary outlays required for the support of children.

Doubling time The number of years required for a population of an area to double its present size, given the current rate of population growth. It is possible to closely approximate the doubling time for a population by dividing the annual rate of population growth into the number 70.

E

Ecological diversity Refers to the diversity of a place at the level of ecosystems.

Ecology The relationships that organisms have with each other and with their natural environment.

Ecumene The inhabited part of the earth.

Elderly In the U.S., usually defined as having reached 65 years of age.

Emigration rate The number of emigrants departing an area of origin per 1000 inhabitants at that area of origin in a given year.

Emigration The process of leaving one country to take up residence in another.

Environmental degradation The deterioration of the environment through depletion of resources such as air, water and soil; the destruction of ecosystems and the extinction of wildlife.

Epidemiologic transition The transition a society goes through in which death rates decline and the major causes of death shift from communicative diseases to degenerative diseases.

F

Family planning Policies and programs designed to help families achieve their desired family size.

Fecundity The physiological capacity of a woman, man, or couple to produce a child.

Fertility The actual reproductive performance of an individual, a couple, a group, or a population.

Fertility trend Any observed or predicted course of fertility over a specified period of time.

Forced migration A movement of individuals or groups of people in which they participated without choice. The Trans-Atlantic slave trade is a major example.

Free-individual migration Movement of people, individual or in families, acting on their own individual initiative and responsibility without official support or compulsion.

G

General fertility rate The annual number of live births per 1000 women in the childbearing age group (ages 15–44).

Genocide The deliberate and systematic destruction, in whole or in part, of an ethnic, religious or national group.

Green Revolution A rapid increase in agricultural output resulting from the application of new agricultural practices and the use of "miracle strains" of wheat and rice.

Gross migration The total sum of all the people who enter and leave an area.

Gross reproduction rate The average number of daughters a woman would bear if she passed through her entire reproductive life at the prevailing age-specific fertility rates.

Group migration Migration of a clan, tribe, or other social group that is larger than a family.

Growth rate The rate at which a population is increasing (or decreasing) in a given year due to natural increase and net migration, expressed as a percentage of the base population.

H

HIV Human Immunodeficiency Virus. HIV is the virus that causes AIDS. *See AIDS*

I

Illegal immigrants Refers to people who have immigrated across national borders in a way that violates the immigration laws of the destination country.

Immigration Act of 1990 Increased the number of legal immigrants allowed into the United States each year.

Immigration The process of entering one country from another to take up residence.

Impelled migration Migration activated by the state or some other political or social institution which allows the migrant some degree of choice.

Infant mortality rate The number of deaths among infants under one year of age in a given year per 1000 live births in that year.

International Planned Parenthood Federation (IPPF) Global non-governmental organization with the broad aims of promoting sexual and reproductive health, and advocating the right of individuals to make their own choices in family planning.

J

J-shaped mortality curve A curve depicting the rates of death and how these increase with age.

K

KAP Surveys to determine the extent of knowledge, attitude, and practice of contraception (KAP surveys).

L

Laissez-faire solution A solution to population issues based on individual choice. Laissez-faire population exponents feel that those best able to decide on the costs and benefits of children are those who are contemplating having them.

Lee's migration model A model developed to explain migration as the result of four sets of factors: (1) factors associated with place of origin, (2) factors associated with place of destination, (3) intervening obstacles, and (4) personal factors.

Less developed countries Countries that have not industrialized and have relatively low rates of per capita income and per capita productivity.

Life expectancy The average number of additional years a person would live if current mortality trends were to continue.

Life table (also called a **mortality table** or **actuarial table**) A table that gives life expectancy and the probability of dying at each age for a given population, according to the age-specific death rates prevailing at that time.

M

Malthusian theory Theory that population growth will outrun the food supply and that the world cannot continue to support a growing population.

Marxism A theory focusing on the mode of production, a form of economic production, from which most social phenomena including demographic phenomena arise.

Mathematical methods Employ some mathematical formula to a base population using an assumed rate of growth over the projection interval.

Median age The age that divides a population into two numerically equal groups with half the people younger and half older than this age.

Medical geography The branch of Human Geography that deals with the geographic aspects of health (status) and healthcare (systems).

Megacity Urban agglomerations with 10 million or more residents.

Migrate Change in residence intended to be permanent.

Migration differentials Characteristics that make migrants different from the general population. Migrants are not random samples of the population from which they came but are "selected" by certain characteristics such as age or educational level.

Migration interval The time interval over which migration is studied or observed.

Migration stream An established movement of people between two places. For example, there is a migration stream between New York and Miami.

Migration The movement of people from place to place, usually across some political boundary, for the purpose of changing their permanent place of residence.

Morbidity The frequency of disease and illness in a population.

More developed countries Countries that have industrialized and have relatively high levels of per capita income and per capita productivity.

Mortality Deaths as a component of population change. The frequency with which deaths occur.

Multiphasic response theory The incorporation of reflexive and behavioral responses (rather than solely the continuous responses and changes) to demographic change as well as the changes these induce in turn.

N

Natality Births as a component of population change.

Natural increase The difference between the number of live births and the number of deaths during a year.

Neo-Malthusian theory A current revival of interest in the ideas of Malthus. Neo-Malthusians believe that population growth will outrun Earth's food and resource supply. Neo-Mathusians, unlike Malthus, advocate birth control as a preventive check.

Neonatal mortality rate The number of deaths to infants under 28 days of age in a given year per 1000 live births in that year.

Net migration The difference between immigration and emigration for an area's population in a given time period; generally expressed as an increase or decrease.

Nonecumene The uninhabited part of the earth.

Nutritional density Ratio of population to arable land.

O

Opportunity costs Measure of opportunities that parents must sacrifice to have and raise children.

Optimists Also called economic optimists, believe that population determines technology and technology determines food supply.

Overgrazing Occurs when plants are exposed to livestock grazing for extended periods of time, or without sufficient recovery periods.

P

Physiological density The ratio of population to arable land.

Pollution Waste that has increased to the point that it can no longer be accommodated by the ecosystem.

Population A group of objects or organisms of the same kind.

Population density Generally refers to the number of people per unit of land.

Population distribution The pattern of settlement and dispersal of a population.

Population equilibrium The level at which the population is in balance with the natural resource base of an area. Because of the changing nature of technology and the resource base, it is difficult to translate population equilibrium into a precise number.

Population geography A division of human geography that studies ways in which spatial variations in the distribution, composition, migration, and growth of populations are related to the nature of places.

Population growth The change in population over time, and can be quantified as the change in the number of individuals in a population per unit time.

Population policy An official government policy specifically designed to affect the size and growth rate of a population, the distribution of a population, or its composition. See *population program*.

Population program The various means and measures that must be utilized in order to achieve the objectives of a population policy. See *population policy*.

Population projection An estimate of the population of an area for some future date, based on a set of assumptions about the future course of births, deaths, and migration.

Population pyramid A graph showing the age and sex structure of a population. It may be constructed for nations, states, cities, or even smaller areas.

Postneonatal mortality rate Deaths occurring after 28 days of life and ending before one year of age per 1,000 live births during a year.

Primate city A city that dominates the state in size and importance. It is commonly at least twice as large as the next largest city, and it is the educational, political, industrial, and commercial center.

Primitive migration The type of migration associated with groups that migrate because they are unable to cope with natural forces related to their physical environment.

Pronatalist policy Population policy that encourage higher birth rates.

"Push-pull" models of migration Derived from the observation that some elements of an origin "push" people to migrate, whereas some elements of a destination "pull" migrants toward it.

R

Rate of natural increase The crude birth rate minus the crude death rate of a population.

Rate of population growth Measure of the average annual rate of increase for a population.

Refugee A person who is outside the country of his or her nationality because he has or had well-founded fear of persecution and is unable or, because of such fear, is unwilling to avail himself or herself of the protection of the government.

Return migration A voluntary or involuntary return of migrants to their place of origin.

Rural-urban migration The moving of people from rural areas into cities.

S

Sample survey A canvass of selected persons or households in a population usually used to infer demographic characteristics or trends for a larger segment or all of the population.

Sex ratio The number of males per 100 females in a population.

T

Total fertility rate The average number of children that would be born to a woman during her lifetime if she were to pass through her childbearing years at the prevailing age-specific fertility rates.

Triage The idea that the world's nations can be put into three categories relative to their food resources: (1) those nations that cannot be saved regardless of food aid from other countries, (2) those nations that will survive without food aid from other countries, and (3) those nations that could feed their people with some aid from the more developed countries.

U

Unmet need for contraception Broadly defined as women who want to delay or stop childbearing but are not using contraception.

Urban A geographical area (including cities and towns) distinct from rural areas.

Urbanism The characteristics or conditions of the way of life of those who live in cities.

Urbanization The percentage of a population that lives in urban areas and the process by which this number increases over time.

V

Vital registrations The recording and compilation of vital statistics, at or near their time of occurrence. They usually include such events as deaths, fetal deaths, births, marriages, divorces, and at times disease and illness.

Vital statistics Demographic data on births, deaths, fetal deaths, marriages, and divorces.

Z

Zero population growth (ZPG) An equilibrium population with a growth rate of zero, achieved when births plus immigration equal deaths plus emigration.

Index

N

O

P

R

S